Library of
Davidson College

UCLA Symposia on Molecular and Cellular Biology, New Series

Series Editor
C. Fred Fox

Volume 1
  Differentiation and Function of Hematopoietic Cell Surfaces, Vincent T. Marchesi and Robert C. Gallo, *Editors*
Volume 2
  Mechanisms of Chemical Carcinogenesis, Curtis C. Harris and Peter A. Cerutti, *Editors*
Volume 3
  Cellular Recognition, William A. Frazier, Luis Glaser, and David I. Gottlieb, *Editors*
Volume 4
  Rational Basis for Chemotherapy, Bruce A. Chabner, *Editor*
Volume 5
  Tumor Viruses and Differentiation, Edward M. Scolnick and Arnold J. Levine, *Editors*
Volume 6
  Evolution of Hormone-Receptor Systems, Ralph A. Bradshaw and Gordon N. Gill, *Editors*
Volume 7
  Recent Advances in Bone Marrow Transplantation, Robert Peter Gale, *Editor*
Volume 8
  Gene Expression, Dean H. Hamer and Martin J. Rosenberg, *Editors*
Volume 9
  Normal and Neoplastic Hematopoiesis, David W. Golde and Paul A. Marks, *Editors*
Volume 10
  Mechanisms of DNA Replication and Recombination, Nicholas R. Cozzarelli, *Editor*
Volume 11
  Cellular Responses to DNA Damage, Errol C. Friedberg and Bryn A. Bridges, *Editors*
Volume 12
  Plant Molecular Biology, Robert B. Goldberg, *Editor*
Volume 13
  Molecular Biology of Host–Parasite Interactions, Nina Agabian and Harvey Eisen, *Editors*
Volume 14
  Biosynthesis of the Photosynthetic Apparatus: Molecular Biology, Development and Regulation, J. Philip Thornber, L. Andrew Staehelin, and Richard B. Hallick, *Editors*
Volume 15
  Protein Transport and Secretion, Dale L. Oxender, *Editor*
Volume 16
  Acquired Immune Deficiency Syndrome, Michael S. Gottlieb and Jerome E. Groopman, *Editors*

Volume 17
  Genes and Cancer, J. Michael Bishop, Janet D. Rowley, and Mel Greaves, *Editors*
Volume 18
  Regulation of the Immune System, Harvey Cantor, Leonard Chess, and Eli Sercarz, *Editors*
Volume 19
  Molecular Biology of Development, Eric H. Davidson and Richard A. Firtel, *Editors*
Volume 20
  Genome Rearrangement, Melvin Simon and Ira Herskowitz, *Editors*
Volume 21
  Herpesvirus, Fred Rapp, *Editor*
Volume 22
  Cellular and Molecular Biology of Plant Stress, Joe L. Key and Tsune Kosuge, *Editors*
Volume 23
  Membrane Receptors and Cellular Regulation, Michael P. Czech and C. Ronald Kahn, *Editors*
Volume 24
  Neurobiology: Molecular Biological Approaches to Understanding Neuronal Function and Development, Paul O'Lague, *Editor*
Volume 25
  Extracellular Matrix: Structure and Function, A. Hari Reddi, *Editor*
Volume 26
  Nuclear Envelope Structure and RNA Maturation, Edward A. Smuckler and Gary A. Clawson, *Editors*
Volume 27
  Monoclonal Antibodies and Cancer Therapy, Ralph A. Reisfeld and Stewart Sell, *Editors*
Volume 28
  Leukemia: Recent Advances in Biology and Treatment, David W. Golde and Robert Peter Gale, *Editors*
Volume 29
  Molecular Biology of Muscle Development, Charles Emerson, Donald Fischman, Bernardo Nadal-Ginard, and M.A.Q. Siddiqui, *Editors*
Volume 30
  Sequence Specificity in Transcription and Translation, Richard Calendar and Larry Gold, *Editors*
Volume 31
  Molecular Determinants of Animal Form, Gerald M. Edelman, *Editor*

UCLA Symposia Published Prior to 1983

(Numbers refer to the publishers listed below.)

1972
Membrane Research (2)

1973
Membranes (1)
Virus Research (2)

1974
Molecular Mechanisms for the Repair of DNA (4)
Membranes (1)
Assembly Mechanisms (1)
The Immune System: Genes, Receptors, Signals (2)
Mechanisms of Virus Disease (3)

1975
Energy Transducing Mechanisms (1)
Cell Surface Receptors (1)
Developmental Biology (3)
DNA Synthesis and Its Regulation (3)

1976
Cellular Neurobiology (1)
Cell Shape and Surface Architecture (1)
Animal Virology (2)
Molecular Mechanisms in the Control of Gene Expression (2)

1977
Cell Surface Carbohydrates and Biological Recognition (1)
Molecular Approaches to Eucaryotic Genetic Systems (2)
Molecular Human Cytogenetics (2)
Molecular Aspects of Membrane Transport (1)
Immune System: Genetics and Regulation (2)

1978
DNA Repair Mechanisms (2)
Transmembrane Signaling (1)
Hematopoietic Cell Differentiation (2)
Normal and Abnormal Red Cell Membranes (1)

Persistent Viruses (2)
Cell Reproduction: Daniel Mazia Dedicatory Volume (2)

1979
Covalent and Non-Covalent Modulation of Protein Function (2)
Eucaryotic Gene Regulation (2)
Biological Recognition and Assembly (1)
Extrachromosomal DNA (2)
Tumor Cell Surfaces and Malignancy (1)
T and B Lymphocytes: Recognition and Function (2)

1980
Biology of Bone Marrow Transplantation (2)
Membrane Transport and Neuroreceptors (1)
Control of Cellular Division and Development (1)
Animal Virus Genetics (2)
Mechanistic Studies of DNA Replication and Genetic Recombination (2)

1981
Immunoglobulin Idiotypes (2)
Initiation of DNA Replication (2)
Genetic Variation Among Influenza Viruses (2)
Developmental Biology Using Purified Genes (2)
Differentiation and Function of Hematopoietic Cell Surfaces (1)
Mechanisms of Chemical Carcinogenesis (1)
Cellular Recognition (1)

1982
B and T Cell Tumors (2)
Interferon (2)
Rational Basis for Chemotherapy (1)
Gene Regulation (2)
Tumor Viruses and Differentiation (1)
Evolution of Hormone-Receptor Systems (1)

Publishers

(1) Alan R. Liss, Inc.
41 E. 11th Street
New York, NY 10003

(2) Academic Press, Inc.
111 Fifth Avenue
New York, NY 10003

(3) W.A. Benjamin, Inc.
2725 Sand Hill Road
Menlo Park, CA 94025

(4) Plenum Publishing Corp.
227 W. 17th Street
New York, NY 10011

Symposia Board

C. Fred Fox, Ph.D., Director
Professor of Microbiology
Molecular Biology Institute
UCLA

Members

Ronald Cape, Ph.D., M.B.A.
Chairman
Cetus Corporation

Pedro Cuatrecasas, M.D.
Vice President for Research
Burroughs Wellcome Company

Luis Glaser, Ph.D.
Professor and Chairman
of Biochemistry
Washington University School
of Medicine

Donald Steiner, M.D.
Professor of Biochemistry
University of Chicago

Ernest Jaworski, Ph.D.
Director of Molecular Biology
Monsanto

Paul Marks, M.D.
President
Sloan-Kettering Institute

William Rutter, Ph.D.
Professor of Biochemistry and Director of
the Hormone Research Institute
University of California, San Francisco,
Medical Center

Sidney Udenfriend, Ph.D.
Member
Roche Institute of Molecular Biology

The members of the board advise the director in identification of topics for future symposia.

MOLECULAR DETERMINANTS OF ANIMAL FORM

MOLECULAR DETERMINANTS OF ANIMAL FORM

Proceedings of the UCLA Symposium
held at Park City, Utah
March 30–April 4, 1985

Editor
Gerald M. Edelman
The Rockefeller University
New York, New York

Alan R. Liss, Inc. • New York

Address all Inquiries to the Publisher
Alan R. Liss, Inc., 41 East 11th Street, New York, NY 10003

Copyright © 1985 Alan R. Liss, Inc.

Printed in the United States of America.

Under the conditions stated below the owner of copyright for this book hereby grants permission to users to make photocopy reproductions of any part or all of its contents for personal or internal organizational use, or for personal or internal use of specific clients. This consent is given on the condition that the copier pay the stated per-copy fee through the Copyright Clearance Center, Incorporated, 27 Congress Street, Salem, MA 01970, as listed in the most current issue of "Permissions to Photocopy" (Publisher's Fee List, distributed by CCC, Inc.), for copying beyond that permitted by sections 107 or 108 of the US Copyright Law. This consent does not extend to other kinds of copying, such as copying for general distribution, for advertising or promotional purposes, for creating new collective works, or for resale.

Library of Congress Cataloging-in-Publication Data
Main entry under title:

Molecular determinants of animal form.

   Includes index.
   1. Morphology (Animals)—Congresses. 2. Morphogenesis—Congresses. 3. Molecular biology—Congresses.
I. Edelman, Gerald M.
QL799.M73   1985   591.3′32   85-18173
ISBN 0-8451-2630-X

Contents

Contributors xiii
Preface
    Gerald M. Edelman xvii

I. INDUCTION AND EARLY EVENTS

Alternative Implementation Mechanisms of Embryonic Induction
    Lauri Saxén 1
Cytoplasmic Localization, Inductions, and the "Organizer" of the Frog Embryo
    Robert L. Gimlich 15
The Activation of Muscle-Specific Actin Genes in *Xenopus* Development
    J.B. Gurdon, T.J. Mohun, S. Brennan, and S. Cascio 37
Retroviruses and Insertional Mutagenesis in Mice
    Rudolf Jaenisch 47
Inductive Interaction and Determination; A New Approach to an Old Problem
    Pieter D. Nieuwkoop 59

II. PRIMARY PROCESSES AND THE BODY PLAN

The Embryonic Cell Lineage of Mammals and the Emergence of the Basic Body Plan
    Michael H.L. Snow 73
Morphogenetic Movements and Fate Maps in the Avian Blastoderm
    L. Vakaet 99
Convergent Extension by Cell Intercalation During Gastrulation of *Xenopus laevis*
    Ray Keller, Michael Danilchik, Robert Gimlich, and John Shih . . . 111
The Cortical Tractor Model for Epithelial Folding: Application to the Neural Plate
    Antone G. Jacobson, Garrett M. Odell, and George F. Oster 143
Cell Migration in the Vertebrate Embryo
    Jean Paul Thiery, Jean Claude Boucaut, and Kenneth M. Yamada . . 167

III. CELL-CELL ADHESION AND CELL-MATRIX INTERACTIONS

The Molecular Bases and Dynamics of Cell Adhesion in Embryogenesis
Gerald M. Edelman, Stanley Hoffman, Cheng-Ming Chuong, and Bruce A. Cunningham 195

Selective Cell Adhesion Mechanism: Role of the Calcium-Dependent Cell Adhesion System
Masatoshi Takeichi, Kohei Hatta, and Akira Nagafuchi 223

Two Cell Adhesion Molecules: Characterization and Role in Early Mouse Embryo Development
Caroline H. Damsky, Margaret J. Wheelock, Ivan Damjanov, and Clayton Buck 235

Detection Using Monoclonal Antibodies of a Structurally Altered Form of Cell-CAM 105 on Rat Hepatocellular Carcinomas
Douglas C. Hixson and Kerry D. McEntire 253

Frog Gastrula Cells Adhere to Fibronectin-Sepharose Beads
Kurt E. Johnson 271

Extracellular Matrix, Cell Polarity and Epithelial-Mesenchymal Transformation
Elizabeth D. Hay 293

IV. HISTOGENESIS

Extracellular Matrix in Skin Morphogenesis
Philippe Sengel, Annick Mauger, Joelle Robert, and Madeleine Kieny 319

Current Concepts of Kidney Morphogenesis
Peter Ekblom and Hannu Sariola 349

The Formation of Microvilli
G.F. Oster, J.D. Murray, and G.M. Odell 365

Monomolecular Induction of Two Components of the Postsynaptic Apparatus in Muscle
Bruce G. Wallace, Noreen E. Reist, Ralph M. Nitkin, Justin R. Fallon, and U.J. McMahan 385

Morphogenesis of the Mouse Motor Endplate
François Rieger 393

V. PATTERN FORMATION

Positional Information and Pattern Formation
L. Wolpert 423

Shaping of the Body Column in Hydra: Is a Pre-Pattern Necessary?
Hans R. Bode, Patricia M. Bode, Lorette C. Javois, and Shelly Heimfeld 435

Positional Maps and Cellular Interactions in Insect Development
Vernon French 455

Position-Specific Antibodies as Probes for the Regional Identity of *Drosophila* Imaginal Disc Cells
Danny L. Brower 477

Molecular Analysis of the Involvement of the *Drosophila engrailed* Gene in Embryonic Pattern Formation
Patrick H. O'Farrell, Claude Desplan, Stephen DiNardo, Judy Kassis, Jerry Kuner, Emily Lim, Elizabeth Sher, James Theis, and Deann Wright 489

VI. NEURAL MAPPING

Cell Patterning in Neural Maps: Specificity and Dynamics in Retinotectal Map Formation
Scott E. Fraser and Nancy A. O'Rourke 521

Selective Stabilization of Retinotectal Synapses by an Activity Dependent Mechanism
John T. Schmidt 539

Selection Mechanisms in Neural Mapping
Leif H. Finkel 571

Index 619

Contributors

Hans R. Bode, Developmental Biology Center and the Department of Developmental and Cell Biology, University of California, Irvine, CA 92717 **[435]**

Patricia M. Bode, Developmental Biology Center and the Department of Developmental and Cell Biology, University of California, Irvine, CA 92717 **[435]**

Jean Claude Boucaut, Laboratoire de Biologie Expérimentale, Université Paris René Descartes, 75270 Paris Cedex 06, France **[167]**

S. Brennan, CRC Molecular Embryology Unit, Department of Zoology, University of Cambridge, Cambridge CB2 3EJ, England **[37]**

Danny L. Brower, Departments of Molecular and Cellular Biology and Biochemistry, University of Arizona, Tucson, AZ 85721 **[477]**

Clayton Buck, The Wistar Institute, Philadelphia, PA 19104 **[235]**

S. Cascio, CRC Molecular Embryology Unit, Department of Zoology, University of Cambridge, Cambridge CB2 3EJ, England **[37]**

Cheng-Ming Chuong, Laboratory of Developmental and Molecular Biology, The Rockefeller University, New York, NY 10021 **[195]**

Bruce A. Cunningham, Laboratory of Developmental and Molecular Biology, The Rockefeller University, New York, NY 10021 **[195]**

Ivan Damjanov, Department of Pathology, Hahnemann Medical College, Philadelphia, PA 19102 **[235]**

Caroline H. Damsky, The Wistar Institute, Philadelphia, PA 19104 and Department of Stomatology and Anatomy, Schools of Dentistry and Medicine, University of California, San Francisco, CA 94143 **[235]**

Michael Danilchik, Department of Molecular Biology, University of California, Berkeley, CA 94720 **[111]**

Claude Desplan, Department of Biochemistry and Biophysics, University of California, San Francisco, CA 94143 **[489]**

Stephen DiNardo, Department of Biochemistry and Biophysics, University of California, San Francisco, CA 94143 **[489]**

Gerald M. Edelman, Laboratory of Developmental and Molecular Biology, The Rockefeller University, New York, NY 10021 **[195]**

Peter Ekblom, Friedrich-Miescher-Laboratory, Max-Planck-Society, Spemannstrasse 37-39, D-7400 Tübingen, West Germany **[349]**

Justin R. Fallon, Department of Neurobiology, Stanford University School of Medicine, Stanford, CA 94305 **[385]**

Leif H. Finkel, The Rockefeller University, New York, NY 10021 **[571]**

The number in brackets is the opening page number of the contributor's article.

Contributors

Scott E. Fraser, Department of Physiology and Biophysics, and The Developmental Biology Center, University of California, Irvine, CA 92717 **[521]**

Vernon French, Department of Zoology, University of Edinburgh, Edinburgh EH9 3JT, Scotland, UK **[455]**

Robert L. Gimlich, Department of Molecular Biology, University of California, Berkeley, CA 94720 **[15,111]**

J.B. Gurdon, CRC Molecular Embryology Unit, Department of Zoology, University of Cambridge, Cambridge CB2 3EJ, England **[37]**

Kohei Hatta, Department of Biophysics, Faculty of Science, Kyoto University, Sakyo-ku, Kyoto 606, Japan **[223]**

Elizabeth D. Hay, Department of Anatomy and Cellular Biology, Harvard Medical School, Boston, MA 02115 **[293]**

Shelly Heimfeld, Developmental Biology Center and the Department of Developmental and Cell Biology, University of California, Irvine, CA 92717 **[435]**

Douglas C. Hixson, University of Texas System Cancer Center, Science Park, Research Division, Smithville, TX **[253]**

Stanley Hoffman, Laboratory of Developmental and Molecular Biology, The Rockefeller University, New York, NY 10021 **[195]**

Antone G. Jacobson, Center for Developmental Biology, Department of Zoology, University of Texas, Austin, TX 78712 **[143]**

Rudolf Jaenisch, Whitehead Institute for Biomedical Research, Cambridge, MA 02142 **[47]**

Lorette C. Javois, Developmental Biology Center and the Department of Developmental and Cell Biology, University of California, Irvine, CA 92717 **[435]**

Kurt E. Johnson, Department of Anatomy, George Washington University Medical Center, Washington D.C. 20037 **[271]**

Judy Kassis, Department of Biochemistry and Biophysics, University of California, San Francisco, CA 94143 **[489]**

Ray Keller, Department of Zoology, University of California, Berkeley, CA 94720 **[111]**

Madeleine Kieny, Unité Associée au CNRS n° 682, Laboratoire de Biologie animale, Université scientifique et médicale de Grenoble, 38402 Saint Martin d'Hères, France **[319]**

Jerry Kuner, Department of Biochemistry and Biophysics, University of California, San Francisco, CA 94143 **[489]**

Emily Lim, Department of Biochemistry and Biophysics, University of California, San Francisco, CA 94143 **[489]**

Annick Mauger, Unité Associée au CNRS n° 682, Laboratoire de Biologie animale, Université scientifique et médicale de Grenoble, 38402 Saint Martin d'Hères, France **[319]**

Kerry D. McEntire, University of Texas System Cancer Center, Science Park, Research Division, Smithville, TX **[253]**

U.J. McMahan, Department of Neurobiology, Stanford University School of Medicine, Stanford, CA 94305 **[385]**

T.J. Mohun, CRC Molecular Embryology Unit, Department of Zoology, University of Cambridge, Cambridge CB2 3EJ, England **[37]**

J.D. Murray, Centre for Mathematical Biology, Mathematics Institute, University of Oxford, Oxford OX1 3LB, England [365]

Akira Nagafuchi, Department of Biophysics, Faculty of Science, Kyoto University, Sakyo-ku, Kyoto 606, Japan [223]

Pieter D. Nieuwkoop, Hubrecht Laboratory, Utrecht, The Netherlands [59]

Ralph M. Nitkin, Department of Neurobiology, Stanford University School of Medicine, Stanford, CA 94305 [385]

Garrett M. Odell, Department of Mathematical Sciences, Rensselaer Polytechnic Institute, Troy, NY 12181 [143,365]

Patrick H. O'Farrell, Department of Biochemistry and Biophysics, University of California, San Francisco, CA 94143 [489]

Nancy A. O'Rourke, Department of Physiology and Biophysics, and The Developmental Biology Center, University of California, Irvine, CA 92717 [521]

George F. Oster, Departments of Biophysics, Entomology, and Zoology, University of California, Berkeley, CA 94720 [143,365]

Noreen E. Reist, Department of Neurobiology, Stanford University School of Medicine, Stanford, CA 94305 [385]

François Rieger, Groupe Biologie et Pathologie Neuromusculaires, INSERM, U 153 17 rue du Fer-à-Moulin, 75005 Paris, France [393]

Joelle Robert, Unité Associée au CNRS n° 682, Laboratoire de Biologie animale, Université scientifique et médicale de Grenoble, 38402 Saint Martin d'Hères, France [319]

Hannu Sariola, Friedrich-Miescher-Laboratory, Max-Planck-Society, Spemannstrasse 37-39, D-7400 Tübingen, West Germany [349]

Lauri Saxén, Department of Pathology, University of Helsinki, S-F 00290 Helsinki, Finland [1]

John T. Schmidt, Department of Biological Sciences and Neurobiology Research Center, State University of New York at Albany, Albany, NY 12222 [539]

Philippe Sengel, Unité Associée au CNRS n° 682, Laboratoire de Biologie animale, Université scientifique et médicale de Grenoble, 38402 Saint Martin d'Hères, France [319]

Elizabeth Sher, Department of Biochemistry and Biophysics, University of California, San Francisco, CA 94143 [489]

John Shih, Department of Zoology, University of California, Berkeley, CA 94720 [111]

Michael H.L. Snow, MRC Mammalian Development Unit, University College, London NW1 2HE, England [73]

Masatoshi Takeichi, Department of Biophysics, Faculty of Science, Kyoto University, Sakyo-ku, Kyoto 606, Japan [223]

James Theis, Department of Biochemistry and Biophysics, University of California, San Francisco, CA 94143 [489]

Jean Paul Thiery, Institut d'Embryologie, 94130 Nogent-sur-Marne, France [167]

L. Vakaet, Department of Anatomy and Embryology, State University, Ghent, Belgium [99]

Bruce G. Wallace, Department of Neurobiology, Stanford University School of Medicine, Stanford, CA 94305 **[385]**

Margaret J. Wheelock, The Wistar Institute, Philadelphia, PA 19104 **[235]**

L. Wolpert, Department of Anatomy and Biology as Applied to Medicine, The Middlesex Hospital Medical School, London, UK **[423]**

Deann Wright, Department of Biochemistry and Biophysics, University of California, San Francisco, CA 94143 **[489]**

Kenneth M. Yamada, Membrane Biochemistry Section, Laboratory of Molecular Biology, National Cancer Institute, Bethesda, MD 20205 **[167]**

Preface

Recent analyses of the primary processes of development using immunological and molecular genetic techniques have opened new possibilities for the understanding of morphogenesis. The advances have occurred, for example, in the arenas of cell division cycle control, morphogenetic cell movements, cell adhesion, and cell death. A number of model systems of great promise have emerged. These include analysis of cell division cycle mutants in yeast, the in vitro culture of neural crest cells, the application of immunological identification methods to fate maps, the molecular genetic analysis of homoeotic mutants in *Drosophila*, and molecular techniques of perturbation of mapping in systems such as the retinotectal projection of frogs. Moreover, great progress has occurred in the analysis of three families of molecules mediating cell contacts at different stages of development—cell adhesion molecules, substrate adhesion molecules, and cell junction molecules.

These diverse sets of experimental approaches have a unitary goal: the understanding of the molecular bases of morphogenesis and pattern formation. This problem, the solution of which would have enormous implications for the fields of development, aging, and disease, also has the most profound implications for evolutionary theory. The central issue is how a one-dimensional genetic code can give rise to a three-dimensional animal. This issue must be framed in cellular terms but ultimately it will be solved only in chemical and molecular terms. The challenge is to discover the molecular determinants of pattern formation. It is of signal importance, therefore, that classical embryologists and developmental biologists make clear to practitioners of the molecular disciplines the main features of pattern specification. Of equal importance is the application of the new techniques of molecular biology within the context of workable developmental systems, ranging from those in tissue culture to the whole animal.

The meeting which formed the basis for this book was designed to encourage such mutual exchange; for continuity's sake, the order of papers in the book differs from the order in which they were delivered during the meeting. This meeting would not have been possible without the generous support of the March of Dimes Birth Defects Foundation, the Sybil and William Golden Foundation, and the Becton and Dickinson Company, as well as the generous help of Messrs. William Golden, Marvin Asnes and Raymond Gilmartin.

Special thanks are also due to Dr. Bruce A. Cunningham of The Rockefeller University and Dr. Jean Paul Thiery of the Institut d'Embryologie, Nogent-Sur-Marne, France for their help in organizing the meeting. Ms. Colette Desbas of the The Rockefeller University and Ms. Lynn Ianni of the UCLA Symposia cheerfully provided the organizational skills that guaranteed the success of the occasion. Finally, gratitude must be expressed to the speakers and chairmen whose papers are presented here for their role in ensuring a lively and scholarly exchange.

Gerald M. Edelman

ALTERNATIVE IMPLEMENTATION MECHANISMS OF EMBRYONIC INDUCTION

Lauri Saxén

Department of Pathology, University of Helsinki, Finland

INTRODUCTION

Over the past 80 years it has become evident that cytodifferentiation and organogenesis are governed by interactive events between cells and cell lineages of different developmental prehistory and different fate. Evidence for the fundamental role of these <u>inductive</u> interactions is largely based on interference experiments in vivo and in vitro; separation of two embryonic cell populations from each other by microsurgery or by inserting physical barriers between them brings their development to a standstill, while their re-conjugation initiates progressive cytodifferentiation and morphogenesis. Such results, while convincing the scientists of the developmental significance of inductive interactions, have frequently led to the obviously erroneous view that the different interactive events might be implemented via similar or identical mechanisms. This unwarranted generalization might be one reason for the failure to unravel these mechanisms and to explain the inductive interactions at a molecular level. Today, when an increasing emphasis is laid on understanding the molecular background of these interactive events, and when the technical ability to do so is available, it seems worthwhile to review some of the biological (supramolecular) features of inductive interactions, i.e. the basis for future molecular approaches. While doing this, I will focus on two reasonably well explored inductive events and show that apart from their developmental significance, they have very little in common as far as our present knowledge reveals. The two examples are the classic neural induction during gastrulation and the induction of the metanephric nephron (Reviews: 1-4).

BIOLOGY OF THE INTERACTIVE EVENTS

Morphogenetically meaningful interactions are known to occur throughout embryogenesis, beginning from early blastula stages up to mature, renewing tissues. Most experimental approaches for the clarification of these interactions focus on one particular stage singled out from a long chain of interactive processes. Consequently, tissues with varying prehistory are explored, and both their developmental options and commitments vary greatly. The induction might thus be either initial, intermediate or terminal. An initial induction acts upon a fully uncommitted, omnipotent target cell population with unlimited developmental options, whereas cells exposed to a terminal inductor might have only two developmental choices, i.e. either to remain at the preinduction stage or to express their ultimate phenotype.

An initial induction is also an example of a directive influence while the terminal induction might represent a permissive event. According to my terminology, a directive influence would instruct multipotent target cells to choose between several developmental options, whereas a permissive interaction merely provides the predetermined cells with the exogeneous clues required for the expression of their already chosen option. Between the initial, directive inductions and the ultimate permissive conditions, varying types of interactions take place and the directive vs. permissive stages of development alternate (5) (Fig. 1).

The neural induction occurs during the invagination of the dorsal blastopore lip during amphibian gastrulation, and the target is the presumptive neuroectoderm. During the prehistory of this ectodermal cell population, its developmental options have apparently not been restricted as shown by numerous experiments where it has been converted to cell types and tissues normally derived from both mesodermal and endodermal germ layers.

In an experimental situation, an isolated fragment of the competent gastrula ectoderm can be directed into these various paths by exposing it to varying inductors, either normal or heterogeneous; a young blastopore lip will induce predominantly forebrain derivatives, while the corresponding region from an advanced gastrula stage converts the target cells into caudal neural structures (6). Similarly, some heterogeneous tissues mimick these varying effects, and by proper choice and combination of such abnormal inductors practically any mature type of

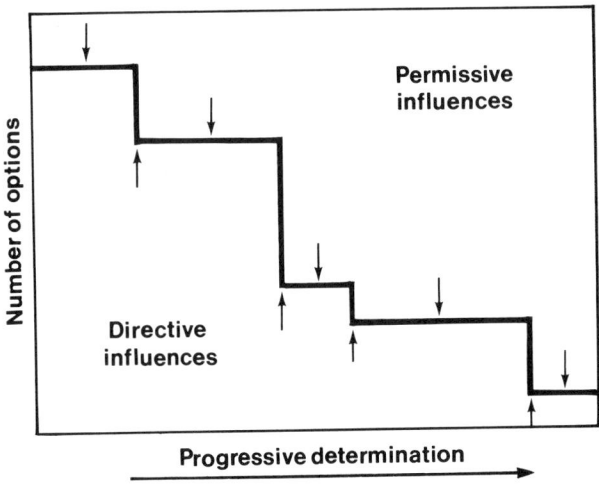

Fig. 1. Schema of the hypothesis of directive vs. permissive inductions (5).

cells can be obtained (7). Hence, there seems to be good reason to consider this induction to be of directive type.

From the complex process of primary induction, an initial neuralizing step can be distinguished experimentally from a secondary action determining the type of the CNS structures to be formed. First, an isolated fragment of the competent ectoderm is exposed either to an inductor known to trigger the formation of the cranial CNS stuctures or to an inductor regularly leading to caudal neural structures. Knowing that the decision to become "neural" is made during the first hours of the inductive contact, the inductor can be removed, and the still fully undifferentiated ectodermal cells will follow the options chosen during the initial period. However, when the inductor has been removed, and the two fragments of the ectoderm are brought together, their ultimate development diverges from that deduced from their inductors and shown in experiments where the two types were subcultured separately. The combined explants show a repertoire of CNS structures both quantitatively and qualitatively different from that expected after the initial neuralization (8).

The initial neuralization can also be distinguished from a subsequent segregation of the CNS by a simple experiment. Ectodermal cells at the cranial, prechordal area of the neural plate are already determined and will form anterior CNS structures when cultivated in vitro after dissection from their natural surroundings. However, when combined experimentally to cells isolated from the caudal axial mesoderm, the ultimate fate of the anterior neural plate cells can be profoundly altered. Depending on the ratio neural/mesodermal cells, a whole array of CNS structures can be obtained (9). It is therefore concluded that the neural induction should be considered directive and initial.

Induction of kidney tubules can be explored with basically similar isolation/conjugation experiments; an isolated metanephric mesenchyme denuded of its epithelial, inductive component can be exposed to either the normal inductor or to a variety of heterogeneous tissues mimicking the action (3, 10). Unlike the gastrula ectoderm, however, this target tissue seems to be strictly restricted in its developmental options; none of the abnormal inductors tested will convert the metanephric mesenchymal cells into anything but epithelial, tubule-forming elements. Conversely, none of the tissues showing this triggering action upon the nephric mesenchyme will cause a similar response in the various embryonic mesenchymes thus far tested (11). Hence, the conclusion seems warranted that the kidney tubule inductors act upon a predetermined target cell population with a definite, epithelial bias to be released by a permissive action.

The above conclusion of the strictly restricted developmental options of the nephric mesenchyme also suggest that we might be dealing with a terminal induction. The epithelial tubules formed after the induction will, though, subsequently segregate into various portions of the nephron. It would therefore be interesting to know whether additional inductive stimuli are required for this - as just shown for the regional specification of the CNS. The experimental design to study such stimuli follows the same lines; the nephric mesenchyme is exposed to an inductor for a limited time after which the induction is known to be completed. For the nephric mesenchyme this critical period is of the order of 24 h whereafter the inductor can be removed, and tubules will form after an additional sub-cultivation of 24 h to 48 h. To examine how far this differentiation, triggered during the 24-h "pulse", will proceed, various segment-specific markers were used: glomeruli were visualized by certain lectins or colloidal iron decorating the anionic surface of the podocytes, the brush border of the proximal tubules was examined in immunofluorescence with a specific antigen or directly visualized in electron microscopy, and, finally, the distal tubules were localized in immunohistology with an antibody against the Tamm-Horsfall protein. The results showed that the initial induction pulse was sufficient to program the segregation of the nephron into its three major portions without any subsequent action of an inductor tissue (12, 13). Hence, with these criteria, induction of the predetermined metanephric mesenchyme would be terminal.

MECHANISM OF TRANSMISSION OF THE INDUCTIVE SIGNALS

As long as the signal substances emitting inductive messages are unknown, we should content ourselves with indirect data on their localization and transmission during the interaction - such information is also essential for the search of the active molecules. Three types of relevant data are available: information about the kinetics of the process (already referred to above), data on the mode of transmission of the signals, and, finally, experimental results related to the spread of the effect within the target tissue. The experimental design to study these components of transmission of the biological effect (induction) falls into different categories.

Kinetics of the interactive process can be explored in various types of conjugation/separation experiments. Johnen (14) used the competent ectoderm and the middle to caudal part of the archenteron roof of Amblystoma embryos, and the removal of the inductor was visually controlled after vital stain of the latter. The results showed that the first neuralized cells were obtained already after 30 min contact with the inductor, and after a total contact of 2 h most of the cells had entered the neural program (Fig. 2). After this initial neuralization stage, a prolonged contact was required to complete the second, regionalizing phase of induction.

In the kidney model-system, use was made of the transfilter technique allowing a complete and easy removal of the inductor tissue. Such experiments by Saxén and Lehtonen (15) revealed a much longer contact time for the tubule induction than for the neural one. Although in our

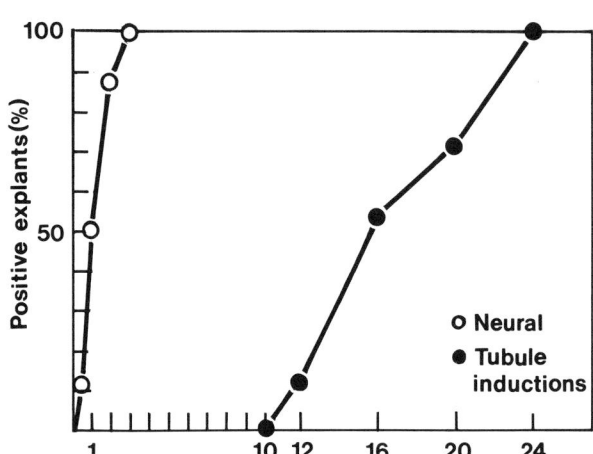

Fig. 2. Kinetics of neural vs. tubule inductions (14, 15).

conditions, an actual inductor-target contact was established through the large pore-size filters in less than one hour, the first cells to become irreversibly programmed towards epithelial direction were detected only after 12 h of contact (Fig. 2). Hereafter new cells gradually entered the program until the response did not differ from that of the controls after some 26 to 28 h (the controls consisted of transfilter explants from which the inductor had not been removed). Thus, though the two investigations cannot be compared directly, the results suggest that the neural induction is of short-term, triggering type while the kidney tubule induction is a time-consuming event. This overt difference might, in part, be explained by the different transmission and spreading mechanisms of the induction in the two model-systems.

Transmission of inductive signals can be investigated both indirectly and directly. Indirect information might be gained by following intercellular transfer of known compounds traceable by various methods (staining, radio-labeling, immunohistology) or by detailed analysis of the inductor/target interphase where the actual exchange of information takes place. Despite many such approaches in the two model-systems discussed here, conclusions remain meager and to a certain extent contradictory (cf. 2, 16, 17). Both transfer of radiolabelled molecules and intimate cell-to-cell contacts have been reported during neural and tubule inductions, but their significance for the actual induction has remained open.

Transfilter experiments have proven more informative: the inductor tissue and the target cells can be separated in vitro by various types of mechanical barriers including filter membranes of known pore diameter (18, 19). By comparing the passage of the inductive signal (recorded by its morphogenetic effect) and the establishment vs. prevention of intercellular contacts through filters of varying pore size, some correlations have been shown (Table 1). The findings suggest that the neural induction is transmitted by signal substances diffusing over an intercellular space of several microns while the kidney tubule induction requires actual intercellular contacts.

TABLE 1

Transmission of neural vs. kidney tubule induction through membrane filters of varying pore diameter as compared with penetration of cytoplasmic processes through the pores (16, 20-22).

Pore diameter µm	Inductions		Cytoplasmic penetration	
	Neural %	Tubule %	Neural	Tubule
0.05	80	0	nd	–
0.1	80	30	–	+
0.2	100	90	nd	++
0.6	100	100	+++	+++

Spread of the "inductive wave" within the target tissue can be thought to be implemented by various mechanisms: (1) by long-range diffusion of the signal substances into the competent target tissue; (2) through an assimilatory, chain-like reaction where one induced cell will carry the message further to the next in contact and so on; or (3) by actual cell movements carrying the induced cells into the deeper layers of the target tissues and leaving space for new cells to approach the inductor.

Both in amphibian embryos and in experimental chick/quail tissue conjugates, an assimilatory mode of spreading has been demonstrated for the neural induction; once induced towards neural direction, the cells can carry the message on to the uninduced cells nearby (23-25).

By an experimental conjugation of induced nephric mesenchymal cells from mouse embryos carrying a chromosomal marker to uninduced, wild-type mesenchymes we have excluded the assimilatory mode of spreading in the kidney tissue (26). Since a long-range transmission was already excluded by the transfilter experiments, spreading based on cell migration became the most plausible mechanism. Interspecies chick/quail conjugates strongly suggested that this is the case (27).

For the time being, I would conclude this paragraph by suggesting that the neural induction is of short duration, carried by diffusible transmitters and further spread by an assimilatory mode. The kidney tubule induction is relatively slow, mediated by direct cell-to-cell contacts and thereafter spread by migratory cells.

TABLE 2

Heterotypic inductors for neural vs. tubule induction (Compiled by Saxén, 34).

Neural induction	Tubule induction
Oocyte nuclei	Spinal cord
Frog retina	Brain tissue
Devitalized liver etc.	Salivary mesenchyme
HeLa-cells	Head mesenchyme
Human serum	Somites
Plant tissues	Neural teratoma

THE SIGNAL SUBSTANCES

The signal substances, still hypothetic but apparent, carry the induction message from the inductor to the target tissue. In fact, both the neural and the tubule inductions have been shown to be reciprocal (10, 28) and, hence, an actual exchange of messages seems to take place. The carrier molecules are not known, but they seem to differ in several respects in the two model-systems under exploration.

Neural induction can be mimicked with a great variety of heterogeneous inductors including adult mammalian tissues, plant tissues, devitalized material, body fluids, etc. (Table 2). Most important, neural induction has also been released by various chemicals, cell-free fractions and even small cations (cf. 29, 30). Some well characterized compounds transmitting neural induction have been isolated from chick embryos, and also to some extent from amphibian gastrulae. This protein is not an integral part of the plasma membrane; its molecular weight is 10.000 to 16.000 daltons, and it is inactivated by trypsin treatment but not by removal of its RNA (30). Covalent binding of this fraction with a neuralizing action to Sephadex beans does not abolish its activity, and this has led to the conclusion that neural induction acts upon receptors at the surface of the target cells (31, 32). Inhibition of neural induction by certain lectins attached to the cell surface have been considered further evidence of the above conclusion (17, 33).

Tubule induction, likewise, can be mimicked with many heterogeneous tissues (Table 2), but in strict contrast with neural induction, not by devitalized tissues or cell-free preparations (15, 37). Nothing is known of the actual molecules obviously acting as transmitters for tubule induction in these tissues. Thus far, their immediate consequences in the target tissue do not permit any far-reaching conclusions on the type of compounds involved. As such early events preceding morphogenesis in the nephric mesenchyme, a stimulation of the DNA synthesis and a probable activation of proteolytic enzymes have been reported (35, 36).

TABLE 3

Some features of neural vs. kidney tubule inductions in comparison.

	Neural	Tubule
Type	Initial Directive	Terminal Permissive
Transmission	Short-term By diffusible molecules Spread by assimilatory induction	Prolonged Via cell contacts Spread by cell migration
Heterogeneous inductors	Devitalized tissues Cell-free fractions Chemical compounds	Living cells only

CONCLUSIONS

The many basic differences in the features of the neural vs. tubule inductions are summarized in Table 3. The data, though still scattered and often superficial, may warrant the conclusion that while refering to "embryonic induction", we include in these events a great variety of interactive events with morphogenetic consequences but with very different implementation mechanisms. This should be kept in mind while exploring these central developmental control processes and, especially, when making an effort to find common denominators and generalizations.

REFERENCES

1. Saxén L, Toivonen S (1962). "Primary Embryonic Induction." London: Academic Press.
2. Saxén L, Toivonen S. Neural induction. In Cowan WM (ed): "Handbook of Physiology. Development of the Nervous System," The American Physiological Society (in press).

3. Saxén L, Koskimies O, Lahti A, Miettinen H, Rapola J, Wartiovaara J (1968). Differentiation of kidney mesenchyme in an experimental model system. In Abercrombie M, Brachet J, King TJ (eds): "Adv. Morphog. Vol. 7," London: Academic Press, p 251.
4. Saxén L, Ekblom P, Lehtonen E (1981). The kidney as a model system for determination and differentiation. In Ritzen M, Hall K, Zetterberg A, Aperia A, Larsson A, Zetterström R (eds): "Biology of Normal Human Growth," New York: Raven Press, p 117.
5. Saxén L (1977). Directive versus permissive induction: a working hypothesis. In Lash JW, Burger MM (eds): "Cell and Tissue Interactions," New York: Raven Press, p. 1.
6. Mangold O, Spemann H (1927). Über Induktion von Medullarplatte durch Medullarplatte im jüngeren Keim, ein Beispiel homöogenetischer oder assimilatorischer Induktion. W Roux'Arch EntwickMech Org 111:342.
7. Toivonen S, Saxén L (1955). The simultaneous inducing action of liver and bonemarrow of the guinea pig in implantation and explantation experiments with embryos of Triturus. Exp Cell Res Suppl 3:346.
8. Saxén L, Toivonen S, Vainio T (1964). Initial stimulus and subsequent interactions in embryonic induction. J Embryol Exp Morph 12:333.
9. Toivonen S, Saxén L (1968). Morphogenetic interaction of presumptive neural and mesodermal cells mixed in different ratios. Science 159:539.
10. Grobstein C (1955). Inductive interaction in the development of the mouse metanephros. J Exp Zool 130:319.
11. Saxén L (1970). Failure to demonstrate tubule induction in a heterologous mesenchyme. Dev Biol 23:511.
12. Ekblom P, Miettinen A, Virtanen I, Dawnay A, Wahlström T, Saxén L (1981). In vitro segregation of the metanephric nephron. Dev Biol 84:88.
13. Lehtonen E, Jalanko H, Laitinen L, Miettinen A, Ekblom P, Saxén L (1983). Differentiation of metanephric tubules following a short transfilter induction pulse. Roux'Arch Dev Biol 192:145.
14. Johnen AG (1961). Experimentelle Untersuchungen über die Bedeutung des Zeitfaktors beim Vorgang der neuralen Induktion. Arch EntwMech Org 153:1.
15. Saxén L, Lehtonen E (1978). Transfilter induction of kidney tubules as a function of the extent and

duration of intercellular contacts. J Embryol Exp Morphol 47:97.
16. Lehtonen E (1976). Transmission of signals in embryonic induction. Med Biol 54:108.
17. Grunz H (1984). Early embryonic induction: the ectodermal target cells. In Duprat A-M, Kato AC, Weber M (eds): "The Role of Cell Interactions in Early Neurogenesis," New York, London: Plenum Press, p 21.
18. Grobstein C (1956). Trans-filter induction of tubules in mouse metanephrogenic mesenchyme. Exp Cell Res 10:424.
19. Saxén L (1961). Transfilter neural induction of Amphibian ectoderm. Dev Biol 3:140.
20. Toivonen S, Tarin D, Saxén L, Tarin PJ, Wartiovaara J (1975). Transfilter studies on neural induction in the newt. Differentiation 4:1.
21. Saxén L, Lehtonen E, Karkinen-Jääskeläinen M, Nordling S, Wartiovaara J (1976). Are morphogenetic tissue interactions mediated by transmissible signal substances or through cell contacts? Nature 259:662.
22. Toivonen S (1979). Transmission problem in primary induction. Differentiation 15:177.
23. Deuchar EM (1971). Transfer of the primary induction stimulus by small numbers of amphibian ectoderm cells. Acta Embryol Exp 2:93.
24. Rasilo M-L, Leikola A (1976). Neural induction by previously induced epiblast in avian embryo in vitro. Differentiation 5:1.
25. Kurihara K, Sasaki N (1981). Transmission of homoiogenetic induction in presumptive ectoderm of newt embryo. Develop Growth Differ 23:361.
26. Saxén L, Saksela E (1971). Transmission and spread of embryonic induction. II. Exclusion of an assimilatory transmission mechanism in kidney tubule induction. Exp Cell Res 66:369.
27. Saxén L, Karkinen-Jääskeläinen M (1975). Inductive interactions in morphogenesis. In Balls M, Wild A (eds): "The Early Development of Mammalian," Cambridge: Cambridge University Press, p 3.
28. Nieuwkoop PD, Weijer CJ (1978). Neural induction, a two-way process. Med Biol 56:366.
29. Barth LG, Barth LJ (1959). Differentiation of cells of the Rana pipiens gastrula in unconditioned medium. J Embryol Exp Morphol 7:210.
30. Tiedemann H (1984). Neural embryonic induction. In Duprat A-M, Kato AC, Weber M (eds): "The Role of Cell

Interactions in Early Neurogenesis," New York, London: Plenum Press, p 89.
31. Tiedemann H, Born J (1978). Biological activity of vegetalizing and neuralizing inducing factors after binding to BAC-cellulose and CNBr-Sephrarose. W Roux Arch 184:285.
32. Born J, Grunz H, Tiedemann H, Tiedemann H (1980). Biological activity of the vegetalizing factor: decrease after coupling to polysaccharide matrix and enzymatic recovery of active factor. W Roux 189:47.
33. Takata K, Yamamoto KY, Takahashi N (1984). A molecular aspect of neural induction in Cynops presumptive ectoderm treated with lectins. In Duprat A-M, Kato AC, Weber M (eds): "The Role of Cell Interactions in Early Neurogenesis," New York, London: Plenum Press, p 83.
34. Saxén L (1985). Morphogenetic interactions: two model-systems in comparison. In Andersson LC, Ekblom P, Gahmberg CG (eds): "Gene Expression During Normal and Malignant Differentiation," Academic Press, p 1.
35. Ekblom P, Lehtonen E, Saxén L, Timpl R (1981). Shift in collagen type as an early response to induction of the metanephric mesenchyme. J Cell Biol 89:276.
36. Saxén L, Salonen J, Ekblom P, Nordling S (1983). DNA synthesis cell generation cycle during determination and differentiation of the metanephric mesenchyme. Dev Biol 98:130.
37. Auerbach R, Grobstein C (1958). Inductive interaction of embryonic tissues after dissociation and reaggregation. Exp Cell Res 15:384.

CYTOPLASMIC LOCALIZATION, INDUCTIONS, AND THE "ORGANIZER" OF THE FROG EMBRYO[1]

Robert L. Gimlich[2]

Department of Molecular Biology, University of California Berkeley, California 94720

ABSTRACT A cortico-cytoplasmic reorgainzation in the uncleaved frog egg ultimately determines the orientation and completeness of the embryonic body axis. This paper describes an experimental study of steps in this determination. The results indicate that a specialization of one quadrant in the vegetal half of the early embryo is both neccessary and sufficient to initiate normal body axis development. This quadrant, centered on the future dorsal midline, contains precursors to the chordamesoderm, prechordal plate, head mesoderm, and anterior endodermal structures. Inductive interactions within this region promote normal development of the prospective axial mesodermal component, and the later inductive influence of the axial mesoderm may be responsible for specification of the remainder of the dorsal axial structures.

INTRODUCTION

The fertilized eggs of many different animal species accomplish morphologically detectable redistributions of their contents early in development. There is evidence in some cases that such rearrangements are important in specializing certain egg regions so that the cells which cleave from these regions have particular morphogenetic behaviors and developmental fates (reviewed in 1). It has

---

[1] This work was supported by USPHS training grant GM07232 and USPHS research grant GM19363 to J.C. Gerhart.
[2] Present address: Department of Cell Biology, Baylor College of Medicine, Houston, Texas 77030.

been proposed that ooplasmic rearrangements localize "morphogenetic determinants" which influence the development of the cell lineages which inherit them (reviewed in 1, 2).

In the amphibian egg, fertilization or activation provoke a reorganization of the cytoplasm and cortex during the first cell cycle (3, 4, 5, 6). This reorganization involves an oriented shift of the endoplasm relative to the egg surface (6, 7, 8), and the geometry of the shift faithfully predicts the position of the future dorsal midline. Interference with the progress of the movement has interesting developmental consequences. For instance, brief cold shock and ultraviolet light irradiation of the vegetal egg surface both cause a dose-dependent inhibition of the pre-cleavage reorganization (6, 9). These same treatments, together with hydrostatic pressure shock (10) and brief nocadazole treatment (S. Scharf, personal communication), cause a syndrome of deficiency in dorsal structures of the embryonic body axis (11, 12, 13). The syndrome may be characterized as a dose-dependent, anterior-to-posterior truncation of the body axis. At high doses, most embryos fail to develop any of the dorsal axial structures--the notochord, somitic mesoderm, and central nervous system. Instead, they develop into radially symmetrical "belly piece" embryos (14), with a ciliated epidermis, red blood cells, mesenchyme, and a rudimentary gut (10).

It is remarkably easy to rescue embryos from the effects of the "axis reducing" treatments, and it is hoped that the means of rescue will reflect the normal steps in axis formation. For example, a simple temporary rotation of the UV irradiated egg 90° away from its normal orientation in the gravitational field is sufficient to rescue complete body axis development (10, 13, 15). The rotation operation displaces the dense vegetal yolk mass, resulting in a cytoplasmic rearrangement similar to that which the egg normally makes (7, 16, 17), and the gravity-driven movement is clearly sufficient to mimic the developmental effects of the normal process completely. Rotation is even effective when applied prior to UV, cold, or pressure shocks (10, 13). These results indicate that the axis-reducing treatments do not simply destroy preformed or prelocalized components needed for axis formation--axial structure "determinants." Instead, the treatments act by interfering with a process involving cytoplasmic displacement whereby determinants are effectively localized or locally activated.

Of course, the possible nature of the axial structure determinants is a mystery. One test for the presence of morphogenetic determinants is transfer of cytoplasm from one site to another in the early embryo (e.g., 18). To date, however, cytoplasm transfers in amphibian embryos have failed to produce ectopic dorsal axial structures reproducibly (D. Gardiner, personal communication; M. Danilchik, personal communication). Curtis (19, 20) was able to obtain secondary body axes in Xenopus embryos by transplanting pieces of the equatorial egg cortex from the future dorsal side to the future ventral side. He concluded that the prospective dorsal "grey crescent" region of the egg cortex is a site of determinants for dorsal structure specification. There are, however, other possible explanations of Curtis' results (reviewed in 21, 22).

Another way to detect the positions of determinants involved in axis formation is to transplant groups of cells from one region to another in the early embryo. This approach yielded the discovery that the prospective chordamesoderm region of the early gastrula stage embryo has unique properties. This group of cells, normally located just above the dorsal lip of the blastopore, can, after transplantation to a new embryonic site, organize around itself an entire ectopic body axis, incorporating surrounding cells whose new fates are set by their proximity to the transplanted tissue (23, 24, 25, 26). The prospective chordamesoderm has therefore been named the embryonic "organizer," and the interactions by which it induces central nervous system formation and patterns the mesodermal layer have been subjects of study for many years (25, 27, 28, 29).

The experiments which I will summarize in this report employ a cell transplantation approach to investigate steps in the development of the organizer in Xenopus laevis embryos. Several questions have been addressed. Does the organizer gain its special properties by virtue of inheriting determinants localized in the equatorial grey crescent cytoplasm or cortex? Alternatively, does some other earlier specialized region induce the equatorial cells of the prospective organizer to develop as chordamesoderm and exhibit organizer activity? What is the normal role of the prospective chordamesoderm in specifying the medio-lateral pattern of morphogenesis in the mesodermal layer? The results confirm and extend the earlier work of others (25, 30, 31), showing that regional interactions are important at several steps in embryonic axis formation. The observations suggest strategies for studying the mechanism of an early inductive interaction, and for determining which aspects of the early ooplasmic rearrangement process are involved in pattern specification in the embryo.

RESULTS

Rescue of axis formation by vegetal cell transplantation.

The recipient embryos in these cell transplantation experiments were derived from eggs irradiated over the vegetal surface with ultraviolet light before first cleavage. As mentioned above, such eggs develop as radially symmetrical, axis deficient embryos. They usually cleave normally, and complete gastrulation by closing the blastopore. They fail in neurulation and also in anterior-posterior extension and medio-lateral convergence within the mesodermal layer (32; and Keller et al., in this volume). It has been suggested that cells of the irradiated embryo are competent to respond to a grafted normal organizer by contributing to the dorsal structures of a graft-induced body axis (33). Thus embryos cleaved from irradiated eggs should provide an excellent environment in which to test the developmental capacities of transplanted normal cells thought to be involved in axis development. To identify the locations of axial determinants in the normal embryo, the following approach was used. A few blastomeres of an irradiated embryo were removed at an early cleavage stage, and replaced with their counterparts from an untreated embryo of the same stage. In this way it was possible to test whether certain donor blastomeres are able to promote normal axis formation in the axis-deficient background of the irradiated recipient. In the first experiments, vegetal cells from various regions of the prospective endoderm were transplanted, to test the proposals of Nieuwkoop (30, 31), regarding the ability of vegetal regions of the midblastula stage embryo to induce the formation of mesoderm by cells of the animal hemisphere.

In fact, as illustrated in Fig. 1a, axis formation can be rescued by transplants of certain of the vegetal-most cells of the 64-cell embryo (34). The vegetal tier of cells at this stage usually contains 6-8 blastomeres. Cells were transplanted from each of the four quadrants at this level, which will be distinguished as follows. The "dorsal-most" quadrant is centered on a meridian (the prospective dorsal

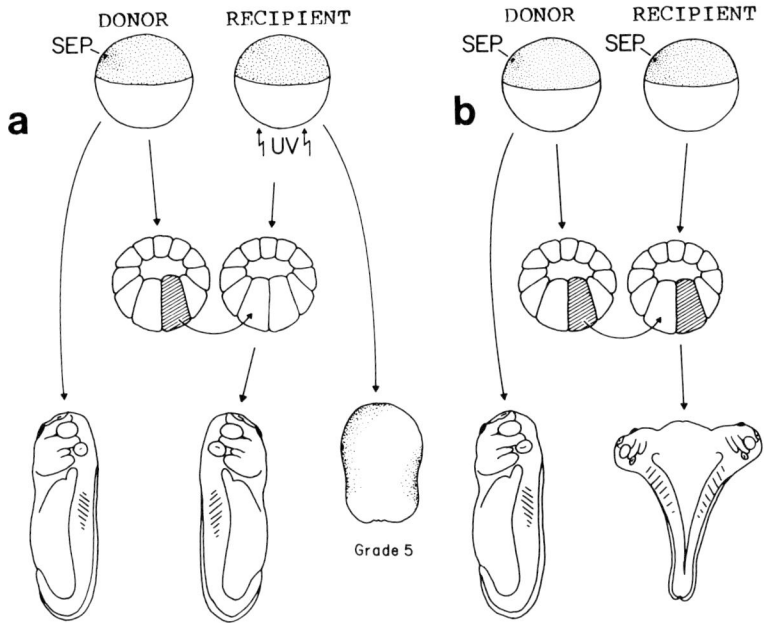

FIGURE 1. a. Schematic diagram of rescue by blastomere transplantation. Eggs were fertilized and the SEP marked with a vital dye spot. Sibling embryos were UV-irradiated as described in (34). At the 64-cell stage (middle, shown in sagittal section) the two blastomeres in the dorsal-most quadrant of the vegetal level (crosshatched) were transplanted into a cavity made in the irradiated recipient by plucking out a pair of cells. Recipients of dorsal-most vegetal blastomeres (bottom) show rescue of axis development, whereas UV-irradiated controls develop into radially symmetrical embryos lacking axial structures. b. Second axis formation resulting from transplantation of dorsal-most vegetal cells (crosshatched) into the ventral-most vegetal quadrant of a normal embryo at the 64-cell stage. Control embryos had ventral-most vegetal blastomeres transplanted into the ventral-most quadrant.

midline) opposite that containing the visible point of sperm entry (SEP; 22). The lateral-most and ventral-most quadrants are centered 90° and 180°, respectively, away from the prospective dorsal midline. When the transplants

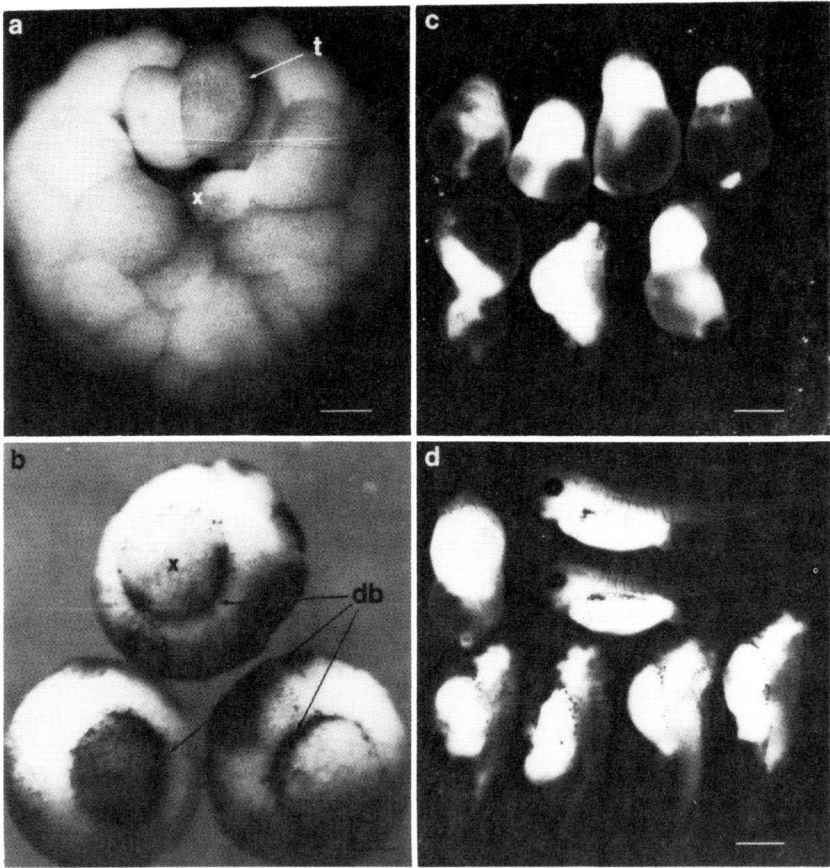

FIGURE 2. a. Vegetal pole (x) view of an irradiated recipient embryo with a cavity in the vegetal hemisphere into which two dorsal-most vegetal blastomeres (t) from an unirradiated donor have been implanted (bar, 0.2 mm). b. Vegetal pole view of dorsal-most blastomere recipients at the early gastrula stage. Like normal embryos, these recipients show a typical dorsal blastopore lip (db) where invagination occurs first (bar, 0.5 mm). c. Control recipients of vegetal blastomeres from synchronous irradiated siblings, shown at an age equivalent to stage 41 of unirradiated embryos (35). d. Batch of recipients of dorsal-most vegetal blastomeres, shown at stage 41. Note the extensive rescue compared to the control embryos of panel c (bar, 1 mm). From (34).

originated in the future dorsal quadrant, recipient embryos displayed a striking degree of body axis development, compared to control recipients of blastomeres from their irradiated siblings (Fig. 2). In contrast, transplants from the lateral-most and ventral-most vegetal regions did not significantly rescue axis formation. As illustrated in Fig. 1b, the same cells often cause a second body axis to develop when transplanted into the ventral-most vegetal region of a normal host embryo at the 64-cell stage (34).

It is important to note that the frequency of complete rescue by vegetal cell transplants is only about 15%. Many of the recipients develop partial body axes, and about 15% are not rescued at all. In fact, the set of morphologies which results from this transplantation conform closely to that which results from mild UV, cold, or hydrostatic pressure doses (10), and the results can be scored according to the "index of axis deficiency" (IAD) of Scharf and Gerhart (13). The IAD assigns to each embryo a grade of 0-5, based on the completeness of body axis development. A grade of 0 denotes normal development, while grades 1-5 denote axis deficiencies ranging from microcephaly through acephaly to complete absence of dorsal axial structures. This scale can be used to relate monotonically the dose of UV, cold, or pressure and the extent to which definitive axial structures are missing in the embryo. Interestingly, the same series of morphologies characterizes the dose-dependent rescue of axis formation by experimental gravity-driven cytoplasmic rearrangement in UV, cold, or pressure treated eggs (13).

To understand how the rescue occurs, it was neccessary to determine the fates of the transplanted vegetal cells in normal development and in the recipient embryos. This was accomplished by labeling the cells with the microinjectable cell lineage tracer fluorescein-dextran-amine (FDA; 36). As suspected on the basis of existing fate maps (37, 38), the 64-cell stage dorsal-most vegetal cells normally contribute only to the endoderm of the tadpole. When FDA-labeled blastomeres are transplanted into unlabeled irradiated host embryos, their progeny also contribute only to the gut in well-rescued embryos (34). Thus the dorsal mesodermal structures and central nervous system in these embryos forms of host cells as a result of the inductive influence of progeny of the transplanted vegetal blastomeres.

These simple experiments, which extend the results of Nieuwkoop and Ubbels (39), show that mesodermal structures can form from competent animal or equatorial cells as the result of an inductive influence from the vegetal core of the embryo. In the best cases, the special properties of these few cells from the prospective endoderm can initiate the formation of a full set of body axis structures with an entirely normal organization.

Developmental autonomy of the equatorial region.

One way to test the involvement of the inductively active vegetal region in the process of axis formation in normal embryos is to remove the vegetal blastomeres at an early stage. It had been previously reported (40) that deletion of all of the eight vegetal-most cells from the 32-cell Xenopus embryo often did not prevent formation of notochord tissue or muscle by the remaining embryonic fragment. I repeated this operation to asses its effects on the organization of the body axis. The eight vegetal cells were removed from 32-cell embryos which displayed a regular

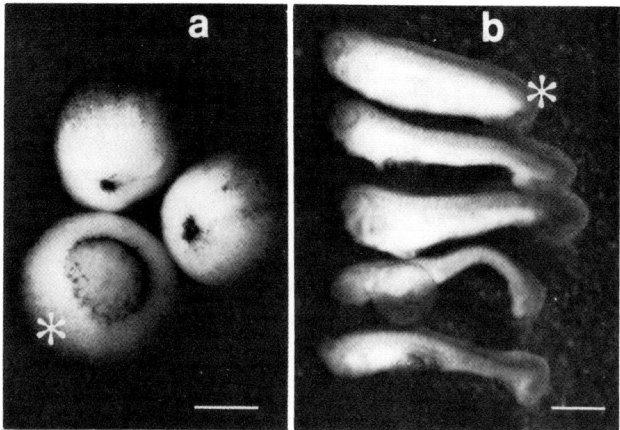

FIGURE 3. The results of vegetal tier deletion. The eight cells of the vegetal-most tier at the 32-cell stage were removed from normal embryos. The gap in these embryos healed within one hour. a. Two such embryos are shown at the early gastrula stage, with a synchronous unoperated embryo (asterisk; bar, 0.5 mm). b. Four operated embryos at control stage 29, with an unoperated control embryo (asterisk). The bottom two embryos are acephalic, with otocysts, somites, and , as histological observation shows, an anteriorly truncated or absent notochord (bar, 1 mm).

cleavage pattern, with the vegetal cells separated from their equatorial neighbors by horizontal cleavage furrows. In such a cleavage pattern, the progeny of the vegetal cells are subblastoporal in position at the early gastrula stage, and contribute primarily to endodermal structures in the tailbud stage embryo (37, 41). Fig. 3a shows two of the experimental embryos at the early gastrula stage. In these embryos the bottle cells of the early blastopore lip are all concentrated near the new vegetal pole, and there is little or no superficial subblastoporal material. Nevertheless, these embryos proceed to gastrulate, and some of them develop a well-organized and complete set of axial structures (Fig 3b). In some cases the only externally obvious defect is the small size of the gut. In others, severe deficiencies in axial structure formation result from vegetal cell deletion. Two such embryos are shown in Fig. 3b. Defects include cyclopy, acephaly, and in 3 cases out of a total of 24, complete axis deficiency.

Clearly an inductive influence from the vegetal core of the embryo is often not essential after the 32-cell stage for at least partial axis formation. This result prompted experiments which test directly the degree of developmental autonomy of cells in the equatorial, prospective mesodermal region of the early embryo. As in the initial experiments, UV-irradiated eggs provided recipient embryos which are normally devoid of dorsal axial structures. Blastomeres from the two middle tiers (tiers 2 and 3) of normal 32-cell embryos were transplanted into the equatorial level of recipients, and the degree of axis formation was scored (Fig. 4a). Transplantation was begun at the 32-cell stage because this is the earliest time at which horizontal cleavage planes have divided an equatorial area which contains much of the prospective mesoderm from a vegetal level which contains mostly prospective endoderm (37, 41).

When these equatorial cell transplants originated in the quadrant centered about the future dorsal midline, the irradiated recipients often developed partial or complete body axes. Again, the set of morphologies which results from such transplantations closely resembles those described by the IAD series (13). Therefore, the result of dorsal equatorial blastomere transplantations will also be termed a rescue of axis formation.

Similar transplants from the two equatorial tiers in the ventral or lateral quadrants did not cause a significant degree of rescue in irradiated recipients (41), even though these cells would have contributed the bulk of the somitic

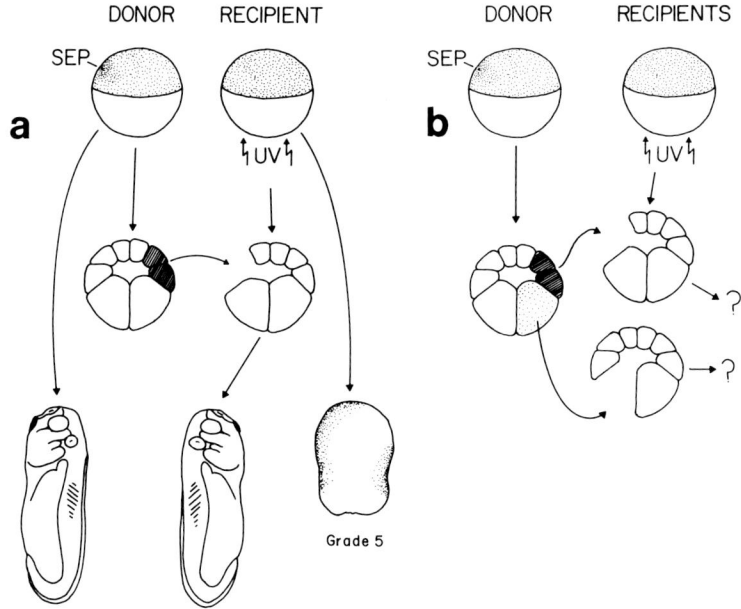

FIGURE 4. a. Rescue of axis formation by equatorial cell transplantation. Donor eggs were marked near the SEP, and sibling eggs were UV-irradiated on the vegetal surface. At the 32-cell stage (middle, in sagittal section) four cells of tiers 2 and 3 from the dorsal quadrant were transplanted into a cavity made in a synchronous recipient by removing four equatorial cells. Recipient embryos often showed rescue of axis formation, conmpared to control irradiated siblings. b. Equatorial and vegetal cell transplantation from single donors. Donor and irradiated recipient embryos were prepared as usual. At the 32-cell stage, the four cells of tiers 2 and 3 in the dorsal quadrant were transplanted into one recipient, and the vegetal cell pair (tier 4) was implanted into another. Recipients were cultured as matched pairs, and the completeness of axis formation was scored at stage 39-41 of controls (41).

mesoderm in normal development (37; A. Smallcombe and J. Cooke, personal communication). Thus at the 32-cell stage, only the dorsal-most equatorial cells have the capacity to initiate body axis formation in the irradiated recipient.

The locus of the rescue activity in the equatorial region is even further restricted. In 32-cell donor embryos with a strictly horizontal plane of third cleavage, only the pair of cells in tier 3 of the dorsal quadrant are able to rescue body axis formation when transplanted into an irradiated recipient. Transplants of tier 2 cells from this quadrant cause partial rescue only when the third cleavage plane is oblique, so that they incorporate some of the unpigmented equatorial egg surface. This result is in accord with observations on the developmental capacities of isolated animal cell quartets of 8-cell embryos, from which the tier 2 cells are derived (42, 43, 44, 45, 46, and Gurdon et al., in this volume).

Rescue of axis formation by the dorsal equatorial blastomeres results from their autonomous ability to develop as axial mesoderm and to organize surrounding host cells into the axial structures. This was shown by clonal marking in normal embryos and in transplant recipients. When the two tier 3 dorsal cells in normal embryos were microinjected with FDA, their progeny were found to populate much of the notochord in the tailbud stage embryo. They also contribute to parachordal somitic mesoderm, head mesenchyme, spinal cord floorplate, archenteron roof, and antero-ventral endoderm (41).

When transplants of these cells cause extensive rescue of axis formation, clonal marking with FDA shows that they make a contribution to the body axis which is strikingly similar to their contribution in normal development. Fig. 5a shows the positions of the transplanted cells' progeny in a case of rescue to IAD grade 1. In this embryo the transplant formed the entire notochord, together with parachordal somite mesoderm, neural tube floorplate, anterior mesenchyme, and branchial and pharyngeal endoderm. Cells of the irradiated host embryo contributed most of the somite mesoderm, central nervous system, and other structures which would not have developed in the absence of the transplanted normal cells (41).

As in the case of rescue by vegetal cell transplantation, the degree of body axis formation following equatorial cell transplantation is quite variable. When rescue is only partial, the clonal marking shows that the transplanted cells fail to develop chordamesoderm autonomously. Fig. 5b shows the distribution of donor cell progeny in a recipient which developed only a grade 3 body axis. Here the transplanted cells contributed only somitic mesoderm, with the somites fused across the dorsal midline.

The failure of many of the 32-cell stage equatorial transplants to develop completely autonomously as chordamesoderm is not attributable to mechanical or physiological damage during transplantation. Dorsal equatorial blastomeres transplanted from one normal embryo to another seldom fail to make a normal contribution to the body axis.

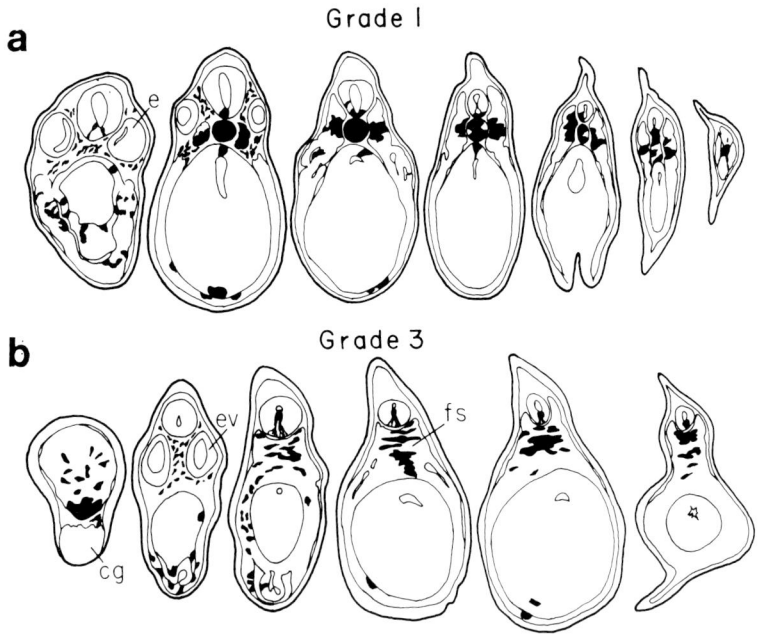

FIGURE 5. Clonal contributions of transplanted dorsal tier 3 blastomeres in rescued recipients. These are camera lucida tracings of the distributions of the progeny of FDA labeled transplant cells. a. Recipient rescued to grade 1. The transplant contributed most of the chordamesoderm, along with some anterior endoderm, head mesenchyme, somitic mesoderm, and ventral neurectoderm (e., optic vesicle). b. Recipient which developed a grade 3 level of axial organization. The transplant cells formed anterior mesenchyme and muscle, and contributed to ventral neurectoderm, lateral mesoderm, and heart. The notochord is absent, and somites are fused across the dorsal midline beneath the neural tube (e.v., otic vesicles; c.g., cement gland; f.s., fused somites).

Organizer Development in *Xenopus* 27

Equatorial cell rescue capacity increases during cleavage.

If the rescue of body axis formation in irradiated recipient embryos is an adequate assay of the specification of cells to form the chordamesoderm, then the average degree of rescue should progressively increase as the transplants are made at successively later cleavage stages. Even in cell explantation experiments the dorsal equatorial region is autonomous in forming the notochord and acting as the organizer by the late blastula stage (47). In fact, this is the case. Only 15% of the 32-cell stage equatorial cell recipients developed as grade 0 or 1 embryos, with an entire complement of dorsal axial structures. Approximately five hours later, at stage 9, the dorsal equatorial region in over 90% of the donor embryos is capable of causing development of a complete body axis in an irradiated recipient (41).

Complete distribution of rescue activity in individual donor embryos.

The results described so far show that the dorsal equatorial region of the early embryo sometimes contains the components needed to initiate body axis development in the absence of the normally adjacent vegetal cells. The frequency with which equatorial region transplants express developmental autonomy increases gradually throughout the blastula period. At the late blastula stage, most dorsal equatorial transplants can rescue complete axis formation in an irradiated recipient embryo. It is reasonable to suppose that developmental autonomy in the region of prospective chordamesoderm is promoted by the inductive activity of vegetal prospective endodermal cells. However, vegetal cell transplants within each batch of donors also show a quite variable capacity for rescue (34). One way to rationalize these results is to propose that the components needed to rescue axis formation are not predictably partitioned by the early cleavage planes. To test this proposal, I transplanted the vegetal cells and the equatorial cells of the dorsal quadrant into separate irradiated recipient embryos, as diagrammed in Fig. 4b.

These paired transplantations were made from 32-cell stage donors into synchronous recipients. Dorsal quadrant cells of tiers 2 and 3 replaced four cells from the same tiers of one recipient. The pair of tier 4 blastomeres

replaced two vegetal cells in another recipient. Donor
embryos were selected on the basis of a very regular pattern
of cleavage, with the four cell tiers separated by
horizontal planes of division. Vegetal or equatorial cells
of matched unoperated embryos in some experiments were
injected with FDA to study their normal fates. The
recipients were cultured as matched pairs and scored for the
completeness of the body axis at a time equivalent to stage
39-41 of controls.

The degree of rescue achieved in these pairs of
recipients is illustrated in Fig. 6. In general, when the
equatorial cell transplant showed extensive developmental
autonomy by causing complete rescue of body axis formation,
the associated vegetal inductive activity was weak.
Conversely, when the vegetal transplant could cause nearly
complete rescue, the dorsal equatorial cells showed limited
autonomy, supporting development of only a grade 3-5 level
of axial organization. In roughly half of the 32-cell stage
donors, both transplants caused an intermediate grade of
rescue (grades 2-3). There were no donors in which both the
equatorial and vegetal cells could completely rescue axis

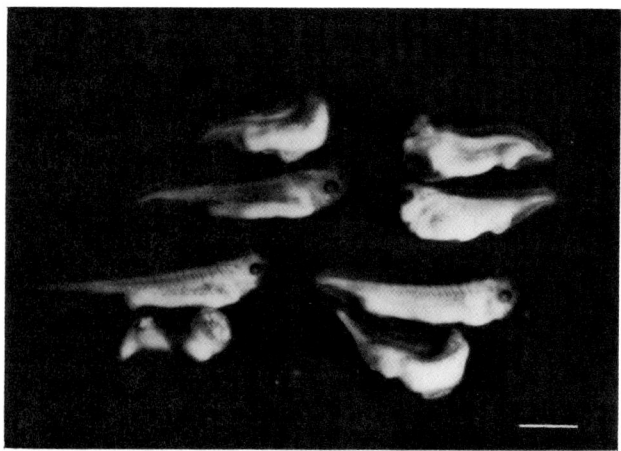

FIGURE 6. Paired recipients of vegetal and equatorial
cell transplants from the dorsal quadrant of the same donor.
Four pairs of recipients are shown. In each pair, the top
embryo was the recipient of a dorsal tier 3 blastomere
transplant, and the bottom embryo received a transplant of
the adjacent dorsal tier 4 cells. Recipients are shown at
an age equivalent to stage 39 of controls (bar, 1 mm).

formation (41).

This experiment shows very clearly that the positioning of "rescue activity" in normal embryos varies without exact regard to the positions of early embryonic cleavage planes. This is true even within a set of embryos selected on the basis of a particular regular cleavage pattern. Correlated clonal marking in such selected embryos shows that the progeny of dorsal equatorial and vegetal blastomeres make a predictable contribution to the various structures of the tailbud stage embryo. The region of overlap between dorsal tier 3 and tier 4 clonal contributions is mostly in the area of the head mesenchyme, the heart, and the pharyngeal and branchial endoderm (41).

DISCUSSION

On the basis of the present results and those of others, a tenative sequence of events in embryonic axis formation will be outlined. A pre-cleavage rearrangement involving a shift of endoplasm relative to the egg surface (3, 6) initially specializes a particular egg region. This specialization becomes incorporated into blastomeres of a unique subequatorial quadrant of the embryo during the early cleavage period. These blastomeres thus inherit determinants enabling them to initiate axis formation, as shown in the transplantation experiments. Cells which inherit these determinants include endoderm precursors in the future dorsal quadrant of the 32-cell embryo, and also equatorial cells which will contribute both endoderm and dorsal axial mesoderm. The latter cells extend deeply within the blastula, and later divide paratangentially to produce a deep cell population and a more superficial layer which, as preliminary experiments show, consists almost entirely of prospective chordamesoderm and pharyngeal endoderm. Yet, as early as these cell layers can be separated for transplantation, the prospective chordamesoderm component shows some frequency of developmental autonomy (Gimlich, unpublished observations). Thus it may be difficult to decide whether axial determinants function strictly be establishing inductive activity in a particular set of cell lineages separate from those which produce axial mesoderm. It is clear that, at least in some embryos, the developmental autonomy of the chordamesoderm rudiment is acquired gradually through the blastula period, probably as a result of the inductive influence of nearby vegetal cells. At gastrulation, the prospective chordamesoderm undertakes a vigorous and autonomous repacking of its constituent cells, and this behavior is thought to be crucial for involution of the mesodermal layer, blastopore closure, and anterio-posterior elongation of the embryo (32; Keller et al., in this volume). As gastrulation proceeds, a dorsal strip of the mesodermal layer acts to induce neural plate formation by the overlying ectodermal sheet (27).

An early specialization of one dorsal-most sector of the embryo is sufficient to account for the generation of medio-lateral and anterio-posterior pattern in the mesodermal layer. In the present experiments, rescue of axis formation by transplanted equatorial cells involved induction of host cells to form axial mesodermal structures, sometimes with entirely normal organization (41). The time course of this induction is not known. Other workers (23, 28, 25) have shown clearly that patterning of the mesodermal mantle can be accomplished by inductions from the chordamesoderm rudiment which occur even after the onset of gastrulation. Since transplants of equatorial or vegetal cells from outside the early dorsal-most quadrant are ineffective in causing a rescue of axis formation (34, 41), it seems likely that similar interactions are involved in normal embryogenesis. In preliminary experiments with 16-cell stage X. laevis embryos, I have deleted or transplanted the two dorsal-most vegetal cells. These blastomeres divide horizontally at the fifth cleavage, giving rise to both the vegetal-most (tier 4) and the equatorial (tier 3) cells of the dorsal quadrant in the 32-cell embryo. Removal of this cell pair often causes complete deficiency in development of axial structures by the remaining fragment, even though the bulk of the somitic mesoderm, for instance, would have formed from the lateral and ventral quadrants in normal development.
Transplantation of the same cell pair into the vegetal level of a UV-irradiated recipient embryo gives substantial rescue of axis formation in most cases. Therefore, the properties of cells in a restricted dorsal vegetal region of the early cleavage stage embryo are both neccessary and sufficient for axis formation in experimental situations.

The results of these cell transplantations are compatible with the recent biochemical observations of Gurdon and his colleagues. This group has reported on the accumulation of messenger RNA transcribed from α-actin genes in Xenopus embryos. They find that these transcripts accumulate specifically in the somitic mesoderm during neurulation, and so they consider α-actin mRNA accumulation as an early marker of cellular commitment to muscle differentiation (48). In a series of experiments involving egg constrictions, blastomere ablation, and in vitro culture of dissociated embryonic cells, this group has obtained evidence that all of the components neccessary for muscle commitment are localized from the earliest stages in the equatorial region (46, 49; and Gurdon et al., in this volume). They suggest that vegetal-equatorial cell interactions during pre-gastrula stages are not neccessary for the proper initiation and accumulation of α-actin gene transcripts. In confirmation of this suggestion, 91% of the dorsal equatorial cell recipients in the present experiments developed some morphologically detectable muscle. However, lateral equatorial cells of the prospective somitic mesoderm were not autonomous in forming muscle when transplanted at the 32-cell stage. Thus the early autonomy of the equatorial zone in initiating α-actin gene transcription may be due to muscle differentiation by the prospective chordamesodermal region, as seen in embryos rescued to a grade 3 level of axial organization by dorsal equatorial cell transplants (Fig. 5b). Alternatively, muscle differentiation in early stage complete equatorial zone isolates would sometimes result from an interaction of the lateral and ventral somite precursors with an autonomous chordamesoderm rudiment. In either case, muscle differentiation by the normal precursors to the somites would result from inductive interactions during blastula and gastrula stages.

The present results give a few clues about the nature of determinants involved in initiating axis formation. First, the experiments illustrated in Fig. 6 suggest that their positions vary without precise regard to the fate map boundaries. Thus it is not likely that they act directly to specify the differentiation of any particular cell type. They can function by conferring mesoderm inducing ability on cells which contain them. They might also function from within prospective mesodermal cell lineages, conferring not only the autonomous capacity to form axial mesoderm, but also the ability to "dorsalize" (50) nearby parts of the mesodermal cell layer. Further clues as to the nature of the regional specialization which the egg makes and its effects on cellular activities may come from a study of the mechanism by which equatorial cell progeny can be induced to form the chordamesoderm. Analysis of the timing of this inductive interaction in experimental situations will be useful (Gurdon et al., in this volume).

A question related to the timing of chordamesoderm induction is, what is the role of new zygotic gene expression in the induction? In preliminary experiments, the dorsal vegetal cell pair in 16-cell embryos has been microinjected in situ with a cell lineage-restricted, amanitin-based inhibitor of RNA synthesis. The injected cells cleave normally, and the inhibited cell clone maintains its integrity until late in gastrulation, when the cells of the clone dissociate from each other and are rejected by the embryo. Interestingly, the remaining parts of the embryo, though not inhibited in RNA synthesis, fail to develop according to their normal fates. Most such embryos are histologically similar to the grade 5, axis deficient embryos derived from UV-irradiated eggs. This result suggests that new transcription within the dorsal vegetal quadrant of the embryo is required for normal specification of the fates of cells in distant embryonic regions. It will be important to determine whether new gene expression is required for the initial inductive specification of the chordamesoderm precursors, or for a later interaction by which these precursors establish the pattern of specifications in the surrounding mesodermal layer. Since detectable new transcription does not begin until about the 4000-cell stage (51, 52), it is likely that at least one of these interactions begins after this stage.

It should also be possible to investigate the mode of intercellular communication which is neccessary for chordamesoderm induction. For instance, using probes which interrupt intercellular communication via the common low resistance electrical pathway, the gap junction (53), the involvement of ion or small molecule exchange in the induction can be determined.

Finally, since the determinants essential for initiating axis formation seem to be localized to just one subequatorial quadrant of the early embryo, it will be interesting to find out what parts of the egg's contents are incorporated into this area. Such information should yield clues as to what aspects of the early cytoplasmic rearrangement process are important in the early steps of axis formation, and may aid in the eventual understanding of the nature of a morphogenetic determinant.

REFERENCES

1. Wilson EB (1925). "The Cell in Development and Heredity," 3rd ed. New York: Macmillan.
2. Davidson EH (1976). "Gene Activity in Early Development." New York: Academic Press, p 249.
3. Elinson RP (1980). The amphibian egg cortex in fertilization and development. Symp Soc Dev Biol 38: 217.
4. Klag JJ, and Ubbels GA (1975). Regional morphological and cytochemical differentiation in the fertilized egg of Discoglossus pictus. Differentiation 3: 15.
5. Palecek J, Ubbels GA, and Rzehak K (1978). Changes in the external and internal pigment pattern upon fertilization in the egg of Xenopus laevis. J Embryol Exp Morph 45: 203.
6. Vincent J-P, and Gerhart JC (1985). Kinematics of grey crescent formation in amphibian eggs: mapping of the displacement of subcortical cytoplasm relative to the egg surface. Submitted to Dev Biol.
7. Ancel P, and Vintemberger P (1948). Recherches sur le déterminisme de la symmétrie bilaterale dans l'oeuf de Amphibiens. Bull Biol Fr Belg (Suppl.) 31: 1.
8. Elinson RP, and Manes ME (1978). Morphology of the site of sperm entry on the frog egg. Dev Biol 63: 67.
9. Manes M, and Elinson RP (1980). Ultraviolet light inhibits grey crescent formation on the frog egg. Wilhelm Roux' Arch 189: 73.
10. Scharf SR, and Gerhart JC (1983). Axis determination in eggs of Xenopus laevis: A critical period before first cleavage, identified by the common effects of cold, pressure, and ultraviolet irradiation. Dev Biol 99: 75.
11. Grant P, and Wacaster JF (1972). The amphibian grey crescent--a site of developmental information? Dev Biol 28: 454.
12. Malacinski GM, Brothers AJ, and Chung H-M (1977). Destruction of components of the neural induction system of the amphibian egg with ultraviolet irradiation. Dev Biol 56: 24.
13. Scharf SR, and Gerhart JC (1980). Determination of the dorsal-ventral axis in eggs of Xenopus laevis: complete rescue of UV-impaired eggs by oblique orientation before first cleavage. Dev Biol 79: 181.
14. Fankhauser G (1930). Zytologische Unterschungen an geschnurten Triton-Eiern. I. Die verzögerte Kernversorgung nach hantelförmiger Einschnürung des

Eies. Wilhelm Roux' Arch 122: 117.
15. Chung H-M, and Malacinski GM (1980). Establishment of the dorsal/ventral polarity of the amphibian embryo: use of ultraviolet irradiation and egg rotation as probes. Dev Biol 80: 120.
16. Pasteels J (1937). Sur l'origine de la symétrie bilatérale des Amphibiens anoures. Arch Anat Micros 33: 279.
17. Black SD, and Gerhart JC (1985). Experimental control of the site of embryonic axis formation in Xenopus laevis eggs centrifuged before first cleavage. Dev Biol 108: 310.
18. Ilmensee K, and Mahowald AP (1974). Transplantation of posterior pole plasm in Drosophila. Induction of germ cells at the anterior pole of the egg. Proc Natl Acad Sci USA 71: 1016.
19. Curtis ASG (1960). Cortical grafting in Xenopus laevis. J Embryol Exp Morphol 8: 163.
20. Curtis ASG (1962). Morphogenetic interactions before gastrulation in the amphibian, Xenopus laevis--The cortical field. J Embryol Exp Morphol 10: 410.
21. Gerhart JC (1980). Mechanisms regulating pattern formation in the amphibian egg and early embryo. In Goldberger RF (ed): "Biological Regulation and Development," New York: Plenum, p 133.
22. Gerhart J, Ubbels G, Black S, Hara K, and Kirschner M (1981). A reinvestigation of the role of the grey crescent in axis formation in Xenopus laevis. Nature 292: 511.
23. Spemann H, and Mangold H (1924). Über Induction von Embryonanlagen durch Implantation artfremder Organisatoren. Wilhelm Roux's Arch 100: 599.
24. Gimlich RL, and Cooke J (1983). Cell lineage and the induction of second nervous systems in amphibian development. Nature 306: 471.
25. Smith JC, and Slack JMW (1983). Dorsalization and neural induction: properties of the organizer in Xenopus laevis. J Embryol Exp Morph 78: 299.
26. Jacobson M (1984). Cell lineage analysis of neural induction: origins of cells forming the induced nervous system. Dev Biol 102: 122.
27. Spemann H (1938). "Embryonic Development and Induction." New York: Hafner, Chapter VII.
28. Cooke J (1982). The relation between scale and completeness of pattern in vertebrate embryogenesis: models and experiments. Amer Zool 22: 91.

29. Slack JMW (1983). "From Egg to Embryo." Cambridge: Cambridge Univ Press, Chapter 3.
30. Nieuwkoop PD (1973). The "organization center" of the amphibian embryo: its origin, spatial organization, and morphogenetic action. Adv Morphog 10: 1.
31. Nieuwkoop PD (1977). Origin and establishment of embryonic polar axes in amphibian development. Curr Top Dev Biol 11: 115.
32. Keller RE (1984). The cellular basis of gastrulation in Xenopus laevis: active, postinvolution convergence and extension by mediolateral interdigitation. Amer Zool 24: 589.
33. Malacinski GM, Allis CD, and Chung H-M (1974). Correction of developmental abnormalities resulting from localized ultra-violet light irradiation of an amphibian egg. J Exp Zool 189: 249.
34. Gimlich RL, and Gerhart JC (1984). Early cellular interactions promote embryonic axis formation in Xenopus laevis. Dev Biol 104: 117.
35. Nieuwkoop PD, and Faber J (1975). "Normal Table of Xenopus laevis (Daudin)," 2nd ed., 1st reprint. Amsterdam: North Holland.
36. Gimlich RL, and Braun J (1985). New fluorescent cell lineage tracers. Dev Biol, in press.
37. Nakamura O, and Kishiyama K (1971). Prospective fates of blastomeres at the 32-cell stage of Xenopus laevis embryos. Proc Japan Acad 47: 407.
38. Jacobson M, and Hirose G (1981). Clonal organization of the central nervous system of the frog: clones stemming from individual blastomeres of the 32- and 64-cell stages. J Neurosci 1: 271.
39. Nieuwkoop PD, and Ubbels GA (1972). The formation of mesoderm in urodelan amphibians. IV. Quantitative evidence for the purely "ectodermal" origin of the entire mesoderm and pharyngeal endoderm. Wilhelm Roux' Arch 169: 185.
40. Nakamura O, and Takasaki H (1971). Analysis of causal factors giving rise to the organizer. I. Removal of polar blastomeres from 32-cell embryos of Xenopus laevis. Proc Japan Acad 47: 499.
41. Gimlich RL (1985). Developmental autonomy and inductive capacity of equatorial cells in the 32-cell Xenopus embryo. Submitted to Dev Biol.
42. Ruud G (1925). Die Entwicklung isolierter Keimfragmente frühester Stadien von Triton taeniatus. Wilhelm Roux's Arch 105: 209.

43. Vintemberger P (1935). Sur les résultats du dévelopment des quatre micromerès isolés au stade de huit blastomères dans l'oeuf d'un amphibien anoure. CR Soc de Biol (Paris) 118: 52.
44. Grunz H (1977). Differentiation of the four animal and the four vegetal blastomeres of the eight-cell-stage Triturus alpestris. Wilhelm Roux' Arch 181: 267.
45. Kaguera H, and Yamana K (1984). Pattern regulation in defect embryos of Xenopus laevis. Dev Biol 101: 410.
46. Gurdon JB, Mohun TJ, Fairman S, and Brennan S (1985). All components required for the eventual activation of muscle-specific actin genes are localized in the subequatorial region of an uncleaved amphibian egg. Proc Natl Acad Sci 82: 139.
47. Nakamura O, Takasaki H, and Mizohata T (1970). Differentiation during cleavage in Xenopus laevis. I. Acquisition of self-differentiation capacity of the dorsal marginal zone. Proc Japan Acad 46: 694.
48. Mohun TJ, Brennan S, Dathan N, Fairman S, and Gurdon JB (1984). Cell type-specific activation of actin genes in the early amphibian embryo. Nature 311: 716.
49. Gurdon JB, Brennan S, Fairman S, and Mohun TJ (1984). Transcription of muscle-specific actin genes in Xenopus development: Nuclear transplantation and cell dissociation. Cell 38: 691.
50. Slack JMW, and Forman D (1980). An interaction between dorsal and ventral regions of the marginal zone in early amphibian embryos. J Embryol Exp Morph 56: 283.
51. Brown DD, and Littna E (1966). Synthesis and accumulation of DNA-like RNA during embryogenesis of Xenopus laevis. J Mol Biol 20: 81.
52. Newport J, and Kirschner M (1982). A major developmental transition in early Xenopus embryos: I. Characterization and timing of cellular changes at the midblastula stage. Cell 30: 675.
53. Warner AE, Guthrie SC, and Gilula NB (1984). Antibodies to gap-junctional protein selectively disrupt junctional communication in the early amphibian embryo. Nature 311: 127.

THE ACTIVATION OF MUSCLE-SPECIFIC ACTIN GENES IN XENOPUS DEVELOPMENT

J.B. Gurdon, T.J. Mohun, S. Brennan, and S. Cascio

CRC Molecular Embryology Unit,
Department of Zoology,
University of Cambridge, Cambridge CB2 3EJ,
England.

ABSTRACT We have used gene-specific probes to determine the time and region, in embryos of Xenopus laevis, when cardiac and skeletal actin genes are first transcribed. Transcripts of both genes are detected in the future body muscle region of late gastrulae. We summarize experiments which have led us to the view that two mechanisms are responsible for the expression of these muscle actin genes in only one region of an embryo. One mechanism depends on the existence of substances localized in the subequatorial region of a fertilized but uncleaved egg. The other mechanism involves animal cells of a blastula responding to an inductive interaction from adjacent vegetal cells.

INTRODUCTION

For many years, two concepts have dominated thoughts on how genes are activated in different cells at the beginning of development, at least in the invertebrates and lower vertebrates. According to the first, molecules which activate genes are localized in different regions of an egg, or become localized during early cleavage divisions of an embryo. Different genes are therefore activated according to the kinds of cytoplasmic components which cells inherit from the egg or early embryo. According to the second concept, inductive interactions between cells provide the principal mechanism by which cell differences arise. Starting with as little as two kinds of cells in an early embryo, it can be argued that all other cell-types arise by a series of cell interactions coupled with the folding of cell layers. In mammals, the outside or inside position of cells in the early embryo is related to their subsequent differentiation as trophectoderm or inner cell mass. An asymmetric distribution of cytoplasmic components seems to be involved in the trophectoderm versus inner cell mass differentiation which therefore resembles in some respects the first concept mentioned above.

Muscle-specific actin genes are the earliest identified genes to be transcribed in a cell-type specific way in Xenopus development (1). It is therefore appropriate to ask whether the expression of these genes in only one part of an early embryo depends on the distribution of materials already localized in the egg or on an inductive interaction between other kinds of cells. In fact it seems that both mechanisms are involved in the initial activation of the same gene. We summarize here recent experiments which have led us to this view.

ACTIN GENE EXPRESSION IN NORMAL XENOPUS DEVELOPMENT

The expression of actin genes at the protein level has been described by Ballantine et al. (2) and by Sturgess et al. (3). Muscle-specific (α) actins are first detected at the end of gastrulation. In our own work (1,4,5), we have used probes specific for cardiac or skeletal actin mRNAs in Northern and S1 nuclease protection analyses. The earliest transcripts, which are also full-length, are detected in late gastrulae (stage 13 of Nieuwkoop and Faber (6). We conclude that, when cardiac and skeletal actin genes are first activated at the late gastrula stage, their earliest transcripts are correctly initiated, terminated, spliced and translated. By analyzing carefully dissected parts of neurula embryos, we found that only the somite or future muscle region of an embryo contains muscle actin transcripts, and that both cardiac and skeletal actin genes are strongly transcribed in the body muscle cells of these early embryos (1). These comments summarize the main facts about Xenopus actin gene regulation which need to be understood. We cannot explain why actin genes are transcribed at a particular time in development, and we are concerned below only with how to account for the activation of muscle actin genes in a particular region of an embryo.

THE ACTIVATION OF MUSCLE ACTIN GENES BY MATERIALS LOCALIZED IN AN UNCLEAVED EGG

Fertilized, but as yet uncleaved, eggs can be divided into two parts by constriction with a hair loop. These ligations can be orientated in respect of the animal-vegetal and dorso-ventral axes of the egg and future larva. Those egg fragments which include at least one pronucleus can be cultured until whole embryo controls reach the late gastrula stage, and then tested with suitable probes for the presence or absence of muscle actin transcripts. From experiments of this design, we have concluded that all components of an uncleaved egg which are required for muscle actin transcription are localized in a subequatorial region (7). When various layers of cells are removed from cleaving embryos, we have found that the materials needed for muscle actin gene transcription are still localized in a subequatorial region, and are largely restricted to the third tier of 8 cells in a 32-cell embryo. The simplest interpretation of these experiments is that the subequatorial cytoplasm of an egg and early embryo contains substances which commit cells acquiring these materials to activate their muscle genes, and that an induction is not required for the activation of these genes.

This conclusion has been strengthened by placing newly fertilized eggs in $Ca^{++}Mg^{++}$-free medium, in which embryonic cells divide apparently as normal, but lose contact with their sister cells after each division. Cell division is allowed to continue under these conditions, until controls have become early gastrulae (stage 10). So long as $Ca^{++}Mg^{++}$ are added at this stage, the loose cells will adhere and form a compact mass in which the activation of muscle genes can be examined. We have found that under these conditions of no normal cell contact from fertilization till stage 10, muscle actin genes are nevertheless transcriptionally activated at the normal time (8). It is possible that an inductive cell interaction takes place after stage 10, but at least no cell interaction involving normal cell contacts is required during the whole of the cleavage phase of development.

Collectively these results suggest that substances already localized in an uncleaved egg may cause the direct activation of muscle actin genes, without the involvement of cell interactions. We think that this is one way in which muscle genes are activated in normal development.

MUSCLE ACTIN GENES CAN BE ACTIVATED EXPERIMENTALLY BY AN INDUCTION MECHANISM

It was originally observed by Nieuwkoop (9) and soon after by Nakamura et al. (10), that mesoderm or organizer activity is created if animal and vegetal regions of a blastula are placed in contact, even though each region cultured on its own forms only ectoderm and endoderm respectively. Subsequent work from each group has documented this effect in more detail, and in particular has argued from detailed histological analysis that mesodermal tissue derives from animal cells which respond to an induction by vegetal cells (11,12,13). The inducing ability of vegetal blastula cells has recently been documented in detail by the blastomere transfer experiments of Gimlich and Gerhart (14).

In our experiments we have asked whether the interaction between animal and vegetal blastula cells results in the activation of muscle actin genes. This is indeed the case. Furthermore the amount of muscle actin gene transcription is substantial, and it takes place in animal not vegetal cells. At the molecular level, this induction can be seen with muscle-specific antibodies, by 2D gel analysis of synthesized proteins, and by various nuclease protection assays with gene-specific probes. The latter are particularly valuable because they are extremely sensitive and they recognize immediate transcripts resulting from induced gene activation. This has made it possible to determine much more accurately than before several aspects of the timing of this induction reaction. The analysis has been facilitated by the demonstration that the inducing vegetal and responding animal cells can be reliably separated after a few hours of contact. The results of all these experiments are described in detail elsewhere (15).

A conclusion from this work which is of special relevance to the present discussion is that the inducing capacity of vegetal cells disappears at stage 9 . We have already mentioned that embryos can be maintained from fertilization until stage 10 in $Ca^{++}Mg^{++}$-free medium, and that so long as cells are allowed to reaggregate at stage 10, muscle actin genes start to be transcribed at the usual time. The main difficulty in interpreting this experiment is that an inductive cell interaction might be thought to take place after reaggregation at stage 10. This now appears to be impossible since vegetal cells have lost their inducing ability by this stage. The only remaining difficulty in assessing these results is the possibility that cells dividing in $Ca^{++}Mg^{++}$-free medium are retarded compared to cells in whole embryos. It might therefore be argued that when controls are at stage 10, dissociated cells are at some earlier stage and may still retain their inducing capacity. Even this reservation seems unlikely, since Miyahara et al. (16) have shown that cells kept in $Ca^{++}Mg^{++}$-free solution synthesize DNA, RNA, and protein in accord with the normal time scale of development up to at least the late blastula stage. The results of our induction experiments have therefore given some support to our argument that muscle actin genes can be activated in normal development by a localization as opposed to induction mechanism.

ARE MUSCLE ACTIN GENES ACTIVATED IN NORMAL DEVELOPMENT BY LOCALIZED SUBSTANCES OR BY INDUCTION?

Since the discovery referred to above that vegetal blastula cells can so effectively induce animal cells to become mesodermal, it has been generally assumed, in agreement with Nieuwkoop (13), that this is the only mode of origin of mesodermal cells in normal development. However, our results with egg ligation, blastomere deletion, and cell dissociation seem to show that muscle gene activation can take place in the absence of an induction mechanism. Since cells from the animal and vegetal poles of a blastula are never normally in contact with each other, it is not self-evident that the mesoderm normally arises by an induction rather than localization mechanism. The matter is of some importance, since the molecular mechanisms by which genes would be activated by the two processes are likely to be very different.

A solution to the problem could be envisaged if it were possible to establish that some of the cells which give rise to muscle by direct lineage do not form muscle if isolated at an early stage but do so if left in contact with other cells. The cells most likely to fall into this class are the animal 4 of an 8-cell embryo, or the second tier of 8 cells in a 32-cell embryo. The first question is whether these cells include direct lineage ancestors of muscle. There is disagreement on this point. Several Japanese workers (17), using vital dyes applied to the surface of certain 32-cell blastomeres, find that most of the axial muscle of larvae traces back to the 3rd tier of a 32-cell embryo; on the other hand, Cooke and Webber (18), using horse-radish peroxidase injection, see most of the axial muscle coming from the second tier of a 32-cell embryo. Probably both second and third tier cells contribute to the muscle lineage, the extent to which each does so being subject to variation among different batches of embryos. The second question we ask is whether cells which can contribute to the muscle lineage do so only if in contact with underlying more vegetal cells, and fail to do so if separated at an early stage. Again, the position is unclear. In our own experience (7), the animal four cells of a typical 8-cell embryo do not form muscle if separated at the 8-cell stage, though Kageura and Yamana (19) find that they sometimes do so. A conclusive result would be one with a single series of embryos in which the animal four eighths of an 8-cell embryo do not form muscle if separated from the rest of the embryo at this stage but do so if left in place in a complete embryo.

The reason for disagreement among laboratories on the questions raised above can probably be explained if we believe that the particular regions of an egg at which cytoplasmic cleavages occur differ from one group of eggs to another, and that the position of cytoplasmic cleavages is in any case unimportant. In <u>Xenopus</u> the pattern of cytoplasmic cleavages seems to be largely unimportant, as is particularly clearly seen from our observations many years ago (unpublished) that nuclear-transplant embryos may undergo radial cleavage (yielding a ring of 8 vertically divided cells), and yet form a normal tadpole. All that appears to matter is that the egg cytoplasm is cleaved into a few thousand cells by the end of the blastula stage. In this case it would not be surprising if the position of the third (equatorial) cleavage varies considerably from one batch of embryos to another, the subequatorial egg cytoplasm being sometimes included in the second as well as third tiers of a 32-cell embryo.

We have to conclude that the origin of the mesoderm, and therefore the mechanism leading to muscle actin gene activation, has not yet been shown to depend exclusively on the distribution of localized egg components, or on inductive cell interactions. At present, it seems most likely that both mechanisms are involved. Indeed it seems to us probable that most cases of induction in animal development are accompanied by some self-differentiating capacity of cells, as is often observed in the absence of induction.

ANALYSIS OF THE MECHANISM OF INDUCTION

Whatever the biological importance of animal-vegetal induction in Xenopus development, it is certainly of great analytical value. This particular induction is instructive, as opposed to permissive (see Saxen and Waartiovaara (20) in the following sense. A separated animal region of a blastula will differentiate as ciliated epidermis if cultured on its own. It can also be very easily diverted into neural differentiation by a wide range of unspecific conditions such as high pH, etc. (21). However the only living material which can induce an animal region to become muscle is vegetal tissue of a blastula. In experiments described elsewhere (15), we have shown that vegetal cells can induce animal cells to express their muscle actin genes within 7 hours of initial contact, and that the minimal contact time sufficient for activation is less than half this period. The rapidity and specificity of this inductive effect should make it fruitful to investigate the nature of the inductive signal. For example, we can ask what kinds of molecules can pass from inducing to responding cells within this time, and whether the inductive effect is eliminated by gap junction antibodies.

ACKNOWLEDGEMENTS

We are indebted to Sharon Fairman for help in the experiments referred to. S.B. and S.C. are in receipt of NIH Fellowships numbers GM.08810 and HD.06593.

REFERENCES

1. Mohun TJ, Brennan S, Dathan N, Fairman S, Gurdon, JB (1984a). Cell type-specific activation of actin genes in the early amphibian embryo. Nature 311: 716.
2. Ballantine JEM, Woodland HR, Sturgess EA (1979). Changes in protein synthesis during the development of Xenopus laevis. J Embryol exp Morph 51: 137.
3. Sturgess EA, Ballantine JEM, Woodland HR, Mohun PR, Lane CD, Dimitriadis GJ (1980). Actin synthesis during the early development of Xenopus laevis. J Embryol exp Morph 58: 303.
4. Mohun TJ, Brennan S, Gurdon JB (1984a). Region-specific regulation of the actin multi-gene family in early amphibian embryos. Phil Trans Roy Soc Lond B 307: 337.
5. Gurdon JB, Brennan S, Fairman S, Dathan N, Mohun TJ (1984b). The activation of actin genes in early Xenopus development. In "Molecular Biology of Development" UCLA Symposia on Molecular and Cellular Biology 19: Alan R. Liss, p 109.
6. Nieuwkoop PD, Faber J (1956). "Normal Table of Xenopus laevis (Daudin)." Amsterdam: North-Holland Publishing Co.
7. Gurdon JB, Mohun TJ, Fairman S, Brennan S (1985a). All components required for the eventual activation of muscle-specific actin genes are localized in the subequatorial region of an uncleaved Amphibian egg. Proc Nat Acad Sci USA 82: 139.
8. Gurdon JB, Brennan S, Fairman S, Mohun TJ (1984b). Transcription of muscle-specific actin genes in early Xenopus development: nuclear transplantation and cell dissociation. Cell 38: 691.
9. Nieuwkoop PD (1969). The formation of mesoderm in urodelan amphibians. Part 1, induction by the endoderm. Wilhelm Roux' Archiv 162: 341.
10. Nakamura O, Takasaki H, Ishihara M (1970). Formation of the organizer from combinations of presumptive ectoderm and endoderm. Proc Jap Acad 47: 313.
11. Sudarwati S, Nieuwkoop PD (1971). Mesoderm formation in the anuran Xenopus laevis (Daudin). Wilhelm Roux' Archiv 166: 189.

12. Nieuwkoop PD (1973). The 'organization center' of the amphibian embryo: its origin, spatial organization, and morphogenetic action. Adv Morphogen 10: 1.
13. Nieuwkoop PD (1977). Origin and establishment of embryonic polar axes in amphibian development. Currt Top Devel Biol 11: 115.
14. Gimlich RL, Gerhart JC (1984). Early cellular interactions promote embryonic axis formation in Xenopus laevis. Devel Biol 104: 117.
15. Gurdon JB, Fairman S, Mohun TJ, Brennan S (1985b). The activation of muscle-specific actin genes in Xenopus development by an induction between animal and vegetal cells of a blastula. Submitted.
16. Miyahara K, Shiokawa K, Yamana K (1982). Cellular commitment for post-gastrular increase in alkaline phosphatase activity in Xenopus laevis development. Differentiation 21: 45.
17. Nakamura O, Takasaki H, Nagata A (1978). Further studies of the prospective fates of blastomeres at the 32-cell stage of Xenopus laevis embryos. Med Biol 56:355.
18. Cooke, Weber (1985). J Embryol exp Morph, in press.
19. Kageura H, Yamana K (1983). Pattern regulation in isolated halves and blastomeres of early Xenopus laevis. J Embryol exp Morph 74: 221.
20. Saxen L, Waartiovaara J (1984). Embryonic induction. In Graham CF, Wareing PF (eds): "Developmental Control in Animals and Plants" Oxford: Blackwell, p 176.
21. Holtfeter J, Hamburger V (1955). Embryogenesis - progressive differentiation in amphibians. In Willier BH, Weiss PA, Hamburger V (eds): "Analysis of Development" p 230.

RETROVIRUSES AND INSERTIONAL MUTAGENESIS IN MICE[1]

Rudolf Jaenisch

Whitehead Institute for Biomedical Research
Cambridge, Massachusetts 02142

ABSTRACT Retroviruses have been shown to cause insertional mutations spontaneously or after infection of early mouse embryos. The most extensively analyzed mutant, the Mov-13 mouse strain, was derived by exposing mid-gestation mouse embryos to the Moloney leukemia virus. In this strain, the provirus has inserted into the first intron of the alpha 1(I) collagen gene, blocking gene activation and resulting in death of homozygous embryos between days 12 and 14 of development. Molecular analysis showed that the provirus prevents the developmentally regulated appearance of a transcriptionally associated DNase I hypersensitive site and induces de novo methylation of collagen gene sequences. The frequency of mutations induced by microinjection of recombinant DNA into mouse zygotes is much higher than that induced by retroviruses, suggesting that these insertional mutagens act by different molecular mechanisms.

INTRODUCTION

The phenotypic analysis of experimentally induced or spontaneous mutations has long been the subject of developmental genetics in the mouse. Lethal mutations have had a significant role in identifying pleiotropic effects of presumably single genes on complex developmental processes. Examples of such genetic systems are the well studied

---

[1] This work was supported by grants from the Deutsche Forschungsgemeinschaft and the Stiftung Volkswagenwerk.

T-locus (1), the albino deletions (2), or mutations affecting hemopoiesis (3). This descriptive approach, however, has its limitations in elucidating the underlying developmental defect on a molecular basis, as a randomly mutated gene or its gene product is difficult to identify. An alternative approach is to induce mutations by insertional mutagenesis and to use the inserted foreign DNA element as a tag to molecularly clone the mutated gene.

In mammals, the experimental insertion of retroviruses and of recombinant DNA are two approaches that have been successful in inducing developmental mutations by insertional mutagenesis. In these experiments, either mouse zygotes were microinjected with DNA (4,5) or mouse embryos were exposed to infectious retroviruses at different stages of development (6,7,8) to derive animals which carried the foreign genetic information in their germ line.

In this report, I will review the derivation of a mouse strain with a recessive lethal embryonic mutation which was induced by provirus insertion into the alpha 1(I) collagen gene. Furthermore, our studies to characterize the mutation on a molecular level and the use of the mutant to investigate the role of type I collagen in embryonic development will be summarized.

RESULTS AND DISCUSSION

Derivation of a Mutant Mouse Strain by Retrovirus Insertion

Mov-mouse strains carrying a single Moloney leukemia proviral copy as a Mendelian determinant were derived by exposing mouse embryos to infectious virus at different developmental stages (Table 1). The great majority of mouse strains was derived from virus-infected preimplantation embryos (8) or from a zygote which was microinjected with proviral DNA (9). The Mov-13 strain which carries a proviral copy in the alpha 1(I) collagen gene was derived from an embryo microinjected with infectious virus at mid-gestation (10,11). While the frequency of germ line integrations is high in animals derived from infected preimplantation embryos, infection of primordial germ cells with virus at mid-gestation is a rare event, and so far Mov-13 is the only strain obtained by this infection protocol.

TABLE 1
INSERTION OF MO-MULV INTO THE GERM LINE OF MICE

Stage of Exposure to Virus	Genetic Locus	Virus Expression	Location (gene)	Ref.
Zygote (microinjection of DNA)	Mov-14	+	X-Chromosome	9
4-8 cell Pre-implantation Stage	Mov-1	+	Chromosome 6	6,7,8
	Mov-2	+		"
	Mov-3	+		"
	Mov-4	−		"
	Mov-5	−		"
	Mov-6	−		"
	Mov-7	−		"
	Mov-8	−		"
	Mov-9	+		"
	Mgpt-1	−		31
Blastocyst	Mov-10	−		8
	Mov-11	−		"
	Mov-12	−		"
Mid-gestation	Mov-13	+	Chromosome 11 $(\alpha_1(I)$ collagen)	12, 13, 34

Heterozygosity at any of the Mov-loci did not interfere with normal embryonic development or result in a detectable phenotype in the postnatal animals. To detect possible recessive mutations caused by insertion of the virus, offspring were derived from heterozygous parents and analyzed to identify animals homozygous at the respective Mov-locus. With the exception of Mov-13, all crosses resulted in homozygous healthy offspring, indicating that proviral insertion at none of the different loci disrupted essential gene functions (11). The Mov-13 substrain, as the other strains, carries a single provirus in its germ line. In contrast to the other strains, however, homozygosity at the Mov-13 locus leads to death of the embryos between day 13 and 14 of gestation (11). The host sequences flanking the insertion site were molecularly cloned by using the virus as a probe. Hybridization of these sequences to RNA extracted from normal embryos or various tissue culture lines detected two species of RNA. The developmentally regulated and cell-specific pattern of Mov-13 locus expression and its sensitivity to transformation by sarcoma viruses led us to screen genes coding for proteins of the extracellular matrix and thus to identify the alpha 1(I) collagen gene as being interrupted by virus insertion in Mov-13 mice (12).

Molecular Analysis of Mutated Collagen Gene.

Sequence analysis of the host DNA flanking the virus showed that provirus insertion had occurred in the first intron of the alpha 1(I) collagen gene with the transcriptional orientation of the provirus opposite to that of the host gene (13). SI mapping experiments using the first exon of the collagen gene as a probe furthermore, demonstrated that virus insertion into these non-coding host sequences completely blocked collagen transcription in vivo as well as in tissue culture cells.

Both DNA hypomethylation (14) and an opened chromatin conformation (15) have been correlated with transcriptional activity of a gene. To understand the mechanisms involved in virus-induced insertional mutagenesis in Mov-13 mice, we compared the methylation pattern and the chromatin conformation of the wild type and mutant alpha 1(I) collagen allele. Our results showed that both the pattern of methylation and the chromatin conformation were changed by virus insertion. The alpha 1(I) collagen gene in fibroblasts which express high concentrations of collagen RNA, displays a DNase I hypersensitive site at 200 bp upstream of the cap site (Fig. 1). This site appears during development, at the time when the alpha 1(I) collagen gene is activated (16). In Mov-13 animals, however, this site is not present in chromatin of the transcriptionally inactive mutant allele (Fig. 1). These results suggest that virus insertion prevents the developmentally regulated appearance of a transcription-associated hypersensitive site.

In addition to inducing a change in chromatin conformation, the virus insertion induces de novo methylation of collagen sequences within 1 kb distance from the integration site (17). DNA methylation has been correlated with gene inactivity in many systems (14). Our results therefore suggest that virus-induced DNA methylation can change DNA-protein interactions and thereby may interfere with correct gene activation during embryonic development.

FIGURE 1

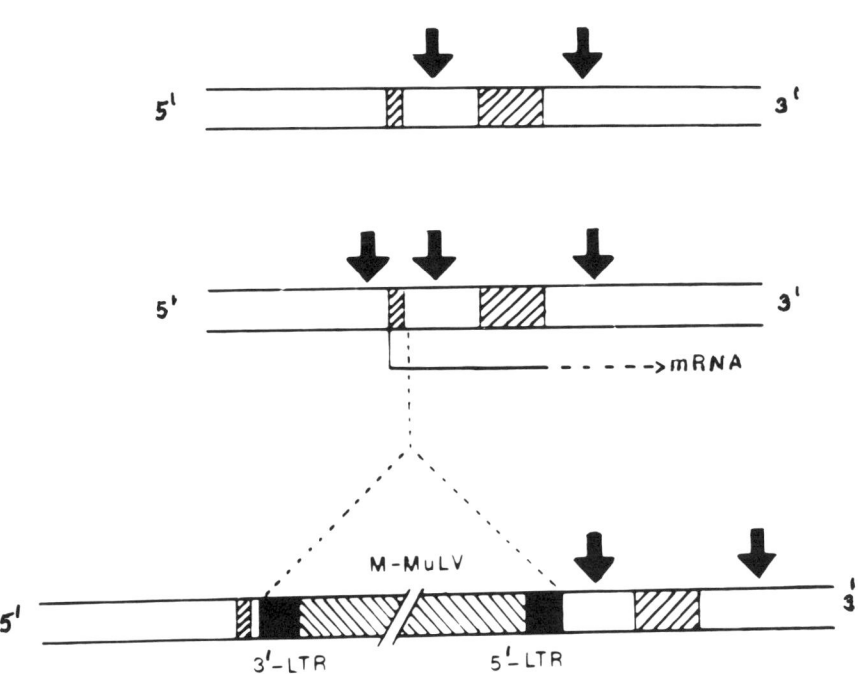

FIGURE 1. DNase-hypersensitive regions at 5' end of the alpha 1(I) collagen gene in wild type and Mov-13 mice (16). The hatched areas represent exons, the open areas, introns, and the black boxes in the lowest panel represent the LTR's of the Mo-MuLV genome. DNase I-hypersensitive areas are marked by arrows.

Upper Panel: wt-cells which do not express collagen have two hypersensitive sites. Middle Panel: wt-cells which express collagen display a third hypersensitive site approximately 200 bp 5' of the cap site. Lower Panel: in Mov-13 mice, this transcription-associated site never appears and its absence correlates with gene inactivity.

Role of Type I Collagen in Early Embryonic Development.

Collagen type I, as the major protein of connective tissues, is likely to have a variety of functions in the organization of the vertebrate body. That it is crucial for the mechanical stability of bones is illustrated in patients with osteogenesis imperfecta, a severe genetic disease caused by mutations in collagen I genes (18). Furthermore, as a major component of the extracellular matrix, collagen I has been proposed to have an important function in mesenchymal-epithelial interactions during organogenesis (19,20). Experimental evidence for such a function includes studies in which morphogenesis in organ culture was shown to be disturbed by collagenase (21) or by drugs that interfere with collagen as well as protein synthesis (22). The availability of a molecularly defined mouse mutant provides the opportunity to study functionally the role of collagen I in mammalian development without having to rely on the use of inhibitors that are likely to have a wide variety of effects on embryo development. Therefore, we analyzed homozygous embryos for pathological events that precede their sudden death (23).

The majority of homozygous embryos die between day 13 and day 14 of gestation. Most organ rudiments are already laid down on or before day 12 of gestation, and growth and histomorphogenesis rather than determinative events are characteristic of the subsequent days of mouse development (24). While type I collagen is already detected in day 8 embryos by immunohistochemical techniques, abundant transcription, which may be related to the formation of mesenchyme, begins at day 12 (12) and thus coincides with the death of homozygous embryos. This suggests that type I collagen has no important role in organ determination but rather is essential for the growth phase subsequent to day 12 of gestation, a period during which its expression increases substantially in normal embryos. Immunohistochemical techniques to detect other components of the extracellular matrix, such as collagen types II, III, IV, laminin and fibronectin, demonstrated that the blockage of type I collagen synthesis does not affect the synthesis or distribution of the other ECM components in the mutant embryo (23).

Histologically, homozygous embryos at day 12 showed two distinct alterations: a progressive necrosis of erythropoietic and mesenchymal cells followed by rupture of major blood vessels which causes sudden death by breakdown of the circulatory system (23). The active hemopoietic proliferation at day 12 of gestation leads to a rapid expansion of the blood volume, and it is likely that the blood vessels in homozygous embryos that lack collagen type I cannot resist the ensuing increase of blood pressure. While collagen type III as the major collagen present in the cardiovascular system is thought to provide strength and elasticity, the Mov-13 phenotype ascribes also to collagen type I an important role for establishing and maintaining the mechanical stability of the circulatory system.

To study the role of collagen type I in epithelial branching morphogenesis during the formation of embryonic organs, organ rudiments from homozygous embryos were explanted in organ culture. In vitro development of all organ types tested, including lung, kidney, salivary gland, mammary buds and skin was perfectly normal despite the absence of collagen type I (25). These observations indicate that collagen I has either no essential role for early morphogenesis or that its function can be replaced by other interstitial collagens.

Insertional Mutagenesis by Retroviruses and Recombinant DNA.

Table 2 summarizes currently known mutations in mice which were caused by insertional mutagenesis. For comparison, some retrovirus-induced mutations obtained in tissue culture cells are included in the table. The published information allows the conclusion that DNA microinjected into the zygote pronucleus is highly mutagenic and mutations can be realistically estimated to occur in 10 to 20% of all transgenic mouse strains generated. In contrast, none of the 14 Mov-mouse strains which were derived from virus-exposed preimplantation embryos showed evidence for a mutant phenotype. The only mutant obtained by experimental manipulation, the Mov-13 mouse strain, was obtained by infecting a primordial germ cell at the midgestation stage. The mode of infection by which the dilute mutation (30) was generated is not known.

Information on the possible mechanism(s) involved in insertional mutagenesis has so far been obtained only for retrovirus-induced mutations. In the most extensively studied mutant, the Mov-13 mouse strain, the virus has inserted into non-coding sequences in the first intron of the collagen gene, which results in a complete block of gene transcription. Molecular analyses demonstrated two virus-

TABLE 2
INSERTIONAL MUTAGENESIS IN MICE

Insertional Mutagen	Developmental Stage of Exposure	Frequency of Mutations Induced	Affected Gene	Mutant Phenotype	Reference
Retrovirus	4-30 cell pre-implant. embryo	0/14	—	—	11, 31
	Midgestation embryo	1/1	$\alpha 1(I)$ collagen	Embryonic lethal	11
	Spontaneous	?	?	Coat color (dilute)	30
	Tissue culture cells	Low	src	Reversion of transformed phenotype	26
		?	p-53	Tumor rejection	27
		?	κ-light chain gene	Decreased IgG synthesis	28
DNA micro-injection	Zygote pronucleus	10-20%	?	Embryonic lethal	4, 32
			?	Transmission distortion	5
			?	Limb deformity	33

induced alterations of the mutated gene: i) The prevention of a DNase I hypersensitive and transcription-associated site appears during embryonic development, ii) de novo methylation of sequences flanking the proviral insertion. The causal relationship between the virus-induced change of chromatin conformation and methylation pattern with gene inactivity remains to be clarified.

It is remarkable that retroviruses have inserted into intron sequences in all known cases of insertional mutagenesis. A Moloney leukemia proviral copy was found in the "intron" of the RSV genome in two revertants of RSV transformed cells (26). Likewise, a provirus copy has mutated the p53 gene (27) and the IgG gene (28) by insertion into the first intron of the respective gene. Furthermore, the provirus which caused the dilute mutation has integrated into non-coding sequences (29). It appears that the proviral copies of spontaneously induced mutations cluster at the 5' end of the affected gene, which may suggest that retroviruses do not randomly integrate as discussed elsewhere (16).

The molecular anatomy of genes that have been mutated by recombinant DNA has not yet been elucidated in a single instance. The remarkably high frequency of mutations induced by DNA injection into the zygote pronucleus, as opposed to retrovirus-induced mutations (table 2), may suggest, however, that introduction of DNA may induce major rearrangements or deletions which cause mutations in the host cell. The use of insertional mutagenesis by retroviruses or by DNA injection is of great potential to dissect molecular mechanisms of mammalian development, and to elucidate the underlying genetic events remains an important challenge.

REFERENCES

1. Silver LM (1981). Genetic organization of the mouse t complex. Cell 27:239.
2. Gluecksohn-Waelsch S (1979). Cell 16:225-237.
3. Russell ES (1979). Hereditary anemias of the mouse: a review for geneticists. Adv Genet 20:357-459.
4. Wagner EF, Covarrubias L, Stewart TA, Mintz B (1983). Prenatal lethalities in mice homozygous for human growth hormone gene sequences integrated in the germ line. Cell 35:647-655.

5. Palmiter RD, Wilkie TM, Chen HY, Brinster RL (1984). Transmission distortion and mosaicism in an unusual transgenic mouse pedigree. Cell 36:869-877.
6. Jaenisch R (1976). Germ line integration and Mendelian transmission of the exogenous Moloney leukemia virus. PNAS USA 73:1260-1264.
7. Jähner D, Jaenisch R (1980). Integration of Moloney leukemia virus into the germ line of mice: correlation between site of integration and virus activation. Nature 287:456-458.
8. Jaenisch R, Jähner D, Nobis P, Simon I, Lohler J, Harbers K, Grotkopp D (1981). Chromosomal position and activation of retroviral genomes inserted into the germ line of mice. Cell 24:519-529.
9. Stewart C, Harbers K, Jähner D, Jaenisch R (1983). X Chromosome-linked transmission and expression of retroviral genomes microinjected into mouse zygotes. Science 221:760-762.
10. Jaenisch R (1980). Retroviruses and embryogenesis: microinjection of Moloney leukemia virus into mid-gestation mouse embryos. Cell 19:181-188.
11. Jaenisch R, Harbers K, Schnieke A, Löhler J, Chumakov I, Jähner D, Grotkopp D, Hoffmann E (1983). Germline integration of Moloney murine leukemia virus at the Mov13 locus leads to recessive lethal mutation and early embryonic death. Cell 32:209-216.
12. Schnieke A, Harbers K, Jaenisch R (1983). Embryonic lethal mutation in mice induced by retrovirus insertion into the α1(I) collagen gene. Nature 304-315-320.
13. Harbers K, Kuehn M, Delius H, Jaenisch R (1984). Insertion of retrovirus into the first intron of α1(I) collagen gene leads to embryonic lethal mutation in mice. PNAS USA 81:1504-1508.
14. Jaenisch R, Jähner D (1984). Methylation, expression and chromosomal position of genes in mammals. Bioch Biophys Acta 782:1-9.
15. Elgin S (1981). DNase I-hypersensitive sites of chromatin. Cell 27:413-415.
16. Breindl M, Harbers K, Jaenisch R (1984). Retrovirus-induced lethal mutation in collagen I gene of mice is associated with altered chromatin structure. Cell 38:9-16.
17. Jähner D, Jaenisch R (1985). Retrovirus induced de novo methylation of flanking host sequences correlates with gene inactivity. Nature, in press.

18. McKusick V (1983). "Mendelian Inheritance in Man" Baltimore: The Johns Hopkins University Press.
19. Bernfield M (1980). Organization and remodeling of the extracellular matrix in morphogenesis. In "Development and Pattern Formation" Conelly, Brinkley, Carlson, eds. New York: Raven Press, pp 139-162.
20. Hay E (1981). Collagen and embryonic development. In "Cell Biology of the Extracellular Matrix" Hay E, ed, New York: Plenum Press, pp 379-409.
21. Grobstein C, Cohen J (1965). Collagenase: effect on the morphogenesis of embryonic salivary epithelium in vitro. Science 150:626-628.
22. Spooner B, Faubion J (1980). Collagen involvement in branching morphogenesis of embryonic lung and salivary gland. Dev Biol 77:84-102.
23. Löhler J, Timpl R, Jaenisch R (1984). Embryonic lethal mutation in mouse collagen I gene causes rupture of blood vessels and is associated with erythropoietic and mesenchymal cell death. Cell 38:597-607.
24. Rugh R (1968). "The Mouse, Its Reproduction and Development" Minneapolis: Burgess Publishing Company.
25. Kratochwil et al. In preparation.
26. Varmus HE, Quintrell N, Ortiz S (1981). Retroviruses as mutagens: insertion and excision of a nontransforming provirus alter expression of a resident transforming provirus. Cell 25:23-26.
27. Wolf D, Rotter V (1984). Inactivation of p53 gene expression by an insertion of Moloney murine leukemia virus-like sequences. Mol Cell Biol 4:1402-1410.
28. Kuff EL, Feenstra A, Lueders K, Smith L, Hawley R, Hozumi N, Shulman M (1983). Intracisternal A-particle genes as movable elements in the mouse genome. PNAS USA 80:1992-1996.
29. Hutchison K, Copeland N, Jenkins N (1984). Dilute-coat-color locus of mice: nucleotide sequence analysis of the d^{+23} and d^{+Ha} revertant allele. Mol Cell Biol 4: 2899-2904.
30. Jenkins N, Copeland N, Taylor B, Lee B (1981). Dilute (d) coat colour mutation of DBA/2J mice is associated with the site of integration of an ecotropic MuLV genome. Nature 293-370-374.
31. Jähner D, Jaenisch R, in preparation.
32. Lacy L, personal communication.
33. Leder P, personal communication.
34. Munke, et al, in preparation.

INDUCTIVE INTERACTION AND DETERMINATION;
A NEW APPROACH TO AN OLD PROBLEM

Pieter D. Nieuwkoop

Hubrecht Laboratory, Utrecht, the Netherlands

The concepts of "inductive interaction" and "determination" date from the beginning of this century, when Spemann studied the interaction between the primary eye vesicle and the overlying head epidermis in the amphibian embryo, leading to the formation of the lens anlage (Spemann 1901, 1902). Later he also studied the action of the mesodermal archenteron roof upon the overlying ectoderm in the late gastrula, which leads to the development of the central nervous system (Spemann & H.Mangold 1924). With these findings he disproved the validity of the old *preformistic* concept of embryonic development, according to which development would be a mere unfolding of a pre-existing, invisible pattern of organ anlagen. His observations demonstrate the essentially *epigenetic* nature of developmental processes, at least in the vertebrates (but see further below).

These important discoveries led to a direct search for the presumed chemical factors involved. When surveying the literature of the last fifty to sixty years it is evident that the development of any new or more refined analytical method led to a new search for the chemical factor responsible for a given induction process. Unfortunately, the outcome of all these studies has time and again been unsatisfactory. Although biochemists succeeded in isolating active fractions from adult or embryonic tissues, it turned out that quite different substances can also release the induction process. For instance, a high--MW protein fraction isolated from 9-day chick embryos causes competent early gastrula ectoderm from an amphibian embryo to develop into meso- and endodermal structures (Tiedemann 1966), but a simple treatment of the same ectoderm with Li^+ ions does exactly the same (Masui 1961). Hoperskaya *et al* (1984) recently showed that extracellular matrix material from Bruch's membrane is also a powerful meso-endodermal inductor. Similarly, a nucleoprotein fraction isolated from 9-day chick embryos induced early gastrula ectoderm from an amphibian embryo to differentiate into neural structures (Tiedemann 1966), but older observations by Holtfreter (1947 a,b,c, 1948) had already shown that neuralisation of such ectoderm can also be achieved by raising or lowering the pH of the culture medium.

The notion of specific inductors may have originated from the discipline of endocrinology. In the older embryo and in the adult specific hormones are formed in particular organs, are released into the blood stream with which they circulate throughout the body, but act only on a specific organ or cell type in another part of the organism. It must be realised that in a system where source and target are far apart both the acting hormone and the responding cells must possess specificity. In early embryonic development, where various, not highly differentiated tissues or organ anlagen come into direct contact with each other, quite different mechanisms may be involved in the transfer of information from one tissue or organ anlage to the other.

Are we perhaps approaching the problem of embryonic induction from the wrong angle? Are there really compelling reasons to assume that specific compounds are responsible for inductive actions? Is it not more likely that the inductive action is hardly specific and that the main specificity in the induction process resides in the responding system? Let us try to approach the problem of embryonic induction from a different point of view by asking ourselves: What are the essential requirements for inductive interaction?

There is good reason to assume that embryonic development of a multicellular organism requires a nearly continuous interaction between its constituent parts for development to be properly coordinated and for its complex three-dimensional organisation to be built up stepwise. Interruption of cellular communication by means of blockage of gap junctions by specific antibodies leads to abnormal development (Warner, Guthrie & Gilula 1984).

Interaction presupposes the exchange of information between the various parts of the egg or developing embryo. Let us therefore assume that cells continuously release *"messages"* which may be recognised by other cells as distinct *"signals"* to which they can react. Although messages are also released by cells in a monoculture, they are not recognised as distinct signals because the messages are identical to those released by the neighbouring cells themselves. Recognition of messages as distinct signals only occurs between cells or tissues which are sufficiently different from each other. This simple assumption already implies that the recognition of messages as distinct signals is the principal event in the exchange of information. It also implies that interaction between two different systems is essentially *reciprocal*. And it leads to yet another important conclusion: development can only occur in an embryonic system which has a minimum of spatial heterogeneity. An entirely homogeneous system cannot develop and cannot acquire a higher complexity because no real interaction occurs. The freshly released *Fucus* egg, which has spherical symmetry, can only develop after this symmetry has been broken and a radial symmetry with regional differences along the polar axis has been established e.g. by unilateral illumination or by the application of an electrical current. From that moment on interaction can occur between the different regions.

The eggs and embryos of higher chordates initially show a single polar axis. The amphibian egg consists of only two different cytoplasmic moieties with different properties, arranged along the animal-vegetal axis: the pigmented animal moiety and the unpigmented vegetal one. Not only the cytoplasmic but also the cell membrane domains of the two moieties are different; the latter are markedly different in composition and membrane fluidity (Dictus *et al* 1984). The presence of only two different moieties holds also for the amniote embryo, where the two primary moieties segregate from the multicellular blastoderm (Eyal-Giladi & Kochav 1976).

The presence of only a minimum of spatial heterogeneity in the chordate egg and embryo implies that its development is almost completely *epigenetic*, the embryo acquiring its increasing spatial and structural complexity by interaction between its constituent parts.

So-called *"mosaic"* eggs seem to possess a greater initial complexity. Their development was once thought merely to represent an unfolding of a preexisting invisible pattern. However, the more we begin to understand about the development of mosaic eggs, the more epigenetic aspects become evident and the smaller the initial heterogeneity turns out to be. I shall return to mosaic development at the end of this paper.

After the egg has acquired its bilateral symmetry as a result of the penetration of the sperm or of rotation under the influence of gravity, and after it has cleaved into a large number of blastomeres, the first large-scale interaction occurs between the cells of the two original moieties. This interaction leads to the formation of the so-called "meso-endoderm". In the amphibians the latter is formed from the totipotent animal moiety of the blastula/early gastrula under an inductive influence emanating from the endodermal vegetal yolk mass, particularly from its dorsal part. A similar process occurs in the avian embryo, where primitive streak formation, representing meso-endoderm formation, is due to the interaction of the (secondary) endodermal hypoblast with the totipotent epiblast of the double-layered blastoderm.

The morphogenetic movements occurring during gastrulation transform the single-layered amphibian blastula into a triple-layered embryo, bringing regions of the embryo into direct contact which were originally far apart, among other things as a result of a reversal of the antero-posterior axis of the endo- and mesoderm. In the triple-layered embryo the three (germ) layers are in contact with each other over their entire inner surfaces, which strongly enhances mutual interaction. This leads in first instance to the induction of the neural anlage in the overlying ectoderm by the endo- and mesodermal archenteron roof; this anlage gives rise to the central nervous system and the neural crest and its derivatives. This induction process consists of two successive steps: the initial activation step determines the spatial extension of the neural anlage and the superimposed transformation step its regional cranio-caudal segregation.

During the morphogenetic movements of the succeeding neurulation process, by which the neural tube is formed, the contact between the neural tube and the axial mesoderm is enhanced, while the subsequent migration of the neural crest cells leads to many different local interactions. Experimental analysis has shown that, apart from the appearance of mesectodermal differentiation tendencies in the cephalic neural crest cells prior to the initiation of migration, the various cell types which the neural crest produces are mainly determined by the surrounding tissues after the arrival of the cells at the ultimate destination sites. However, their development already seems to be influenced to some extent by tissues with which they came into transient contact along their migration routes (Le Douarin *et al* 1979).

In all interactions a distinction must be made between an *action system* and a *reaction system*. During embryonic development both the action and the reaction system show time-dependent changes and may go through successive phases of inductive activity on the one hand and responsiveness or *competence* on the other. The phases of competence are generally more sharply delimited than those of inductive activity. For instance, the neural inductive activity of the archenteron roof, which starts at the beginning of gastrulation, extends at least till an early tail-bud stage, while neural competence of the ectoderm, which may already start before gastrulation (Chuang 1955), drops sharply at the mid-gastrula stage (Nieuwkoop 1958, 1960). This again points to a higher specificity in the responsiveness of the reacting cells than in the nature of the inductive action.

A clear distinction should be made between *"inductive"* and *"supporting"* interactions. Only those interactions which switch the reacting cells into a *new* developmental pathway should be called *"inductive"*, while those which only support a developmental pathway already in progress should be considered as non-inductive and should be classified as *"supporting"* interactions. An inductive interaction involves a *pluripotential* reaction system which has a *choice* from among several different potential pathways. For instance, the animal moiety of the amphibian blastula and the epiblast of the avian blastoderm show weak differentiation tendencies towards epidermal development in the form of, respectively, ciliation (Grunz *et al* 1975) and a characteristic epithelial cell arrangement (Eyal-Giladi & Kochav 1976), but can still be switched into the meso-endodermal or the neural developmental pathway. It seems likely that the first effect of an inductive action is the suppression of the developmental pathway already in progress, after which a new developmental pathway can be opened. The most fundamental property of embryonic tissues in an epigenetically developing system is therefore their pluripotentiality.

From a theoretical point of view it seems more likely that the recognition of messages as distinct signals by the reacting cells represents a much more specific event than the continuous release of messages by the action system. Competence for a given developmental pathway appears at a specific developmental stage, increases with time and reaches a maximum, after which it declines rather sharply. It may be replaced by a different competence. For instance, the meso-endodermal competence of the blastula and early gastrula ectoderm shows an earlier maximum than its neural competence (Leikola 1963, 1965). The latter is replaced by competence for placodal ectoderm formation in the ageing ectoderm (Nieuwkoop 1963). After the loss of placodal competence the ectoderm can no longer form anything but epidermis; this pathway only needs supporting influences from the underlying mesoderm.

I have already mentioned that very different agents can induce the early gastrula ectoderm to develop first into meso-endodermal and slightly later into neural tissues. This indicates that the competent reaction system is already thoroughly prepared for these new pathways of differentiation, so that a rather unspecific stimulus suffices to effect the necessary switch. This places strong emphasis on the particular properties of the reaction system. It should however be realised that a certain specificity cannot be denied to the action system: for instance, the meso-endodermal inductive action is most probably not identical to the neural inductive action, while the reaction system is the same, with the sole exception that it is slightly older at the maximum of neural competence.

Unfortunately, little is known about the nature of competence. Clayton (1982) found that various eye tissues from which lenses can regenerate under experimental conditions, in the embryo contain low levels of specific lens antigens. This induced her to suggest that lens competence may imply the presence of low levels of transcription of specific mRNAs. This would mean that induction represents an enhancement of a process already in operation, rather than the initiation of entirely new syntheses. This seems to be a valuable suggestion but requires careful confirmation.

In my opinion there is no compelling evidence that the messages released by the action system in inductive interaction represent special products of biosynthesis, as do the hormones that act in later phases of development and during adult life. It seems much more likely that the released messages are nothing else than *transient products of cellular differentiation* of the tissues in question. During development both the action and the reaction system show a progressing cellular differentiation along different developmental pathways, so that the transient products of differentiation may change with time, both quantitatively and qualitatively. These products which are released into the intercellular spaces may vary enormously in complexity, from ions and simple metabolites to high-MW components of extracellular matrix material. Moreover, physico-chemical changes in the apposing cell membranes or extracellular matrices of the action and reaction system may also lead to inductive interactions, a possibility that is supported by the effect of Li^+ and H^+ ions in meso-endodermal and neural induction, respectively.

On the one hand, these considerations make it unlikely that a direct search for specific inductive compounds, such as has been tried repeatedly in the past decades, will be successful. On the other hand, they stress the desirability of a thorough study of the physiology of both the action and the reaction system during the critical period of inductive interaction. It seems particularly promising to look for important differences between the two systems, since these must in my opinion be responsible for the interaction. Moreover, we urgently need a better insight into what constitutes the cellular nature of competence, since that is the principal prerequisite for inductive interaction.

"Determination" represents the step-wise restriction or potential developmental pathways in a pluripotential embryonic system. Switching of the reacting system into a particular pathway implies the suppression of other pathways, both the one already in progress and others for which special competences are present. For instance, the determination of gastrula ectoderm to neural tissue implies the suppression of the epidermal pathway of differentiation already in progress as well as the meso--endodermal pathway. Subsequent inductive interactions may further restrict neural development, e.g. to neural crest formation as it occurs in the peripheral region of the neural anlage. Neural crest formation may be due to the loss of competence of the neuralised cells for the caudalising, transforming action of the archenteron roof, which is slowly spreading medio-laterally through the activated neural ectoderm (Nieuwkoop, Johnen & Albers 1985). As I have already mentioned, the process of determination then continues for particular cells of the neural crest during and after their migration to their ultimate destination sites (Le Douarin *et al* 1979). The step-wise

determination of a particular cell type may therefore cover a considerable period of development and may consist of many successive restricting steps, each of which seems to be preceded by a period of corresponding competence.

However, determination is not only restrictive. It is accompanied by a progressive differentiation process: the cells of a particular part of the embryo begin to differentiate early, long before their terminal state of differentiation is reached. The various phases of this long process no doubt require various determinative stimuli specific for the differentiation pathway in question. We know nothing about the interplay of these stimuli with the suppression of the other potential pathways.

The appearance of a given competence often is the consequence of a preceding inductive action, as e.g. the appearance of lens competence in the placodal ectoderm under the influence of the pharyngeal endoderm (A.G.Jacobson 1966), or that of liver competence in the gut endoderm under the influence of the cardiac mesoderm (Le Douarin 1974). However, competence may also arise spontaneously during the normal ageing process of the embryonic system, as e.g. the successive appearance of meso-endodermal, neural and placodal competences in the totipotent animal ectodermal moiety of the amphibian embryo. As far as we know no inductive action can be made responsible for the appearance of these competences.

Since we do not know the biochemical nature of competence we do not know either what determination means in biochemical terms. We can only say that somehow a particular set of genes which is involved in a particular pathway of differentiation must ultimately be derepressed, while other sets of genes for alternative pathways must be repressed.

In my opinion certain objections can be made to the title of this symposium: "Molecular Determinants of Animal Form". The term "determinant" was introduced by August Weismann in the late 19th century for species-specific characteristics (later called genes) localised in the nucleus. However, later the term "determinant" was also used for cytoplasmic factors, e.g. for specific cytoplasmic inclusions such as the germ plasm in anuran amphibians (Smith 1966; Blackler 1970) and in higher insects (Mahowald 1977). Recently Whittaker (1979) has used the term for components of the yellow ooplasm of the ascidian egg, which possibly contains specific mRNAs or unknown gene regulation factors. Although the term "molecular determinants" says nothing about their localisation, it is evident that in this title it does not refer to genetic factors but to cytoplasmic factors or components of the extracellular matrix. In my opinion the use of the term "determinant" for cytoplasmic factors may be rather confusing.

Also, I cannot refrain from making some critical remarks on the use of the term "determinant" in the current sense. First of all, the term implies high specificity: it indicates a factor thought to be responsible for the determination of a given developmental pathway or a specific cell type. In this contribution I have tried to provide arguments in favour of the thesis that the role of highly specific inductors in embryonic development is unlikely. As I see it, the messages released by the inducing system very probably represent ordinary, transient products of cellular differentiation with little specificity, while the main specificity of the inductive interaction resides in the reaction system, which at a particular stage of development manifests a high competence for a specific pathway of differentiation. Although mRNAs may be involved in the appearance of competence, we simply do not know what competence means in biochemical terms. The term "determinant" might more likely be applicable to factors that characterise competence than to those characterising inducing factors. We know too little of the processes involved to use a term like "determinant" with its implication of high specificity; it may actually be misleading.

I want to end my contribution by giving some examples which, I think, should caution us against using the term "determinant" for special cytoplasmic factors. The first example concerns the germ plasm of the amphibian primordial germ cells and the second the yellow ooplasm of the ascidian and cephalochordate egg.

It is at present widely accepted that in both the anuran amphibians and the higher insects the primordial germ cells develop from cells which contain part of the so-called germ plasm. The latter is localised in the subcortical cytoplasm in the vegetal hemisphere of the amphibian egg and in the vegetal pole region of the insect egg, and during cleavage ends up in certain blastomeres. During further development of the anuran embryo the descendants of these blastomeres migrate towards the genital ridges situated on either side of the dorsal mesentery, where they differentiate into oocytes or spermatocytes during sexual differentiation. A somewhat similar process takes place in the insect embryo. Therefore the germ plasm

is generally considered as the "determinant" of the germ
cells (Blackler 1970, Mahowald 1977). However, Mahowald
and coworkers, who carefully analysed the protein spectrum
of the germinal granules of the insect egg, found no
specific proteins in them, all proteins of the germinal
granules also being present in other parts of the egg cytoplasm (personal communication). This makes their role
rather doubtful. My next objection is the following. It
has been proved experimentally that in the urodeles the
primordial germ cells develop from the lateral plate mesoderm as part of the process of meso-endoderm induction
(Nieuwkoop, 1973, Nieuwkoop & Sutasurya 1979). In the
urodeles the primordial germ cells (PGCs) apparently
develop from ordinary cells of the totipotent animal moiety
of the blastula, since any portion of the animal moiety
representing either presumptive lateral plate, somites,
notochord, epidermis or neural plate can form primordial
germ cells when combined with ventral yolk mass material
(Sutasurya & Nieuwkoop 1974). The primordial germ cells
of the urodele larva ultimately also contain germ plasm,
which looks nearly identical to that of anuran PGCs.
However, the urodele PGCs acquire this germ plasm only at
a late stage of development, around stage 40 (Harrison),
when they have long been determined as germ cells, as
shown by their typical migratory behaviour (Ikenishi &
Nieuwkoop 1978). It is evident that in the urodeles the
germ plasm cannot be considered as a determinant but only
represents a product of cellular differentiation. It thus
becomes doubtful whether the same structure may be called
a germ cell determinant in the anuran amphibians, while it
represents only a product of cellular differentiation in
the urodele amphibians.

In both the ascidians (Urochordata) and Amphioxus
(Cephalochordata) a process of cytoplasmic segregation
occurs between fertilisation and the beginning of cleavage.
During this process a "yellow ooplasm", consisting of
clusters of mitochondria with adhering yellow pigment
granules, accumulates in the ventral equatorial region of
the egg, while a grey ooplasm accumulates in the dorsal
equatorial region (Conklin 1905a,b, 1932). In later development the yellow ooplasm is almost exclusively found in
the presumptive muscle cells of the larva, and the grey
plasm in the notochordal and neural anlagen. In ascidians
components of the yellow ooplasm are considered to be the
determinant of the muscle cells (Whittaker 1979). Experiments by Tung *et al* (1960) on Amphioxus embryos have shown

that the mesoderm, including the somites, can also be formed in recombinates of only the most animal blastomere tier with the most vegetal tier of the four rows of cells of the 32- to 64-cell stage, although neither of these two tiers contains the above-mentioned ooplasms. In our opinion Amphioxus shows an epigenetic development of the mesoderm which is very similar to mesoderm formation in the amphibians. Notwithstanding the early cytoplasmic segregation of particular ooplasms these do not seem to play an essential role in mesoderm formation in Amphioxus. Whether this conclusion also holds for the ascidians is unknown but should at least be taken into serious consideration.

These examples make me rather reluctant to ascribe much significance to the role of cytoplasmic inclusions or organelles in both epigenetic and mosaic development, and certainly keep me from qualifying them as "determinants". It seems more likely that the particular cell organelles, localised in specific cells, such as the yellow pigment in presumptive muscle cells of ascidians and Amphioxus and the germ plasm in amphibian and insect PGCs, are *products of cell-specific differentiation* and do not represent determinants of particular pathways of development. It seems plausible to assume that these differentiation phenomena have been preceded by processes of determination. Determination is probably much more complex in nature than this. I would like to repeat that to understand the processes of induction and determination we need a deeper insight into the no doubt complex physiology of the action and reaction systems during their inductive interaction. For a much more extensive treatment of these problems the reader is referred to Nieuwkoop & Sutasurya (1979 and 1981) and Nieuwkoop, Johnen & Albers (1985).

REFERENCES

Blackler AW (1970). Current Topics in Developmental Biology. 5:71.
Chuang H-H (1955). Chinese Journal of Experimental Biology. 4:151.
Clayton RM (1982). In Clayton RM, Truman DES (eds): "Stability and Switching in Cellular Differentiation" New York: Plenum Press, p 327.
Conklin EG (1905a). Journal of the Academy of Natural Sciences of Philadelphia 13:1.

Conklin EG (1905b). Biological Bulletin 8:205
Conklin EG (1932). Journal of Morphology 54:69.
Dictus WJAG et al (1984). Developmental Biology. 101:201.
Eyal-Giladi H, Kochav S (1976). Developmental Biology 49:321.
Grunz H et al (1975). W.Roux's Archives of Developmental Biology 178:277.
Holtfreter J (1947a). Journal of Experimental Zoology 106:197.
Holtfreter J (1947b,c). Journal of Morphology 80:25, 57.
Holtfreter J (1948). Symposium of the Society of Experimental Biology 2:17.
Hoperskaya OA et al (1984). Differentiation (in press).
Ikenishi K, Nieuwkoop PD (1978). Development, Growth and Differentiation 20:1.
Jacobson AG (1966). Science 152:25.
Le Douarin N (1974). Année de Biologie 13:101.
Le Douarin N et al (1979). In N.Le Douarin (ed): "Cell Lineage, Stem Cells and Cell Differentiation", Amsterdam: Elsevier/North Holland Biomedical Press, p 353.
Leikola A (1963). Annales Zoologici Societatis Zoologico--Botanicae Fennicae, Vanamo 25:1.
Leikola A (1965). Experientia 21:458.
Mahowald AP (1977). The American Zoologist 17:551.
Masui Y (1961). Experientia 17:458.
Nieuwkoop PD (1958) Acta Embryologiae et Morphologiae Experimentalis 2:13.
Nieuwkoop PD (1960). Archives Néerlandaises de Zoologie 13:588.
Nieuwkoop PD (1963). Developmental Biology 7:255.
Nieuwkoop PD (1973). Advances in Morphogenesis 10:1.
Nieuwkoop PD, Johnen AG, Albers B (1985). "The Epigenetic Nature of Early Chordate Development: Inductive Interaction and Competence", Cambridge University Press 373 pp.
Nieuwkoop PD, Sutasurya LA (1979). "Primordial Germ Cells in the Chordates: Embryogenesis and Phylogenesis" 187 pp Cambridge University Press.
Nieuwkoop PD, Sutasurya LA (1981). "Primordial Germ Cells in the Invertebrates: from Epigenesis to Preformation" 258 pp Cambridge University Press.
Smith LD (1966). Developmental Biology 14:330.
Spemann H (1901). W.Roux' Archiv für Entwicklungsmechanik der Organismen 13:224.
Spemann H (1902). W.Roux' Archiv für Entwicklungsmechanik der Organismen 15:448.
Spemann H, Mangold H (1924). Archiv für mikroskopische

Anatomie und Entwicklungsmechanik der Organismen 100:599.
Sutasurya LA, Nieuwkoop PD (1974). W.Roux' Archiv für Entwicklungsmechanik der Organismen 175:199.
Tiedemann H (1966). Current Topics in Developmental Biology 1:85.
Tung TC, Wu SC, Tung YFY (1960). Scientia Sinica 9:119.
Warner AE, Guthrie SC, Gilula NB (1984). Nature, London 311:127.
Whittaker JR (1979). In Subtelny S, Konigsberg IR (eds): "Determinants of Spatial Organization", New York: Academic Press, p 29.

THE EMBRYONIC CELL LINEAGE OF MAMMALS AND THE EMERGENCE OF THE BASIC BODY PLAN

Michael H.L. Snow

MRC Mammalian Development Unit
University College, London NW1 2HE

ABSTRACT Lineage studies have been carried out on rodents. In the mouse the fetus is derived from about 15 epiblast cells present in the 4d preimplantation blastocyst. The events of cleavage that are associated with the delineation of these cells are described. The epiblast increases to about 600 cells before the axes of the future body are apparent. The emergence of the primary germ layers is coincident with a dramatic increase in cell proliferation such that the neural plate stage embryo formed in about 36h contains 40-100,000 cells. The evidence that substantial lineage restriction occurs early in this phase of development is reviewed. Although not conclusive this data strongly suggests allocation and perhaps determination of neural and epidermal ectoderm, neural crest, somites, heart, blood and mesenchyme, primordial germ cells and possibly endoderm when the embryo contains about 2000 cells, i.e. at about 7 days of development.

INTRODUCTION

With the exception of monotreme marsupials the embryos of mammals live in the comfortable protection of the maternal reproductive tract with which they make intimate trophic connections. One adaptation to this environment is the almost complete loss of yolk from the egg. Whilst this removes a mechanical restraint to cleavage it carries with it a requirement that supply lines for nutrition of the embryo are established fairly swiftly. Mammals share this strategic necessity with the other amniote embryos, of reptiles and birds who have to process their yolk. Mammals achieve it by the early formation, at the end of cleavage, of a blastocyst in which all but a few cells form temporary nutritive tissues which interact with the uterine environment. These tissues are lost later in development or at birth. The blastocyst is a convenient starting point from which to consider embryonic cell lineages.

The blastocyst is created from a ball of cells by formation of a fluid-filled cavity within it, and bearing somewhere in its periphery a localised accumulation of cells, the inner cell mass (ICM). In regions away from the ICM the wall enclosing the fluid-filled cavity is one cell thick and called the trophectoderm. All mammals make a blastocyst of this general description and cavitation occurs at around the 16-32 cell stage in most species, some $2\frac{1}{2}$ to 5 days after fertilization of the egg. Cleavage therefore is very slow in comparison to other vertebrates. Experimental lineage studies on the trophectoderm and ICM have been carried out on rodents, predominantly the mouse. Since post-implantation development of rodents pursues perhaps the oddest course of all mammalian groups a brief comparison of the anatomy of implantation and embryo formation will be useful.

Figure 1 illustrates the three different morphological pathways to embryo formation found in mammals. The first event is the definition of a primary endoderm layer, one cell thick, which forms on the surface of the ICM facing the blastocyst cavity. This layer seems to behave like the hypoblast of the avian embryo and ought really to be so termed. Attachment to the uterus is effected by the trophectoderm, either that overlying the ICM, as is the case in the primates group, or the mural and abembryonic tissues not in contact with the ICM, such as is seen in the other groups. The embryoblast, blastoderm or embryonic disc is formed from other cells of the ICM. In the rodents this tissue, called the epiblast, becomes suspended on a cylindrical column of tissue which protrudes into the expanding blastocyst cavity. This structure is referred to as the egg cylinder and the distal half, which contains the epiblast is referred to as the embryonic part. The outer surface of this cylinder is covered by the primary endoderm.

Three types of amniotic cavity formation are shown. In the primates group it is formed by a cavitation process occurring in the epiblast; it is reminiscent of blastocyst cavitation. In rodents it is unclear whether it forms by cellular necrosis in the epiblast or by epithelialisation and curling up of the tissue. In the remaining group a

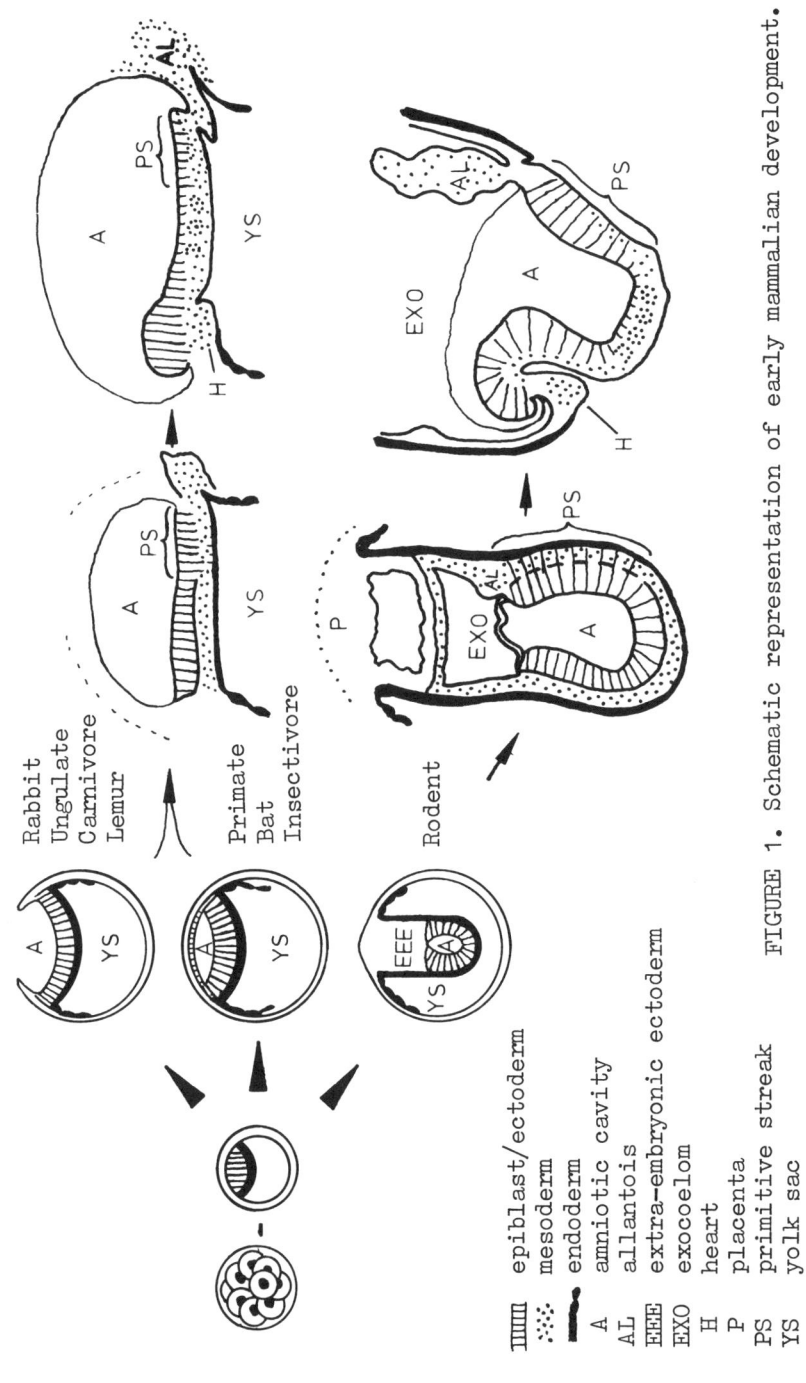

FIGURE 1. Schematic representation of early mammalian development.

ridge of tissue is raised at the junction between the disc of epiblast and the trophectoderm which extends up and over the epiblast to enclose a cavity.

THE EMBRYONIC CELL LINEAGE

Microsurgical techniques developed by Richard Gardner and his colleagues have allowed the construction of chimeric blastocysts in which either individual cells or the whole ICM may be genetically distinguishable from the remainder of the blastocyst either according to pigmentation, isozyme differences or immunological differences (1-4). More recently in situ hybridization with a species specific DNA probe has been used to distinguish lineages in chimeras between Mus musculus and Mus caroli (5).

These studies show clearly that the embryo and fetus is derived entirely from the inner cell mass. Furthermore they show that the primary endoderm, which is differentiated early from the ICM and covers the egg cylinder does not persist into the fetus. The fetus is derived entirely from the few epiblast cells sandwiched between the polar trophectoderm and primary endoderm but no further restriction of lineage into e.g. ectoderm, mesoderm, germ cells etc. can be detected at this stage. The extra-embryonic ectoderm which forms the core of the proximal part of the rodent egg cylinder is derived from polar trophectoderm.

Thus the segregation of the embryonic lineage depends on those processes during cleavage which establish the ICM, in particular those few epiblast cells in its centre. The chimera studies show that irreversible determination of the embryonic lineage does not occur until quite late, in the fully expanded blastocyst (2), so it is facets of the progressive restriction of developmental pathways that need to be sought.

The physical isolation of an inside cell lineage is associated with compaction of the embryo at the 8- to 16-cell stage (6). This process involves a polarised redistribution of mitochondria and cytoskeletal components within the blastomeres and the formation of tight junctions between cells on the periphery of the embryo. If this process does not isolate a cell which is internal, with no surface in contact with the surrounding environment then the next cycle of cell division will provide such a cell(s). The polarisation of cells in the morula that is formed is demonstrable in many ways; localisation of microvilli (7,8), ligand binding sites (9), nuclear displacement (10) and distribution of alkaline phosphatase (11). The establishment of polarity in the blastomeres precedes compaction (12) and will occur if compaction is prevented (13). The overall impression is that these changes occur gradually over the 2- to 8-cell period (14,15).

The eventual differentiation of ICM and trophectoderm is associated with the production of tissue specific proteins which are first detectable at about the 32-cell stage (16,17). The production of these molecules is not dependent upon cell division or those features of compaction which are prevented by cytochalasin D (18,19). Their synthesis is however prevented by α-amanitin provided this transcription inhibitor is applied before 80h after induction of ovulation (20). This corresponds to a stage after compaction, during the initial phase of fluid accumulation leading to cavitation and blastocyst formation, i.e. about the 32-cell stage. Blastocyst formation itself may depend on the production of these specific gene products (20). If so then a number of possible mechanisms whereby the timing of their production is controlled can be eliminated since blastocyst cavitation is not interfered with by reduction in the number of rounds of cytokinesis (13,18, 21-23), by changes in absolute cell number or time (24) and to some extent may be independent of the number of rounds of DNA replication (15,25).

Johnson (15) briefly describes a model whereby three cellular properties interact to finally define the committed cell lineages. These properties, the initial generation and stabilization of polarity in individual blastomeres, the consequential orientation of cleavage planes, and a final segregation of lineage by cell sorting according to differential adhesive properties are seen to act sequentially from the 2-cell stage onwards. These events are regarded as labile up to the 8 - 16-cell stage, and studies in which blastomeres are isolated or rearranged within the embryo support this view (26-29).

In addition to changes in the spectrum of genes expressed during this lineage restriction (15,30) there are indications of other changes in DNA and chromatin. In female embryos one X chromosome becomes inactivated in the trophectoderm lineage at about the time of its segregation (31). Trophectoderm, and later, primary endoderm DNA is relatively undermethylated (32,33), and at the 16-cell stage there would seem to be a change in the DNA and/or chromatin of the emerging ICM lineage which renders it highly susceptible to damage and fragmentation caused by radiactive decay of incorporated tritium (34), even if that label is only available during the S phase of the 4-cell stage (35).

DEVELOPMENT OF THE EPIBLAST

At the time of implantation the epiblast of the mouse ICM consists of around 15 cells and for a few days maintains the leisurely cleavage rate seen earlier, with a cell population doubling time of about 10h. When it contains about 600 cells there is an abrupt acceleration in cell proliferation (36,37), which occurs at about $6\frac{1}{2}$d post coitum and is coincident with formation of the primitive streak and onset of mesoderm formation. The first formed mesoderm appears at the most posterior end of the embryo and expands away from the embryo to form the mesodermal tissue lining the exocoelom from which embryonic blood cells will emerge to recolonise the embryo at a later stage. A small mesodermal protrusion at the posterior end of the embryo will subsequently expand to form the allantois. This ultimately forms part of the placenta and the connection between it and the embryo. Concomitant with the development of those mesodermal components, over a period of about 24h, other cells come to surround the cup-shaped epiblast, lying between it and the primary endoderm. The endoderm that will line the gut of the fetus and from which endodermal organs such as liver, pancreas and thyroid will form has not yet appeared. Mitotic activity is high throughout the embryo at this time.

Within the next 12-15h, i.e. within two further cell doublings, an embryo with identifiable neural plate, epidermal ectoderm, heart primordium, gut invaginations, somite(s), primordial germ cells and mesenchyme to hold it together is formed.

This brief description of epiblast development is generally accepted but there is some controversy over the kinetics of development and the segregation of the various cell lineages. Serial reconstructions of sectioned embryos and measurement of cell and tissue volumes suggest that within 24h of primitive streak formation the 600 or so epiblast cells have generated some 8,000 ectoderm cells and 6,000 mesoderm cells within the embryo itself (36,38). These figures do not include mesoderm which lines the exocoelom, nor that which is in the allantoic bud. There are fewer than 1,000 such cells, maybe only a few hundred. Such an increase in cell number requires a dramatic increase in cell proliferation to provide a population doubling time of about 5h - or double the previous rate. The mitotic index is not uniform throughout the epiblast showing a patch in the midline with an index $2\frac{1}{2}$ times higher than elsewhere, and which may therefore contain cells with a cycle time of 3h or less (36). However, autoradiographic studies (39,40) and analyses of shape changes in parts of the embryo (40) fail to provide evidence for such rapid cell division. These latter studies advocate cell cycle times of around 7h and Poelmann (40) further suggests that between 15 and 45 percent of the cell population in the epiblast does not incorporate H^3-thymidine and hence is non-dividing. Such tissue kinetics clearly will not permit the generation of over 14,000 cells in one day's development from a starting population of as few as 600. The autoradiographic study by Solter et al. (39) was carried out on thick wax sectioned material and it took no account of the autoradiographic artefact this introduces, leading to a serious under-estimate of labelling index and hence overestimate of cell cycle parameters (41). Poelmann's study (40) examines only a part of the epiblast, takes no account of mesoderm production and is difficult to reconcile with studies done elsewhere in which labelling indices of over 90% are routinely obtained (37,42-45). He acknowledges a requirement for more cells than his kinetic studies would account for and accepts that they may be generated from the epiblast at the distal end of the egg cylinder. This is a piece of tissue he does not analyse and corresponds to the area of high cell proliferation reported by Snow (36,39). It also seems likely however that Poelmann's criteria for recording a cell as labelled may not be adequate. His acceptance as labelled of a nucleus showing only two silver grains suggests difficulty in achieving a satisfactorily high grain count and could reflect difficulty in getting the precursor through the maternal/embryo barrier at these stages in development. Many laboratories have encountered this as yet unexplained phenomenon. It can be shown that the lower the grain count the greater is the chance of failing to detect a cell capable of incorporating the chosen precursor (46) so I think it likely that the <u>in vivo</u> autoradiographic data generally gives a false impression of cell kinetics in these embryos. Estimates of changes of tissue volumes during primitive streak stages of other mammals (human, pig, baboon) suggest that they also approximately double their cell proliferation rate (47) and that the embryos that emerge from this phase - the neural plate stage embryos - probably contain between 40 and 100 thousand cells, irrespective of species. Thus, in contrast to lower vertebrates, mammals seem to have developed the ability to combine the phase of rapid cell proliferation with that of the tissue rearrangements of gastrulation that yield the basic body plan, correctly spatially organised. What do we know of this regionalisation?

THE ORIGIN OF FETAL CELL LINEAGES

Epiblast cells isolated from $4\frac{1}{2}$d mouse blastocysts show no evidence for assignment and restriction to lineages such as ectoderm, mesoderm etc. within the embryo and it has not yet been possible to get early postimplantation epiblast to colonise a host blastocyst (2,44), so the status of the cells therein in terms of their developmental potential is not known. However the extensive chimerism found after transfer of a single cell (1) suggests that lineage restriction cannot occur until that cell has undergone considerable multiplication. Primitive streak stages and later, which can be cultured in vitro have been the subject of direct analysis by microsurgery (48,49), or grafting (41,42), and of indirect assessment using cytotoxic agents (49-51), retrospective clonal analysis using chimeras or X-inactivation mosaics (52 for review and discussion of methodology, 53,54), or by the capacity of embryonic parts to make teratomas after ectopic grafting (55-57). Other direct techniques using ingested or injected cell markers have been developed recently but have not yet been used to study fetal cell lineages. They have been employed in a non-regionalised examination of mesoderm formation in the rat (58) and, more elegantly to follow single cell fates in the primary endoderm during that period of development when it is believed a definitive endoderm will emerge (59). The species-specific DNA probe previously mentioned (5,33) has been used to study lineage from the blastocyst stage (60) and would be adaptable as a lineage marker for use in grafting experiments. Similarly the null mutation for cytoplasmic malic enzyme has potential as a cell marker in chimeras (61).

The direct approaches to the developmental capacities

of various regions within the embryo probe somewhat different aspects of lineage restriction. The microsurgery experiments, whereby the embryo is cut into pieces, or has specific parts removed, but which causes minimal disturbance to the relationships of the tissue components within the parts allows the construction of a fate map confining the developmental potential to form certain organs to certain regions. The physical separation of parts of the embryo potentially allows the detection of any necessary contribution or interaction between disparate regions. Of course failure for certain structures to appear does not necessarily imply missing interaction but could be due to destruction of that potential by the surgical procedure. Similar fate maps, but on a finer scale can be obtained by marking cells in situ and following their movement and development. Neither of these techniques necessarily gives information about the state of determination of cells at the time of isolation or marking. Grafting of tissue into heterotopic sites on the other hand is designed to test such commitment, by analysing the structures made or contributed to by the graft.

    An indirect analysis is generally fairly speculative without considerable supportive evidence. This is because unverified assumptions usually need to be made about the tissues or developing systems being analysed. For example in the retrospective clonal analysis it is a necessary condition of the statistics that the original cell population from which the precursors of a particular lineage are selected or segregated is a random assortment with respect to the cell marker used in the analysis. In the case of aggregation chimeras, or injection chimeras in which there is not parity in numbers of epiblast cells between host and donor, this condition is unlikely to be met (52,62). Another condition is that there should not be differential expansion of one cell population at the expense of the other. Since cell proliferation rate is not uniform throughout the embryo during crucial primitive streak stages this condition also seems unlikely to be met. Teratoma studies presume that the developmental capacity of transplanted pieces is not modified by the new environment, e.g. testis or kidney, in which it is placed. Finally the effects of reduction in cell number produced by cytotoxic agents can only be assessed in the light of other information on the development of these early embryonic stages.

    Nevertheless the data accrued from these various

82 Snow

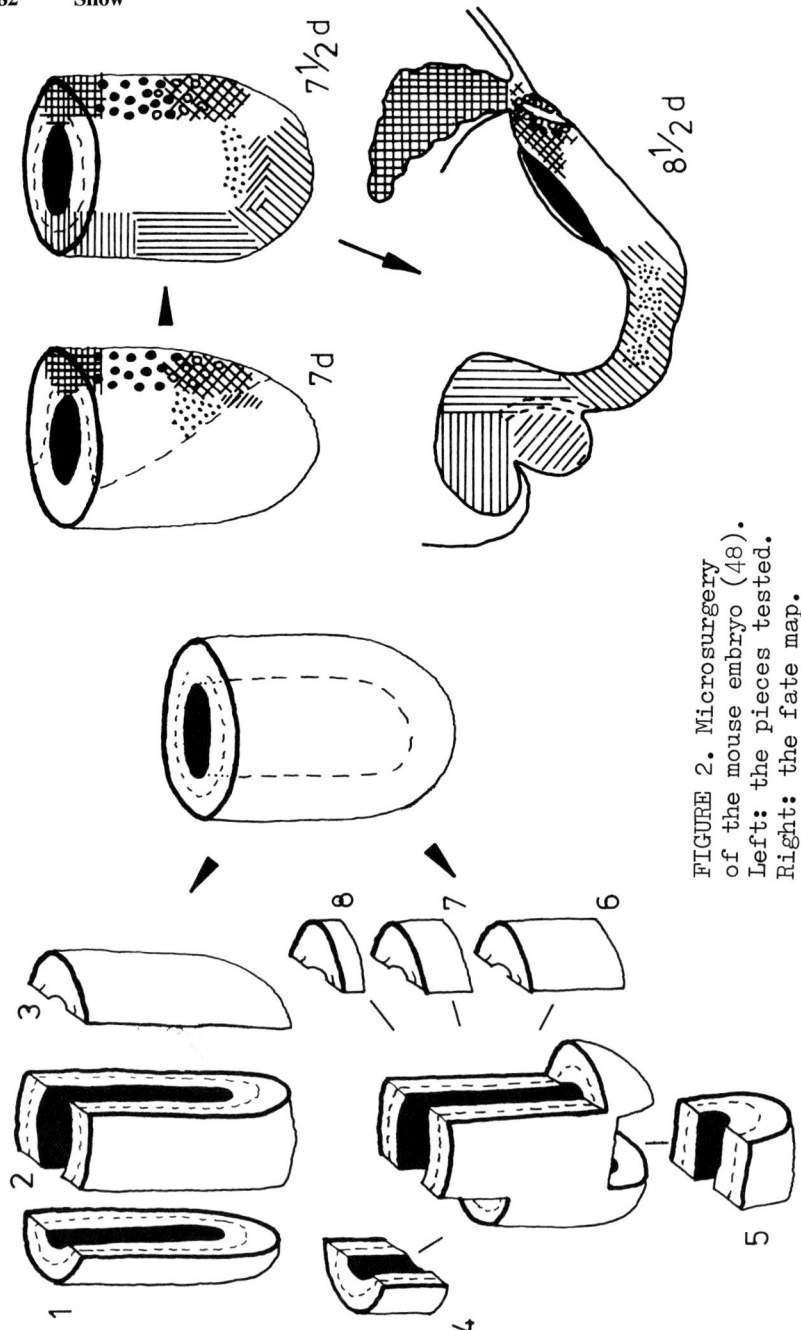

FIGURE 2. Microsurgery of the mouse embryo (48). Left: the pieces tested. Right: the fate map.

	n	HEAD	FORE-GUT	HEART	SP.CORD SOMITES	TAIL-BUD	HIND-GUT	GERM CELLS	ALLANTOIS
PIECES									
1	49	+	+	+	–	–	–	–	–
2	24	–	–	11	+	–	–	–	–
3	58	–	–	–	–	+	+	+	+
6	18	–	–	–	–	+	8	ND	–
7	18	–	–	–	–	–	3	+	–
8	14	–	–	–	–	–	–	9	+
DEFICIENT EMBRYOS									
–5	37	+	+	4	+	+	+	+	+
–6	18	+	+	+	+	–	5	+	+
–7	19	+	+	+	+	+	+	–	+
–8	14	+	+	+	+	+	+	+	–

TABLE 1. The development of pieces (Figure 1) isolated from primitive streak stage mouse embryos, or of embryos lacking pieces. +,–: present or absent from all cultures, Nos.: number of positive recordings. ND= not done.

approaches, coupled with the obvious anatomical and cytological development that takes place provide a reasonably compelling argument for substantial lineage restriction occurring in early to mid-primitive streak stages when the mouse embryo contains around 2,000 cells.

Direct Investigation

Figure 2 illustrates the microsurgical experiments and the fate map that can be derived from them (48). The results (Table 1) mostly bear out what one would predict from embryonic anatomy but there are several unexpected and important features in the data, concerning in particular heart development and the nature of the primitive streak.

The anatomy of embryonic development illustrates heart development from a small group of mesoderm cells located at the most anterior extremity of the embryonic axis (63). It is tacitly assumed that this mesoderm originates early from the primitive streak and reaches this site by migration around the proximal part of the embryonic region of the egg cylinder. However heart development is most profoundly disturbed if a piece is removed from the distal tip of the egg cylinder (Piece 5, Figure 2). If this operation is done on 7d old embryos heart formation is abolished completely while in $7\frac{1}{2}$d old embryos only 4 out of 32 developed a heart after such surgery. It is not known what essential contribution to heart development lies in this region. Piece 2 in isolation (which incorporates piece 5) develops neural tube, somites and <u>sometimes</u> beating hearts despite the complete absence of the presumptive heart region located in Piece 4.

The effect of surgical removal of pieces from the posterior 75% of the primitive streak is the same whether this is done at 7d or $7\frac{1}{2}$d of embryonic development, and the development of those pieces in isolation is also the same. This part of the streak is thus seen to contain, arranged in an anterior to posterior sequence, cells from which the tailbud, hind gut, primordial germ cells and allantois will be formed. If small pieces are removed from the streak then commonly the wound will heal yet the ensuing embryo will remain deficient in respect of the tissue formed by the removed part. It is difficult to reconcile these data with the widely held view of the streak as a channel through which cells generated in the epiblast or in the streak itself are passing to emerge as mesoderm cells. That view could be maintained if there

Embryonic/Fetal Cell Lineage

Origin of donor cells	Anterior				Distal				Posterior			
No. of chimeras	30				35				38			
Tissue colonised	n.	E	M	En	n.	E	M	En	n.	E	M	En
Site of graft												
Anterior	10	8	2	1	11	9	3	0	9	8	1	4
Distal	12	12	0	0	11	1	10	3	14	0	14	0
Posterior	8	3	6	0	13	0	13	0	15	0	15	0
total	30	23	8	1	35	10	25	3	38	8	30	4
%		77	27	3		28	72	8		19	77	11

TABLE 2. A summary of the results of grafting embryonic ectoderm in mouse embryos (42,43). E = ectoderm, M = mesoderm, En = endoderm.

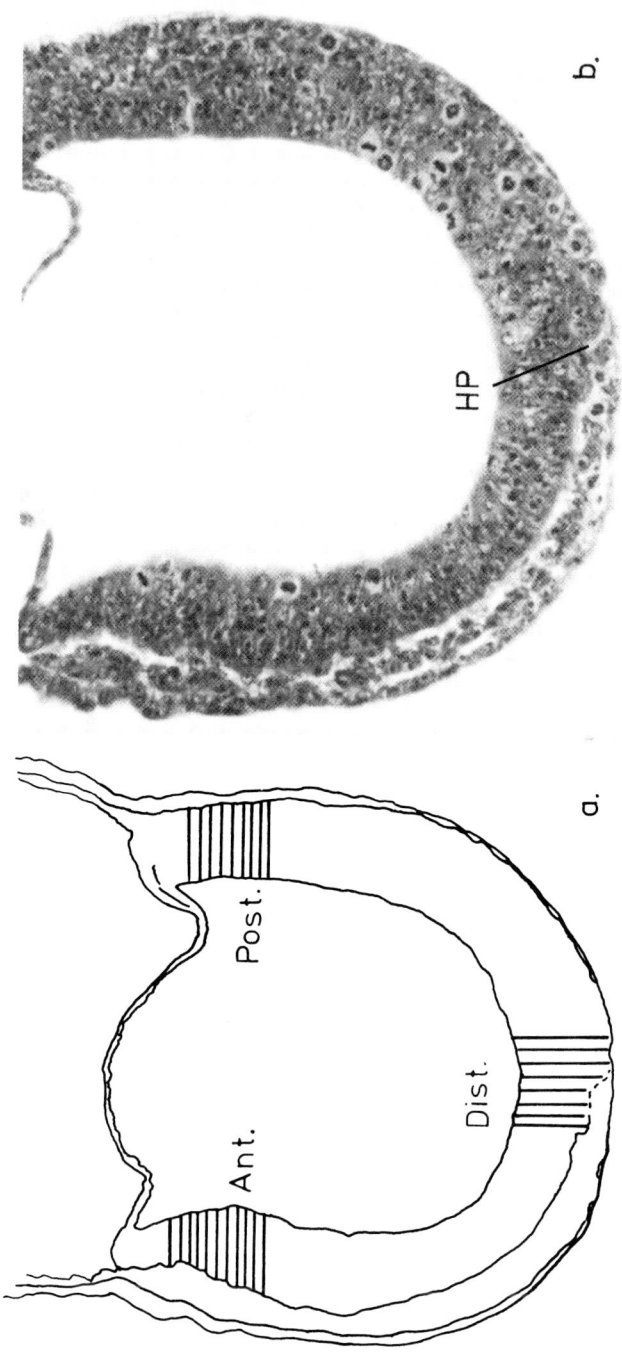

FIGURE 3. (a) Outline drawing of a late primitive-streak stage mouse embryo illustrating the sites involved in the grafting experiments (42,43). They all come from the mid-line. (b) Photograph of the embryonic region of egg cylinder drawn in (a). Note the good definition of ectoderm at the anterior site, poor discrimination at the posterior site and potentially ambiguous definition at the distal site, where the head process (HP) forms.

was evidence for the replacement (regeneration?) of the tissues surgically removed. In the absence of such evidence it seems much more likely that the cells that congregate in the posterior part of the streak during its elongation simply remain there thus confining the continued generation of mesoderm cells to passage through the anterior end of the streak. (This is somewhat similar to the chick primitive streak). It also seems reasonable to regard the sequestration of these few tissues into the posterior streak as allocation to these lineages in normal development (48). The expression of their varied differentiated phenotypes is not dependent upon any further inter-relationships with other parts of the embryo, nor, when isolated from the embryo, do they express any other phenotype.

It was pointed out by Snow (48) that the pieces removed by microsurgery contain elements of all tissue layers of the embryo (primary endoderm, mesoderm and epiblast/ectoderm) and that the fate maps derived give no firm indication of determination, or irreversible restriction of lineage. Grafting experiments (42,43) should provide that information and at first sight they seem to indicate quite clearly that there is considerable developmental lability in cells derived from different parts of the embryo. Table 2 summarises the data and the sites of the grafts are indicated in Figure 3a. Overall 103 chimeras were analysed and the grafted 'ectoderm' cells were found in ectodermal structures in 41, in mesodermal structures in 63 and endodermal (fore- or hind-gut epithelium) in 8. There is some regional variation in tissue colonisation such that cells taken from the anterior site preferentially colonise ectoderm whereas distal and posterior cells are usually found in the mesoderm. Of particular note (43) is the observation that both distal and posterior tissue readily colonises surface ectoderm when placed in an anterior site but do not do so when grafted orthotopically. In only 4 of the 20 chimeras obtained from these particular grafts was neural ectoderm colonised. The simplest interpretation of this data is that anterior ectoderm shows some evidence for a restricted developmental capacity but other ectoderm is quite uncommitted and capable of forming many different tissues. However that conclusion is almost certainly an overstatement. There are several features of the grafting results that are not adequately accounted for and which could indicate fairly widespread restriction in developmental capacity of cells from different regions.

The most successful graft is the orthotopic posterior one with an 83% success rate, whereas the success rate with all other grafts is in the 40-65% range. Most grafts, even the most successful, are found to contain dead donor cells the reason for whose demise can only be guessed at, but could possibly be that they failed to relocate in sites compatible with their developmental commitment. In the orthotopic posterior grafts the integrated donor cells are found to be harmoniously interspersed with the host cells. This is also true of posterior cells grafted into the distal region. In all other grafts, including the remaining orthotopic grafts, donor cells are invariably found as discrete coherent patches within the host tissue. In the first paper (42), concerning the orthotopic anterior and distal grafts, Beddington argues that the coherence of the donor tissues is evidence of health and true tissue colonisation, yet then fails to discuss the obvious implications this has for the diffuse colonisation observed when posterior cells are grafted (43). Since she accepts that the strong tendency for anterior ectoderm to form ectodermal derivatives when grafted elsewhere is probably an indication of lineage restriction the claims for developmental lability rest heavily on the fate of distal and posterior tissues, and on the acceptance that their presence in host embryos reflects real integration into the development of tissues in which they reside. An alternative explanation for the coherence observed in heterotopic grafts and non-posterior orthotopic grafts could simply be that close coherence is a property of like cells and when bits of this tissue find themselves in a strange environment they associate more readily with one another than with their neighbours whose developmental specification is different. If this were the case then the diffuse colonisation would reflect true participation in development and all other types of distribution would be aberrant. In reality, neither extreme is likely to be correct. It is interesting to note however that in only 10% of graft chimeras was colonisation observed in two germ layers. In half of these posterior cells were grafted.

There must also be some doubt about the true nature of the distal and posterior cells used in the grafts. The site from which the posterior cells are removed is in the primitive streak and there is no distinction in the tissue of this area that permits the identification of an ectodermal or mesodermal component. Histologically, and in the electron microscope, the tissue is homogeneous. The same might be true of the distal tissue since it is at this site that the head process lies. This structure also shows no clear demarcation between an internal ectodermal surface, an intermediate mesodermal component and an outer endoderm, although it is believed such a distinction exists. Figure 3b shows a light micrograph of a section through the midline of an embryo of the stage used in the grafting experiments, which illustrates the regions in question.

Although it is claimed by the author (43) that the grafting experiments show lability and contradict the microsurgery data it is possible to draw the opposite conclusion.

Piece 1 (Figure 2) on its own makes brain, foregut and heart; the grafting of anterior ectoderm seems to confirm a strong commitment or determination toward neural ectoderm formation. Removal of Piece 5 interferes with heart development; in the 3 grafts of distal ectoderm to an anterior site in which a mesodermal component was formed that component was found in the heart (43). It was the only mesodermal derivative of those grafts. Removal of Piece 5 also interferes with somite formation (48), and Piece 2 (which includes Piece 5) will generate all the trunk mesoderm and ectodermal structures of the embryo; distal tissue grafts yield mesodermal components in over 70% of cases and neural ectoderm in 14%. The microsurgery of the primitive streak locates tailbud, hind gut and maybe primordial germ cells in its posterior part at the developmental stages at which grafting was done. During development the tailbud will generate ectodermal and mesodermal structures of the posterior end of the embryo. Grafts of posterior tissue colonise ectoderm and endoderm (foregut) in an anterior site and mesodermal components in distal or posterior sites. In these cases the identity of cells which die after grafting is crucial in understanding whether the different colonisation in anterior and other sites reflects developmental lability or cell selection. Teratoma studies using tailbuds could indicate that it is cell selection (57).

Indirect Analysis

Clonal analysis of X-inactivation mosaics. This lends indirect support for the view that some lineage segregation takes place at about 7 dpc in the mouse. It has been pointed out that random sampling and coherent clone growth are prerequisites for the statistical approach to be valid (52). It is rare for these conditions to be met but the studies of McMahon et al. (63,64) present data indicating that the first condition at least is satisfied. Their investigation was consequent upon the observation that preferential inactivation of the paternally derived X chromosome was coincidental with the differentiation of early, extraembryonic, lineages whereas in later developmental stages either the maternal or paternal chromosome seems randomly inactive (65). It was postulated that the availability of the maternally derived X chromosome for inactivation might be progressive and coincide with the sequential segregation of cell lineages, such as ectoderm, mesoderm, endoderm and germ cells. It was argued that this would be reflected in derivatives of those tissues by differences in the proportions of active paternal and maternal X-linked genes (66). Quantitation of the activity of the enzyme phosphoglycerate kinase (PgK) was done on $12\frac{1}{2}$d female mouse embryos heterozygous for the electrophoretic variants PgK-1^a and PgK-1^b. Brain ectoderm, heart mesoderm and liver were taken as representative of the primary germ layers and primordial germ cells were surgically removed from the developing gonad. Extraembryonic mesoderm and ectoderm were included in the study (64).

The results show that in the extraembryonic ectoderm the paternally derived chromosome is inactive, i.e. only the maternally derived isozyme was detected. In all other tissues both alleles are expressed. The relative isozyme contributions are very similar in all tissues from the same embryo but strikingly different between siblings. Thus it seems that the embryonic tissues examined are derived by simultaneous random sampling of a common pool of cells after X chromosome inactivation has occurred, rather than sequentially and as a separate event in the germ layers and germ cells.

Statistical analysis of these data (64,67) allows an estimate of a minimum of 47 cells from which the fetus is derived and of 193 cells (95% confidence limit, 120 $-\alpha$) for the precursor cell pools of the tissues analysed. This latter figure is somewhat higher than the 20-60 cells estimated for several tissues using cytological techniques and an X-autosome translocation to identify the source of the X chromosome (68). The use of an abnormal karyotype may have contributed to the higher independent variance

of tissues which was observed and which generates the
lower estimate.

The data indicates that lineage allocation follows
X-inactivation, which is known to occur sometime before
6 dpc in the mouse (see 64 for references and discussion).
Lineage definition must precede the appearance of a
differentiated phenotype. If the tissues analysed in the
PgK study are representative of the germ layers then
segregation of those tissues is known to occur anatomically
at $6\frac{1}{2}$ to 7 dpc i.e. when the fetal lineage consists of 600
to 2,000 cells. Primordial germ cells are segregated
anatomically and by enzyme activity at about 8 dpc (54)
and microsurgery locates them in the primitive streak at
7 dpc.

Teratoma studies. If the epiblast of primitive
streak stages or the ectoderm of headfold stages of rat or
mouse embryos is divided into e.g. anterior and posterior
parts and then grafted into an ectopic site teratomas will
develop which contain differentiated tissues representative
of all germ layers (55,56). There is no obvious disprop-
ortionate representation of anterior or posterior type
tissues in either sort of graft. Indeed, the caudal fragments are observed to form both cranial and caudal tissues.
It should be noted however that the course of development
of the ectoderm into the teratoma involves an initial disorganization of the epithelial structure of the ectoderm
(55) prior to the emergence of different tissue types.
This process is reminiscent of the dedifferentiation and
tissue reorganisation that is characteristic of regeneration. Grafting of the tailbud of later stage mouse embryos,
up to the hind limb-bud stage show this caudal tissue to
retain considerable histogenic potential, generating
teratomas with a whole range of tissues. Eventually the
tailbud loses some of this capacity by about the 40-somite
stage (57).

Cytotoxic studies. Treatment of primitive streak
stages of mouse embryos with Mitomycin C reveals that
further embryonic development is maintained even if the
cell number at $7\frac{1}{2}$ dpc is reduced by cell death and mitotic
delay to 10% of normal (49-51). Development proceeds
along a curious pathway (69) but does not result in major
malformation nor in any parts being missing. This could
have been expected if cell populations were irreversibly
determined at 7 or $7\frac{1}{2}$ dpc. In this context small cell
populations, such as the emerging germ-line or the early
somites, would have been particularly at risk. These

tissues are never found to be missing (49,50), nor is heart formation, another early forming organ probably derived from few cells, ever abolished (Tam, unpublished observations).

CONCLUSIONS AND COMMENT

Segregation of the dozen or so epiblast cells which constitute the embryonic cell line is progressive through cleavage and most likely brought about by a combination of changes in polarity and adhesiveness of cells from the 2-cell stage onwards, aided by small differences in the cell cycle times of the descendants of the first two blastomeres. Irreversible commitment to the embryonic lineage does not occur until the expanded blastocyst stage.

Establishment of different lineages within the epiblast probably does not start until there are 600 or so cells. It is likely that segregation of these lineages also is progressive. Considerable autonomy of development of isolated parts of primitive streak stage embryos (2,000-15,000 cells) can be demonstrated and grafting of tissues between embryos reveals some regional restriction of developmental capacity. However there is also evidence from these grafting experiments that suggests that some developmental lability is present. This is supported by data from teratomas and from cytotoxicity studies.

However there is reason to consider that some of the demonstrated lability of development may reflect cell selection (in the grafting experiments) or dedifferentiation and regeneration (in the teratoma studies).

Whatever the true status of determination of cells within primitive streak stage mouse embryos it is clear that within a very short time (less than 12h, allowing on average less than two cell divisions) an embryo will be formed which, for development to proceed normally, must possess in a reasonably differentiated state, neural ectoderm, epidermal ectoderm, neural plate mesoderm, heart mesoderm, a blood cell line, somitic mesoderm, mesenchyme, perhaps an endothelial cell line, endoderm and germ cells. It is not unrealistic to think that allocation to these lineages takes place during primitive streak stages, perhaps coincidentally with the definition of the germ layers, and that each would show the X inactivation characteristics and variance already demonstrated for some of them. This would suggest a necessary requirement for about 10 tissues, in each of which the precursor cell pool is estimated at around 150. This would require an embryo of at least 1,500 cells. The 7 dpc mouse embryo, for which a fate map can be made, consists of about 2,000 cells.

The emergence and differentiation of the tissues which form the somite stage embryo is not dependent upon the number of cell or nuclear divisions in the lineage since it is not interfered with by large changes in cell number, either upwards or downwards (69). On balance therefore I feel that commitment of cells to those lineages needed to construct a neural plate/somite stage embryo takes place about 24h earlier, in early primitive streak stages, in an embryo of 1,000-2,000 cells. Only when specifically challenged by providing them with the strange environment of a kidney or testis in an ectopic graft is there good evidence that the normal pathway of development can be altered, and it would then appear to involve a dedifferentiation step prior to formation of new cell and tissue types.

REFERENCES

1. Gardner RL, Lyon MF (1971). X chromosome inactivation studied by injection of a single cell into the mouse blastocyst. Nature, London 231: 385.
2. Gardner RL, Papaioannou VE (1975). Differentiation in the trophectoderm and inner cell mass. In Balls M, Wild AE (eds): "The Early Development of Mammals," 2nd Symp Brit Soc Devl Biol, Cambridge University Press, p 107.
3. Gardner RL, Johnson MH (1975) Investigation of cellular interaction and deployment in the early mammalian embryo using interspecific chimeras between rat and mouse. In "Cell Patterning," Ciba Foundn Symp 29: 183. Excerpta Medica, North Holland, Amsterdam.
4. Gardner RL (1982). Investigation of cell lineage and differentiation in the extraembryonic endoderm of the mouse embryo. J Embryol exp Morph 68: 175.
5. Rossant J, Vijh KM, Siracusa LD, Chapman VM (1982) Identification of embryonic cell lineages in histological sections of Mus musculus ↔ Mus caroli chimeras. J Embryol exp Morph 73: 179.
6. Ducibella T, Anderson E (1975). Cell shape and membrane changes in the eight-cell mouse embryo: prerequisite for morphogenesis of the blastocyst. Devl Biol 47: 45.

7. Ducibella T (1977) Surface changes of the developing trophoblast cell. In Johnson MH (ed) "Development in Mammals," Vol 1, Amsterdam: Elsevier, p 5.
8. Reeve WJD, Ziomek CA (1981) Distribution of microvilli on dissociated blastomeres from mouse embryos: evidence for surface polarity at compaction. J Embryol exp Morph 62: 339.
9. Handyside AH (1980) Distribution of antibody- and ligand-binding sites on dissociated blastomeres from mouse morulae: evidence for polarisation at compaction. J Embryol exp Morph 60: 99.
10. Reeve WJD (1981) Cytoplasmic polarity develops at compaction in rat and mouse embryos. J Embryol exp Morph 62: 351.
11. Izquierdo L, Lopez T, Marticorena P (1980) Cell membrane regions in preimplantation mouse embryos. J Embryol exp Morph 59: 89.
12. Ziomek CA, Johnson MH (1980) Cell surface interaction induces polarization of mouse 8-cell blastomeres at compaction. Cell 21: 935.
13. Kimber SJ, Surani MAH (1981) Morphogenetic analysis of changing cell associations following release of 2-cell and 4-cell mouse embryos from cleavage inhibition. J Embryol exp Morph 61: 331.
14. Johnson MH (1981) The molecular and cellular basis of preimplantation mouse development. Biol Rev 56: 463.
15. Johnson MH, McConnell J, Van Blerkom J (1984) Programmed development in the mouse embryo. J Embryol exp Morph 83, Suppl: 197.
16. Van Blerkom J, Barton SC, Johnson MH (1976) Molecular differentiation in the preimplantation mouse embryo. Nature, Lond 259: 319.
17. Handyside AH, Johnson MH (1978) Temporal and spatial patterns of the synthesis of tissue specific polypeptides in the preimplantation mouse embryo. J Embryol exp Morph 44: 191.
18. Surani MAH, Barton SC, Burling A (1980) Differentiation of 2-cell and 8-cell mouse embryos arrested by cytoskeletal inhibitors. Exp Cell Res 125: 275.
19. Pratt HPM, Chakraborty J, Surani MAH (1981) Molecular and morphological differentiation of the mouse blastocyst after manipulations of compaction with cytochalasin D. Cell 26: 279.
20. Braude PR (1979) Time-dependent effects of α amanitin on blastocyst formation in the mouse.

J Embryol exp Morph 52: 193.
21. Snow MHL (1973) Tetraploid mouse embryos produced by cytochalasin B during cleavage. Nature 224: 513.
22. Eglitis MA, Wiley LM (1981) Tetraploidy and early development: effects on developmental timing and embryonic metabolism. J Embryol exp Morph 66: 91.
23. Petzoldt U, Burki K, Illmensee GR, Illmensee K (1983) Protein synthesis in mouse embryos with experimentally produced asynchrony between chromosome replication and cell division. Wilhelm Roux Arch devl Biol 192: 138.
24. Smith R, McLaren A (1977) Factors affecting the time of formation of the mouse blastocoele. J Embryol exp Morph 41: 79.
25. Dean WL, Rossant J (1984) Effect of delaying DNA replication on blastocyst formation in the mouse. Differentiation 26: 134.
26. Wilson IB, Bolton E, Cuttler RH (1972) Preimplantation differentiation in the mouse egg as revealed by microinjection of vital markers. J Embryol exp Morph 27: 467.
27. Hillman N, Sherman MI, Graham CF (1972) The effect of spatial arrangement on cell determination during mouse development. J Embryol exp Morph 28: 263.
28. Kelly SJ (1977) Studies of the developmental potential of 4- and 8-cell stage mouse blastomeres. J exp Zool 200: 365.
29. Graham CF, Lehtonen E (1979) Formation and consequences of cell pattern in preimplantation mouse development. J Embryol exo Morph 49: 277.
30. Magnusson T, Epstein CJ (1981) Genetic control of very early mammalian development. Biol Rev 56: 369.
31. Monk M, Harper MI (1979) Sequential X chromosome inactivation coupled with cellular differentiation in early mouse embryos. Nature 281: 311.
32. Manes C, Menzel S (1981) Demethylation of CpG sites in DNA of early rabbit trophoblast. Nature 293: 589.
33. Chapman V, Forrester L, Sanford J, Hastie N, Rossant J (1984) Cell lineage specific undermethylation of mouse repetitive DNA. Nature 307: 284.
34. Snow MHL (1973) The differential effect of H^3-thymidine upon two populations of cells in pre-implantation mouse embryos. In Balls M, Billett FS (eds) "The Cell Cycle in Development and Differentiation," 1st Symp Brit Soc Devl Biol,

Cambridge: Cambridge University Press, p 311.
35. Snow MHL (1973) Abnormal development of preimplantation mouse embryos grown in vitro with H^3-thymidine. J Embryol exp Morph 29: 601.
36. Snow MHL (1977) Gastrulation in the mouse: growth and regionalisation of the epiblast. J Embryol exp Morph 42: 293.
37. Lewis NE, Rossant J (1982) Mechanism of size regulation in mouse embryo aggregates. J Embryol exp Morph 72: 169.
38. Snow, MHL, Bennett D (1978) Gastrulation in the mouse: assessment of cell populations in the epiblast of t^{w18}/t^{w18} embryos. J Embryol exp Morph 47: 39.
39. Solter D, Skreb N, Damjanov I (1971) Cell cycle analysis in the mouse egg cylinder. Exp Cell Res 64: 331.
40. Poelmann RE (1980) Differential mitosis and degeneration pattern in relation to the alterations in the shape of the embryonic ectoderm of early postimplantation mouse embryos. J Embryol exp Morph 55: 33.
41. Snow MHL (1976) Embryonic growth during the immediate postimplantation period. In "Embryogenesis in Mammals," Ciba Found Symp 40, Amsterdam: Elsevier, p 53.
42. Beddington RSP (1981) An autoradiographic analysis of the potency of embryonic ectoderm in the 8th day postimplantation mouse embryo. J Embryol exp Morph 64: 87.
43. Beddington RSP (1982) An autoradiographic analysis of tissue potency in different regions of the embryonic ectoderm during gastrulation in the mouse. J Embryol exp Morph 69: 265.
44. Beddington RSP (1983) The origin of the foetal tissues during gastrulation in the rodent. In Johnson MH (ed) "Development in Mammals," Vol 5, Amsterdam: Elsevier, p 1.
45. Tam PPL (1980) The regulation of growth and morphogenesis of mouse embryos during the postimplantation period. PhD Thesis, University College, London.
46. England JM, Rogers AW, Miller RG (1973) The identification of labelled structures on autoradiographs. Nature 242: 612.
47. Snow MHL (1985) Control of embryonic growth rate

and fetal size in mammals. In Falkner F, Tanner JM (eds) "Human Growth," Vol 1, New York: Plenum Press (in press).
48. Snow MHL (1981) Autonomous development of parts isolated from primitive-streak-stage mouse embryos. Is development clonal? J Embryol exp Morph 65, Suppl: 269.
49. Tam PPL, Snow MHL (1981) Proliferation and migration of primordial germ cells during compensatory growth in mouse embryos. J Embryol exp Morph 64: 133.
50. Tam PPL (1981) The control of somitogenesis in mouse embryos. J Embryol exp Morph 65, Suppl: 103.
51. Gregg BC, Snow MHL (1983) Axial abnormalities following disturbed growth in Mitomycin C treated mouse embryos. J Embryol exp Morph 73: 135.
52. McLaren A (1976) "Mammalian Chimaeras", Cambridge: Cambridge University Press, p 92.
53. McMahon A, Fosten M, Monk M (1983) X chromosome mosaicism in the three germ layers and the germ line of the mouse embryo. J Embryol exp Morph 74: 207.
54. Snow MHL, Monk M (1983) Emergence and migration of mouse primordial germ cells. In McLaren A, Wylie CC (eds) "Current Problems in Germ Cell Differentiation," Brit Soc Devl Biol Symp 7, Cambridge: Cambridge University Press, p 115.
55. Svajger A, Levak-Svajger B, Kostovic-Knesevic L, Bradamante Z (1981) Morphogenetic behaviour of the rat embryonic ectoderm as a renal homograft. J Embryol exp Morph 65, Suppl: 243.
56. Beddington RSP (1983) Histogenetic and neoplastic potential of different regions of the mouse egg cylinder. J Embryol exp Morph 75: 189.
57. Tam PPL (1984) The histogenetic capacity of tissues in the caudal end of the embryonic axis of the mouse. J Embryol exp Morph 82: 253.
58. Smits-Van Prooije AE, Poelmann RE, Vermeij-Keers C (1984) Mesoderm formation in the rat embryo, cultured in vitro, analysed with a lectin-coated colloidal gold marker. J Embryol exp Morph 82, Suppl: 76.
59. Lawson KA, Meneses JJ, Pedersen RA (1984) Fate mapping of the endoderm in pre-somite mouse embryos by intracellular microinjection of horseradish peroxidase. J Embryol exp Morph 82, Suppl: 67.
60. Rossant J, Sanford J, Chapman VM, Andrews G (1984) Cell lineage analysis in the mouse embryo using

molecular probes. J Embryol exp Morph 82, Suppl: 73.
61. Gardner RL (1984) An *in situ* cell marker for clonal analysis of development of the extraembryonic endoderm in the mouse. J Embryol exp Morph 80: 251.
62. Garner W, McLaren A (1974) Cell distribution in chimaeric mouse embryos before implantation. J Embryol exp Morph 32: 495.
63. McMahon A, Fosten M, Monk M (1981) Random X chromosome inactivation in female primordial germ cells in the mouse. J Embryol exp Morph 64: 251.
64. McMahon A, Fosten M, Monk M (1983) X-chromosome mosaicism in the three germ layers and the germ line in the mouse embryo. J Embryol exp Morph 74: 207.
65. Monk M, Harper MI (1979) Sequential X chromosome inactivation coupled with cellular differentiation in early mouse embryos. Nature 281: 311.
66. Monk M (1981) A stem-line model for cellular and chromosomal differentiation in early mouse development. Differentiation 19: 71.
67. Stone M (1983) A general statistical model for clone-tissue studies using X-chromosome inactivation data. Biometrics 39: 395.
68. Nesbitt MN (1971) X chromosome inactivation mosaicism in the mouse. Devl Biol 26: 252.
69. Snow MHL, Tam PPL, McLaren A (1981) On the control and regulation of size and morphogenesis in Mammalian embryos. In Subtelny S, Abbott UK (eds) "Levels of Genetic Control in Development," 39th Symp Soc Devl Biol, New York: Alan R Liss, p 201.

MORPHOGENETIC MOVEMENTS AND FATE MAPS IN THE AVIAN BLASTODERM

L. Vakaet

Department of Anatomy and Embryology, State University, Ghent, Belgium

ABSTRACT Using a combination of marking techniques (vital dyes, iron oxide grains and especially chick-quail grafts after N. Le Douarin) the disposition of the Anlage Fields of the avian blastoderm has been investigated. Fate maps have been constructed of ten stages from the just laid blastoderm through the head fold stage. While the fate maps of stages 6 through 9 mainly confirm Nicolet's findings, a more precise description is given of the final placement of the gut endoderm by ingression through the anterior part of the primitive streak between stages 4 and 5. Early marking has made it possible to draw fate maps of just laid stage 0 blastoderms. A hypothetic fate map of a radially symmetrical avian blastoderm (about five hours before laying) is presented. The bilateral symmetrization of the blastoderm is discussed in relation with external (gravity) and internal (extra-cellular matrix) factors. The role of a band of extra-cellular fibers on the basal lamina of the upper layer in the marginal zone of the young blastoderm is considered.
The description of the movements correlated with the evolution of the disposition of the Anlage Fields is updated using cinemicrographic data. Possible mechanisms for these movements are presented based on transmission electron microscopic studies.

This work was supported by the Belgian Nationaal Fonds voor Wetenschappelijk Onderzoek (FGWO Grant N° 3.9001.81).
Present address : Department of Anatomy, State University, Ledeganckstraat, 35, B-9000 Ghent, Belgium.

INTRODUCTION

Morphogenetic movements need to be cut in artificial stages for analysis. This enables us to follow the trajectory of the Anlage Fields between successive stages. Therefore, each stage has a different Fate Map. Drawing these maps is relatively easy in stages shortly before segregation of the Anlage Fields into primitive organs. It is increasingly difficult as the stages considered are younger.

The fate maps of the avian blastoderm I present here are mainly based on results obtained with cinemicrographic and histological observation of a combination of marking techniques. Vital dye and iron oxide marks have been applied onto the deep layer of the blastoderm (1). This has led to distinguish ten stages, from the just laid blastoderm (stage 0) to headfold (stage 9). More recently the quail-chick xenograft technique (2) has rendered it possible to describe more precisely the morphogenetic movements in the upper layer. While the data about stages 6 to 9 mainly confirm Nicolet's (3) findings, a more precise description can be given of the final placement of the gut endoderm between stages 4 and 5 (4).

In this paper I start from a hypothetical fate map of a radially symmetrical avian blastoderm about 5 hours before laying. This blastoderm which is not symmetrized, is said to be at stage -1. I propose to compare successively pairs of stages from -1 to 6. The weak part of this description lies of course in the extrapolation from observations to movements. Only Anlage fields that will ingress are mentioned here.

DESCRIPTION OF MAPS AND MOVEMENTS

At stage -1, on which no marking experiments have been done, a radial symmetry is suggested by circles. The darker circle shows the margin of the area centralis which constitutes the intraembryonic part of the blastoderm. Outside this margin, the extra-embryonic mesoblast or area vasculosa is drawn. Inside the margin of the area centralis, beginning from the periphery, we find circular Anlage Fields corresponding respectively to pre-gut endoderm, pre-lateral plates, pre-somites and pre-axial mesoblast. Only half of the Anlage Fields is shaded, to indicate that about half of the upper layer of the blastoderm will eventually ingress during gastrulation.

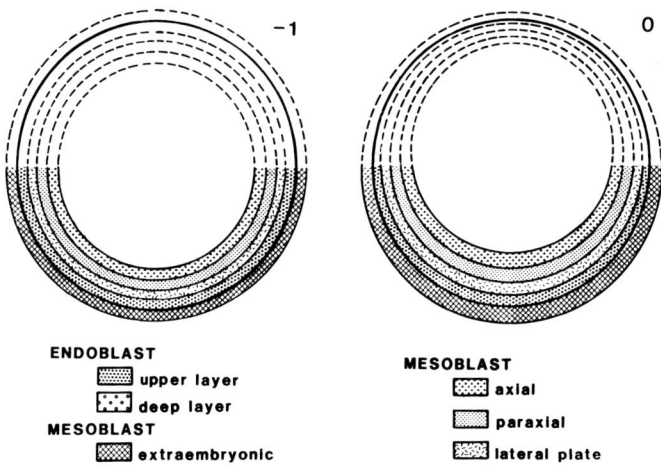

FIGURE 1. Stage -1 and 0. The legends of the shading are the same in Figs. 2,3,5,6.

The shaded area might have been oriented in any direction; the limit of ingression would always cross the center of the area centralis which I consider as the punctum fixum of the blastoderm during gastrulation. The other borders have been drawn in concentric dash lines outside the limit of ingression. This is to point out that it is possible to evoke experimentally a secondary primitive streak in this area up to stage 6 (5). After normal symmetrization no blastoporal ingression will take place in the anterior half of the blastoderm, outside the shaded area. In this outline the future posterior end of the embryo is at the bottom.

Between stage -1 and stage 0, symmetrization is imposed by gravity (6, 7). This orientation is not visible in the freshly laid blastoderm but has been proven by Lutz to exist (8). In my theory it occurs through a shifting of the Anlage Fields towards the rear part of the blastoderm, where they are represented broader than in the anterior part. Neither their total surface nor the position of the limit of ingression have changed. From now on the Anlage Fields have acquired their temporo-spatial disposition on their path of development to the corresponding organs. This disposition will guide each of them into determination during development. The mechanism by which the quantitative data of disposition are transduced into qualitative differences is unknown.

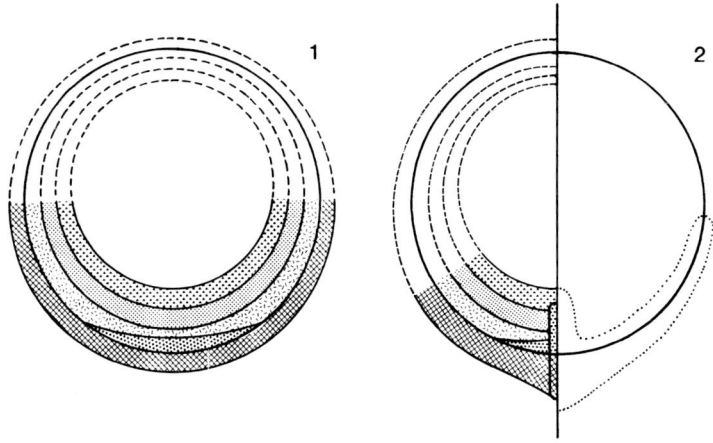

FIGURE 2. Stage 1 and stage 2.

The transition between stage 0 and stage 1 is brought about by the beginning of morphogenetic movements in the plane of the upper layer, due to the temperature of incubation. These initial movements may be compared, after Spratt (9) with the slowly clapping together of the posterior half of the upper layer like a folding fan. The rivet of the fan would be in the center of the area centralis while its sticks would correspond to the limit of ingression. These movements lead to some sort of compression of the Anlage Fields. This first results in an inward bending of the upper layer of the area centralis along its margin at the rear end, leading to the formation of a sickle that has been first described by Koller (10). Part of the pre-endoblast is thus shifted into the middle layer.

From stage 1 to stage 2 the clapping together continues and leads to the appearance of the primitive streak. The streak arises as a linear piling up, perpendicular to the most posterior tangent to the margin of the area centralis. This tangent crosses the primitive streak at its midpoint. The anterior half of the primitive streak is built up by pre-gut endoblast, the posterior half contains extra-embryonic mesoblast of the area vasculosa. The anterior tip of the primitive streak lies within the Anlage Field of the axial mesoblast. This situation will not change until stage 6.

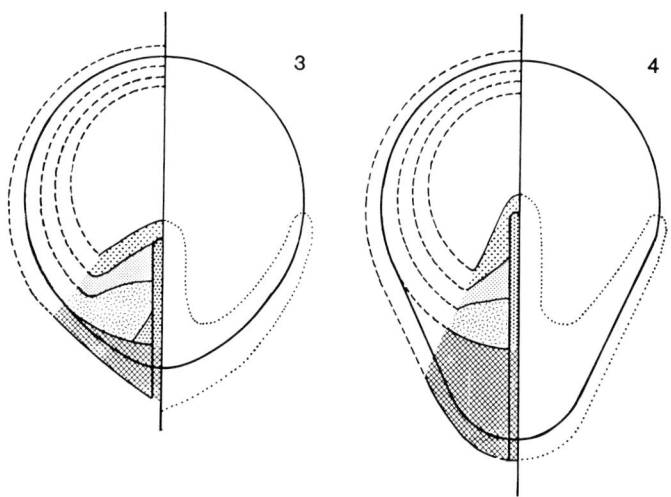

FIGURE 3. Stage 3 and 4.

From stage 2 to stage 3 the primitive streak elongates to the front as well as to the rear (11). It is still built up by the Anlage Fields of the gut endoblast and the area vasculosa. The extent of the Anlage Fields alongside the primitive streak broadens but they are not yet incorporated in the streak.

During the transition to stage 4, the whole Anlage Field of the endoblast is incorporated in the anterior half of the primitive streak, where it is present in the upper as well as in the middle layer. Together with the further broadening of the Anlage Fields alongside the primitive streak, the limit of ingression comes to lie closer to the primitive streak.

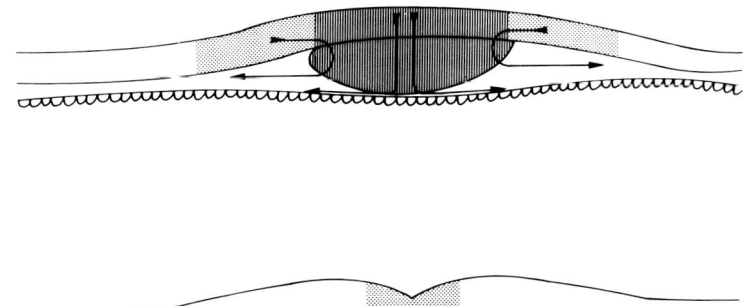

FIGURE 4. Outline of the processes of ingression of the definitive endoblast (vertical striations) and the mesoblast (dotted) between stage 4 and stage 5.

By the end of stage 4 grooving of the primitive streak starts. A groove appears at the anterior end of the primitive streak when it reaches the center of the area centralis and rapidly extends over the entire length of the streak. By that time, all pre-endoblast has disappeared from the primitive streak and has attained its final placement in the lower layer. This means that the pre-endoblast is the first to ingress, and that it sinks directly into the deep layer, dragging along the intra-embryonic mesoblastic Anlage Fields. The more medial parts of these fields now build up the lips of the primitive streak. They are the next to ingress. Their destination is the middle layer.

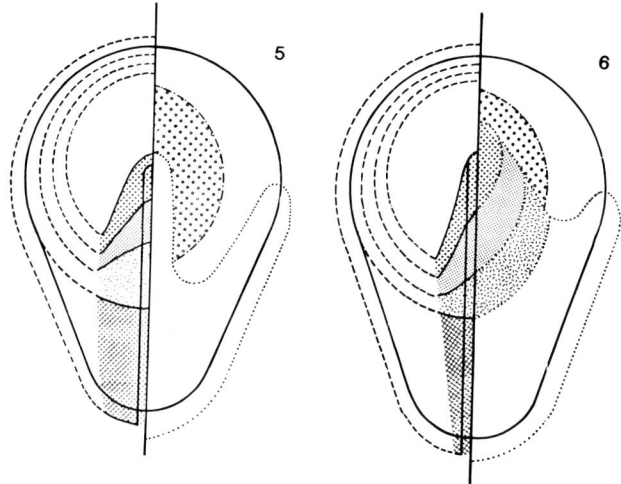

FIGURE 5. Stage 5 and stage 6. Shading as in Fig.1.

The ingression of the pre-mesoblast fields abides by some rules. The first ingressing mesoblast cells will end most foreward in the germ, the following more and more to the rear. On the other hand, the initially most centrally disposed intraembryonic mesoblastic Anlage Fields will form axial structures, the adjoining ones somites and the most peripheral ones, down to the middle of the primitive streak, lateral plate. Outside the margin of the area centralis, in the extraembryonic part of the primitive streak, lies the area vasculosa.

From stage 5 to 6 the primitive streak no longer elongates by convergence but by cutting its groove forward into the Anlage Field of the axial mesoblast. Hensen's Node extends up to the anterior end of this field and attains its most forward position at stage 6. The median cells of the preaxial mesoblast ingress during this cutting process. They form the prechordal plate in the middle layer. The Anlage Field of the notochord is situated in the anterior part of the primitive streak, behind it lie successively those of the somites and the lateral plate. The limit of ingression approaches the primitive streak.

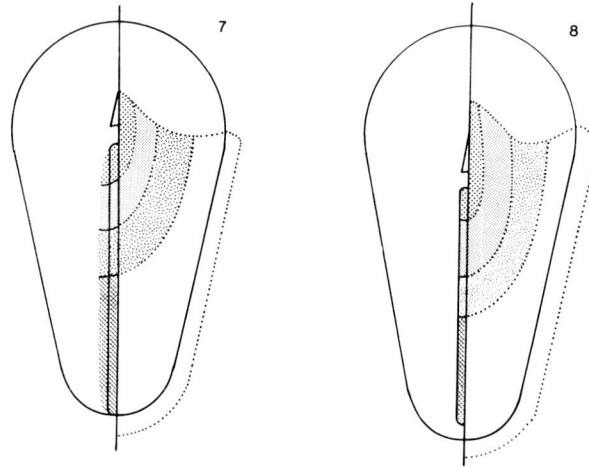

FIGURE 6. Stage 7 and stage 8.

After stage 6 regression of the anterior half of the primitive streak starts, while the ingressed mesoblastic Anlage Fields spread away from their source in the middle layer : the primitive streak. The combination of these two movements leads to the parallel arrangement, side by side, of notochord, somites and lateral plate in the situation as found in the basic type of any vertebrate embryo, the pharyngula.

DISCUSSION

For the discussion on the mechanisms of avian symmetrization and the beginning of gastrulation movements I will limit myself to pointing to the literature and some unpublished observations on which my assumptions are based.
The symmetrization of the avian blastoderm by its rotations in utero have been studied by Clavert (6). Symmetrization has been attributed by Vakaet (12) to the effect of the intrauterine rotations, leading to an inclination of the germ and this has been proven experimentally by Eyal-Giladi (7). The mechanisms by which inclination of the germ leads to symmetrization is explained in diverging ways by these authors. All authors agree, however, that bilateral symmetrization of the avian blasto-

derm occurs about five hours before laying.

The result of symmetrization is not visible macroscopically. In my opinion, a band of fibers situated at the ventral surface of the upper layer in the marginal area of the blastoderm, might play a role in determining the area where the primitive streak will arise. First described by Wakely and England (13) with Scanning electron microscopy and interpreted as a migratory pathway, this band appears on the contrary to inhibit cell locomotion (14). It might constitute a barrier to ingression as well, because it is best developed in the anterior three quarters of the marginal area. We are currently using Scanning electron microscopy to examine younger blastoderms than have been studied so far in search of this structure.

We are going to make use of the possibility to induce secondary primitive streaks in the marginal area. That this is possible by rotating the deep layer of the young blastoderm in its plane, was discovered by Waddington (15) and confirmed by Vakaet (12). However, both were probably in error in their interpretation, as it appears that induction of a secondary primitive streak is brought about by ingressing pre-area vasculosa cells (5). Cells of this sort may have contaminated earlier experiments. The mechanism of this induction is still unknown. A verifiable theory is that it may operate by dissolution of the fiber zone at the site where the streak arises.

The formation of the early primitive streak by convergence and extension has been shown by explanting fragments of the posterior half of the area centralis (11). The evolution of the limit of ingression and the final elongation of the streak have been followed with precision using cinematography combined with vital marking on the upper layer (16). The trajectory of the Anlage Fields (stage -1 excepted) has been followed after quail-chick xenografting (4). Using Transmission electron microscopy, the presence of microfilament bundles in the upper layer and their versatility have been demonstrated (17). They are possible driving organelles for the extensive locomotory activity of the gastrulating avian blastoderm.

ACKNOWLEDGMENTS

The author thanks Dr. N. Goossens for helpful discussions, Mrs. E. Bijtebier for typing the manuscript and the technical staff of the Laboratory for their kind help in preparing the illustrations.

REFERENCES

1. Vakaet L´(1970). Cinephotomicrographic investigations of gastrulation in the chick blastoderm. Arch Biol 81:387.
2. Le Douarin N (1969). Particularités du noyau interphasique chez la caille japonaise (Coturnix coturnix japonica). Bull Biol 103;435.
3. Nicolet G (1970). Analyse autoradiographique de la localisation des différentes ébauches présomptives dans la ligne primitive de l'embryon de poulet. J Embryol Exp Morph 23:79.
4. Vakaet L (1984). Early development of birds. In McLaren A, Le Douarin N (eds): "Chimeras in Developmental Biology", London: Acad Press, p 71.
5. Vakaet L (1984). The initiation of gastrular ingression in the chick blastoderm. Amer Zool 24:555.
6. Clavert J (1962). Symmetrization of the egg of vertebrates. Adv Morph 2:27.
7. Kochav S, Eyal-Giladi H (1971). Bilateral symmetry in chick embryo, determination by gravity. Science 171:1027.
8. Lutz H (1955). Contribution expérimentale à l'étude de la formation de l'endoblaste chez les oiseaux. J Embryol Exp Morph 3:59.
9. Spratt NT (1946). Formation of the primitive streak in the explanted chick blastoderm marked with carbon particles. J Exp Zool 103:259.
10. Koller C (1982). Untersuchungen über die Blätterbildung im Hühnerkeim. Arch f Microsc Anat 20:174.
11. Vakaet L (1960). Quelques précisions sur la cinématique de la ligne primitive chez le poulet. J Embryol Exp Morph 8:321.
12. Vakaet L (1982). "Pregastrulatie en gastrulatie der vogelkiem", Thesis, Brussels: Arscia, p. 185.
13. Wakely J, England M (1979). Scanning electron microscopical and histochemical study of the structure and function of basement membranes in the early chick embryo. Proc R Soc Lond B 206:329.

14. Harrisson F, Andries L, Van Hoof J, Vakaet L (1985). Arrest of cell migration along a fibrillar band in the early chicken embryo. Eur J Cell Biol suppl 7:24.
15. Waddington CH (1933). Induction by the endoderm in birds. W Roux Arch Entw Mech Org 128:502.
16. Vakaet L (1967). Contribution à l'étude de la prégastrulation et de la gastrulation de l'embryon de poulet en culture "in vitro". Mém Acad Roy Med Belg II serie 5:231.
17. Vanroelen C, Verplanken P, Vakaet L (1982). The effects of partial hypoblast removal on the cell morphology of the epiblast in the chick blastoderm. J Embryol Exp Morph 70:189.

CONVERGENT EXTENSION BY CELL INTERCALATION DURING GASTRULATION OF XENOPUS LAEVIS

Ray Keller, Michael Danilchik, Robert Gimlich, and John Shih

Department of Zoology and Department of Molecular Biology University of California, Berkeley, CA 94720

ABSTRACT Microsurgical manipulation and explantation studies, analyzed by time-lapse video and cine microscopy, show that the autonomous morphogenetic movement of convergent extension of the circumblastoporal region accounts for the major tissue displacements during Xenopus gastrulation, including spreading of the marginal zone toward the blastopore, involution of the marginal zone, and closure of the blastopore. Tissue recombination experiments show that the deep cells of the dorsal involuting marginal zone (IMZ) bring about onvergent extension. Labeling of small populations of these cells with a cell lineage tracer shows that convergent extension involves intercalation of deep cells to form a longer, narrower array. Direct time-lapse videomicrography and cinemicrography of deep cells in cultured explants shows convergent extension to involve radial and circumferential intercalation of cells. Removal of the entire blastocoel roof of the early gastrula shows that convergent extension of the circumblastoporal region alone can bring about its involution and blastopore closure.

INTRODUCTION

The Function of Specific, Regional Processes

Amphibian gastrulation appears to involve all parts of the gastrula in a complex integration of autonomous, local processes that together bring about the distortion of the whole (reviewed in 1 and 2). These include epiboly of the animal cap region (3, 4, 5, 6), the change in shape or the active migration of bottle cells (7, 8, 9, 10, 11), migration of involuted cells toward the animal pole (3, 9, 12, 13, 14, 15, 16, 17), and convergence and extension of the marginal zone (18, 19, 20, 21, 22, 23).

Convergence and extension has largely been ignored in recent studies, with some exceptions (23, 24, 25), but the classical literature shows it to be important (see 2). Vogt (18, 19) recognized the capacity of the marginal zone to constrict and close the blastopore, and the isolated dorsal marginal zone autonomously undergoes the narrowing (convergence) and lengthening (extension) (20, 21, 22) that is characteristic of this region of the gastrula in vivo (3). Moreover, convergence and extension appears to be the only one of the several processes identified as playing a role in gastrulation that can account for blastopore closure. Migration of mesodermal cells animal-ward might result in shear at the interface between themselves and the overlying gastrular wall (17) but it is unlikely that this could explain either epiboly, involution, or blastopore closure (see 2). Bottle cell formation may be important in reorienting the convergent extension machinery in the dorsal sector of the early gastrula (see 2) but removal of bottle cells after their formation does not stop gastrulation (see 26, 27). These facts suggested that we do the following experiments in order to define the role and the mechanism of convergent extension in gastrulation of Xenopus.

The following work demonstrates: 1) that convergent extension plays a major role in gastrulation; 2) that it results from active intercalation of deep cells; 3) that the role of other processes in gastrulation are different from what was previously supposed; and 4) that our notions of the nature of regional morphogenetic processes and their interactions should be revised. A history of the investigation of convergent extension, its neglect during the more recent interest in the role of bottle cells and in the migration of mesodermal cells, and arguments for its reconsideration are reviewed in 2.

MATERIALS AND METHODS

Obtaining and Handling Embryos

Embryos the African clawed frog, Xenopus laevis, were obtained, dejellied, and held for experiments as previously described studies (see 27). Staging was done according to Nieuwkoop and Faber (28). Staging designations for the explants represent the stage of companion, intact embryos.

Microsurgery, Labeling of Cells, and Culture of Explants.

Microsurgery was done in modified Steinbergs solution (Steinberg's salts with 5 mM HEPES buffer, pH 7.4) on a Permoplast base, using hair loops and knives of eye-brow hairs. Embryos were transferred to agarose-coated culture dishes after the operations. Local populations of cells in the early gastrula were labeled by injecting precleavage embryos with fluorescein-lysine dextran [FDA] (see 29). At the early gastrula stage, selected regions of labeled embryos were grafted to the corresponding regions of unlabeled embryos. These embryos were then allowed to develop to the appropriate stage, fixed in 4 percent formaldehyde, serially sectioned at 10 micrometers in paraffin, and the degree of intercalation of labeled and unlabeled cells determined by fluorescence microscopy under epi-illumination. "Sandwich explants" of circumblastoporal regions of the gastrula were made and cultured in modified Steinberg's as described in the Results. "Open-faced" sandwich explants and embryos without blastocoel roofs were made and cultured as described in the Results. Both these preparations were cultured in Danilchik's medium (Appendix I), which was designed after the measurements of the ionic composition of the blastocoel fluid of Xenopus (30). This medium allows normal morphogenesis and development of the deep cells and permits direct observation of the activities of the deep cells of the gastrula that were previously hidden within the opaque embryo.

Time-lapse Micrography, Scanning Electron Microscopy (SEM), and Morphometric Analysis.

Time-lapse cinemicrography was done with an Arriflex or Bolex 16 mm camera and intervalometer on Kodak Plus X or Ektachrome Reversal film; videomicrography was done with a Panasonic time-lapse recorder, camera, and monitor. Standard or inverted microscopes were used with objectives from 2.5x to 40x (water immersion) magnification under epi-illumination from heat-filtered tungsten or from fiber-optic sources. SEM was done as described previously (17), using an ISI microscope at 10 to 15 Kv. Morphometrics were done with a Numonics digitizer and Apple II computer.

RESULTS

Prospective Areas and Morphogenetic Movements

We shall briefly review the patterns of morphogenetic movements of Xenopus gastrulation. A fate map of the early gastrula (Fig. 1a, b) and the subsequent distortions of each of the prospective areas (Fig. 1c-f) are shown diagramatically in exploded views of the embryo (see 31, 32). The early Xenopus gastrula (Fig. 1a) consists of the animal cap (AC), the noninvoluting marginal zone (NIMZ), the involuting marginal zone (IMZ), and the subblastoporal endoderm (SBE). These are shown in an exploded view of the early gastrula (Fig. 1b), and each is shown separately at the end of gastrulation (Fig. 1c-f). Dorsal is to the left on all figures and the layers of the neurula are shown from the outside (Fig. 1c) to the inside (Fig. 1f). During gastrulation the AC and NIMZ expand to form the entire outer wall of the gastrula, consisting of prospective neural and epidermal ectoderm (Fig. 1c). The dorsal sector of the NIMZ expands greatly, lengthens, and converges toward the dorsal midline, whereas the AC spreads uniformly (Fig. 1c). The deep sector of the IMZ is an annulus containing prospective notochord (N), somitic mesoderm (SM), head mesoderm (HM), and lateral-ventral mesoderm (LM). During gastrulation, the annulus turns inside out and thus forms the mesodermal mantle (Fig. 1d). HM and LM spread toward the animal and ventral regions whereas N and SM converge dorsally and elongate to form the axial, dorsal structures (Fig. 1d). The superficial, epithelial layer of the IMZ consists of prospective suprablastoporal endoderm (SPE); it lies outside the deep mesodermal annulus (Fig. 1b) and moves with it during gastrulation to form the endodermal archenteron roof (AR) at the end of gastrulation (Fig. 1e). The major distortion of gastrulation involves this double-annulus of deep mesoderm and superficial endoderm in the IMZ. It turns inside out, converges dorsally, and elongates. The SBE is pushed inside, covered by the IMZ, and thus forms the archenteron floor (AF). Note that we have divided the

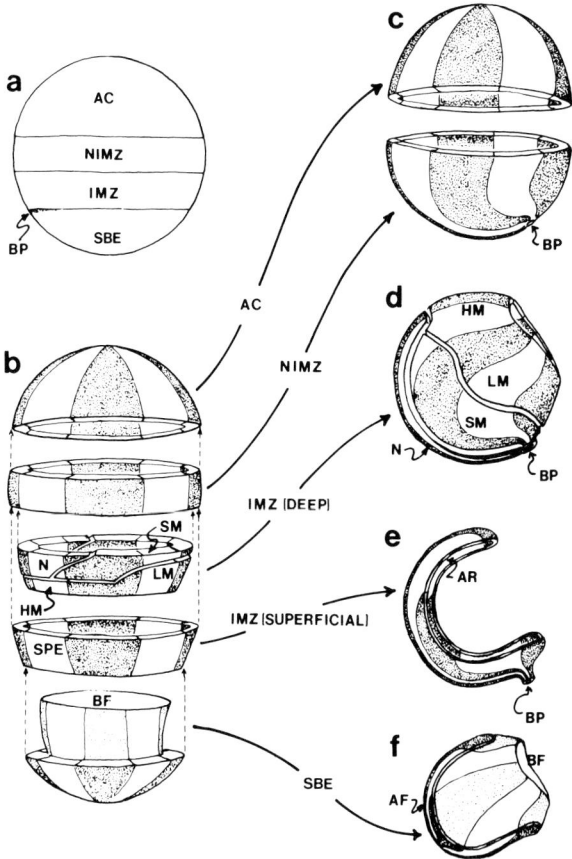

FIGURE 1. The fate map (a,b) and the morphogenetic movements (c-f) of Xenopus laevis are shown diagramatically. From 2, with permission.

marginal zone into two regions, the IMZ and NIMZ. The
IMZ undergoes involution during gastrulation. The NIMZ
lies adjacent to the IMZ and bounds it at the limit of
involution. The limit of involution comes to lie at the
rim of the closed blastopore at the end of gastrulation.
Note that both the IMZ and NIMZ undergo convergence
(narrowing) and extension (lengthening) in closing the
blastopore, with the maximum convergence and extension
occurring in dorsal sector of each. Also, the degree of
convergence and extension increases symmetrically in each
region, toward a maximum in their prospective posterior
ends, represented by the limit of involution (Fig. 1).
Henceforth we shall refer to convergence and extension as
convergent extension, because the two movements occur
together in the events dealt with in this study.

Convergent Extension in Explants of the Gastrula.

Two major, morphogenetic movements are required to
bring the prospective areas from their positions at the
onset of gastrulation (Fig. 1b) into their positions at
the neurula stage (Fig. 1c-f). These are: 1) involution
of the IMZ, and 2) the coordinate convergent extension of
the both IMZ and NIMZ, particularly of their dorsal sectors. How do these movements occur? Previous microsurgical rearrangement experiments showed that the dorsal IMZ
undergoes autonomous convergent extension (23, 2) but that
it occurs largely in the second half of gastrulation and
after involution (25). It was proposed (25) that the
dorsal sector of the mesodermal annulus is turned under
(involuted) in the first half of gastrulation, probably
in part by the mechanical forces generated during bottle
cell formation and in part as the result of migration of
the leading head mesodermal cells (see 2). Then, at the
midgastrula stage, this region uudergoes convergence to
form a constriction ring that acts on the inner surface
of the blastoporal lip and tends to roll the remaining,
uninvoluted IMZ over the lip. The convergent extension,
which occurs predominantly on the dorsal side, results in
movement of the lip across the yolk plug. These coordinate
actions result in involution of the IMZ remaining outside
the blastopore, eccentric closure of the blastopore, and
the dorsal convergence and extension observed in fate maps.
Culture of explants of the dorsal sector of the early
gastrula confirms our notion of how convergent extension

functions in gastrulation. The dorsal sectors of two early
gastrulae, consisting of a superficial epithelium and 4 to
6 layers of deep cells, were excised and sandwiched with
their inner surfaces together, and allowed to heal (Fig.
2c, 3a). Four regions can be recognized in the explants

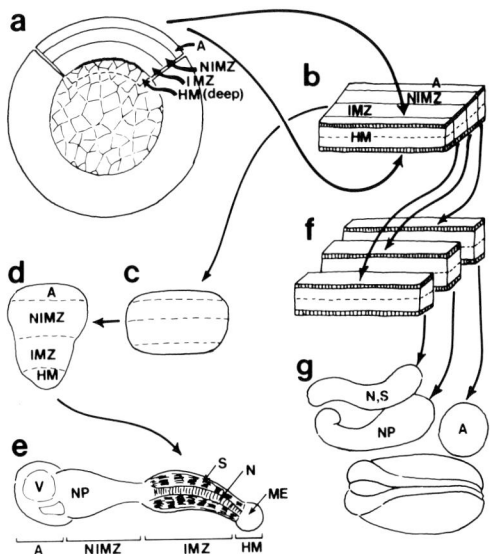

FIGURE 2. Development of the sandwich explants of
the dorsal sector of the gastrula is shown diagramatically.
See text for explanation. From (2), with permission.

(Fig. 2b), based on their development and morphogenesis
(Fig. 2c-d): 1) the animal cap (AC); 2) the NIMZ; 3) the
IMZ; and 4) the head mesoderm (HM). After explantation,
epithelial edges heal together (Fig. 3a) and no major
change in the shape of the explant occurs until the mid-
gastrula stage. Then the IMZ begins to narrow and
lengthen as does the NIM[(Fig. 2c-d; Fig. 3b). Both
regions continue their convergent extension through the
late gastrula stage and into the late neurula stage (Fig.
2e; Fig. 3c). The AC and HM become spherical and knob-
like (Fig. 2e, 3c). The IMZ differentiates into a central
notochord (N), flanked by somitic mesoderm (S)(Fig. 2e;
Fig. 3d; Fig. 4a, b). The NIMZ forms an elongated struc-

ture of small pleiomorphic cells (Fig. 4c). This region
is similar in behavior to the notoplate (NP), which is
that part of the neural anlage in contact with the notochord (33). The AC invariably inflates one or more fluid-
filled vesicles (V), surrounded by cells histologically
similar to those of the notoplate (Fig. 4d). The HM region
forms a ball of mesenchyme cells (ME) by the early tail bud
stage (Fig. 2e). These regions appear to have some measure
of autonomy with respect to one another and the rest of the
gastrula by the early gastrula stage; they can be cut apart
from one another at a stage equivalent to stage 10.25 (28),
and each piece will follow its usual pattern of behavior
and differentiation (Fig. 2f, g). Note that both IMZ and
NIMZ may form the full length of the notochordal-
notoplate region of the corresponding control neurula (Fig.
2g). Similar explants of the lateral and ventral sectors
show progressively decreasing capacity for convergent
extension during the gastrula stages. In these sectors,
convergent extension appears to occur only in the IMZ,
and it does not continue into the neurula stages (data
not shown). Similar behavior of dorsal, lateral and ventral sectors of the IMZ were observed by Schechtman (23)
in Hyla regilla embryos.

Tracings from time-lapse cinemicrographic frames of
explants filmed simultaneously from the top (tangential or
surface view; Fig. 5) and from the side (not shown), show
the distortion involved in convergent extension. When
convergent extension begins, the explant thickens in the
posterior IMZ, the lateral regions move toward the midline,
and the anterior region begins extension (Fig. 5, t = 0 -1
hr). Likewise, the NIMZ begins extension, without thickening, in the opposite direction from the IMZ (Fig. 5, t = 0
-1 hr). Extension continues through the late gastrula
stage (Fig. 5, t = 1.5 -3 hr) and into the late neurula
stage (Fig. 5, t = 6 hr). The mean rate of extension of
the IMZ is 2.6 µm/min and that of the NIMZ is 3.6 µm/min
(mean of 8 and 6 explants respectively).

Gastrulation Without the Blastocoel Roof

The independence and the timing of convergent extension in the gastrula (25), the pattern of distortion of
the gastrula (Fig. 1), and other evidence in the literature
(2), argue strongly that convergent extension alone can

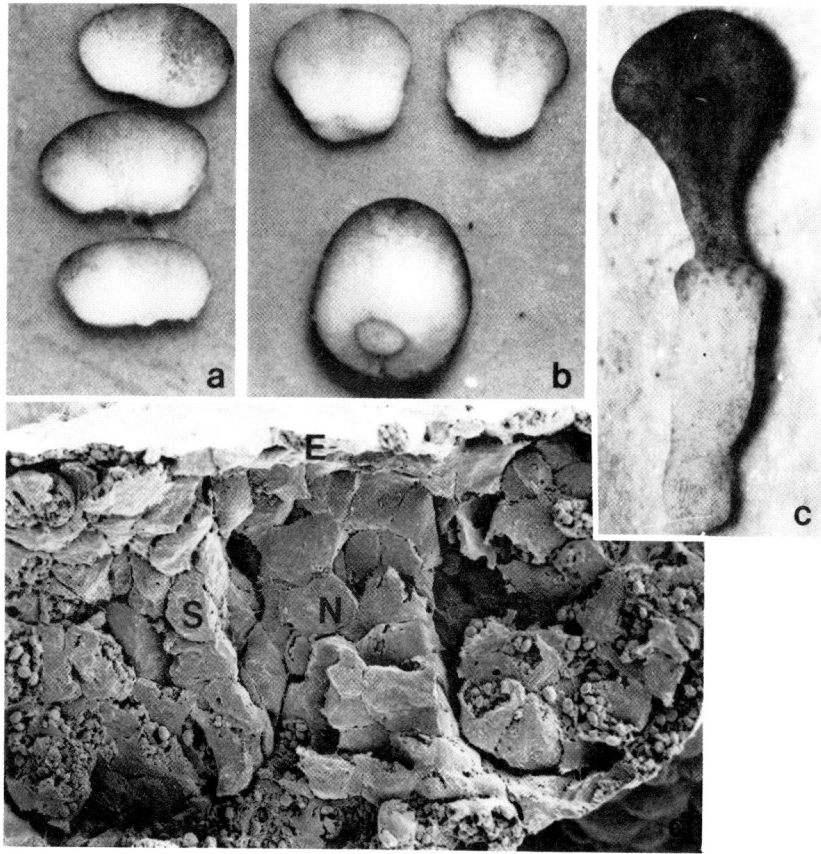

FIGURE 3. Sandwich explants of the dorsal sector of the gastrula are shown a half hour after construction (a), at the late midgastrula stage (b), and at the late neurula stage in external view (c) and in cross section by SEM (d). Figures c and d from 51, with permission.

account for most of the distortion of the gastrula and that other processes, such as bottle cell formation and mesodermal cell migration on the roof of the blastocoel, play ancillary roles. This notion was verified by studying gastrulation in the absence of the blastocoel roof.

The blastocoel roof was removed from early gastrulae by cutting circumferentially, parallel to the prospective blastoporal pigment line (BPL) through the wall of the gastrula and lifting off the roof (Fig. 6a-b). The

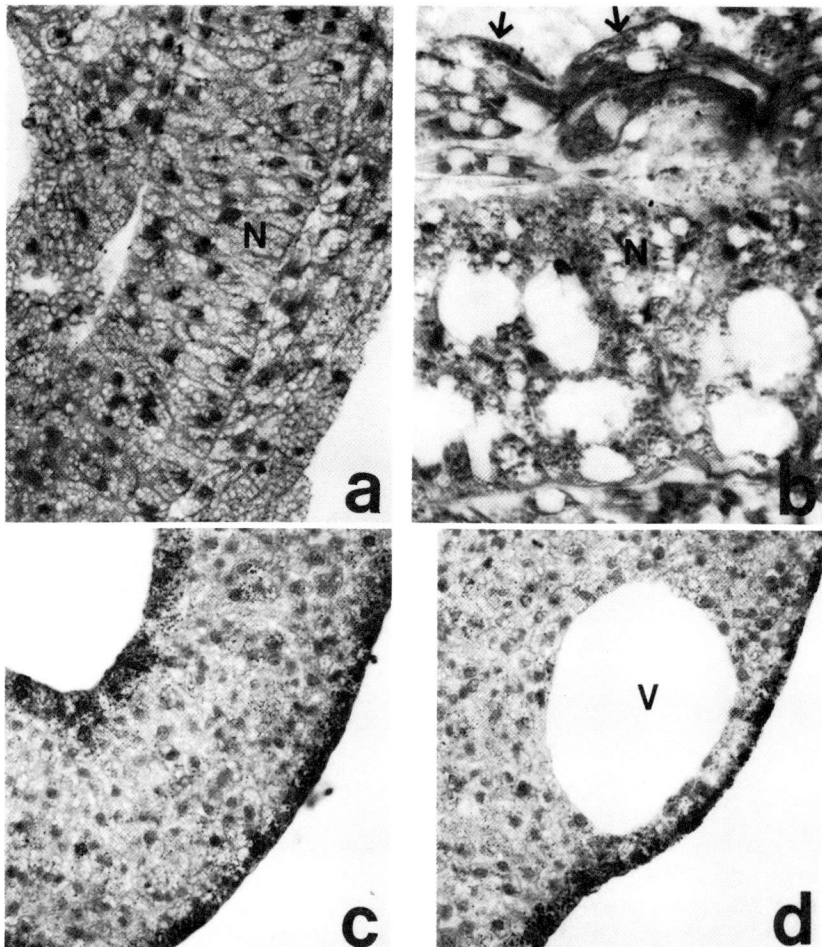

FIGURE 4. Histological sections through the IMZ at the late neurula (a) and tailbud (b) stages show the notochord (N) and somites (pointers, b). Similar sections of show the NIMZ (c) and AC (d), with its characteristic vesicle (V), at the late neurula stage.

animal-vegetal level of the cut was varied to include only the AC, the AC and part of the NIMZ, the AC, NIMZ and part of the IMZ. The resulting embryos consisted of the central endoderm (CE), bottle cells (BC), and the sub-

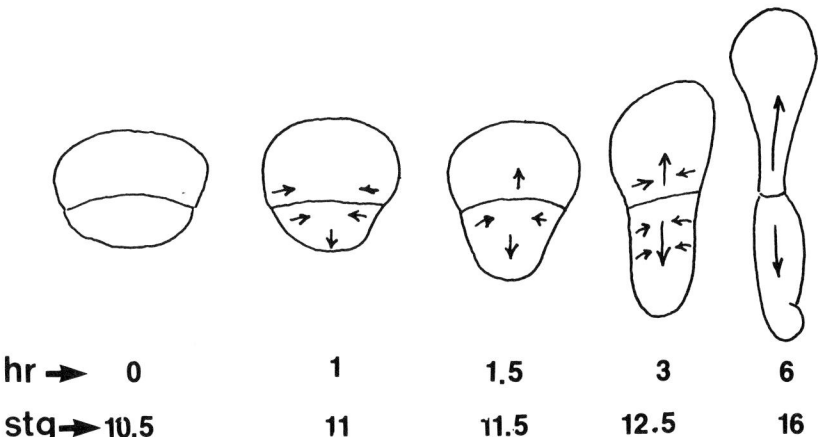

FIGURE 5. Tracings of sandwich explants show the movements (indicated by arrows) of convergent extension.

blastoporal endoderm (SBE), and what is left of the IMZ and NIM[(shaded, Fig. 6a, b). The embryos were then cultured in Danilchik's medium. Time-lapse videomicrography of the vegetal regions of these embryos shows that involution of the IMZ and closure of the blastopore occur as fast or faster than in the controls (Fig. 7, 8).

Embryos with the NIMZ remaining show involution of the IMZ, closure of the blastopore (Fig. 7, stages 10.5-12.5), and convergence and extension, particularly in the dorsal sector (Fig. 7, stages 11.5-19). Moreover, if part of the animal-most (prospective posterior) IMZ is also removed, constriction of the blastopore stalls as soon as all the IMZ is involuted (Fig. 8, stage 12.5). Figure 9 shows a control midgastrula (a) and a gastrula of the same stage but lacking the AC, the NIMZ, and about half the IMZ (b); all the available IMZ has involuted and blastopore closure is stalled. The removed blastocoel roof is shown in Figure 9c. From the late gastrula stage onward, the course of development of these embryos depends on whether or not there is dorsal NIMZ available for interaction with the dorsal involuted IMZ. If dorsal NIMZ is left on the embryo, it comes into apposition with the involuted IMZ in the late gastrula and the two continue convergent

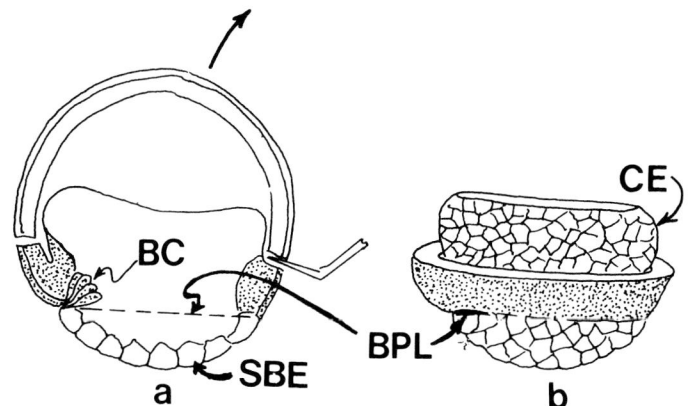

FIGURE 6. The removal of the blastocoel roof is shown diagramatically.

extension together in the neurual stages (Fig. 7, stages 11.5-19). Under these conditions, they form a stable array of dorsal, axial structures, including notochord, somites, an archenteron, and the ventral part of the neural tube. A control neurula (stage 17) is shown (Fig. 9d) alongside such an embryo (Fig. 9e). The elongated dorsal NIMZ can be seen between the somitic mesodermal masses on either side (arrows, Fig. 9e). A section (Fig. 9f) through an embryo similar to the one shown in Figure 9e, shows the notoplate (NP), notochord (N), and somitic mesoderm (S). Without the NIMZ, the IMZ involutes and undergoes convergent extension to form notochord (N) and other axial mesodermal structures (Fig. 9g), but its position and direction of extension may vary considerably.

Mechanism of Convergent Extension

The convergent extension and involution of the IMZ depends on the activities of its deep, nonepithelial cells (27). In the sandwich explants, described above, the superficial epithelium can be replaced with epithelium from other nonextending regions or it can be reoriented without effect (see 25, 2). In contrast, replacement or reorientation of the native, deep cells affects the pro-

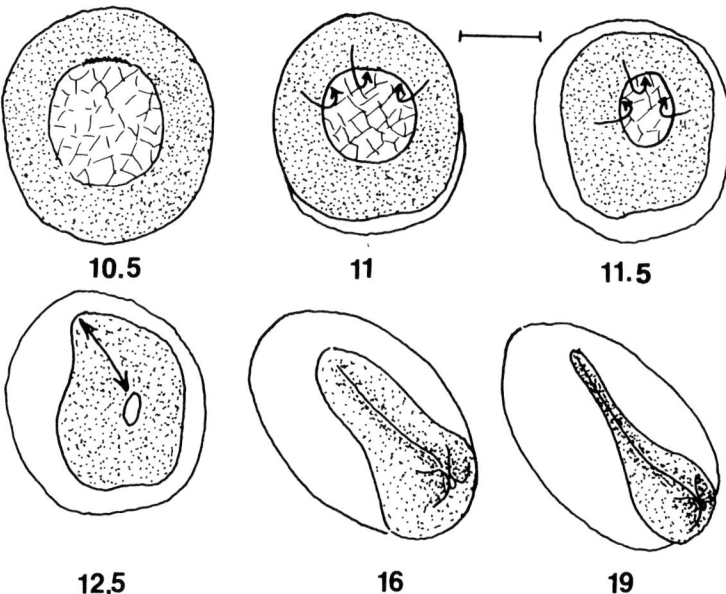

FIGURE 7. Tracings of an embryo from which the blastocoel roof has been removed show the involution of the IMZ and closure of the blastopore in its absence. The numbers indicate the stage of development of companion, control embryos. The bar represents 0.5 mm.

cesses of involution, extension, and convergence (Keller, 1981, 1984).

What do the deep cells do to bring about convergent extension? Waddington (34, p. 109) pointed out that since the elongation of the dorsal sector of the amphibian embryo is not accompanied by overall change in cell shape, cell rearrangement must occur during this distortion. Such is the case. Upon grafting a small patch of IMZ cells from an embryo labeled with FDA to an unlabeled embryo (Fig. 10a-b) and examining the sectioned embryo with fluorescence microscopy at later stages, the circumferential intercalation of labeled graft and unlabeled host cells is apparent. After grafting a labeled patch to an unlabeled embryo, the patch boundaries are continuous, with few

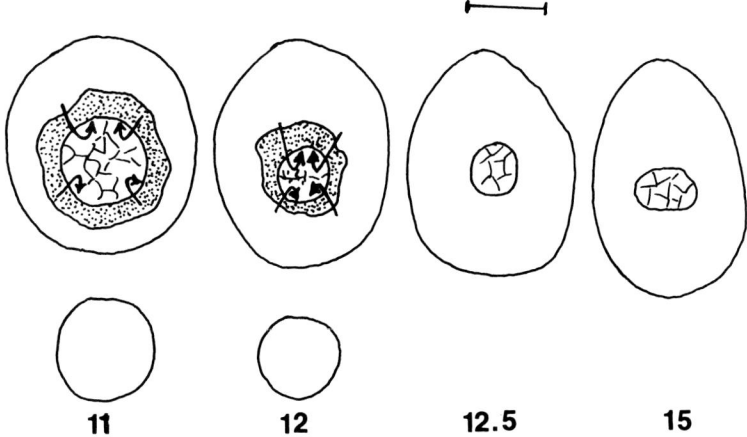

FIGURE 8. Tracings show involution and partial blastopore closure in an embryo lacking the blastocoel roof and half of its IMZ. The numbers indicate the stage of development of companion, control embryos. The bar represents 0.5 mm.

separated cells (Fig. 10c). In contrast, by the early neurula stage, extensive mixing has occurred in the circumferential direction such that labeled cells are spaced over nearly the full length of the notochord and have intercalated with many unlabeled cells (Fig. 10d). This evidence shows that intercalation of many rows cells to form fewer rows of greater length is coincident with convergent extension. A graft of a labeled patch of the ventral IMZ into the corresponding position of an unlabeled embryo shows no such intercalation, but the labeled cells spread out as a sheet of ventral mesoderm (Fig. 10e).

More evidence comes from direct observation of cell behavior during convergent extension. The dorsal sector of the gastrula is explanted as shown for the construction of sandwich explants. But it is cultured between an agarose sheet and an agarose base, as an "open-faced" sandwich in Danilchik's medium (Fig. 11). In this preparation, the behavior of the deep cells can be seen directly and recorded by video or cine methods, using a compound microscope with a 40x water immersion objective and

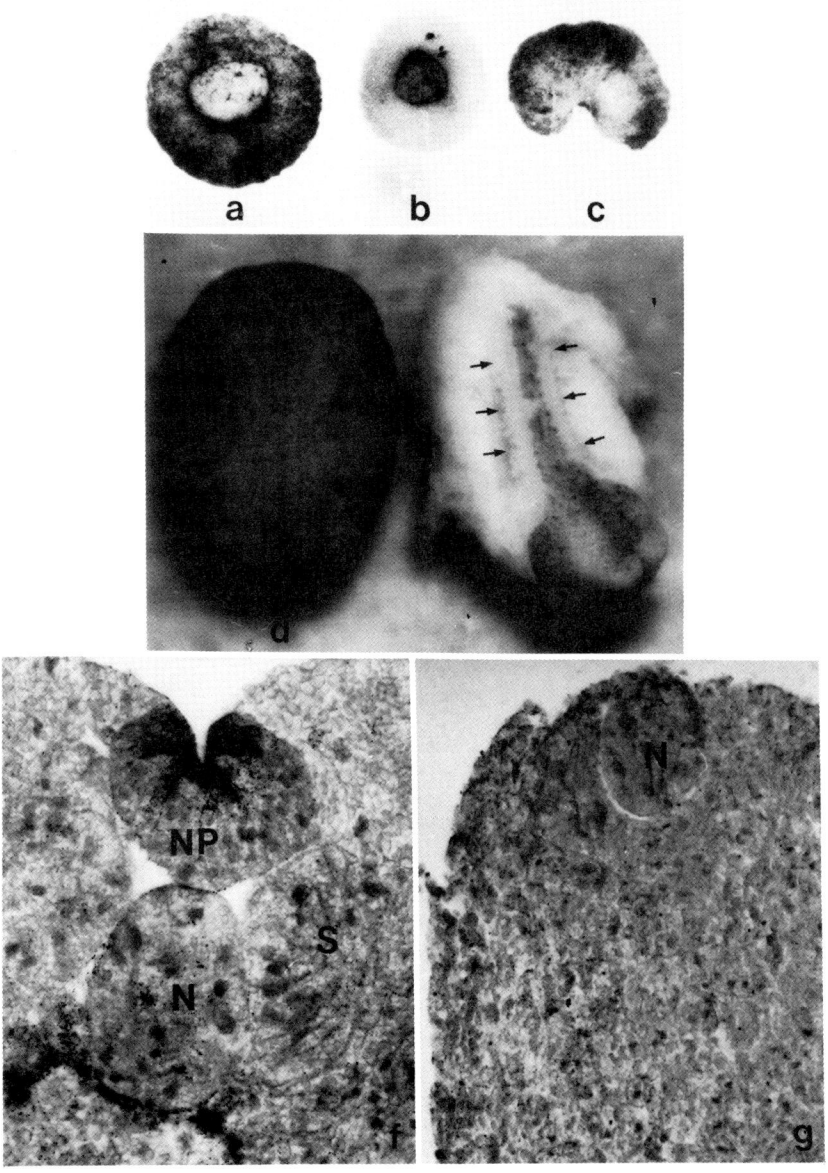

FIGURE 9. Legend appears on following page.

epi-illumination at low angles from one side (EP, Fig. 11). In this preparation, the epithelial wouud-healing response is held at bay and the explant is not distorted by attempts of the free edge to heal with itself. The IMZ sector of the explant converges (dashed arrows, Fig. 11), extends (solid arrows, Fig. 11), and differentiates into notochord and blocks of somitic mesoderm under full view of the experimenter. As suggested by the surface distortion of the sandwich explants, deep cells of "open-faced" explants move medially at the posterior IMZ and forward at the vegetal end of the explant. This is accompanied by repeated breaking and reforming of contacts between cells and their separation to form fractures that extend between a few to ten or more cells. Individual cells intercalate between one another in the circumferential (Fig. 12a) and radial (Fig. 12b) directions. In the neurula stage, the boundaries of the notochord become apparent and the notochordal cells undergo circumferential intercalation to form the "stack of coins" array (Fig. 12c) characteristic of this tissue (35).

The mechanism of intercalation and its relationship to production of a force sufficient to produce elongation and narrowing of the explant is not yet clear. The fracturing observed may be simply a mechanism of releasing internal stresses; alternatively, it may be organized and produce spaces into which cells can move. All intercalation events that could be clearly resolved in the gastrula stage explants involved dramatic cytoplasmic flow. Such flow often passed distally, through the core of a protrusion from the cell leading the intercalation. At the distal end of the protrusion it turned peripherally in a "fountain-

FIGURE 9. The vegetal view of a midgastrula control embryo (a) and an embryo lacking the blastocoel roof (the AC, the NIMZ, and part of the IMZ) are compared; the excised blastocoel roof is shown in c. A control embryo (d) and one lacking its AC (e) are compared at the late neurula stage. Transverse sections through neurulae that developed without the AC (f) and both the AC and the NIMZ (g) show the notochord (N); the notoplate (NP); and the somitic mesoderm (S).

Gastrulation of *Xenopus* 127

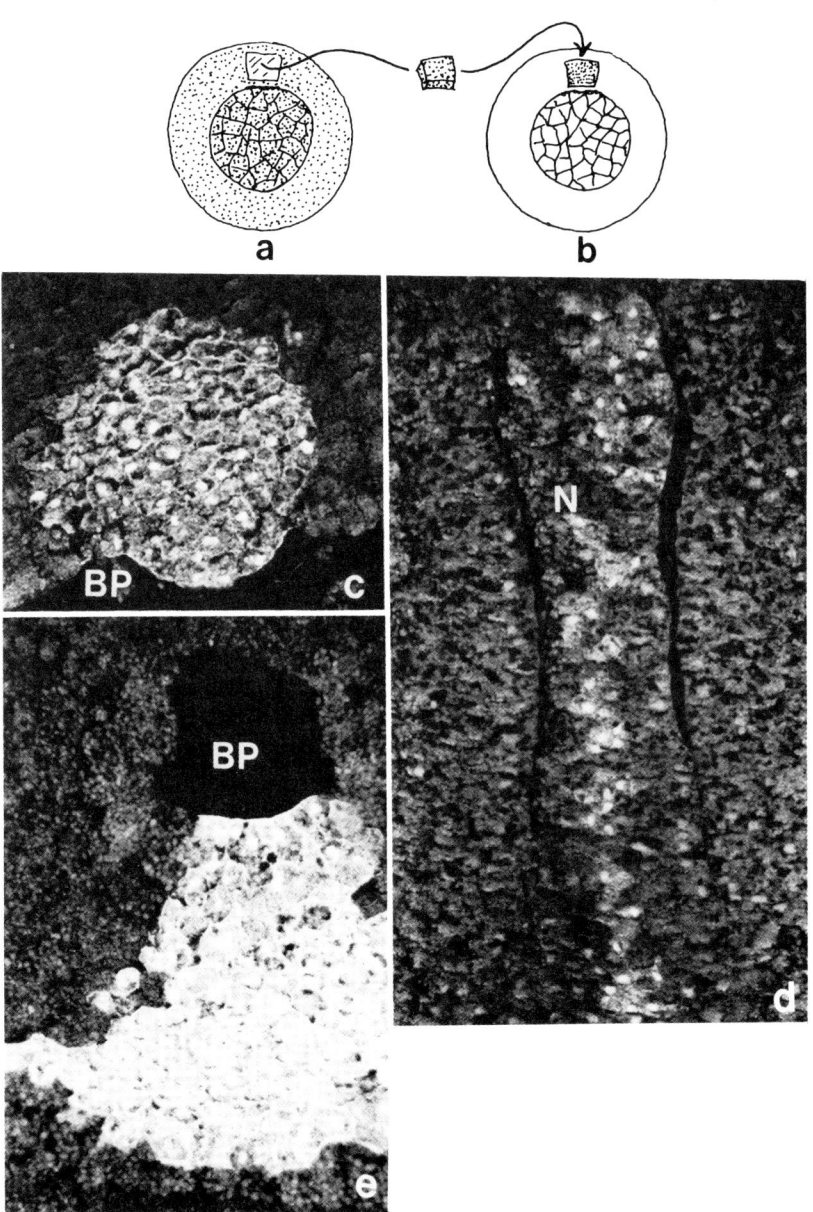

FIGURE 10. Legend appears on following page.

head" and joined the cortical region of the protrusion, which appeared to be attached to the adjacent cells. Cell behavior in explants that show progressive loosening of cell contacts in the culture medium suggest that cytoplasmic flow and traction on adjacent cells drives intercalation. Explants from different spawnings show differences in behavior in Danilchik's buffer. In a few the epithelium attempts to heal and in others the contacts between deep cells loosen. In the latter case, cytoplasmic flow no longer results in movement of the cell, but slippage occurs on one or several sides of the cell such that the cell either remains stationary or gyrates around what appears to be its attached side.

It is clear that the convergent extension of these open-faced explants occurs as a result of forces generated within themselves and without traction on an external substratum. They do not attach to the agarose base. The notochord apparently elongates with some force. In some preparations left overnight, the notochord continues to elongate, forms its characteristic sheath, and may throw itself into bends, some of which may push laterally through the somites to one side.

DISCUSSION

Classical Evidence Bearing on Convergent Extension

The classical literature supports a major role for convergent extension in gastrulation, and suggests that a rekindling of interest in this morphogenetic process is long overdue. Spemann (20) and Vogt (18) showed the capacity for the marginal zone to constrict. Schechtman (23) showed that the dorsal marginal zone has the capacity

FIGURE 10. A graft of dorsal IMZ from an embryo labeled with FDA (a) to an unlabeled embryo (b) shows no intercalation of labeled and unalbeled cells after healing (c) and much intercalation in the notochord of the early neurula (d). Corresponding grafts to th ventral IMZ show little intercalation at the same stage (e). BP, blastopore; N, notochord.

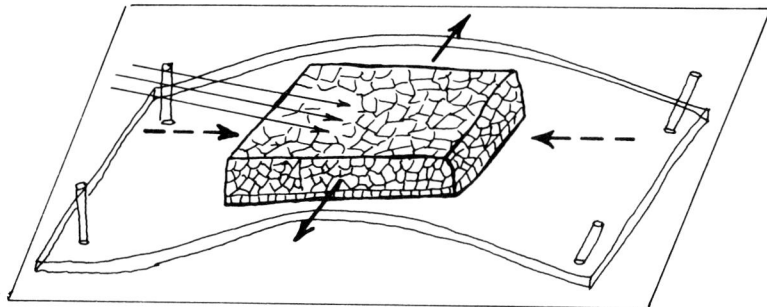

FIGURE 11. "Open-faced" explants are made in Danilchik's medium (Appendix I) and covered with a thin agarose sheet pinned to an agarose base at its corners, as shown. Extension (solid arrows) and convergence (dashed arrows) occurs and cell behavior can be viewed directly with the compound microscope under low-angle epi-illumination (EL).

to extend and converge and argued that this behavior plays a major role in closure of the blastopore. A variety of microsurgical manipulations (21, 5, 22, 6, 23) and explantations (Holtfreter, 36, 37, 23, 38, 39, 40) of the dorsal marginal zone all result in formation of elongated, narrowed structures. From older works, it is difficult to determine when (gastrula or neurula stage) and where (pre- or post-involuxion) convergent extension takes place because of the lack of continuous records of shape changes and because of distortions due to woundhealing. In these early works, the common view seems to have been that convergent extension of the marginal zone pushes material toward the lip from the outside. However, Holtfreter (41, p. 187) clearly suggested that it might also occur in the involuted mesoderm.

Several facts argue that convergent extension plays the major role in gastrulation. Those regions of the gastrula that distort greatly lie in a narrow annulus immediately above the blastopore (the IMZ and NIMZ), and it is the dorsal sector of this region that shows the greatest change in shape (31, 33). Traction of post-involution cells on the roof of the blastocoel could

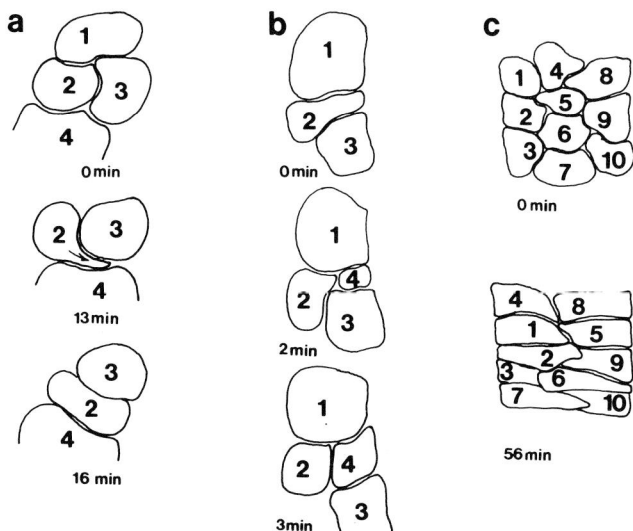

Figure 12. Tracings of numbered cells from video-recordings of open-faced explants show intercalation in the circumferential (a) and radial (b) directions in the gastrula stage and circumferential intercalation in forming the notochord in the neurula stage (c). From 2, with permission.

produce shearing at the interface of the two tissues and perhaps move the outside vegetally and the inside toward the animal pole (see 17). This would not account for blastopore closure nor for the convergent extension movements in the circumblastoporal region. Lastly, removal of one member of the pair of tissues involved in this traction, the blastocoel roof, as described here and by Holtfreter (6), fails to hinder involution, blastopore closure, or convergence and extension. In contrast, several experimental manipulations of the circumblastoporal deep region results in immediate and local blockage of all these processes (see 23, 27, 25). These facts suggest that the "main engine" of gastrulation lies in the region

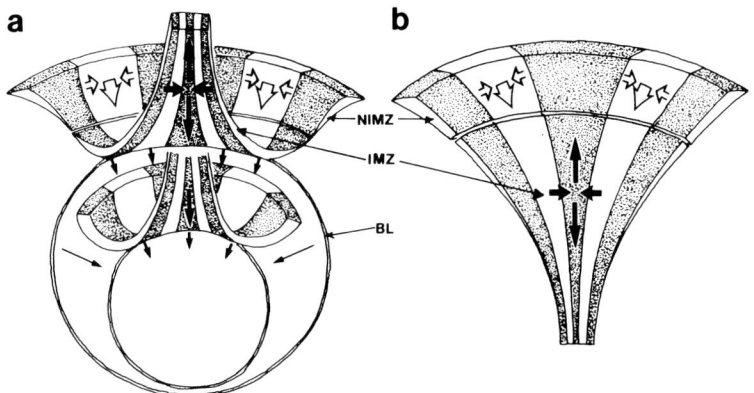

FIGURE 13. The mechanics of the function of convergent extension in gastrulation is shown diagramatically for normal gastrulation (a) and exogastrulation (b). See text for explanation. From 2, with permission.

just above the blastopore and involves the autonomous capacity of this sector to converge and extend.

Function of Convergent Extension in Gastrulation

In contrast to other mechanisms, convergent extension of the circumblastoporal region can account for the bulk of the distortion of the gastrula. The proposed mechanism can best be visualized by considering a diagram of the dorsal sector of the gastrula, viewed from inside the blastopore, looking out (Fig. 13). The vegetal edge of the dorsal sector of the mesodermal annulus is turned over the lip in the early gastrula stage, probably due to distortions associated with formation of the bottle cells (see 2) and perhaps due to the invasive, migratory behavior of the leading mesodermal cells in this region (23). At the midgastrula stage, the involuted part of the IMZ begins convergent extension (large solid arrows) posteriorly and thuo, in one stroke, constricts and pushes vegetally on the inside of the blastoporal lip (BL, Fig. 13a). This tends to roll preinvolution cells of the IMZ over the lip where they join in the convergent extension movement. At some point, the NIMZ also begins convergent extension (open arrows, Fig. 13a) and feeds cells vegetally to take up positions vacated by the involution of the IMZ. The maximum constriction of the blastopore, and, thus, the limit of involution, is fixed by the maximum degree of convergent extension of the adjacent, prospective posterior regions of the IMZ and NIMZ. If the mesodermal annulus is not turned under by the onset of convergent extension, the convergent extension machinery is directed outward and produces an exogastrula (Fig. 13b).

Convergent extension is characteristic of only the dorsal sector of the NIMZ that normally becomes the central part of the posterior neural plate (the notoplate; see 33). However, the conditions under which convergent extension will occur autonomously in this region are more limited than for the dorsal IMZ. The dorsal NIMZ does not show convergent extension when the dorsal IMZ-NIMZ is rotated or repositioned in the embryo, whereas the dorsal IMZ does (25). The NIMZ may show convergent extension but it is too weak to overcome the mechanical resistance of the uncooperative tissues around it. Alternatively, some tissue relationship necessary for the exercise of convergent extension may be disturbed in these circumstances. Also, convergent extension of the dorsal NIM[does not appear to occur in the "open-faced" explants, described below. Lastly, in embryos lacking an animal cap region, the dorsal NIM[does not extend if it is not in contact with the underlying involuted part of the IMZ.

The common feature in all these situations in which convergent extension of the dorsal NIMZ fails to occur is absence of contact of its basal region with itself or with the IMZ. Is convergent extension of the dorsal NIM[regulated by basal contact with the involuted IMZ (notochordal mesoderm)? Dissections of living Xenopus gastrulae do show the dorsal NIMZ (prospective notoplate) and underlying notochordal region becoming tightly connected and nearly inseparable from stage 11 onward. Dependence of convergent extension of the NIMZ on basal contact would ensure coordination of its convergent extension with that of the IMZ. The evidence gathered thus far suggests that the basal stimulus sufficient to produce elongation of the posterior, central part of the neural plate may not be limited to that normally associated with chordamesodermal induction of neural development, since the prospective notoplate is capable of elongation and narrowing similar to that seen in the embryo as a result of basal contact with itself rather than with the underlying chordamesoderm.

Cellular Behavior During Convergent Extension

It is yet not clear how intercalation is related to generation of the force underlying convergent extension. It is clear that the explants, both the sandwiches and the "open-faced" sandwiches, do not require an external substratum and therefore must generate the force from within themselves. It is also clear from our direct and indirect observations and from Waddington's argument (34) that intercalation of cells to form a longer, narrower array is the geometrical basis of convergent extension. Although cells of the superficial layer (42) and the deep region rearrange, those in the former do so passively (42, 27). Therefore, deep cells must provide the motive force for convergent extension. We have two notions about how they might do so. The act of intercalation might be directly involved in generating the forces producing elongation. In this mechanism, cells would actively force their way between their lateral neighbors (circumferential intercalation) or between their deep neighbors (radial intercalation), probably using the pattern of cytoplasmic flow seen in many of the cells. Alternatively, intercalation might be of secondary importance. Local contractions might pull cells apart from each other, producing microfractures between them. Adjacent cells might then invade the free space and thus narrow the array; once intercalated, all cells would then forcefully adopt their original shape and produce the elongation. In the first mechanism, the intercalation would presumably occur against the mechanical resistance of adjacent cells and directly produce convergent extension. In the second, intercalation occurs in an island of tissue where compressive forces have been relieved locally by active contraction. Here, intercalation occurs unresisted. A subsequent active change in cell shape produces the driving force for convergent extension.

The cellular motile events underlying intercalation are not yet understood. As described above, cytoplasmic flow is seen in at least some of the intercalation events, and the explants in which cell contacts loosen show a continuous and progressive increase in slippage of cells on one another and a commensurate decrease in effectiveness of intercalation. These facts suggest that intercalation occurs by an amoeboidal type of translocation, involving cytoplasmic flow and traction on adjacent cells. Dissociated, cultured cells of the gastrula and neurula of amphibians show vigorous cytoplasmic flow in several patterns, depending on the origin of the cell (see 9, 41). The most common of these patterns is the "limnicola" or "circus" movement in which a hemispherical bleb forms and rotates around the cell (9). This behavior was thought to be an aberrant behavior caused by loss of contact with adjacent cells (9). The present study supports this view, in that the circus movement was the extreme behavior of cells in semi-dissociated explants that also ceased their convergent extension. But prior to loss of contact and traction, more directed cytoplasmic flow occurred, and this was correlated with effective cell displacement. It is possible that related cell behavior patterns, such as those displayed by the "sausage" cells (41), vermiform cells (9), and several others (see 43, 44) might be involved in intercalation.

The Blastocoel Roof and the Role of Mesodermal Cell Migration

The removal of the blastocoel roof shows that active migration of involuted material contributes little, directly, to the major distortions of the embryo. Holtfreter did this experiment in 1933 on Hyla, with results similar to those reported here (6, see 2). This does not mean that mesodermal migration has no role in gastrulation. The invasive migration of the head mesoderm on the blastocoel roof may function in the initial involution of the powerful convergent extension machinery and thus prevent exogastrulation or formation of an excrescence (see 23). The mesoderm that does not undergo convergent extension, such as head mesoderm, heart mesoderm, and the ventral mesoderm that forms the circulatory system over the yolk, may not move, maintain position, or differentiate properly without an overlying substratum of blastocoel roof. Lastly, the association of the dorsal involuted IMZ (differentiating notochord) with the overlying dorsal NIMZ (notoplate) is necessary for stabilizing the position of dorsal, axial structures.

What are the relative roles of convergent extension and mesodermal migration in urodeles? The urodele marginal zone shows convergent extension (18, 3, 36, 41). Moreover, a large area of lateral (somitic) and ventral mesoderm leaves the superficial layer during gastrulation and migrates away from the circumblastoporal region (see 3, 41). The removal of the lateral mesodermal cells from the circumblastoporal array, along with the narrowing of the notochordal region, would presumably generate constriction forces that might act to bring about involution, even in the absence of the blastocoel roof of the Ambystoma (see 45, 46, 2). Thus migration might indirectly assist convergence of the lateral and ventral sectors of the circumblastoporal regions of the urodele gastrula, and thus contribute to constriction of the blastopore. Migration may also make a greater direct contribution to gastrulation movements in urodeles than in anurans, such as Xenopus. Mesodermal cells have been observed to migrate in opened gastrulae (47). Their morphology and distribution during dissections suggests that they have a stronger association with the overlying blastocoel roof than corresponding cells in Xenopus, and thus perhaps depend on migration as a means of displacement (see 16). The fibrillar matrix on the roof of the urodele gastrula is better developed than it is in anuran gastrulae studied thus far (48). Injection of anti-fibronectin antibodies (49) or a synthetic peptide perhaps containing the cell recognition sequence of fibronectin (50) result in failure of mesodermal cells to adhere and migrate on the blastocoel roof, but it also results in collapse and thickening of the blastocoel roof and other abnormalities of gastrula shape that are difficult to associate solely with failure of mesodermal cells to migrate. Experiments similar to the work by Lewis (45, 46) should be done to elucidate the contribution of mesodermal cell migration in urodeles.

Remaining Questions About Convergent Extension

The major unanswered question concerning convergent extension is, what protrusive activity and contact behavior is involved in bringing about intercalation? Our working hypothesis is that intercalating protrusions form and extend by cytoplasmic flow through the core of the protrusion, cortical gelation, and exertion of traction on adjacent cells. It is not clear what elicits such protrusive activity nor is it clear how the underlying directional property of convergent extension is established. Does it arise from anisotropy of protrusive activity, of adhesion, or of some other property? Holtfreter (41) suggested that the original tissue intergity is necessary for elongation, based on his observation that dissociated and reaggregated cells of the dorsal marginal zone would form notochord tissue but would not converge and extend. Lastly, there may be several mechanisms of convergent extension operating in early development. Differences in cell morphology and arrangement between the dorsal NIMZ and IMZ suggest that different cell behaviors may lead to convergent extension in these two regions. The same applies to the early, gastrula stage convergent extension of the IMZ and the convergent extension of the same region later in neurulation.

ACKNOWLEDGDMENTS

We wish to thank Paul Tibbetts for his technical assistance and Cathy Lundmark and Ann Sutherland for their comments on the manuscript. This work was supported in part by NSF Grant 81-10985 and NIH Biomedical Grant 84-39 to Ray Keller, NIH GM 08738 to M.V.D., and, for R.L.G., NIH GM19363 to John Gerhart.

REFERENCES

(1) Spemann H (1938). "Embryonic Development and Induction." Yale University Press. Reprinted 1962, Hafner Publishing Company, Inc., New York.
(2) Keller RE (1985). The cellular basis of amphibian gastrulation. In Browder L (ed): "Developmental Biology: A Comprehensive Synthesis," Plenum Press. In press.

(3) Vogt W (1929). Gestaltungsanalyse am Amphibienkeim mit ortlicher Vitalfarbung. II. Teil. Gastrulation und Mesodermbilduug bei Urodelen and Anuren. Wilhelm Roux Arch Entwicklungsmech Organismen 120:384-706.
(4) Mangold O (1923). Transplantationsversuche zue Frage der Spezefitat und der Bildung der Keimblatter bei Triton. Arch f mikr Anat Entw Mech 100:198-301.
(5) Spemann H (1931). Uber den Anteil von Implantat und Wirtskeim an der Orientierung und Beschaffenheit der induzierten Embryonalange. Arch Entw mech 123:3z0-517.
(6) Holtfreter J (1933). Die totale Exogastrulation, eine Selbstablosung des Ektoderms von Entomesoderm. Arch f Entw mech 129:669-793v
(7) Rhumbler L (1902). Zur Mechanik des Gastrulationsvorganges, insbesondere der Invagination. Eine entwicklungsmechanishe Studie. Wilhelm Roux's Arch Entwicksluugsmech Org 14:401-476.
(8) Holtfreter J (1943a). Properties and function of the surface coat in amphibian embryos. J Exp Zool 93:251-323.
(9) Holtfreter J (1943t). A study of the mechanics of gastrulation. Part I. J Exp Zool 94:261-318.
(10) Baker P (1965). Fine structure and morphogenetic movements in the gastrula of the treefrog, Hyla regilla. J Cell Biol 24:95-116.
(11) Perry M, Waddington CH (1966). Ultrastructure of the blastoporal cells in the newt. J Embryol Exp Morph 15:317-330.
(12) Nakatsuji N (1974). Studies on the gastrulation of amphibian embryos; pseudopodia in the gastrula of Bufo bufo japonicus and their significance to gastrulation. J Embryol Exp Morph 32:795-804.
(13) Natasuji N (1975a). Studies on the gastrulation of amphibian embryos: Light and electron microscopic observations of a urodele Cynops pryyhogaster. J Embryol Exp Morph 34:669-685.
(14) Nakatsuji N (1975b). Studies on the gastrulation of amphibian embryos: Cell movement during gastrulation in Xenopus laevis embryos. Wilhelm Roux's Archives 178:1-14.
(15) Nakatsuji N (1976). Studies on the gastrulation of amphibian embryos: Ultrastructure of the migrating cells of anurans. Wilhelm Roux's Archives 180:229-240.

(16) Nakatsuji N (1984). Cell locomotion and contact guidance in amphibian gastrulation. Amer Zool 24: 615-627.
(17) Keller, RE, Schoenwolf GC (1977). An SEM study of cellular morphology, contact, and arrangement, as related to gastrulation. Wilhelm Roux Arch 182: 165-186.
(18) Vogt W (1922a). Die Einrollung und Streckung der Urmundlippen bei Triton nach Versuchen mit einer neuen Methode embryonaler transplantation. Verh d D Zool Ges 27:49-51.
(19) Vogt W (1922b). Opertiv bewirkte "Exogastrulation" bei Triton und ihre Bedeutung fur die Theorie der Wirbeltiergastrulation. Anat Anz Erg 55:53-64.
(20) Spemann H (1902). Entwicklungsphysiologische Studien am Triton-Ei II. Arch f Entw Mech 15:448-534.
(21) Mangold O (1920). Fragen der regulation und Determination an umgeordneten Furchungsstadien und verschmolzenen Keimen von Triton. Roux's Arch 47:250-301.
(22) Lehman FE (1932). Die Beteiligung von Implantats-und Wirtsgewebe bei der Gastrulation und Neurulation induzierter Embryonalanlagen. Arch Ent mech 125:566.
(23) Schechtman AM (1942). The mechanism of amphibian gastrulation. I. Gastrulation-promoting interactions between various regions of an anuran egg (Hyla regilla), Univ Calif Publ Zool 51:1-39.
(24) Phillips H (1984). Physical analysis of tissue mechanisms in amphibian gastrulation. Amer Zool 24:567-672.
(25) Keller RE (1984). The cellular basis of gastrulation in Xenopus laevis: Active, postinovolution convergence and extension by mediolateral interdigitation. Am Zool 24:589-603.
(26) Cooke J (1975). Local autonomy of gastrulation movements after dorsal lip removal in two anuran amphibians. J Embryol Exp Morph 33:147-157.
(27) Keller RE (1981). An experimental analysis of the role of bottle cells and the deep marginal zone in gastrulation of Xenopus laevis. J Exp Zool 216: 81-101.
(28) Nieuwkoop P, Faber J (1967). "Normal Table of Xenopus laevis (Daudin)." Second edition, North Holland Publishing Company, Amsterdam.

(29) Gimlich R, Cooke J (1983). Cell lineage and the induction of second nervous systems in amphibian development. Nature 306:471-473.
(30) Gillespie JI (1983). The distribution of small ions during the early development of Xenopus laevis and Ambystoma mexicanum embryos. J Physiol 344:359-377.
(31) Keller RE (1975). Vital dye mapping of the gastrula and neurula of Xenopus laevis. I. Prospective areas and morphogenetic movements of the superficial layer. Develop Biol 42:222-241.
(32) Keller RE (1976). Vital dye mapping of the gastrula and neurula of Xenopus laevis. II. Prospective areas and morphogenetic movements in the deep region. Develop Biol 51:118-137.
(33) Jacobson A (1982). Morphogenesis of the neural plate and tube. In Connelly et al. (eds): "Morphogenesis and Pattern Formation," Raven, New York, pp. 223-263.
(34) Waddington CH (1940). "Organizers and Genes." Cambridge University Press, Cambridge.
(35) Mookerjee S, Deuchar E, Waddington CH (1953). The morphogenesis of the notochord in Amphibia. J Embryol Exp Morph 1:399-409.
(36) Holtfreter J (1938a). Differenzierungspotenzen isolieter Teile der Urodeleangastrula. Wilhelm Roux's Archives 138:522-656.
(37) Holtfreter J (1938b). Differenzierungspotenzen isolierter Teile der Anurengastrula. Wilhelm Roux's Archives 138:657-738.
(38) Ikushima N (1959). The formation of two independent notochords in an explant taken from the dorsal blastoporal area of the early gastrula of Amphibia. Experientia 15:475-476.
(39) Ikushima N (1961). Formation of notochord in an explant derived from the dorsal marginal zone of the early gastrula of Amphibia. Jap J Zool 13:117-140.
(40) Ikushima N, Maruyama S (1971). Structure and developmental tendency of the dorsal marginal zone in the early amphibian gastrula. J Embryol Exp Morph 25:263-276.
(41) Holtfreter J (1944). A study of the mechanics of gastrulation. Part II. J Exp Zool 95:171-212.
(42) Keller RE (1978). Time-lapse cinemicrographic analysis of superficial cell behavior during and prior to gastrulation in Xenopus laevis. J Morph 157:223-248.

(43) Satoh N, Kageyama T, Sirakami K-I (1976). Motility of embryonic cells in Xenopus laevis. Develop Growth, Diff 18:55-67.
(44) Kubota H (1981). Creeping locomotion of the endodermal cells dissociated from the gastrula of the Japanese newt, Cynops pyrrhogaster. Exp Cell Res 133:1 137-148.
(45) Lewis WH (1948). Mechanics of Ambystoma gastrulation. Anat Rec 101:700.
(46) Lewis WH (1952). Gastrulation of Ambystoma punctatum. Anat Rec 112:473.
(47) Kubota H, Durston AJ (1978). Cinematographical study of cell migration in the opened gastrula of Ambystoma mexicanum. J Embryol Exp Morph 44:71-80.
(48) Nakatsuji N, Johnson K (1983). Comparative study of extracellular fibrils on the ectodermal layer in gastrulae of five amphibian species. J Cell Sci 59:61-70.
(49) Boucaut J-C, Darribere T, Boulekbache H, Thierry J-P (1984). Antibodies to fibronectin prevent gastrulation but do not perturb neurulation in gastrulated amphibian embryos. Nature 307:364-367.
(50) Boucaut J-C, Darribere T, Poole TJ, Aoyama H, Yamada KM, Thiery JP (1984). Biologically active synthetic peptides as probes of embryonic development: A competitive peptide inhibitor of fibronectin functions inhibits gastrulation in amphibian embryos and neural crest cell migration in avian embryos. J Cell Biol 99: 1822-1830.
(51) Keller RE, Danilchik M, Gimlich R, Shih J (1985). The function of convergent extension in gastrulation of Xenopus laevis. J Embrbryol Exp Morph, in press.

Appendix I

Ionic concentrations in Danilchik's medium were chosen to reflect the mean free intercellular ionic activaties in Xenopus embryos reported by Gillespie (30). Final ionic concentrations are (mM): Na^+, 95.0; K^+, 4.5; Ca^{2+}, 1.0; Mg^{2+}, 1.0; Cl^-, 55.0; bicarbonate, 18 - 19; sulfate, 1.0; isethionate, approximately 20; gluconate, 4.5; bicine (N,N-bis[2-hydroxyethyl]-glycine), 5.0. pH at $21^{o}C$ = 8.30. Salt concentrations used to make the medium are (mM): NaCl, 53.0; $NaHCO_3$, 15.0 mM; K-gluconate, 4.5; $MgSO_4$, 1.0;

$CaCl_2$, 1.0. These salts are dissolved in glass-distilled water (about 90% of the final volume). The pH is then adjusted to 8.30 with a measured volume of 1.0 M Na_2CO_3, the total Na concentration is then calculated, and the final Na concentration is brought to 95.0 mM with Na-isethionate. The final volume is then completed with glass-distilled water.

THE CORTICAL TRACTOR MODEL FOR EPITHELIAL FOLDING: APPLICATION TO THE NEURAL PLATE[1]

Antone G. Jacobson, Garrett M. Odell, and George F. Oster

Center for Developmental Biology, Department of Zoology, University of Texas, Austin, Texas 78712 (A.G.J.) Department of Mathematical Sciences, Rensselaer Polytechnic Institute, Troy, New York 12181 (G.M.O.), and Departments of Biophysics, Entomology, and Zoology, University of California, Berkeley, California 94720 (G.F.O.)

ABSTRACT We propose a "cortical tractor model" to account for the shaping and folding of epithelial tissues. We suggest that a particular type of cortical flow common to both mesenchymal and epithelial cells allows epithelial cells to interdigitate with one another while maintaining the integrity of their apical seals. Indeed we propose that the cortical tractoring builds these apical seals by cycling intracellular adhesive junctions. The boundaries between cellular domains with differing adhesive characteristics may organize movements amongst the epithelial cells to produce some of the characteristic changes of shape of the tissue. We apply the model to the phenomenon of neurulation and demonstrate that it can account for neural plate elongation and rolling into a tube.

INTRODUCTION

During early development, embryonic cells appear to fall into two broad morphological classes; epithelial and mesenchymal. Epithelial cells array themselves into sheets and may present a very regular "paving stone" appearance

---

[1] This work was supported by NIH grant NS 16072 to AGJ, by NSF grant MCS 8301460 to GMO, and by NSF grant MCS 8110557 to GFO.

when viewed from the apical side. Epithelia form the external and internal boundary layers of embryos. Mesenchymal cells are more amorphous, migratory cells that move about in spaces between and on epithelial layers. Form-shaping movements of the embryo involve the coordinated activities of both cell types. An epithelial layer folds, rolls, invaginates and deforms, while retaining its integrity as a connected sheet. Mesenchymal cells migrate independently and aggregate into patterned collections of cells.

Classification of embryonic cells into these two catagories is based largely on the gross differences in appearance between epithelia and mesenchyme. These distinctions have proven so conceptually appealing that it is easy to downplay the fact that cells frequently interconvert between the two forms. For example, from the boundary of the epithelial epidermis and neural plate, mesenchymal neural crest cells emerge. As another example, epithelial cells of the epiblast of amniote embryos converge to the primitive streak at the midline where many epithelial cells detach during focal contractions (C. Stern, personal communication) and spread out below the epiblast as migrating mesenchymal cells. Some of these condense into somites and become epithelial again, and then later the epithelial somite again breaks up into mesenchyme (1).

In this paper we shall blur the distinction between the two types of cells still further by proposing a model for epithelial morphogenesis that imparts to epithelial cells certain "mesenchymal" behavioral characteristics. In particular, we shall propose a mechanism by which epithelial cells can fold, roll, and change their neighbors. This mechanism not only ensures maintenance of the all-important apical seal that insulates the embryo from the external environment, but actually constructs that seal at the cell apices.

The sorts of phenomena to which this model may apply include epiboly in amphibian embryos, many movements during amphibian gastrulation, convergence and ingression of the primitive streak of birds, neurulation, the formation and invagination or evagination of epidermal placodes, and eye-cup morphogenesis. In this paper we shall use elongation and tube formation in the neural plate of the newt embryo as our example of how active cortical tractoring of cells in an epithelial tissue may be organized to produce large-scale morphogenetic deformations.

THE CORTICAL TRACTOR MODEL

The model that we present here departs from previous views of how epithelial layers are constructed. Rather than treating the cells that constitute the layer as static entities, we propose that they are constantly cycling their surfaces in a certain way. It is only on the short time scale of visual observations - or in the frozen rictus of the scanning electron microscope - that epithelial cells convey the appearance of stationary cobblestones that can at most contract and relax their apical surfaces. We shall assume that epithelial cells are rather more "mesenchymal-like" than heretofore presumed. To make clear exactly what we mean, we briefly discuss mesenchymal cell motions as we present the postulates of our model of epithelial cell behavior.

Motile Cells Cycle Their Cortex

A cell moving across a substratum clearly exhibits several features. First, the cell puts out protuberances (eg. filopodia, lamellipodia) at its leading edge. When one of these appendages attaches to the substratum, contraction of the cortex pulls the cell forward a small increment. This cycle of spreading, or extension, followed by attachment and contraction, is accomplished by internal flows of cytoplasm, generally forward through the center of a lamellipodium and rearward in the cortex. If a time-lapse movie could be taken of the cortical flow, the net average motion would be the double fountain movement of cytoplasm shown schematically in Fig. 1. Oster (2) has previously presented a model for lamellipodial motion based on a cycle of events involving solation of the actomyosin cortical gel followed by osmotic expansion and active contraction of the cortex. However, our discussion here does not depend on the detailed assumptions of that model, but only on the average cortical flow pattern, whatever the driving mechanism. Therefore, the first postulate of our model is:

I. *Cell motion is characterized by a "cortical tractor" that produces a time-averaged cortical flow from the leading to the trailing surface, as shown in Fig. 1.*

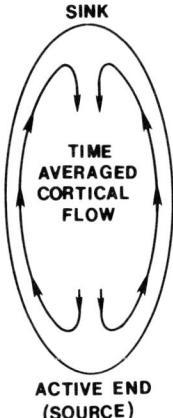

FIGURE 1. Schematic of the cortical tractor mechanism. The actomyosin cortex flows as indicated by the arrows, and cell-to-cell adhesion structures are dragged along by the flow of the cortex.

This sort of fountainoid motion of the cortex has been suggested by a number of workers (eg. 3-7). Marking experiments have shown that the cortical flow is reflected on the surface of the cell, indicating an intimate connection between the flowing cortex and the plasma membrane (8). It is also necessary for us to postulate that the adhesion structures that anchor the cell to the substratum and to its neighboring cells are cycled along with the cortex (9):

II. *Adhesion and junctional molecules are inserted at the site of the cortical "source" and flow with the cortex to be resorbed at the cortical "sink", unless stabilized by bonding with the substratum, or with another cell surface.*

Thus a moving cell attaches to its surroundings via attachment sites that are continuously being inserted at the leading edge, dragged posteriorly, and there resorbed.

Cortical Activity Can Be Triggered by Ionic Stimuli

As shown by experiments with ionophores and channel-plugging molecules, cell motility can be stimulated by ionic leaks, especially calcium, and inhibited by blocking ionic channels (10, 11). The model suggested by Oster (2) provides one possible mechanism for this behavior, however, we need not be more specific here than that motility can be ionically triggered. Thus our third postulate is:

III. *Local ionic conditions can "activate" any face of a cell, which thence becomes the leading edge; that is, the cortical source in Fig. 1.*

This postulate accords with the observation that cells can "contact inhibit" one another. When the leading membrane of a cell becomes closely attached to the membrane of another cell, the ionic leak, which stimulates motility, may become sealed, thus paralysing cell movement. However, we will not be specific about the cellular or molecular mechanisms that regulate the intercellular ionic conditions; that such mechanisms exist is amply evident, but how they are mediated is yet unclear. Furthermore, the proximal signal for activating a cell's surface may be other, nonionic, chemical messengers. For example, chemotactic agents that bind to specific receptors on a cell's surface may initiate the ionic leak that triggers motile activity. For the purposes of our discussion here we can remain vague about the specific extracellular molecular trigger that activates a cell's surface.

Epithelial Layers Are Dynamic Structures

For the final ingredient in our model we reconsider the nature of an epithelial layer. Microscopic examination of the surface of an epithelium reinforces the impression of a static paving of cells. The individual cells of an epithelium are bonded firmly along their apical edges to form a molecular seal that insulates the interior of the embryo from the exterior environment. The epithelial layer as a whole can deform by folding, bending, rolling into tubes, invaginating or evaginating. The occasional evidence of apical bundles of microfilaments has led to the postulate that constriction of these filaments could generate the force that drives these deformations (12-14). Here we augment that model dramatically by postulating the following:

IV. *Epithelial cells continuously cycle their cortical cytogel in a manner analogous to motile mesenchymal cells, with the added constraint that their apical boundaries remain firmly attached.*

The implications of Postulate IV can be appreciated by considering the sequence of events shown schematically in Fig. 2. An epithelial layer consists of an array of cells, all joined at their apical boundaries. The basal surfaces of each cell may be active, so that each cell's cortex undergoes the tractor motion. This means that each cell is, in effect, attempting to crawl downward on its neighbors by adhesion of its surface and contraction of its cortex. However, if each cell undergoes the same motion, then no net traction can be developed between adjacent cells, and thus no net motion transpires. That is, if adjacent cells attempt to crawl on one another with equal intensity in the same direction, then no relative motion can take place. Moreover, since the adhesion structures are being continuously added in the basal region, they will be carried apically by the cortical flow, finally piling up at the apical seal.

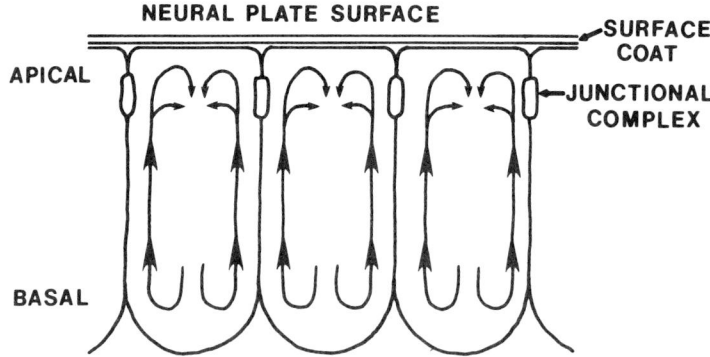

FIGURE 2. Diagram of a section of the neural epithelium. Each cell undergoes the cortical tractor motion (arrows). In this case, the cells are active basally.

Microfilament bundles, usually found attached to surface adhesion structures, linking them together, would be concentrated at the apical end of a cell by cortical flow. Contraction of these filament bundles could then rapidly contract the apex of the cell. In this way, our cortical tractor model subsumes the previous models of Baker and Schroeder (12), Burnside (13), Jacobson and Gordon (19), and Odell, et al. (14).

Thus according to the cortical tractor model, an epithelial layer is a dynamic structure, each cell continually cycling its cortical cytogel in a basal-to-apical fountain flow. When each cell cycles with nearly equal intensity, there is no net relative motion, and the appearance of a static structure is maintained.

The Cortical Tractor Model Allows Epithelial Cells to Change Neighbors Without Breaking the Apical Seal

The cortical tractor model can explain how cells can interdigitate while still maintaining the integrity of the apical seal. Consider the situation illustrated in apical view in Fig. 3a: cells A and C are in contact with each other, and cells B and D are separated. If at some later time we observe that cells B and D are in contact, then the bonds between A and C must have been severed. How could this neighbor shift take place without breaking the apical seal? According to the cortical tractor model, the answer lies below the surface. Suppose that the cortex source of cell B shifts laterally toward cell D. Thus cell B puts out lamellipodia on its basal and lateral surfaces and the lateral lamellipodia contact the neighboring cells, eventually contacting cell D near its basal end. This lamellipodial surface area, along with its adhesive structures, is swept apically by the cortical tractor. Thus cell B interposes itself between its neighbors from underneath as shown in Fig. 3b, and new attachments between cells B and D are swept

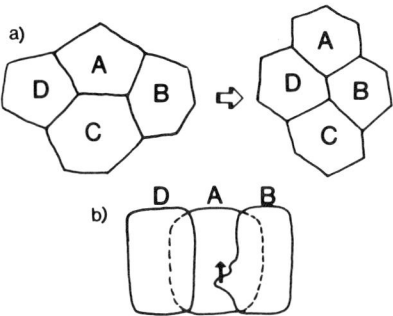

FIGURE 3. (a) Apical view of four epithelial cells as they change neighbors. (b) Lateral view of the cortical tractor mechanism showing cell B interdigitating between cells A and C. Cell B is laterally and basally active.

from their basal to apical ends. Viewed from the top, it appears as if cell B is actively crawling between cells A and C, giving the appearance that A-C bonds are being broken. However, the basal to apical flow of cortical cytogel - which carries the junctional structures - inserts junctions as fast as they are recycled at the apex. The net result is that cell B can appear to interdigitate itself between A and C, and bring about a neighbor exchange without breaking the apical seal.

A cell population, wherein each cell tractors its cortex in a fountainoid such as we have described, can behave quite counterintuitively. Recently, Odell and Bonner (15) have shown that the crawling of the *Dictyostelium discoideum* grex can be explained by the concerted action of a population of amoebae, each of which crawls on its neighbors. Indeed, much of the morphogenesis of *D. discoideum* can be interpreted in terms of the cortical tractor model, including the formation of the grex, its motion, the distribution and orientation of the stalk and spore cells in the grex, and the erection of the fruiting body. Here we apply the same physical analysis to model the shaping and rolling of the neural plate. We show next how the type of cell dynamics described by the cortical tractor model could drive the process of neurulation in newt embryos.

NEURULATION AND THE CORTICAL TRACTOR MODEL

The details of newt neurulation have been reviewed recently by Jacobson (16, 17), so we shall present only a brief summary here.

Two populations of cells compose the neural plate. One population lies directly above the notochord and is called "notoplate" (16). The other population comprises the rest of the neural plate. We will refer to these two populations of cells as "notoplate" and "neural plate" respectively (Fig. 4). The prospective notoplate in the early gastrula is a crescent of tissue between the prospective neural plate and the prospective notochord; all of these parts lying in sequence above the forming lip of the blastopore (Fig. 4). As gastrulation proceeds, the notochord involutes around the blastopore lip, converges toward and elongates along the midline. The notoplate does not involute, but rather converges toward and elongates along the midline in concert with the notochordal tissue below it. As this happens, the neural plate is shaped around the notoplate, which comes to occupy the midline of the emerging spinal cord and brain up through the mesencephalon (Fig. 4).

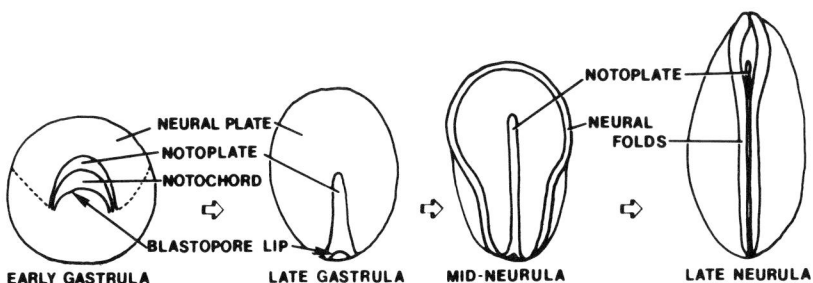

FIGURE 4. The parts referred to in the text are labelled on these drawings of a rear view of an early gastrula, and dorsal views of late gastrula to late neurula of newt embryos. The notochord shown above the dorsal lip of the blastopore of the early gastrula has involuted around the blastopore lip and lies beneath the notoplate in subsequent stages. Figures are approximately to scale. An early gastrula is about 2.5 mm in diameter.

Besides the profound differences in behavior that have been described between notoplate and the rest of the neural plate (19), there are other observations that suggest the notoplate is quite a different population of cells from the rest of the neural plate. For the most part, the notoplate does not differentiate into nervous tissue. It occupies the same position as the floor plate of the spinal cord and brain. The fate of the floor plate is to become a raphe (a seam) between the basal plates of the spinal cord and the brain stem (which is precisely what one would expect to be the ultimate consequence of the cell behavior proposed here for the notoplate).

The notoplate may not be a part of the induced neural plate, but may be more closely related to the chordamesoderm, and be induced along with that tissue. As shown in Fig. 4, the position of the notoplate in the early gastrula is contiguous with the prospective notochord in the dorsal lip of the blastopore. In the famous experiments of Spemann and Mangold (20) in which a piece of dorsal lip of the blastopore of an early gastrula of *Triton cristatus* was transplanted into the ventral ectoderm of a *Triton taeniatus* host embryo, a second neural plate was induced in the ventral ectoderm by the implanted dorsal lip material. Pigmentation differences between the donor and host species allowed the tissues of the two species to be distinguished from one another. At least in some cases, the induced neural plate formed from host tissue except for the midline region (the notoplate) that came from the grafted implant (Spemann (21), his Fig. 78, p. 144). In illustrations of sections through the spinal cord of later stages when the secondary nervous system had formed into a tube, the donor notoplate material appears as the floor plate of the spinal cord (Spemann (21), his Fig. 80, p. 146). The donor notoplate cells must have insinuated themselves down the midline of the induced host neural plate.

When neurulation begins, several things occur simultaneously. The surface of the neural plate reduces as the apical surfaces of the cells that compose the plate contract, the cells of the neural plate get taller (i.e. more columnar), and the cells of the notoplate rearrange to elongate the midline of the neural plate. These activities reshape the disc-shaped neural plate into a keyhole shape (19). At the keyhole stage, neural folds appear at the boundary between the neural plate and the epidermis. The length of this boundary remains constant up to the keyhole stage, but the boundary elongates extensively as the neural plate subsequently folds into a tube. During the period of neural tube formation, which commences at the keyhole stage, the neural plate elongates very rapidly, and the cells of the plate continue to reduce their apical surfaces and to get taller (the plate gets thicker) (22).

Cell behavior, as described by the cortical tractor model, could be responsible for most of the activities that are observed during the shaping of the neural plate and the rolling of the plate into a tube. If the cells that compose the single-layered neural plate in urodele and amniote embryos are active at their basal ends, then the cortical tractors of the cells would function as shown in Fig. 2. Since the apical surfaces of the cells are bound together by junctional complexes, and some embryos, such as the newt, have a surface coat, the cells are not free to move, but are held in place by their apical surfaces. If the rate of tractoring in the apical junctions is the same as in the basal ends of the cells, then no net motion will occur. If the rate of tractoring in the apical junctions is slightly lower than the rate of tractoring toward the basal end of the cells, then the end result of cortical tractoring could be to elongate the cells, and this could account for the observation in newt embryos that neural plate cells get taller during neurulation. If the rate of tractoring of the apical junctions is much slower than the rate of tractoring toward the basal ends of the cells, or if there is no recycling of the junctional structures, then the end result could be drawing out of the apical end, bottle-cell formation, or the traction of the basal end could be enough to break the cells free of the epithelium. This scenario does not eliminate specific contraction of microfilament bundles arranged as pursestrings around the apical surfaces of the cells, as suggested by others (12, 13), but it better accounts for the observation that cell elongation and apical constriction keep pace with each other during much of neurulation (19).

Elongation of the Notoplate

In contrast to the rest of the neural plate, the cells of the notoplate do not get taller during neurulation through to the midneurula keyhole stage (19), but they elongate thereafter. There is intimate contact between the basal surfaces of the notoplate cells and the cells of the notochord (Fig. 5), while more lateral neural plate cells hang free in the extracellular space, which is probably filled with extracellular matrix.

We propose that the notoplate cells would have their basal faces inactivated by contact with the notochord cells, and that their apical surfaces, which are covered by the surface coat, are normally inactive. Only the lateral faces of the notoplate cells could have active cortical tractors.

In accordance with this assumption, tangential sections through the midline of the neural plate (through the notoplate) of a newt neurula reveal numerous long lamellipodia extending up to three or four cell ranks from the lateral faces of notoplate cells (Fig. 6). Thus we suggest that notoplate cells can interdigitate between one another according to the cortical tractor mechanism described above. (It may be relevant to note that Cooper and Keller (23) have shown in tissue culture that it is not unusual for cells to move perpendicular to their long axes.)

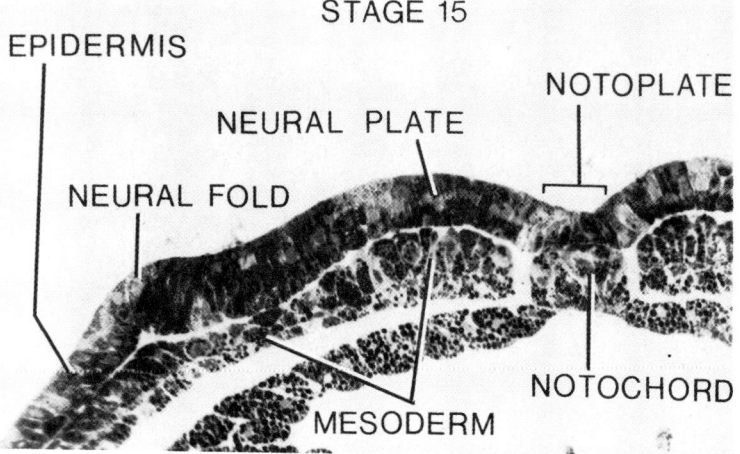
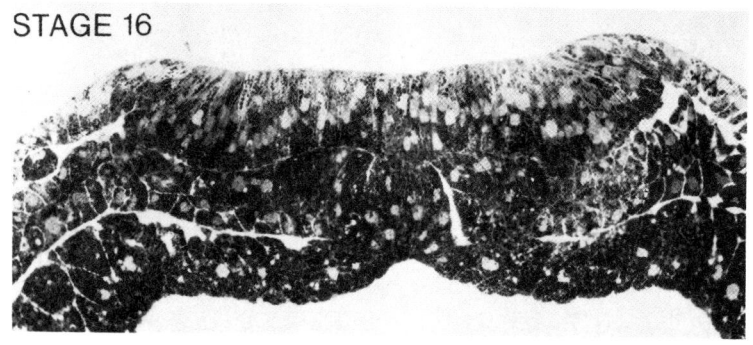
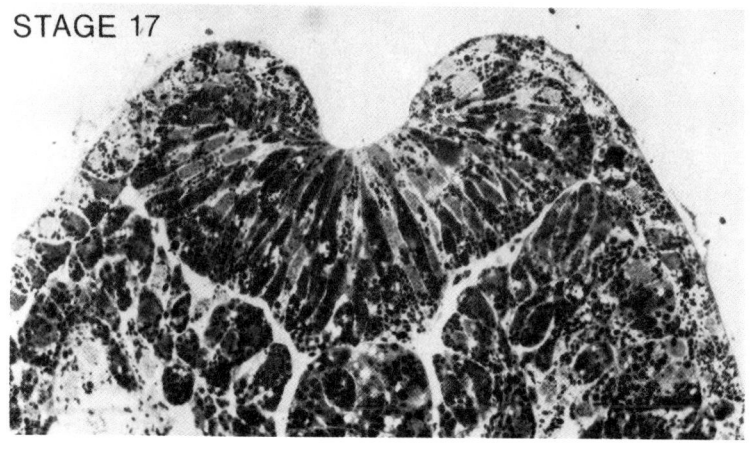

FIGURE 5. (Continued on the next page.)

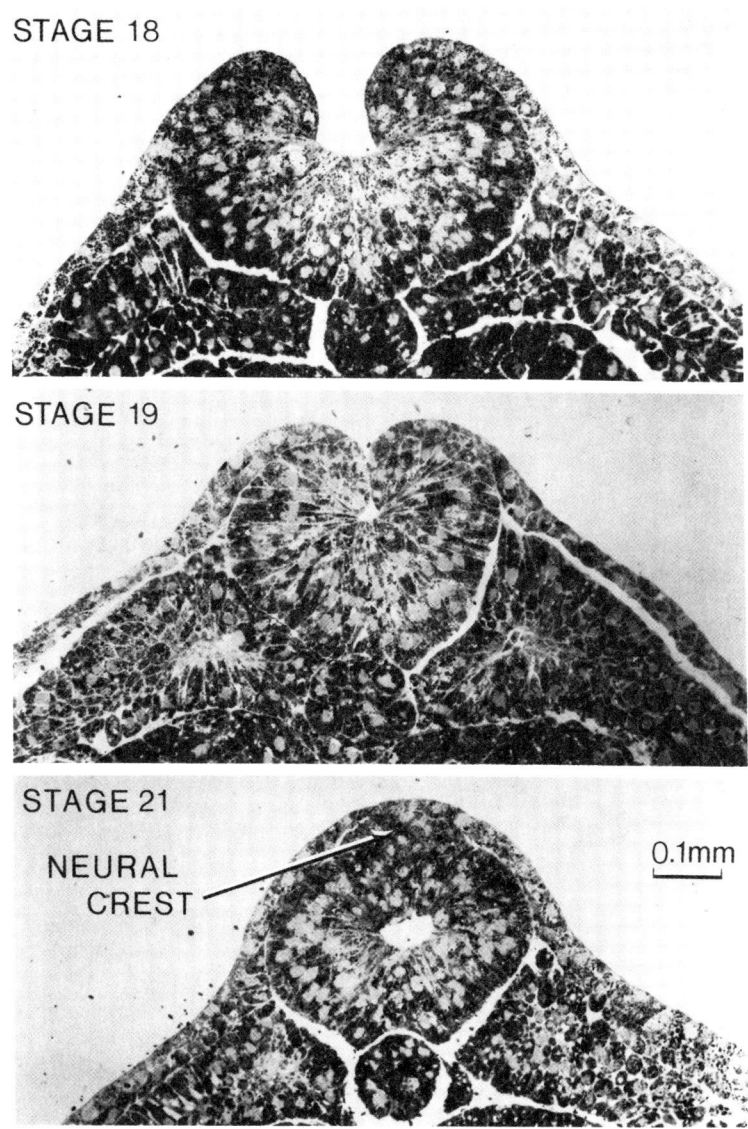

FIGURE 5. The neural plate, and the tissues that underlie it, are shown in cross section at stages through the period of neural fold formation and rolling of the plate into a tube, in newt embryos. These plastic sections (2μm thick) are from the region where spinal cord joins the brain. All are at the same magnification.

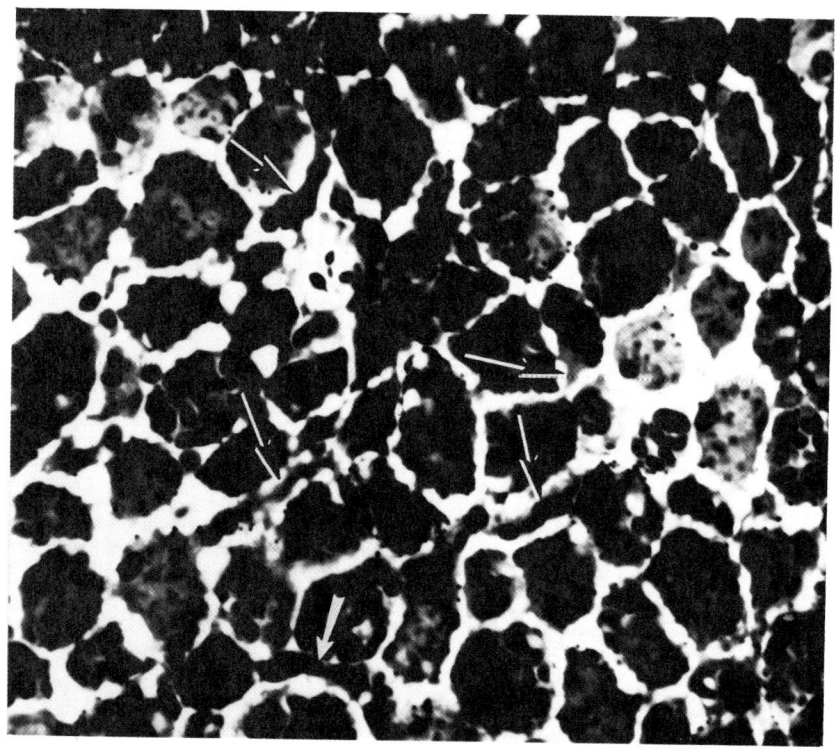

FIGURE 6. Tangential section through the neural plate of a stage 17 newt embryo, about half way down through the thickness of the notoplate. Numerous lamellipodia (arrows) can be seen extending laterally several cell diameters from the bodies of the notoplate cells. The neural plate-notoplate boundary runs across the top of the figure. The plate was removed from underlying tissues (which opened the spaces between the cells some) then fixed immediately in aldehydes, embedded in plastic, and sectioned at 2μm thickness.

If the direction of tractoring were random on the lateral faces of the notoplate cells, the net effect should be a random walk, with cells changing neighbors, but with no net distortion of the notoplate in any direction. However, there are at least two mechanisms by which the boundary between the notoplate and the rest of the neural plate could play a role in organizing these lateral interdigitations amongst the notoplate cells: (a) notoplate cells that contact

neural plate cells stick to them more than they stick to
each other (differential adhesion), or (b) contact with a
neural plate cell inhibits the activity of that face of the
notoplate cell (contact inhibition). Either mechanism
would restrict the tractoring to faces not abutting the
neural plate. Thus directed tractoring by notoplate cells
stuck up against the neural plate boundary can elongate the
boundary by cell interdigitation, as shown in Fig. 7.

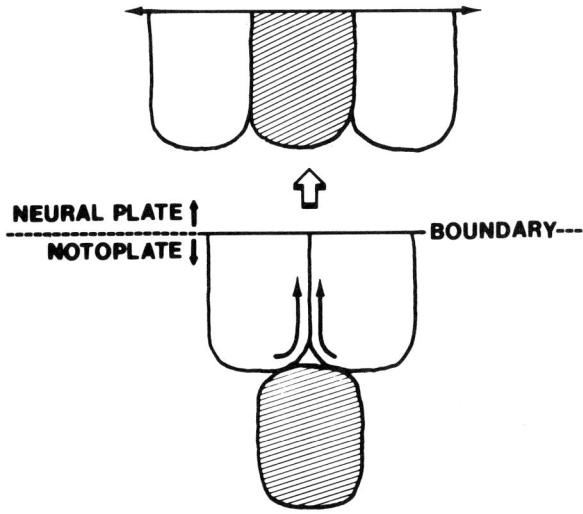

FIGURE 7. This diagram (top view, looking down on the
cell apices) illustrates how interdigitation of cells can
elongate the boundary at the neural plate/notoplate
interface.

Elongation of the notoplate boundaries would stress the
adjacent notoplate and neural plate tissue, and would
facilitate interdigitation of these adjacent ranks of cells
along the long axis of the embryo. The spreading of this
effect would elongate the notoplate and neural plate to
conform to the elongating notoplate boundaries. Meanwhile,
the boundary cells would continue to tractor more cells to
the boundary. If no countermanding forces intervene, this
process could continue until the notoplate has been reduced
to a single rank of cells (i.e. becomes a raphe).
We have measured about a four-fold increase in length

of the notoplate region during neurulation. The minimum number of ranks of notoplate cells required to accomplish this by interdigitation is four (Fig. 8).

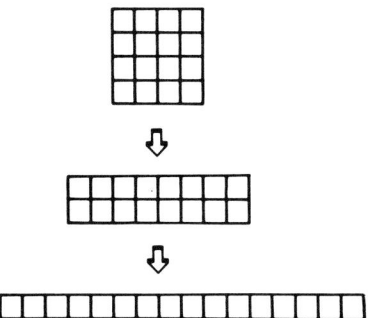

FIGURE 8. This diagram illustrates how a four-fold increase in length results from the complete interdigitation of four ranks of cells.

Jacobson and Gordon (19) found notoplate elongation to be responsible for most of the shaping of the neural plate up to the keyhole stage. After the keyhole stage, the plate rolls into a tube, and exactly during that time the rate of notoplate elongation increases ten-fold (19). Chick embryos have a similar rapid rate of neural plate elongation during tube closure (16-18). As suggested by Jacobson (16, 17, 22), notoplate elongation can help to roll the neural plate into a tube, much as a rubber sheet folds into a tube when stretched along a line (16, 17 22). However, analyses discussed below suggest that this is not the only mechanism that drives the folding of the plate into a tube.

Jacobson and Tam (24) found that while the brain plate of the mouse embryo rolls into a tube, the neural folds elongate considerably, but the midline does not. This led us to look more closely at the neural folds of the newt embryo during neural tube formation. We find that the neural folds begin to elongate considerably just at the stage when the plate begins to fold, and increase in length by 30% by the time the neural tube is completely formed. During this same period, the midline increases by 35%.

The boundary of the neural plate and epidermis, which becomes the neural fold and crest, could be another organizing boundary for elongation of the neural plate. We assume

that the epidermal cells have inactive basal and apical
surfaces, and so can move only tangentially amongst one
another. The adjacent neural plate cells are basally (and
perhaps laterally) active, and as they tractor, their faces
that contact the epidermal cells adhere to them. This
would cause the neural plate cells to crawl beneath the
epidermal cells, and as their adhering faces become inactived,
to preferentially tractor more neural plate cells after
them, elongating the plate edge, and drawing epidermal cells
along the edge. This would crowd even more adhering plate
cells into the forming neural folds. Since the plate cells
are tethered apically, their tractoring would tend to pull
them under the epidermis, while buckling that epidermis up
into an arch, and stretching the plate cells doing the
crawling. As the plate cells stretch beneath the epidermis,
their apical surfaces will constrict much as the trailing
tail of a migrating fibroblast is constricted. Some of the
plate cells at the edge of the plate may have their apical
surfaces reduced to a point, then pull free of the epithelium and become neural crest cells. Compare Fig. 9, which
illustrates these proposed events schematically, to the
sections in Fig. 5.

Preliminary Simulations of the Cortical Tractor Model

The assumptions of the cortical tractor model seem to
lead quite naturally to the above scenario for neurulation.
However, it is important to verify the verbal descriptions
we have given by mathematical analyses and/or numerical
simulation. Otherwise, we have no way to prove that our
verbal scenarios will actually evolve from the mechanical
assumptions of the model. These simulations are exceedingly
complex, and we have approached them piece-meal, simulating
only restricted aspects of the model. A complete simulation
of the cortical tractor model is in progress, however, we
report here a number of related simulations that give us
confidance that the verbal descriptions we have given are
indeed valid.

The Mechanical Properties of Cytogel

The basis for the cortical tractor model lies in the
physicochemical properties of the cortical cytogel in each
epithelial cell. We have constructed a series of detailed
models of actomyosin gels, and have applied them to a number
of cell motility phenomena (25-31). We feel confident that

our models of cytogel behavior, while certainly not yet a complete description, nevertheless capture enough of the relevant physics to warrent the assumptions of the cortical tractor model. In particular, the basic cycle of ionically triggered solation, osmotic expansion and active contraction appears well verified in a number of experimental systems.

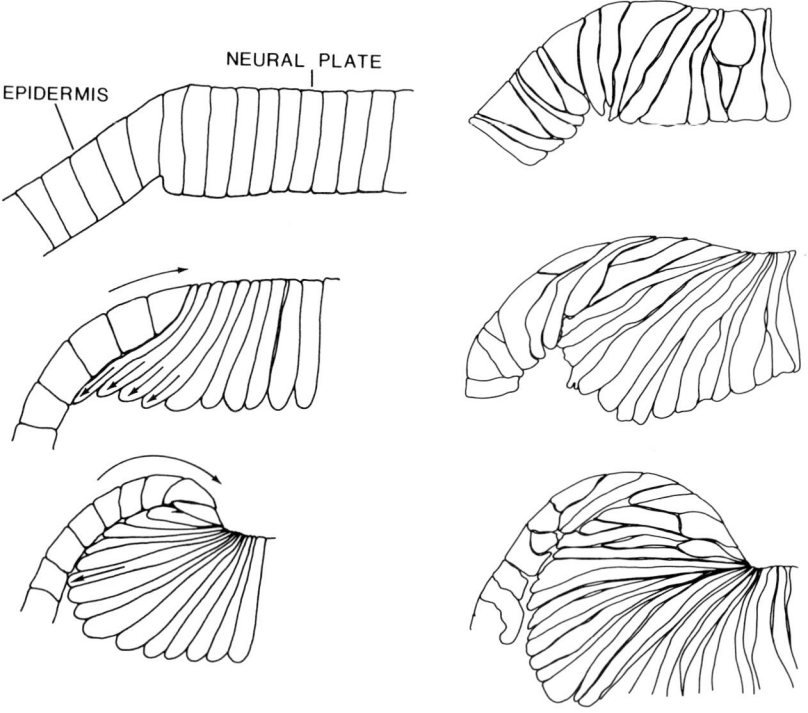

FIGURE 9. The diagrams at left illustrate how we interpret events at the epidermis/neural plate boundary. Neural plate cells tractor on the bottoms of the epidermal cells, pulling them into a fold, and at the same time stretching the neural plate cells until their apical surfaces are points, or even become released. Neural plate cells interdigitate along the boundary (not shown) to elongate the neural folds, and their tractoring produces a rolling moment toward the midline and lifts the folds up out of the plane. The drawings at right are tracings of cells from cross sections of newt neurulae at stages 15, 16, and 17. Actual cell shapes can be compared to the diagrams.

Motile Cells Can Exert Considerable Forces on Their Surroundings

A basic assumption of the cortical tractor model is that the motile behavior of epithelial cells is identical to that of mesenchymal cells, except for the constraint on their motion by their apical junctional structures. The cortical tractor we have imputed to epithelial cells is identical to that observed in motile mesenchymal cells *in vitro*. That motile cells can exert considerable forces on their own is well established (32, 33), and we have assumed that this property is retained by motile cells that are tethered into an epithelial configuration. Thus the neural plate cells should be amply strong to roll the tube, as shown in Fig. 9.

Simulations of Epithelial Sheet Folding

Odell, et al. (14) simulated the rolling of the neural tube employing a model for epithelia wherein the only force generated by the cells was developed by apical microfilament bundles. These simulations demonstrated that apical contraction was sufficient to mimic many aspects of epithelial folding phenomena. In this paper, we have relegated apical contraction to a secondary role in contrast to active crawling; however, it is likely that both mechanisms operate during neurulation. Indeed, our cortical tractor model provides a mechanism for assembling and concentrating microfilament contractile machinery at cell apices, and thus subsumes and enhances our previous work.

Does Elongation of the Neural Folds and/or Notoplate Roll the Neural Plate into a Tube?

During neurulation, not only do the neural folds close into a tube, but they rise out of the plane of the neural plate a considerable distance. Previously, Jacobson and Tam (24) and Jacobson (16, 22) have suggested that both rising and rolling can result from the elongation of the neural folds and/or the notoplate. In order to investigate the relative importance of elongation of the neural folds in comparison to the rolling action of the crawling plate cells, we employed a finite element program developed by L. Chang (to appear). Fig. 10 shows a simulation wherein the folds elongated, but no other forces acted. The elongation develops transverse compressive stresses (Poisson buckling) sufficient to raise the folds upward. However, we were unable to cause the folds to close into a tube. We concluded that an active bending moment must be developed by the plate in order to roll completely into a tube. These bending moments can be generated by crawling, according to the cortical tractor model, or by apical constraction, as previously assumed by Odell, et al. (14) in their simulations, or by both mechanisms acting in concert.

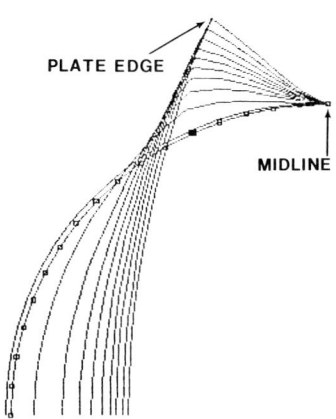

FIGURE 10. Results of simulations in which the only force applied is the elongation of the plate edge (neural fold). The simulations were done using only one quadrant, as shown, since the other quadrants may be reconstructed by symmetry.

Does the Neural Tube Roll From the Centerline Outward, Or From the Edge of the Plate Inwards?

There is also the question of whether the neural tube rolls from the centerline outwards, or from the neural folds inward, or both. Fig. 11 shows a comparison between the two possibilities. The simulations wherein rolling commences at the folds and proceeds inward look more realistic, reinforcing our suspicion that the bending moments that roll the neural plate mostly commence at the plate edge and propagate toward the centerline. However these simulations do not address the issue of whether the bending moments are generated by activation of the basal and lateral faces of the plate cells, or by apical contraction, or both.

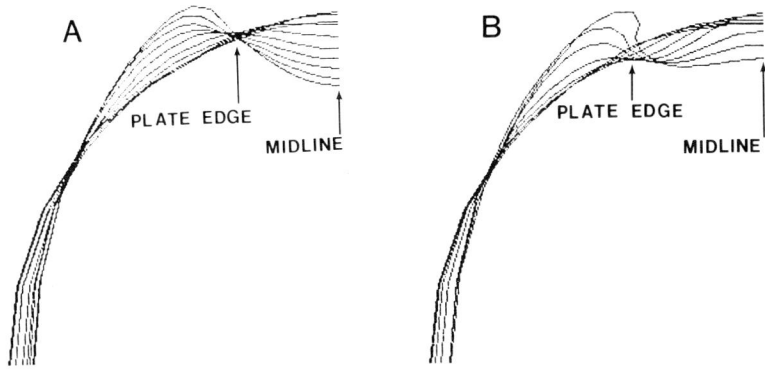

FIGURE 11. A) Results of simulations in which the imposed rolling moment is from the midline to the plate edge. B) Results of simulations in which the imposed rolling moment is from the plate edge to the midline. In neither case is any elongation imposed. These simulations were done using only one quadrant, as shown, since the other quadrants may be reconstructed by symmetry.

It may be difficult to distinguish whether the contraction proceeds from the centerline to the edge, or from the margins inward, because in the previous simulation of Odell et al. (14), the cells were triggered at the centerline, yet rolling commenced at the margins. This was due to purely geometrical effects.

The tractor motion suggested in the cortical tractor model for epithelial morphogenesis can create quite complex epithelial foldings, and many other epithelial movements may be generated by this mechanism that both epithelial cells and mesenchymal cells hold in common. Key features of the cortical tractor model, as applied to epithelia, are that cell movements and neighbor changes can occur amongst the cells of an epithelium without breaking the apical seals; when different populations of cells exist within the tissue layer, each with distinctive adhesive characteristics (or with the ability to contact inhibit one another), then the tractor movements of the cells may be organized to produce orderly epithelial morphogenesis.

ACKNOWLEDGMENTS

We thank L. Chang for adapting a general purpose finite element program (Ph.D. dissertation, University of California, Berkeley) to generate the simulations seen in Figs. 10 and 11. We appreciate Raymond Keller's contributions to discussions of the ideas in this paper.

REFERENCES

1. Hay ED (1968). Organization and fine structure of epithelium and mesenchyme in the developing chick embryo. In Fleischmajer R, Billingham RE (eds): "Epithelial-Mesenchymal Interactions," Baltimore: Williams & Wilkins Co., p 31.
2. Oster GF (1984). On the crawling of cells. J Embryol Exp Morph 83(Suppl.):329.
3. Allen RD (1961). A new theory of amoeboid movement and protoplasmic streaming. Exp Cell Res Suppl 8:17.
4. Harris AK (1973). Cell surface movements related to cell locomotion. In Porter R, Fitzsimons DW (eds):"Locomotion of Tissue Cells," Ciba Foundation Symposium 14(new series), Amsterdam: Elsevier/North Holland, p 3.
5. Odell GM, and Frisch HL (1975). A continuum theory of the mechanics of amoeboid pseudopodium extension. J Theor Biol 50:59.
6. Abercrombie M (1980). The crawling movements of metazoan cells. Proc Roy Soc Lon B 207:129.
7. Bretscher MS (1984). Endocytosis: Relation to capping and cell locomotion. Science 224:681.
8. Dembo M, and Harris AK (1981). Motion of particles adhering to the leading lamellae of crawling cells. J Cell Biol 91:528.
9. Campbell RD, and Campbell J (1971). Origin and continuity of desmosomes. In Reinert J, and Ursprung H (eds): "Results and Problems in Cell Differentiation," Vol 2, "Origin and Continuity of Cell Organelles", New York: Springer-Verlag, p 261.
10. Zigmond S (1978). Chemotaxis by polymorphonuclear leukocytes. J Cell Biol 77:269.
11. Snyderman R, and Goetzl E (1981). Molecular and cellular mechanisms of leukocyte chemotaxis. Science 213:830.

12. Baker PC, and Schroeder TE (1967). Cytoplasmic filaments and morphogenetic movement in the amphibian neural plate. Dev Biol 15:432.
13. Burnside B (1971). Microtubules and microfilaments in newt neurolation. Dev Biol 26:416.
14. Odell GM, Oster G, Alberch P, and Burnside B (1981). The mechanical basis of morphogenesis. I. Epithelial folding and invagination. Dev Biol 85:446.
15. Odell GM, and Bonner JT (1985). How the Dictyostelium discoideum grex crawls. Phil Trans Roy Soc, B(in press).
16. Jacobson AG (1981). Morphogenesis of the neural plate and tube. In Connelly TG, Brinkley LL, and Carlson BM (eds): "Morphogenesis and Pattern Formation," New York: Raven Press, p 233.
17. Jacobson AG (1985). Adhesion and movement of cells may be coupled to produce neurulation. In Edelman GM, and Gall WE (eds): "The Cell in Contact: Adhesions and Junctions as Morphogenetic Determinants," New York: John Wiley and Sons,(in press).
18. Jacobson AG (1984). Further evidence that formation of the neural tube requires elongation of the nervous system. J Exp Zool 230:23.
19. Jacobson AG, and Gordon R (1976). Changes in the shape of the developing nervous system analysed experimentally, mathematically, and by computer simulation. J Exp Zool 197:191.
20. Spemann H, and Mangold H (1924). Über Induktion von Embryonalanlagen durch Implantation artfremder Organisatoren. Arch f Mikr Anat u Entw Mech 100:559.
21. Spemann H (1938). "Embryonic Development and Induction," New Haven:Yale University Press (Reprinted by Hafner, New York, 1962.)
22. Jacobson AG (1978). Some forces that shape the nervous system. Zoon 6:13.
23. Cooper MS, and Keller RE (1984). Perpendicular orientation and directional migration of amphibian neural crest cells in dc electrical fields. Proc Natl Acad Sci USA 81:160.
24. Jacobson AG, and Tam PPL (1982). Cephalic neurulation in the mouse embryo analysed by SEM and morphometry. Anat Rec 203:375.
25. Oster GF, and Odell GM (1984). The mechanochemistry of cytoplasmic contractility. Physica 12D:333.
26. Oster GF, and Odell GM (1984). Mechanics of cytogels 1: Oscillations in Physarum. Cell Motility 4:469.

27. Odell GM (1984). A mathmatically modelled cytogel cortex exhibits periodic Ca^{++}-modulated contraction cycles seen in Physarum shuttle streaming. J Embryol Exp Morph 83 (Suppl.):261.
28. Murray JD, and Oster GF (1984). Generation of biological pattern and form. IMA J Math Med and Biol 1:51.
29. Oster GF, Murray JD, and Odell GM (1985). The formation of microvilli. (this volume).
30. Oster GF, Murray JD, and Harris AK (1984). Mechanical aspects of mesenchymal morphogenesis. J Embryol Exp Morph 78:83.
31. Cheer A, Nuccitelli R, Oster GF, and Vincent J-P (1985). Cortical waves in vertebrate eggs: The activation wave. Dev Biol (submitted).
32. Harris AK, Stopak D, and Wild P (1981). Fibroblast traction as a mechanism for collagen morphogenesis. Nature 290:249.
33. Harris AK, Wild P, and Stopak D (1980). Silicone rubber substrata: A new wrinkle in the study of cell locomotion. Science 208:177.

CELL MIGRATION IN THE VERTEBRATE EMBRYO

Jean Paul Thiery°, Jean Claude Boucaut+ and Kenneth M. Yamada*.

° Institut d'Embryologie, 49 bis Avenue de la Belle Gabrielle, 94130 Nogent-sur-Marne, France. + Laboratoire de Biologie Expérimentale, Université Paris René Descartes, 45, rue des Saints Pères, 75270 Paris Cedex 06, France. * Membrane Biochemistry Section, Laboratory of Molecular Biology, National Cancer Institute, Bethesda, Maryland 20205.

INTRODUCTION

A great diversity in the developmental program is observed in the animal kingdom. In the vertebrates, early embryogenesis proceeds differently in different species even within the same class. The structure of the egg and the amount and localization of maternally derived molecular determinants must contribute to the formation of these unique patterns in early embryos.
In addition, cells of the blastoderm which have been identified as precursors of defined territories are considerably displaced throughout the early stages of development. Extensive morphogenetic movements including individual cell migration, cell sheet deformation and remodeling are essential in the establishement of the basic structures of the embryo. These movements allow cells of different area of the blastoderm to interact transiently to complete the process of primary induction. These inductive events are responsible for the segregation of the endoderm, mesoderm and ectoderm and most importantly in the latter, of the nervous system.
There is now an urgent need to establish some general principles of morphogenesis particularly in order to better define the genetic and epigenetic controls of developmental processes.

At any given time, cells can be engaged in one or more of the five primary processes of division, migration, adhesion, differentiation and death. Cell division is necessary but far from sufficient for the construction of the body plan. For instance, gastrulation proceeds similarly in the chick embryo, which has 60000 cells and in the mouse embryo, which has only 500-1000 cells at this stage. Distinct cell lineages appear within different tissues involved in a similar morphogenetic event. Therefore cell division, cell differentiation and cell death could be considered as secondary compared to cell migration and cell adhesion in the shaping of the embryo. In fact, cell adhesion has a paramount influence on the mechanism of cell migration and more generally in epithelium-mesenchyme interconversion and in tissue remodeling.

In this chapter, we shall examine some of the adhesive behavior of cells during migration and their final localization after its completion. Mechanisms providing proper directionality to the migrating cells will also be mentionned. Recent experimental analyses of the neural crest will be integrated into a tentative model defining the molecular repertoire required for normal development.

MOLECULAR ANALYSIS, LOCALIZATION AND FUNCTIONS OF CELL-CELL AND CELL TO SUBSTRATE ADHESION MOLECULES.

Cell adhesion molecules-CAMs

The structures and functions of CAMs are already described in several chapters of this volume. So far, three molecules have been identified (1, 2). N-CAM is a heavily glycosylated glycoprotein containing as much as 30% carbohydrate, most of which consists of sialic acid. N-CAM amino terminal domains from two cells bind directly to each other in a calcium-independant process. The binding strength depends on the surface density of the molecule and on the amount of sialic acid. The binding strength is likely to be altered through cis-interaction with other molecules such as Ng-CAM (3). Adhesive interactions could also be modulated by other cell surface glycoproteins (4). L-CAM, a cell surface glycoprotein acting via a calcium dependent mechanism, is also directly involved in cell adhesion (5). These two molecules are found on all cells of the chick blastoderm (6). L-CAM (also termed cadherin, uvomorulin, or gp 80/120 in mammals) has been localized at the cleavage stage and on teratocarcinoma cells (7-10). In the latter cases, L-CAM was clearly shown to be directly involved in the mechanism of compaction and in the formation of epithelial sheets of teratocarcinoma cells.

In birds (11), both N-CAM and L-CAM are maintained on cells of the superficial layer, except at the level of the primitive streak where L-CAM and to a lesser degree N-CAM disappear. The newly formed middle layer (or mesoderm) has lost L-CAM but rapidly reacquires N-CAM. The deep layer constituting the definitive endoderm contains predominantly L-CAM. Most interestingly, L-CAM also progressively disappears from the surface of the newly induced neural plate. The last cells to contain L-CAM in the neural plate are located at the dorsal border, which is the presumptive territory of neural crest cells.

Metamerization of the axial mesoderm is accompanied by a local increase of N-CAM in the somitic mesoderm; N-CAM remains on dermomyotomal and later on myotomal cells (12). L-CAM permanently remains localized in the ectoderm and in the endoderm. All of the endodermal structures derived from the gut are stained with equal intensity by anti L-CAM antibodies, and the same glycoprotein is recognized in all of these tissues. In the mesoderm, the urogenital tract epithelium is also rich in L-CAM. The mesonephric inductor, the Wolffian duct, elongates as an epithelium containing L-CAM. The newly induced mesenchymal cells aggregate locally along the duct concomittantly with the appearance of N-CAM. The mesonephric rudiment develops into a S-shaped tube that expands later; during this process, L-CAM appears at the surface of the newly formed epithelium and persists until adulthood.

A third glycoprotein, Ng-CAM (13), was recently characterized on the surface of post-mitotic neurons both in the central and in the peripheral nervous sytem (14). Ng-CAM has been shown to mediate adhesion between neurons and glia. It is noteworthy that Ng-CAM is found in vivo mostly at the surface of neurites at a stage where neurons or their processes are migrating along glial cells. Axons emerging from the neural tube traverse a defined compartment of the somite; this compartment is already occupied by neural crest cells that will give rise to Schwann cells. In the developing cerebellum, the external granular cells descending along Bergman radial glia have leading processes covered with Ng-CAM. The ganglion cell axons leaving the optic cup also contain Ng-CAM during their migration on or within the optic stalk (14).

In contrast to N-CAM and L-CAM, Ng-CAM appears late and has a restricted distribution; this molecule has therefore been designated as a secondary CAM.

Although it is not yet possible to define the minimal number of primary and secondary CAMs required for the assembly of cells into functional organs, the results so far indicate that the same two primary CAMs are found either transiently or permanently in many different tissues belonging to the three primary germ layers. On the fate map of the chick blastoderm, both CAMs cover more than 2/3 of the presumptive tissues (6). It is therefore tempting to consider that no more than two or possibly three primary CAMs may be found in all vertebrates. On the other hand, it is likely that heterophilic interactions of the type mediated by Ng-CAM will be controlled by different molecules. For instance in T lymphocyte subpopulations, a unique glycoprotein was found to be involved in adhesion to the high endothelial venules of lymph nodes (15). More specialized molecules of this kind are likely to be found in the near future.

Cell to substrate adhesion molecules

The extracellular matrix (ECM) produced by mesenchymal cells and by epithelial tissues contain several adhesive glycoproteins (16). Laminin, a high molecular weight glycoprotein contains several distinct binding domains for attachment and spreading of cells and for binding to other basal lamina components including type IV collagen (17); a high-affinity receptor for laminin has been isolated and characterized (18). Laminin appears very early during development; one polypeptide chain is already synthesized during oogenesis, while two and ultimately all three polypeptides appear in the cytoplasm of the 2 to 8 cell-stage mouse embryo. Laminin is finally expressed on the surface of blastomeres at the 16 cell-stage (19-20). In the chick embryo, laminin appears at the epiblast basal surface. Most epithelia, whether transient such as the somites or permanent such as the skin, are underlain by a laminin-rich basal lamina (21).

Laminin becomes detectable at sites where mesenchymal cells are organizing into epithelia (22). Its synthesis and assembly in a basal lamina, however, appears to be delayed compared to the synthesis of N-CAM (21); in the case of the somites and the mesonephros, laminin as well as other components of the basal lamina and its associated ECM must be important in the maintenance of epithelia, as has been shown for other epithelia (23). It is noteworthy that laminin disappears from sites where cells are egressing from epithelia (21).

Fibronectin (FN) is another well-characterized adhesive glycoprotein of the ECM. FN is composed of two nearly identical polypeptide subunits, each containing approximately 2350 amino acids. FN is encoded by a single gene (24-25) containing 48 exons in the chick embryo (26). Alternative splicing has been described during the processing of the primary transcript of this gene. One class of mRNA completely lacks sequences corresponding to the extra-domain exon (28,29). This splicing occurs, in the liver, where plasma FN is synthesized (30), but other cell types appear to contain mRNA with this sequence. Interestingly, incom- plete or complete splicing occurs for another exon (IIICS) (25). As many as ten distinct mRNA molecules can be produced by the FN gene (Kornblihtt, personal communication). Furthermore, post-translational modifications contribute to the heterogeneity of FN (31). However, all of the FN are multifunctional proteins with similar binding domains along their polypeptide chains (32).

Plasma FN lacking peptide sequences corresponding to the ED exon are soluble, in contrast to the cellular FN which readily assemble into fibrils. However, so far, the sequences responsible for fibrillogenesis have not been identified with certainty. Fibrillar FN can be found at the cell surface or in the ECM. Studies on the binding of FN to cell surfaces have revealed surface receptors with a fairly low affinity constant (33 and references therein).

A leading candidate for the FN receptor is a glycoprotein complex of 140 000 dalton component that can be detected by monoclonal antibodies (34-37) and affinity chromatography (38). This cell surface complex was found to colocalize with microfilaments and with extracellular FN fibrils (39). These putative FN receptors are expressed early during development in most cell types (Duband et al. in preparation).

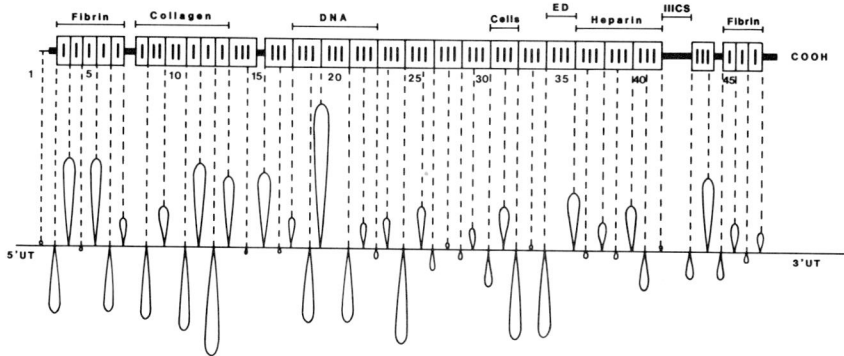

Figure 1 : Schematic representation of the primary transcript and one of the polypeptide chains of FN. The size of each exon and intron is deduced from the work of Hirano et al. in the chick (26), whereas the arrangements of the different types of homologies derive mostly from data obtained in mammals including sequence analyses from Kornblihtt et al. in press. Type III homology units are probably generally encoded by 2 exons each. The materials bound at distinct regions by functional domains of the protein are indicated at the top, as well as the regions modified by splicing termed ED (extra domain) and IIICS. The bottom of the figure depicts the organisation of exons and introns according R-loop analysis; the loops are introns and the exons are shown linearly relative to the protein structure. There are 5' and 3' untranslated (UT) regions on the mRNA, and as many as 10 mRNA variants.

Recently, the region of FN involved in interactions with the receptor was shown to contain a unique peptide sequence Arg-Gly-Asp-Ser (RGDS) that is required for the attachment of cells to intact FN or FN fragments (38, 40-43). Variants of this sequence have been tested for their ability to inhibit binding of cells to FN or of the FN cell-binding region to the cell surface. Most substitutions except at the carboxy terminus inactivate the peptide (41).

Interestingly, specific spacing between two charged amino acids (Arg and Asp) is required; these changes must be located at a specific distance from the peptide backbone. In addition, adjacent sequences can also modify the binding capacity of the RGDS peptide (44).

CELL MIGRATION IN EARLY EMBRYOGENESIS

Gastrulation in amphibians.

In amphibians, after the phase of segmentation, the blastula appears as a hollow sphere containing a blastocoelic cavity. The first sign of gastrulation is a slit, the blastopore, that appears on the future dorsal side of the embryo, corresponding to the site of invagination of the mesodermal and endodermal cells.

Scanning electron microscopy, time-lapse microcinematography and cultures of tissue explants have shown that gastrulation involves epiboly of the superficial layer, the formation of bottle cells and active movements of mesodermal cells along the blastocoelic roof (45-48).

The surface of the blastocoelic roof to which mesodermal cells adhere is covered by a network of fibrils. Biochemical studies and immunocytochemistry have revealed only the presence of FN as a major component of this ECM.

FN is synthesized at a low rate from maternally derived mRNA during oogenesis, then translation increases rapidly at the late blastula and early gastrula stages (49,50). FN assembles specifically along the ectoderm lining the roof of the blastocoel, even though most cells are able to synthesize FN. Immunolabeling with fluorescent or gold-coupled antibodies has clearly shown that the fibrils described by scanning electron microscopy (51) contain FN (52).

The role of FN in the adhesion and migration of mesodermal cells has been assessed by three types of perturbation experiments, which have led to the following conclusions: i, when part of the blastocoelic roof is inverted, mesodermal cells avoid the area now lacking an ECM. ii, microinjection of monovalent anti-FN antibodies into the blastocoelic cavity of late blastulae or early gastrulae blocks gastrulation (53).

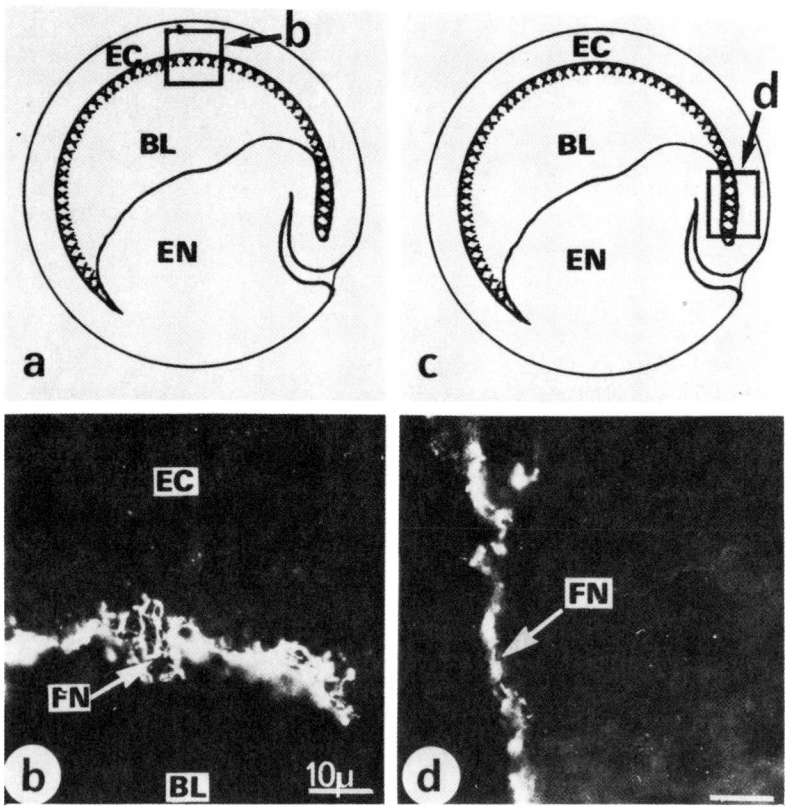

Figure 2 : Gastrulation in Pleurodeles waltlii
c,d: Mesodermal cells migrate along the blastocoelic roof in a FN-rich matrix. a,b : Note that the matrix is already assembled in regions not yet occupied by mesodermal cells. EC : ectoderm; BL : blastocoelic cavity; EN : endoderm.

iii, similarly when Arg-Gly-Asp-Ser- containing peptides are injected, gastrulation is also arrested (54). These peptide sequences contained in the cell-binding site of FN have been shown to be directly involved in the mechanism of binding of fibroblasts to FN (41-43).

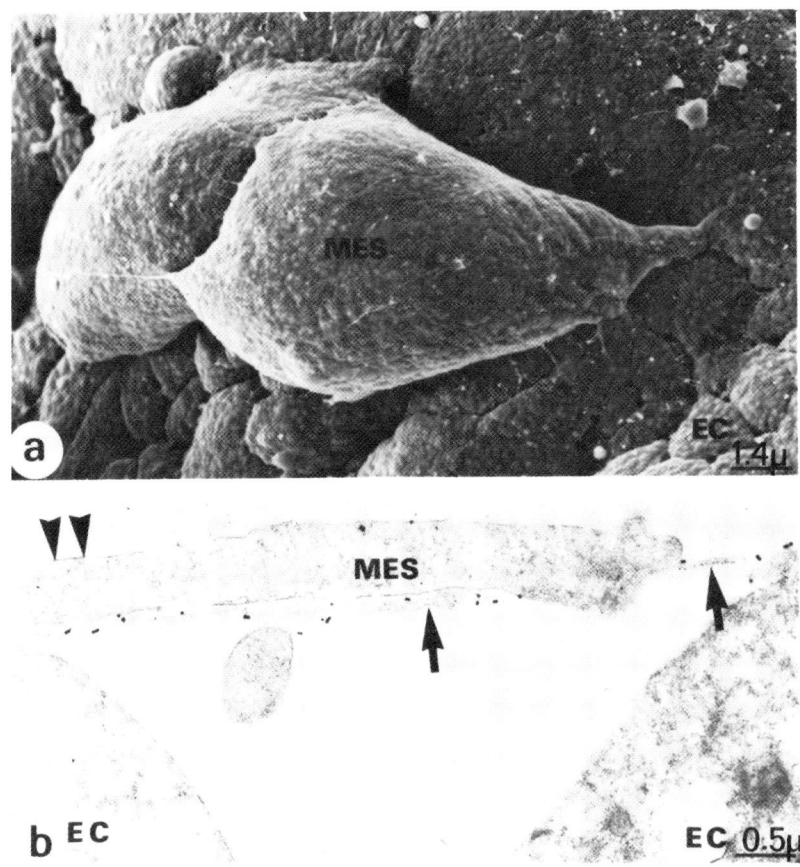

Figure 3 : a: Mesodermal cells (MES) adhere to the blastocoelic roof; b: Mesodermal cell processes bind to FN-rich fibrils attached to the ectoderm as detected by anti-FN coupled to colloidal gold. Note that FN is present only at the site of contact (arrows). The mesodermal cell surface facing the cavity lacks FN (double arrowhead).

Thus, competitive inhibition for the receptors and an immunological steric hindrance effect, preventing the interaction between the cell surface and FN, both interfere with the movement of mesodermal cells during gastrulation.

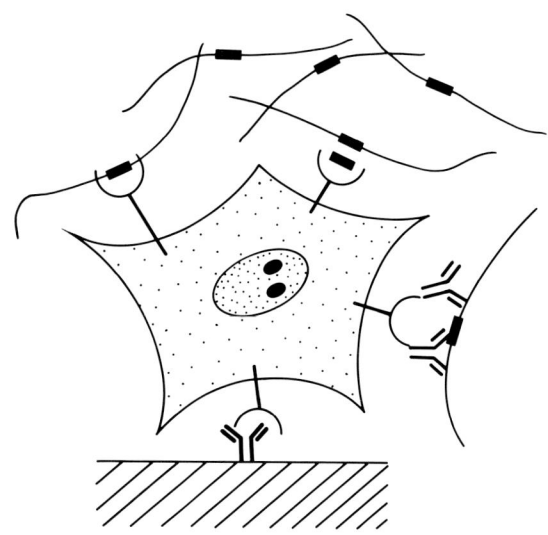

Figure 4 : Cell attachment and migration can be altered by two different procedures. If one assumes that specific surface receptors interact directly with the cell binding sequence of FN, peptides containing such a sequence should saturate receptor sites and antibodies to FN should prevent access to the binding sites. Antibodies directed against the receptor could also block migration. However, if the antibodies are adsorbed to a surface they can provide a substrate for adhesion.

In contrast to the requirement for FN in the formation of the mesoderm, FN is unlikely to be involved in the mechanism of neural induction; in vitro, mesodermal cells associated with the apical surface of the ectoderm devoid of ECM can induce the appearance of neural elements (55). In vivo, when either the antibodies or the peptides are introduced during or at the end of gastrulation, a partial or a complete neural plate still forms (53, 54).

Embryonic Cell Migration 177

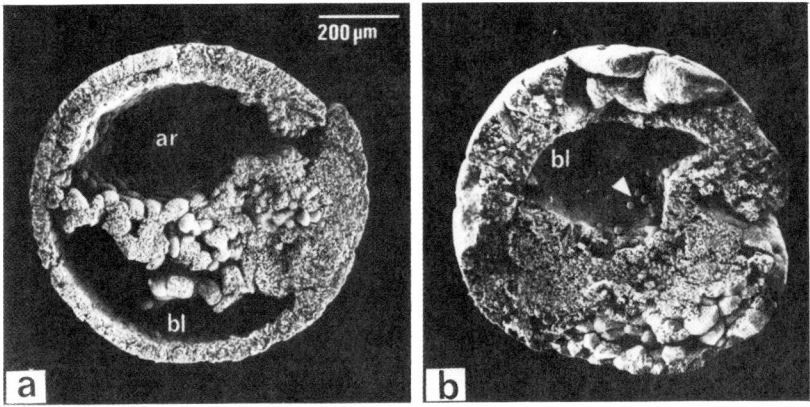

Figure 5 : a: Control embryos gastrulate normally when injected with several peptides including a peptide contained in the collagen binding domain. b: Pleurodeles waltlii embryos injected with the cell-binding peptide fail to gastrulate. A few cells are found in the blastocoelic cavity (arrowhead), whereas most mesodermal cells remain on the outside. The ectoderm does not undergo proper epiboly and instead becomes highly convoluted.

The neural crest.

Definition. The neural crest is a transient embryonic structure occupying the dorsal border of the entire neural axis. In the chick embryo, crest cells detach from the neural tube during and after its closure and migrate to various sites where they undergo differentiation. Crest cells give rise to a large number of cell types as diverse as pigment cells, most types of peripheral neurons and glia, myoblasts, chondrocytes, and connective tissues in the head (for reviews see 56, 57). A variety of experiments involving transplantation of neural tubes into host embryos (57-59) have suggested that the morphology of the embryo plays a major role in the pattern of the neural crest migration and ultimate fate. In addition, the behavior of crest cells appears to depend on the control of adhesion between themselves and to the ECM. From this point of view, the natural history of crest cells can be described as a series of phases involving different adhesive systems.

Separation from the neural tube. Prior to their migration, the presumptive crest cells are integrated in the neuroectodermal epithelium. Their release from the neural epithelium can be compared to gastrulation; it involves local disruptions of the basal lamina and the disappearance of intercellular junctions. Gap junctions are lost among the premigratory crest cells, as shown by the absence of the electrical coupling (60). Viewed by transmission and scanning electron microscopy, presumptive neural crest cells are irregular in shape, do not show tight junctions and are frequently separated by acellular spaces (61-63). The basal lamina overlying the neural tube is interrupted and then completely disappears (21, 61-63). Thereafter, crest cells send projections out of the epithelium and are progressively surrounded by FN (62-64).
The factors triggering the disruption of the neural epithelium have not yet been identified; however, it has been shown that the disappearance of the basal lamina induces epithelial cell destabilization (65), whereas direct access to a three-dimensional ECM can promote emigration (66). The numerous morphogenetic movements that occur in the head (67), accompanied by intense cell proliferation, could generate mechanical forces responsible for damaging the basal lamina. Alternatively, plasmin and

collagenases produced locally could digest the lamina components; interestingly, crest cells synthesize large amounts of plasminogen activator (68) and plasminogen stored in the yolk can diffuse into the embryo before the blood circulation is established (69). The quantity of adhesive molecules, i.e., L-CAM and subsequently N-CAM, diminishes at the surface of newly formed crest cells (11, 12). Furthermore, other molecules inserted into the plasma membrane may alter the binding properties of the residual N-CAM (4).

Patterns of migration. The migration of neural crest has been followed by means of various markers for crest cells themselves, such as the quail nucleolar marker (56-58), acetylcholinesterase (70), and more recently the monoclonal antibody, NC-1 (71-73); migration pathways have been characterized by electron microscopy and immunohistology for FN and laminin (21, 62-64, 74). These studies have shown that morphogenesis of tissues becoming populated by crest cells is a major element controlling the pattern of crest cell migration. Structures adjacent to the neural tube are metamerized; the somitomeres in the head (75) and the somites in the trunk (74) thereby provide several distinct pathways.

In the trunk, one can identify a transient pathway offered to the crest cells between two consecutive somites, and a second one between the somite and the neural tube. The first pathway leads the crest cells to the aortic and mesonephric area, where autonomic differentiation occurs. Cells within the other pathway give rise to the dorsal root ganglia and to the Schwann cells lining motor and sensory nerves. Subsequently, the metameric pattern in the trunk is rapidly modified, as two adjacent half-somites fuse to form a vertebra. This extensive reorganization of the mesoderm surrounding the neural tube contributes to the formation of new pathways, but primarily creates physical barriers achieving a complete separation of different crest cell populations.

Substrate of migration. The crest cell migratory pathways contain an ECM limited by one or two laminin-rich basal lamina, which provide defined channels. FN, type I and III collagens, hyaluronate and small amounts of chondroitin sulfate are constituents of the three dimensional network of fibers (21, 64 75-77).

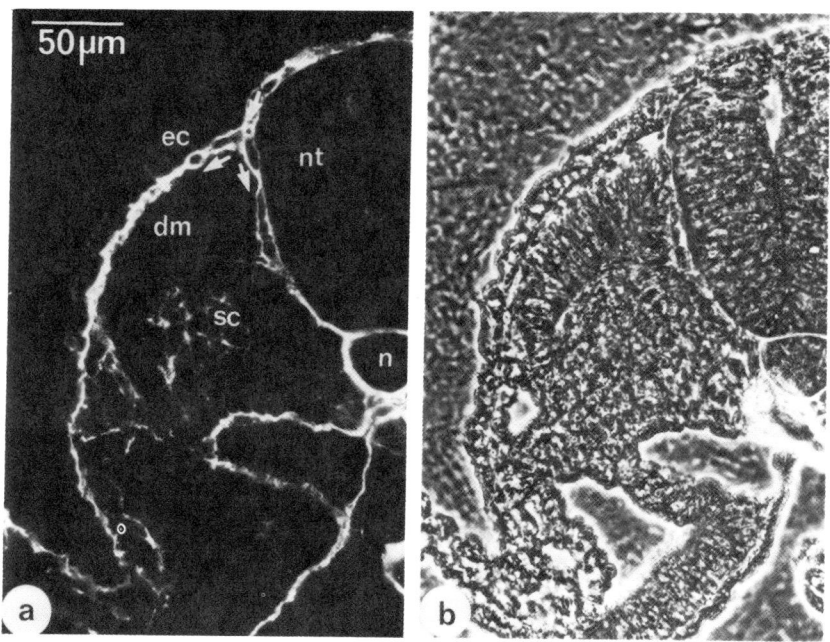

Figure 6 : Lateral and ventral pathways of neural crest cell migration at vagal level; pathways (arrows) are identified after immunolabeling for FN. Crest cells are located between ectoderm (ec) and dermomyotome (dm) and between neural tube and dermomyotome and sclerotome (sc). a: immunofluorescence; b: phase constrast.

The ECM is deposited prior to crest migration. However, so far there is no direct evidence for its role in triggering neural crest emigration from the neural tube. Furthermore, potential pathways are not always used by crest cells. For example, the presence of high levels of chondroitin sulfate around the notochord (76) and of a non-defined factor in the skin of the white axolotl (78) seem to prevent crest cell migration. Therefore, in addition to defined pathways, crest cell migration may be further restricted according to the chemical composition of the matrix.

FN alone or associated with other ECM components greatly promotes the attachment, spreading and motility of crest cells. In contrast, serum proteins, collagens, hyaluronate, chondroitin sulfate and LN are very poor substrates for crest cell attachment and movement and frequently induce their aggregation (79-81). While collagens may provide a scaffold for the organization of the ECM, hyaluronate is thought to expand the space and indirectly enhances the speed of locomotion (82).

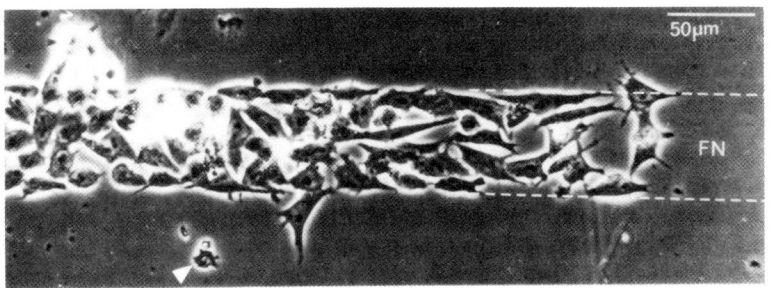

Figure 7 : In vitro crest cells preferentially migrate on FN-rich substrata (dotted area) cells that wander from the FN-coated region round up and cease migration (arrowhead). Directionality of migration is provided by the narrow pathway and population pressure.

The essential role of FN in crest migration has been confirmed by in vivo and in vitro perturbation experiments using either monovalent antibodies directed against FN or a decapeptide which competes for the cell binding sequence of FN. Both agents reversibly block the migration of crest cells (54, 81).

Behavior of crest cells. The dispersal of crest cells is also the result of their specific motile behavior, which differs strikingly from that of other embryonic cells (83-84). In contrast to somitic and notochordal fibroblasts which are polarized, crest cells on a two dimensional substrate are stellate with numerous filopodia. In addition they do not exhibit typical contact inhibition of movement (81).

Crest cells contain very little organized cytoskeleton and exert a weak tractional force on their substratum (85). Moreover, the cell surface glycoprotein complex that is likely to be the receptor for FN has a uniform distribution on the neural crest cell surface, in contrast to other mesenchymal cells, where it is concentrated in the cell-to-substratum contact sites (Duband et al., unpublished). In addition, most crest cells lack the ability to synthesize and deposit FN as a matrix in their immediate environment (86). In contrast, they synthesize large amounts of hyaluronate (87), a property that may favor their displacement (82). A similar situation was observed for mesenchymal cells emigrating from embryonic heart explants on exogenous FN (88); these cells become stationary when they begin secreting and assembling FN fibrils at their surface. These observations prompt the hypothesis that the ability to migrate is directly linked to a particular distribution of FN receptors and the inability to synthesize and organize FN fibrils at the cell surface.

Mechanisms for directional migration. Haptotaxis (89), positive (90) and negative chemotaxis (91-92), galvanotaxis (93), contact guidance, (94) and contact inhibition of movement (80, 84) have all been considered as possible mechanisms for oriented migration (95). However, it was shown recently that the "negative chemotaxis" thought to be exhibited by crest cells is caused by an artefact due to cell death (96). At least in the early phase of migration, haptotaxis is also unlikely, since crest cells transfered to the middle of the pathways can migrate in both directions (83, 84).

On a suitable substrate, isolated crest cells move very actively but randomly; their effective displacement is very small (97, 81). However, within a dense cell population, crest cells acquire persistence in their direction of movement. A type of contact inhibition of movement may be involved in the mechanism controlling directional migration, although such a phenomenon is difficult to observe in vivo (80, 84). Crest cells proliferate actively in the narrow pathways of migration, thus creating a population pressure responsible for unidirectional migration, at least in vitro (81).

However, neurite outgrowth from ganglia and Schwann cell migration may not follow the same principle. Melanocyte precursors, another crest cell derivative, also behave quite differently; considering their widespread distribution in the embryo, their capacity to invade tissues and even to traverse blood vessels under experimental conditions (98), melanocytes may respond to different cues including chemotactic signals.

Migration within the gut is also an example where several distinct mechanisms may be involved in the directed movement of enteric precursors. At first, crest cells remain in close contact with a delaminating epithelium ; subsequently they intermingle within the mesenchyme before regrouping into plexuses. The analysis of their movement is complicated by the intrisic displacement of the gut during its closure and by the formation of a progressively more differentiated and complex environment (99).

Final localization and aggregation. The final loss of motility of crest cells may result from a sudden modification of the cytoplasmic motile machinery, such as increased formation of microfilament bundles. Cells may interact more tightly with the ECM, and particularly with FN if FN receptors become organized in clusters or in strands; on the other hand, the loss or the polarized distribution of FN receptors could induce rounding up of cells and consequently enhance cell-cell adhesion. The expression of new adhesive properties involved in the aggregation of crest cells can be triggered by a chemical modification in the ECM. As crest cells migrate, their local environment is progressively transformed: epithelia dissociate into mesenchymes, which expand and, in consequence, obstruct the pathways. In some cases, the ECM itself is modified. This latter alteration is particularly true in the area where the spinal ganglia form; chondroitin sulfate increases in amount, FN, hyaluronate and type I and III collagens disappear, and finally, LN appears among crest cells (21, 74, 76). In vitro, LN can induce crest aggregation; when crest cells are cultured for long periods, they develop a greater capacity to bind to LN than to FN (81). In vivo, LN appears transiently within the dorsal root ganglion rudiments.

However, modifications of the environment are not solely responsible for the arrest of crest cells migration; autonomic ganglia form in regions where FN is still very abundant and where LN is totally absent. Crest cells forming clusters very early have N-CAM on their surface. These observations suggest that the arrest of crest cells in autonomic ganglia is induced by a change of their cell surface binding properties. If they exist, environmental factors inducing this modification remain to be defined. In contrast, the crest cell precursors of spinal ganglia do not express N-CAM until after they are maintained at high cell density for some time.

By varying the substrate of migration of crest cells in vitro, it is possible to obtain two-and three-dimensional clusters of cells; these clusters contain N-CAM. An in vitro micro-aggregation assay revealed that crest cells, according to their dissociation conditions, express either a calcium-dependent or a calcium-independent mechanism of aggregation. Anti-N-CAM monovalent antibodies strongly inhibit the calcium-independent aggregation of crest cells (100).

MOLECULAR SPECIFICITIES REGULATING THE MIGRATORY PROCESS

The neural crest offers a remarkable model system of morphogenesis. A very dynamic pattern of cell adhesion is observed throughout the development of crest cells. Furthermore, their progression and their final distribution is greatly influenced by the surrounding tissues which themselves undergo intense morphogenetic processes.

It is remarkable, however, that the adhesive properties of crest cells may be ascribed to only a few molecules. Figure 8 summarizes the developmental time course of expression of cell-cell and cell-ECM interactions. The loss of the two major cell-cell adhesion mechanisms is accompanied by the increasing importance of FN-FN receptor interactions.

At the time of cessation of migration, crest cells progressively reacquire intercellular adhesion mediated by only one of the two primary CAMs and lose the FN adhesion system. Subsequently a secondary adhesion molecule is expressed on the surface of a subpopulation of crest cells. Although other molecules could be involved in this process, localisation and perturbation experiments indicate that these adhesion systems probably provide the basic molecular regulatory determinants for this migratory event.

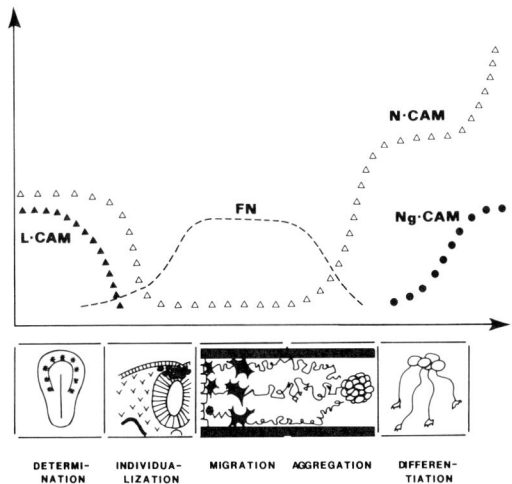

Figure 8 : Neural crest cells precursors (*) arise at the boundary between neural plate and lateral ectoderm. Their dissociation from the neural epithelium (individualization) occurs concommittantly with the disappearance of L-CAM and the N-CAM from their surface. Binding to FN deposited in the ECM predominates during migration. FN binding then diminishes, while intercellular adhesion mediated by N-CAM again increases, resulting in aggregation. Subsequently, newly born neurons express a secondary adhesion molecule (Ng-CAM) that controls adhesion between neurons and glial or Schwann cells. Both N-CAM and Ng-CAM remain on differentiated neurons.

ACKNOWLEDGMENTS

Research by the authors is supported by grants from INSERM (CRL 824018), CNRS (ATP 3701), MRT (84C1312), the Ligue Nationale Française contre le Cancer and the Fondation pour la Recherche Médicale and the National Cancer Institute. Excellent technical assistance was provided by Monique Denoyelle and Dorothy Kennedy. The authors thank Lydie Obert for typing, Sophie Tissot and Stephane Ozounoff for illustrations. We are particularly grateful to our colleagues involved in the original work we describe.

REFERENCES

1. Edelman GM (1983). Cell adhesion molecules. Science - 219: 450.
2. Edelman GM (1985). Cell adhesion and the molecular processes of morphogenesis. Ann Rev Biochem 54: 135.
3. Grumet M, Hoffman S, Chuong CM, Edelman, G.M (1984). Polypeptide components and binding functions of neuron-glia cell adhesion molecules. Proc Natl Acad Sci USA., 81: 7989.
4. Tucker GC, Thiery JP (1984). Surface molecules associated with cell migration and tissue remodelling during embryogenesis. In cellular and pathological aspects of glycoconjugate metabolism" Dreyfus, H. INSERM, in press
5. Gallin WJ, Edelman GM, Cunningham BA (1983). Characterization of L-CAM, a major cell adhesion molecule from embryonic liver cells. Proc Natl Acad Sci 80: 1038.
6. Edelman GM, Gallin WJ, Delouvée A, Cunningham BA, Thiery JP (1983). Early epochal maps of two different cell adhesion molecules. Proc Natl Acad Sci USA 80: 4334.
7. Hyafil F, Babinet C, Jacob F (1981). Cell-cell interactions in early embryogenesis : A molecular approach to the role of calcium. Cell 26: 455.
8. Damsky CH, Richa J, Solter D, Knudsen K, Buck CA (1983). Identification and purification of a cell surface glycoprotein mediating intercellular adhesion in embryonic and adult tissue. Cell 34: 455.
9. Peyrieras N, Hyafil F, Louvard D, Ploegh HL, Jacob F (1983). Uvomorulin : A non integral membrane protein of early mouse embryo. Proc. Natl. Acad. Sci. 80: 6274.
10. Yoshido-Noro Susuki N, Takeichi M (1984). Molecular nature of the calcium dependent cell-cell adhesion system in mouse teratocarcinoma and embryonic cells studied with a monoclonal antibody. Dev. Biol. 101: 19.
11. Thiery JP, Delouvée A, Gallin W, Cunningham BA, Edelman GM (1984). Ontogenetic expression of cell adhesion molecules : L-CAM is found in epithelia derived from the three primary germ layers. Develop. Biol. 102: 61.

12. Thiery JP, Duband JL, Rutishauser U, Edelman GM (1982). Cell adhesion molecules in early chicken embryogenesis. Proc. Natl. Acad. Sci. USA, 79: 6737.
13. Grumet M, Edelman GM (1984). Heterotypic binding between neuronal membrane vesicles and glial cells is mediated by a specific cell adhesion molecule. J. Cell Biol. 98: 1746.
14. Thiery JP, Delouvée A, Grumet M, Edelman GM (1985). Initial appearance and regional distribution of the neuron-glia cell adhesion molecule (Ng-CAM) in the chick embryo. J Cell Biol 100: 442.
15. Gallatin WM, Weissman IL, Butcher EC (1983). A cell-surface molecule involved in organ-specific homing of lymphocytes. Nature 104: 30.
16. Hay ED (1981). Cell biology of extracellular matrix. Plenum Press, New York.
17. Timpl R, Johansson S, Van Delden V, Oberbaümer I, Hook M (1983). Characterization of protease-resistant fragment of laminin mediating attachement and spreading of rat hepatocytes. J. Biol. Chem. 158: 8922.
18. Liotta LA, Horan-Hand P, Rao CN, Bryant G, Barsky SH, Schlom J (1985). Monoclonal antibodies to the human laminin receptor recognize structurally distinct sites. Exp Cell Res 156: 117.
19. Leivo I, Vaheri A, Timpl R, Wartiovaara J (1980). Appearance and distribution of collagens and laminin in the early mouse embryo. Dev Biol 76: 100.
20. Cooper AR, Mc Queen H (1983). Subunits of laminin are differentially synthesized in mouse eggs and early embryos. Dev Biol 96: 467.
21. Duband JL, Thiery JP (1985). Laminin in the early chick embryo : correlation with epithelium-mesenchyme interconversion. Dev Biol Submitted.
22. Ekblom P (1984). Basement membrane proteins and growth factors in kidney differentiation. "In the role of extracellular matrix in development" Trelstad ed Alan R. Liss, p 173.
23. Sugrue SP, Hay ED (1981). Response of basal epithelial cell surface and cytoskeleton to solubilized extracellular matrix molecules. J Cell Biol 91: 45.
24. Kornblihtt AR, Vibe-Pedersen K, Baralle FE (1983). Isolation and characterization of cDNA clones for human and bovine fibronectins. Proc Natl Acad Sci USA. 80: 3218.

25. Tamkun JW, Schwarzbauer JE, Hynes RO (1984). A single rat fibronectin gene generates three different mRNAs by alternative splicing of complex exon. Proc Natl Acad Sci USA. 81: 5140.
26. Hirano H., Yamada Y, Sullivan M, De Crombrugghe B, Pastan I, Yamada KM (1983). Isolation of genomic DNA clones spanning the entire fibronectin gene. Proc Natl Acad Sci USA 80: 46.
27. Petersen TE, Thogersen HC, Skortengaard K, Vibe Pedersen K, Sahl P, Sottrup-Jensen L, Magnusson S (1983). Partial primary structure of bovine plasma fibronectin : three types of internal homology. Proc. Natl Acad Sci USA 80: 137.
28. Kornblihtt AR, Vibe-Pedersen K, Baralle FE (1984). Human fibronectin : molecular cloning evidence for two mRNA species differing by an internal segment coding for a structural domain. EMBO J 3: 221.
29. Vibe Pedersen K, Kornblihtt AR, Baralle FE (1984). Expression of a human globin/fibronectin gene hybrid generates two mRNAs by alternative splicing. EMBO J 3: 2511.
30. Tamkun JW, Hynes RO (1983). Plasma fibronectin is synthesized and secreted by hepatocytes. J Biol Chem 258: 4641.
31. Paul JI, Hynes RO (1984). Multiple fibronectin subunits and their post translational modifications. J Biol Chem 259: 13477.
32. Hynes RO, Yamada KM (1982). Fibronectins : multifunctional modular glycoproteins. J Cell Biol 95: 369.
33. Akiyama SK, Yamada KM (1985). The interaction of plasma fibronectin with fibroblastic cells in suspension. J Biol Chem in press.
34. Greve JM, Gottlieb DI (1982). Monoclonal antibodies which alter the morphology of cultured chick myogenic cells. J Cell Biochem 18: 221.
35. Neff NT, Lowrey C, Decker A, Tover C, Damsky C, Buck CA, Horwitz AF (1982). A monoclonal antibody detaches embryonic skeletal muscle for extracellular matrices. J Cell Biol 95: 680.
36. Hasegawa T, Hasegawa E, Chen WT, Yamada KM (1985). Characterization of a membrane glycoprotein implicated in cell adhesion to fibronectin. J Biol Submitted.
37. Knudsen KA, Horwitz AF and Buck CA (1985). A monoclonal antibody identifies a glycoprotein complex

involved in cell substratum adhesion. Exp Cell Res in press.
38. Pytela R, Pierschbacher MD and Ruoslahti E (1985). Identification and isolation of a 140 kd cell surface glycoprotein with properties expected of a fibronectin receptor. Cell 40: 191.
39. Chen WT, Hasegawa E, Hasegawa T, Weinstock C., Yamada KM (1985). Development of cell surface linkage complexes in cultured fibroblasts. J Cell Biol 100: 1103.
40. Pierschbacher MD and Ruoslahti E (1984). Cell attachment activity of fibronectin can be duplicated by small synthetic fragments of the molecule. Nature 309: 30.
41. Pierschbacher MD, Ruoslahti E (1984). Variants of the cell recognition site of fibronectin that retain attachment-promoting activity. Proc Natl Acad Sci 81: 5985.
42. Yamada KM, Kennedy DW (1984). Dualistic nature of adhesive protein function : fibronectin and its biologically active peptide fragments can auto inhibit fibronectin function. J Cell Biol 99: 29.
43. Akiyama SK and Yamada KM (1985). Synthetic peptides competitively inhibit both direct binding to fibroblasts and functional biological assays for the purified cell binding domain of fibronectin. J Biol. Chem in press.
44. Yamada KM, Kennedy DW (1984). Aminoacid sequence specificities of an adhesive recognition signal. J Cell Biochem in press.
45. Cooke J, (1975). Local autonomy of gastrulation movements after dorsal lip removal in two anuran amphibians. J Embryol exp Morph 33: 147.
46. Keller RE (1978). Time-lapse cinematographic analysis of superficial cell behaviour during and prior to gastrulation in Xenopus laevis. J Morph 157: 223.
47. Keller RE (1980). The cellular basis of epiboly : an SEM study of deep-cell rearrangement during gastrulation in Xenopus laevis. J Embryol. Exp Morph 60: 201.
48. Nakatsuji N (1975). Studies on the gastrulation of amphibian embryos : light and electron microscopic observation of an urodele Cynops pyrrhogaster. J Embryol exp Morph 34: 669.

49. Darribère T, Boucher D, Lacroix JC, Boucaut JC (1984). fibronectin synthesis during oogenesis and early development of the amphibian Pleurodeles waltlii. Cell Diff 14: 171.
50. Lee G, Hynes R, Kirschner M (1984). Temporal and spatial regulation of fibronectin in early Xenopus development. Cell 36: 729.
51. Nakatsuji N, Johnson KE (1983). Comparative study of extracellular fibrils of the ectodermal layer in gastrulae of five amphibian species. J Cell Sci 59: 61.
52. Darribère T, Boulekbache H, De Li Shi, Boucaut JC, (1985) Immuno-electron-microscopic study of fibronectin in gastrulating amphibian embryos. Cell Tissue Res 239: 75.
53. Boucaut JC, Darribère T, Boulekbache H, Thiery JP, (1984). Antibodies to fibronectin prevent gastrulation but do not perturb neurulation in gastrulated amphibian embryos. Nature 307: 364.
54. Boucaut JC, Darribère T, Poole TJ, Aoyama H, Yamada KM, Thiery, JP (1984). Biological active synthetic peptides as probes of embryonic development: a competitive peptide inhibitor of fibronectin function inhibits gastrulation in amphibian embryos and neural crest cell migration in avian embryo. J Cell Biol 99: 1822.
55. Duprat AM, Gualandris L (1984). Extracellular matrix and neural determination during amphibian gastrulation. Cell Diff 14: 105.
56. Le Douarin NM, (1982). The Neural Crest. Cambridge University press
57. Noden DM, (1984). The use of chimeras in analyses of cranio facial development. In "Chimeras in Developmental Biology": Le Douarin, McLaren, London, Academic Press, p 241.
58. Le Douarin NM, Teillet MA, Fontaine-Perus J (1984). Chimeras in the study of the peripheral nervous system of birds. In "Chimeras in Developmental Biology, Le Douarin, McLaren. London, Academic Press, p 313.
59. Weston JA (1963). A radioautographic analysis of the migration and localization of trunk crest cells in the chick. Dev Biol 6: 279.
60. Revel JP, Brown SS, (1975). Cell junctions in development with particular reference to the neural tube. Symp Quant Biol 40: 433, Cold Spring Harbor.

61. Newgreen DF, Gibbins IL (1982). Factors controlling the time of onset of the migration of neural crest cells in the fowl embryo. Cell Tiss Res 224: 145.
62. Tosney KW (1978). The early migration of neural crest cells in the trunk region of the avian embryo. An electron microscopic study. Dev Biol 62: 317.
63. Tosney KW (1982). The segregation and early migration of cranial neural crest cells in the avian embryo. Dev Biol 89: 13.
64. Duband JL Thiery JP (1982). Distribution of fibronectin in the early phase of avian cephalic neural crest cell migration. Dev Biol 93: 308.
65. Sugrue SP, Hay ED (1981). Response of basal epithelial cell surface and cytoskeleton to solubilized extracellular matrix molecules. J Cell Biol 91: 45.
66. Greenberg G, Hay ED (1982). Epithelia suspended in collagen gels can loose polarity and express characteristics of migrating mesenchymal cells. J Cell Biol 95: 333.
67. Karfunkel P (1974). The mechanism of neural tube formation. Intern. Rev Cytol 38: 245.
68. Valinsky JE, Le Douarin NM (1985). Production of plasminogen activator by migrating neural crest cells. EMBO J in press.
69. Valinsky JE, Reich E (1981). Plasminogen in the chick embryo. Transport and biosynthesis. J Biol Chem 256: 12470.
70. Cochard P, Coltey P, (1983). Cholinergic traits in the neural crest : acetylcholinesterase in crest cells of the chick embryo. Dev Biol 98: 221.
71. Vincent M, Duband JL, Thiery JP (1983). A cell determinant expressed early on migrating avian neural crest cells. Dev Brain Res 9: 235.
72. Vincent M, Thiery JP (1984). A cell surface marker for neural crest and placodal cells : further evolution in peripheral and central nervous system. Dev Biol 103: 468.
73. Tucker GC, Aoyama H, Lipinski M, Tursz T and Thiery JP (1984). Identical reactivity of monoclonal antibodies HNK-1 and NC-1: conservation in vertebrates on cells derived from the neural primordium and on some leukocytes. Cell Diff 14: 223.
74. Thiery JP, Duband JL, Delouvée A (1982). Pathways and mechanism of avian trunk neural crest cell migration and localization. Dev Biol 93 324.

75. Anderson CB, Meier S (1981). The influence of the metameric pattern in the mesoderm on migration of cranial neural crest cells in the chick embryo. Dev Biol 85: 385.
76. Derby MA (1978). Analysis of glycosaminoglycans within the extracellular environments encountered by migrating neural crest cells. Dev Biol 66: 321.
77. Duband JL, Thiery JP (1985). Distribution of type I and III collagenes during chick neural crest cell migration. Cell Diff submitted.
78. Spieth J, Keller RE (1984). Neural crest cell behavior in white and dark larvae of Ambystoma mexicanum : Differences in cell morphology, arrangement and extracellular matrix as related to migration. J exp Zool 229: 91.
79. Erickson CA, Turley EA (1983). Substrata formed by combinations of extracellular matrix components alter neural crest cell motility in vitro. J Cell Sci 61: 299.
80. Newgreen DF, Gibbins IL, Sauter J, Wallenfels B, Wütz R (1982). Ultrastructural and tissue-culture studies on the role of fibronectin, collagen and glycosaminoglycans in the migration of neural crest cells in the fowl embryo. Cell Tissue Res 221: 521.
81. Rovasio RA, Delouvée A, Yamada KM, Timpl R, Thiery JP (1983). Neural crest cell migration : Requirement for exogenous fibronectin and high cell density. J Cell Biol 96: 462.
82. Tucker RP, Erickson CA (1984). Morphology and behavior of quail neural crest cells in artificial three dimensional extracellular matrices. Dev Biol 104: 390.
83. Erickson CA, Tosney KW, Weston JA (1980). Analysis of migratory behavior of neural crest and fibroblastic cells in embryonic tissues. Dev Biol 77: 142.
84. Erickson CA (1985). Control of neural crest cell dispersion in the trunk of the avian embryo. Dev Biol in press.
85. Tucker RP, Edwards BF, Erickson CA (1985). Tension in the culture dish : microfilament organization and migratory behavior of quail neural crest cells. Submitted.
86. Newgreen D, Thiery JP (1980). fibronectin in early avian embryos: Synthesis and distribution along the migration pathways of neural crest cells. Cell and Tiss Res 211: 269.

87. Greenberg JH, Pratt, RM (1977). Glycosaminoglycan and glycoprotein synthesis by cranial neural crest cells in vitro. Cell Diff 6: 119.
88. Couchman JR, Rees DA, Green MR, Smith CG (1982). fibronectin has a dual role in locomotion and anchorage of primary chick fibroblasts and can promote entry into the division cycle. J Cell Biol 93: 402.
89. Weston JA (1970). The migration and differentiation of neural crest cells. Adv Morphogen 8: 41.
90. Greenberg JH, Seppä S, Seppä H, Tyl Hewitt A (1980). Role of collagen and fibronectin in neural crest cell adhesion and migration. Dev Biol 87: 259.
91. Twitty VC, Niu NC (1948). Causal analysis of chromatophore migration. J Exp Zool 108: 405.
92. Twitty VC, Niu NC (1954). The motivation of cell migration studied by isolation of embryonic pigment cells singly and in small groups in vitro. J Exp Zool 125: 541.
93. Nuccitelli R, Erickson CA, (1983). Embryonic cell motility can be guided by physiological electric fields. Exp Cell Res 147: 195.
94. Löfberg J, Ahlfors K, Fällstrom C (1980). Neural crest cell migration in relation to extracellular matrix organization in the embryonic axolotl trunk. Dev Biol 75: 148.
95. Weston JA (1982). Mobile and social behavior of neural crest cells in "Cell Behaviour" Bellairs, R. Curtis, A. and Dunn, G.A. p 429 Cambridge University Press.
96. Erickson CA, Olivier, KR (1983). Negative chemotaxis does not control quail neural crest cell dispersion. Dev Biol 96: 542.
97. Newgreen DF, Ritterman M, Peters E.A, (1979). Morphology and behaviour of neural crest cells of chick embryo in vitro. Cell Tissue Res 203: 115.
98. Weiss P, Andres G (1952). Experiment on the fate of embryonic cells (chick) disseminated by the vascular route. J Exp Zool 121: 449.
99. Tucker G, Thiery JP (1985). Avian neural crest cell migration in the developping gut. Submitted.
100. Aoyama H, Delouvée A, Thiery JP (1985). Cell adhesion mechanisms in gangliogenesis studied in avian embryo and in a model system. Cell Diff in press.

THE MOLECULAR BASES AND DYNAMICS OF CELL ADHESION IN EMBRYOGENESIS[1]

Gerald M. Edelman, Stanley Hoffman,
Cheng-Ming Chuong, and Bruce A. Cunningham

Laboratory of Developmental and Molecular Biology
The Rockefeller University
1230 York Avenue
New York, New York 10021

ABSTRACT   In this paper, we review the structure, molecular mechanism, and distribution during development of three cell adhesion molecules (CAMs). N-CAM, the neural cell adhesion molecule, contains two major polypeptide components that have similar structures and appear to be synthesized from distinct mRNAs differentially spliced from a single gene. Both of these components contain a variety of post-translational modifications. One of these modifications, polysialylation, is developmentally controlled with the highest levels of sialic acid found in embryonic tissue and much lower levels in adult tissue. N-CAM binds homophilically in Ca^{++}-independent fashion; the efficacy of this adhesion is strongly modulated by the local concentration of N-CAM and is inversely related to the sialic acid content of the molecule. L-CAM, the liver cell adhesion molecule, contains a single polypeptide chain that is synthesized as a larger precursor and is encoded by a single mRNA species. Like N-CAM, L-CAM contains multiple N-linked oligosaccharides and can be phosphorylated on serine and threonine residues. In contrast to N-CAM, L-CAM-mediated adhesion is Ca^{++}-dependent. Ng-CAM, the neuron-glia cell adhesion molecule, contains three components; the smaller two of these components are antigenically related to the largest but not to each other. Ng-CAM is found on

---

[1] This work was supported by USPHS Grants HD-09635, HD-16550, and AM-04256.

neurons but not glia indicating that Ng-CAM mediated adhesion is heterophilic.

The functions of N-CAM, L-CAM, and Ng-CAM are expressed differentially in space and time and they all have characteristic cellular distributions during development. N-CAM and L-CAM are present from the earliest stages of chicken morphogenesis; one or both is present on the rudiments of almost all organs. During primary induction and secondary inductions (e.g. feather morphogenesis), borders between N-CAM bearing and L-CAM bearing cell collectives are formed and are altered in highly dynamic patterns. The fact that these borders form before overt cell differentiation suggests that these borders are necessary for induction. In contrast to N-CAM and L-CAM, Ng-CAM first appears only after 4 days of embryonic development and only in neural tissues. Levels of Ng-CAM change dramatically during development with the greatest expression in regions of active cell and neurite movements.

The accumulated data suggest that the expression of CAMs and CAM genes in a defined temporal schedule reflecting the occurrence of local cell surface modulation with alterations in the prevalence, cellular distribution, and structure of CAMs is a major mode of regulation of morphogenetic events.

INTRODUCTION

How can the one-dimensional genetic code specify the development of the three-dimensional shape of animals? This unanswered question points up morphogenesis as one of the great unsolved riddles in modern biology. During regulative development, morphogenetic movements bring the appropriate cells together in a complex, continuous set of transactions involving both cell contact and molecular signals (1). The interactions of these cells at particular times and positions lead to milieu-dependent differentiations or embryonic inductions and eventually to the establishment of form and of cytodifferentiation within particular organs.

Each of the primary processes of development--cell division, cell movement, cell adhesion, inductive cell differentiation, and cell death--is the result of many molecular events, which are regulated in parallel not only by interacting genes or gene products but also by concomitant epigenetic events. One of the key tasks in understanding these

processes is to identify proteins and their corresponding
genes that are directly involved in morphogenesis. In our
laboratory, we have focused our attention on the primary
process of cell adhesion. Cell adhesion must take place at
every stage of development to establish form, and it is the
process most susceptible to direct chemical assault. Moreover, it is the process most obviously related to the maintenance of final form.

Our working hypothesis is that cell patterning may be mediated by a dynamic mechanism that is parsimonious with
respect to the number of cell surface gene products required
to establish cell patterns (2,3). In this hypothesis, a relatively small number of different specific cell surface molecules are assumed to be responsible for cell adhesion and can
generate intricate morphogenetic patterns, provided that they
undergo certain changes in expression or in chemical properties during development. These changes, collectively termed
local cell surface modulation (4), consist of modifications in
the number, distribution, or chemical structure of each particular CAM. By changing the binding behavior of CAMs at the
cell surface, the various modulation events would directly or
indirectly alter the dynamics and interactions of the other
primary processes of development. Alterations of these
processes would in turn result in a change in form (Figure 1).
The effects of modulation upon selective cell-cell patterning
would thus depend upon three variables: 1) the number of CAMs

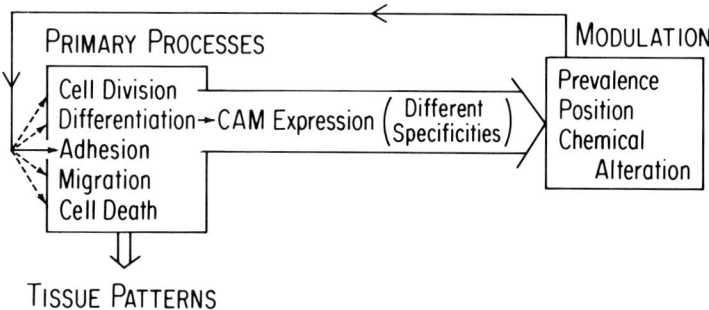

Figure 1. Regulatory effects of CAM expression and modulation. Although the major effect is on the adhesion process,
CAM expression and modulation may alter the sequence and extent of each of the other primary processes, either indirectly
or (as suggested by the dotted arrows) directly. These effects may lead, in turn, to different tissue patterns.

of different specificities; 2) the particular binding mechanisms of these different CAMs; and 3) the change of both of these variables with time (prevalence modulation) or cellular position (polarity modulation) during embryonic development. In the remainder of this paper, we describe data on the structure, function, and distribution of CAMs that support this hypothesis.

STRUCTURE OF CAMS

Using an immunological approach (5), we have isolated three independent cell adhesion molecules (CAMs) (6,7,8): N-CAM (neural CAM), L-CAM (liver CAM), and Ng-CAM (neuron-glia CAM). In each case, polyspecific antisera were prepared that inhibited adhesion between the appropriate cells in a short-term assay. The CAMs were isolated as the specific purified proteins that, when preincubated with the antibodies, neutralized their ability to inhibit adhesion. Monoclonal antibodies were prepared against the purified CAMs and when immobilized on solid supports, used to purify milligram quantities of CAMs for structural analyses.

N-CAM isolated from chicken embryo brains migrates on SDS gel electrophoresis in a microheterogeneous zone of apparent Mr 180-250,000 (Fig. 2). This behavior appears to be due to the unusually high content of sialic acid in embryonic N-CAM (30g/100g of amino acids); much of this carbohydrate is present in the form of polysialic acid (9,10). Following the removal of sialic acid, embryonic N-CAM migrates as two discrete components (Mr=170,000 and 140,000). Similarly, adult N-CAM (which contains only 1/3 as much sialic acid as embryonic N-CAM) migrates primarily as two discrete components in the chicken (11); an additional minor band of 120,000 molecular weight is found in the mouse (12,13). Inasmuch as the amino acid compositions, known sequences, and peptide maps of the embryonic (E) and various adult (A) forms are similar or identical (9,11,14), these observations indicate that a major change in the carbohydrate structure occurs as a function of development. Recent experiments suggest that the A-form of N-CAM is synthesized de novo rather than being produced by the removal of sialic acid from pre-existing embryonic N-CAM (15).

N-CAM can be prepared free of N-linked carbohydrate by synthesis in the presence of tunicamycin (14) or by digestion with endoglycosidase F (16). Both treatments result in N-CAM polypeptides of Mr 160,000 and 130,000. The observation that N-CAM synthesized in the presence of tunicamycin never reaches

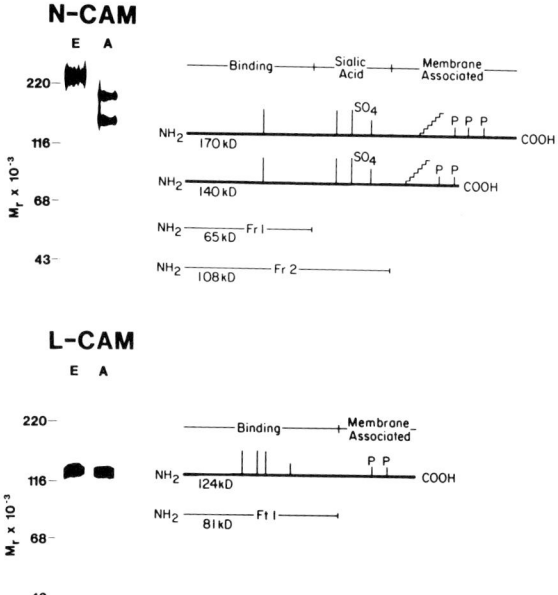

Figure 2. Summary of the chemical features of N-CAM and L-CAM. On SDS gel electrophoresis, N-CAM isolated from 14-day chicken embryo brains (E) migrates as a broad zone of Mr 200,000 to 250,000; N-CAM from adult brains (A) migrates as discrete components of Mr 150,000 and 180,000. In contrast, L-CAM isolated from adult chickens migrates at the same position as L-CAM isolated from embryos. Linear models of N-CAM and L-CAM were deduced from studies of the intact molecules and purified fragments: The amino-terminal region of N-CAM contains a specific binding domain and one oligosaccharide group (vertical line); the middle region contains essentially all of the sialic acid in N-CAM on three oligosaccharides, at least one of which is sulfated (SO_4); the carboxyl-terminal region of N-CAM is associated with the plasma membrane, contains bound fatty acid (staircase), and is phosphorylated (P) at different sites and to different extents in the two components. The amino-terminal region of L-CAM contains a specific binding domain, four oligosaccharide groups (three complex type and one high mannose), and can be isolated intact by trypsinization in the presence of Ca^{++}. The carboxyl-terminal region of L-CAM is associated with the plasma membrane and is phosphorylated.

the cell surface indicates that the smaller component of N-CAM is generated intracellularly and is unlikely to be merely a degradation product of the larger component.

Structural and functional analyses of the two components of N-CAM and two purified fragments of N-CAM have allowed the construction of the linear model shown in Fig. 2. Both components of N-CAM contain an amino-terminal region involved in cell-cell adhesion (14), a middle region in which the sialic acid is localized (14), and a carboxyl-terminal region involved in anchoring the molecule to the plasma membrane. Amino acid sequence analyses and peptide mapping experiments indicate that in the amino-terminal and middle regions the two components are identical or at least very similar in structure. All bound oligosaccharides appear to be in these regions; one attachment site has been detected in the "binding" region (unpublished observations) and three attachment sites (17) have been determined to be present in the "sialic acid-rich" region. At least one of these oligosaccharides contains covalently-bound sulfate (16). The adult (A) form of N-CAM also contains three attachment sites in the "sialic-acid rich" region (17) suggesting that the developmental decrease in sialic acid content results from the presence of shorter chains of polysialic acid on otherwise similar attachment sites.

Structural differences have been detected between the two components of N-CAM in their carboxyl-terminal region. The larger component is phosphorylated to a greater extent than the smaller and this post-translational modification is found only in the carboxyl-terminal region of the molecule (16). While both components contain phosphoserine and phosphothreonine (but no detectable phosphotyrosine), the phosphoserine/phosphothreonine ratio is four times higher in the larger component than in the smaller component. Furthermore, when ^{32}P-labelled tryptic phosphopeptides are prepared from the components of N-CAM and resolved by two-dimensional thin layer chromatography and electrophoresis, phosphopeptides that are unique to each component are detected in addition to other phosphopeptides that may be common to the two components. Both components of N-CAM also contain bound fatty acid within their carboxyl-terminal region; however, the precise location of this modification remains to be determined.

In order to extend our analysis of N-CAM structure and expression, we have prepared cDNA probes that hybridize to N-CAM mRNA (18). The behavior of probe pEC001 on Northern and Southern blot analysis is shown in Figure 3A, B. The probe hybridized to two large discrete poly (A)$^+$ RNA species from

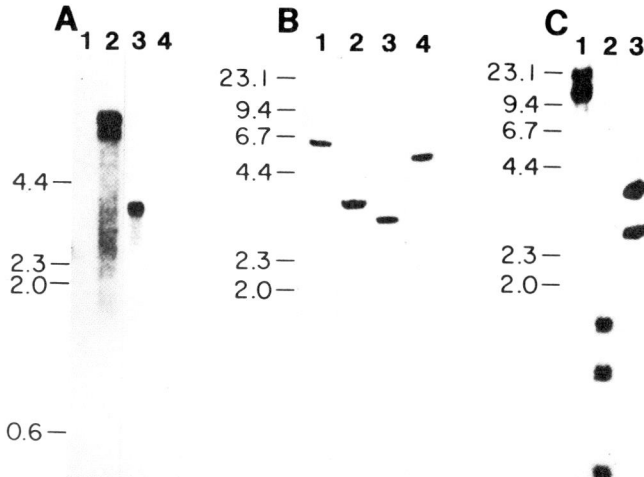

Figure 3. Hybridization analyses of mRNA and genomic DNA. N-CAM and L-CAM cDNA probes were ^{32}P-labeled by nick translation, were hybridized to embryonic chicken liver and brain mRNA which had been resolved on agarose-formaldehyde gels (panel A) or to chicken liver DNA which had been digested with restriction enzymes and resolved on agarose gels (panels B and C), and were detected by autoradiography. Positions of migration of Hind III fragments of lambda DNA of indicated size (in kilobases) are indicated to the left of each panel. (A) Probes: N-CAM (lanes 1 and 2), L-CAM (lanes 3 and 4). mRNA: liver (lanes 1 and 3), brain (lanes 2 and 4). (B) Probe: N-CAM. Restriction enzymes: EcoRI (lane 1), Pst I (lane 2), Sst I (lane 3), EcoRV (lane 4). (C) Probe: L-CAM. Restriction enzymes: EcoRI (lane 1), Pst I (lane 2), Sst I (lane 3).

brain; variable amounts of polydisperse, faster migrating material, assumed to be degradation products of the larger species were also seen. No components were detected in liver RNA, consistent with earlier observations that N-CAM is not detected in liver. The size of the two RNA species detected (6-7 kilobases) is sufficient to code for the two polypeptide chains of N-CAM indicating that these chains are probably derived from separate mRNAs.

To estimate the number of N-CAM genes, adult chicken liver DNA was digested with four restriction enzymes that did not cleave the pEC001 insert. After electrophoretic separation, the digests were tested for the presence of components homologous to the pEC001 insert. Each digest gave only one hybridizing fragment (Fig. 3B). These results suggest that there may be only one N-CAM gene and that the two mRNAs might arise by differential splicing. Such differential splicing of a single gene has been described recently for fibronectin (19).

Our understanding of the structure of L-CAM is summarized in the linear model shown in Figure 2. Like N-CAM, L-CAM is a large glycoprotein with an amino-terminal region that is external to the cell and probably includes the region involved in cell-cell adhesion. Its carboxyl-terminal region is also associated with the plasma membrane and is phosphorylated on some serine and threonine residues (20,21). In other respects, the two CAMs differ. L-CAM consists of a single $Mr=124,000$ polypeptide, the structure of which does not appear to change during development. The structural integrity of the L-CAM polypeptide chain and its function in cell adhesion require Ca^{++}. Digestion of L-CAM with trypsin in the presence of Ca^{++} generates an amino-terminal 81Kd fragment (called Ft1, Fig. 2) in high yield (20). In the absence of Ca^{++}, only smaller polypeptides are generated by trypsin. Uvomorulin, cadherin, and cell-CAM 120/80 (22,23,24) are cell adhesion molecules of mammalian origin that are the same size as L-CAM and have similar Ca^{++} dependencies for function and structural integrity, strongly suggesting that they are equivalents of L-CAM.

L-CAM lacks the unusual amounts and polymeric form of sialic acid that are found in N-CAM, but it is nevertheless glycosylated at multiple sites. Endoglycosidase digestion experiments suggest that both L-CAM and Ft1 contain one high mannose and three complex oligosaccharides. L-CAM which is deglycosylated enzymatically or by synthesis in the presence of tunicamycin has an $Mr=110,000$ (21). A high molecular weight precursor of L-CAM ($Mr=132,000$) has been detected following short pulse labelling of cultured liver cells and it appears to differ from mature L-CAM in both its polypeptide chain length and glycosylation.

We have also succeeded in preparing an L-CAM cDNA probe (pEC301) (25). This probe hybridized to a 4 kilobase species found in chicken liver poly $(A)^+$ mRNA but recognized no component in embryonic brain (Fig. 3A), in accord with earlier observations that L-CAM concentrations are high in liver and undetectable in brain. The mRNA recognized is large enough to code for the L-CAM polypeptide and the fact that only a single mRNA species was detected is consistent with our observations that L-CAM is synthesized as a single polypeptide.

On the other hand, attempts to define the number of L-CAM genes indicated that more than one stretch of DNA hybridizes to the L-CAM probe; nonetheless, the number of possible genes does not exceed three. Pooled DNA from 14-day embryonic chicken livers was digested with restriction enzymes known not to cleave the pEC301 insert. The products were separated on agarose gels, transferred to nitrocellulose, and hybridized with the pEC301 insert (Figure 3C). Three bands of approximately equal intensity were detected in the PstI digest, while two bands, one more intense than the other, were detected in both the EcoRI and SstI digests. The data indicate, therefore, that there may be at least two copies of the L-CAM gene. Because the DNA used for Southern blot analyses was obtained from multiple, outbred animals, the multiplicity could be due to allelic genes; the possibility of two closely related functioning genes or of pseudogenes must also be considered.

Having determined the basic structures of these two primary CAMs, we set out to isolate CAMs uniquely involved in histogenesis, or secondary CAMs. One of these is Ng-CAM. N-CAM and L-CAM were originally identified using assays in which the aggregation of a single cell suspension was monitored. Because Ng-CAM mediates the adhesion between two distinct classes of cells, neurons and glia, more complex assays were necessary. In one type of neuron-glia binding assay, the Ca^{++}-independent adhesion of neurons in suspension to glial cells in a monolayer was quantitated (26). A more easily manipulable neuron-glia binding assay was used in the identification of Ng-CAM. In this assay (8), the binding of neuronal membrane vesicles to glial cells in suspension was monitored and Ng-CAM was purified using the immunological approach described above.

Ng-CAM consists of three components (27), a major component of Mr=135,000 and two minor components (Mr=200,000 and 80,000) (Fig. 4). Immunological analyses of the components of Ng-CAM have demonstrated that the Mr=135,000 and 80,000 components are each antigenically related to the Mr=200,000 component, but that nevertheless, the Mr=135,000 and 80,000 are not antigenically related to each other (28). These results suggest that the two smaller components of Ng-CAM are post-translationally derived from the largest component, but there is, as yet, no direct evidence in favor of this conclusion. Consistent with the idea that there is a region of the molecule common to the Mr=200,000 and 80,000 components but absent from the Mr=135,000 component, these two minor components of Ng-CAM are phosphorylated while the Mr=135,000 component is not (Fig. 4).

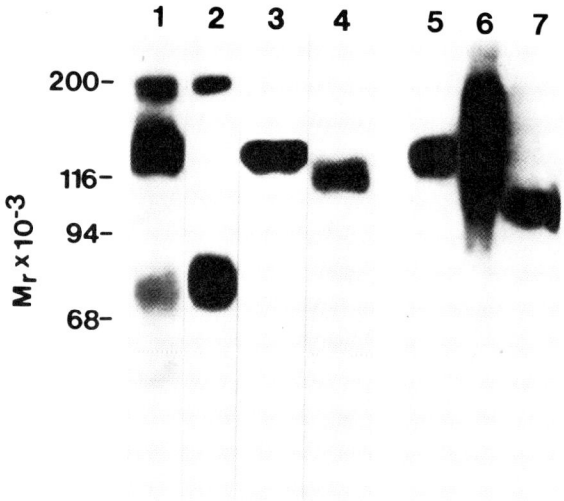

Figure 4. Phosphorylation and glycosylation of Ng-CAM. Ng-CAM was purified, resolved on SDS gels, and detected by Coomassie blue staining (lane 1), by autoradiography of material intrinsically labeled with $^{32}PO_4$ (lane 2), or by immunoblotting (lanes 3-5). The Ng-CAM in lanes 3 and 4 was preincubated with no enzyme (lane 3) or endoglycosidase F (lane 4) prior to electrophoresis and was detected using a monoclonal antibody specific for a polypeptide antigen in Ng-CAM. In lanes 5-7, purified Ng-CAM (lane 5), purified N-CAM (lane 6), and an SDS extract of adult chicken myelin (lane 7) were resolved on SDS gels and immunoblotted using monoclonal antibody anti-N-CAM no. 5. This antibody recognizes a carbohydrate antigen shared by Ng-CAM, N-CAM, and the myelin-associated glycoprotein (additional components in the myelin sample were detected by Coomassie blue staining).

Ng-CAM is highly glycosylated as evidenced by the fact that endoglycosidase F treatment reduces the molecular weight of the $M_r=135,000$ component to 115,000 (Fig. 4). Although Ng-CAM oligosaccharides do not appear to contain large amounts of polysialic acid nor do they change in structure during development as do N-CAM oligosaccharides, Ng-CAM, nevertheless, shares a carbohydrate antigen with N-CAM and the myelin-associated glycoprotein (29) (Fig. 4) as detected by monoclonal anti-N-CAM No. 5. The exact structure of the common antigen is unknown, but it is clear that the antigen does not include sialic acid (27). It remains unknown whether N-CAM, Ng-CAM, and the myelin-associated glycoprotein share a polypeptide sequence that may be the site of addition of this oligosaccharide antigen.

MOLECULAR BINDING MECHANISMS

As suggested above, it is our working hypothesis that the necessary selectivity observed during embryological pattern formation may be obtained using a limited number of CAMs of unique specificity (at most, tens of CAMs), provided that the adhesive function of each CAM can be modulated. Potential modulatory mechanisms include chemical modification and differential expression both at the subcellular level (polarity modulation) and in amount at the cellular level (prevalence modulation). In this section, we will describe several examples of the modulation of CAM function and demonstrate that CAM modulation can alter the efficacy of adhesion.

Several lines of evidence suggest that N-CAM-mediated adhesion is homophilic, i.e., the adhesive receptor for an N-CAM molecule on one cell is an N-CAM molecule on an apposing cell. Adhesion inhibitable by anti-N-CAM occurs between pairs of cells bearing surface N-CAM (brain, retina, and muscle cells), but does not occur between these cells and cells lacking N-CAM (liver, fibroblast, and trypsinized retina cells). Furthermore, even when only one of two cells that would otherwise form an N-CAM bond is coated with anti-N-CAM Fab' fragments, adhesion is completely inhibited (30). In other words, the N-CAM molecule on the uncoated cells does not form adhesive bonds with any of the available cell surface molecules on the anti-N-CAM coated cells. The sialic acid present in the middle domain of the molecule does not participate directly in the binding of N-CAM to N-CAM (14,31) although, as described below, it plays an important modulatory role.

More direct evidence for the homophilic nature of N-CAM mediated adhesion was obtained when purified N-CAM was incorporated into artificial lipid vesicles. These vesicles exhibited the same binding properties as N-CAM-positive cells (30): they bound to N-CAM-positive cells but not to cells lacking N-CAM. Furthermore, these vesicles self-aggregated and this aggregation was inhibited by antibodies to N-CAM. Vesicles lacking N-CAM did not self-aggregate or bind to cells.

Table 1. Aggregation of reconstituted vesicles containing adult or embryonic forms of N-CAM

Vesicles (form of N-CAM)	Neuraminidase Treatment	N-CAM/lipid (μg/mg)	Relative Rate of Aggregation[a]
E		14	1.0
E		17	2.3
E		19	4.3
E		28	36.0
E		17	1.0
A		17	3.5
E	−	14	1.0
E	+	14	3.7
A	−	11	1.0
A	+	11	1.1

[a] In each experiment, the rate of aggregation of the slowest aggregating sample has been normalized to 1.0.

Using these artificial N-CAM vesicles (31), it was possible to evaluate the effect on rates of adhesion caused by alterations at the membrane surface of N-CAM concentration (as observed in vivo particularly during early embryogenesis) or by changes in N-CAM sialic acid content (which decreases greatly during the conversion of N-CAM from E to A forms). A 2-fold increase in N-CAM concentration within vesicles led to a greater than 30-fold increase in rate of vesicle aggregation (Table 1). Smaller but still significant changes in the rate of aggregation were also associated with changes in N-CAM sialic acid content. Vesicles containing the A-forms of N-CAM (with lower sialic acid) aggregated almost 4 times as fast as E-form N-CAM vesicles (Table 1). Neuraminidase treatment of vesicles containing the E-form of N-CAM gave a similar enhancement of aggregation. On the other hand, neuraminidase treatment of vesicles containing the A-form of N-CAM only marginally increased their rate of aggregation. Although the sialic acid in N-CAM clearly affects the rate of vesicle aggregation, it does not alter the specificity as evidenced by the fact that vesicles containing E-form N-CAM and vesicles containing A-form N-CAM co-aggregate (31).

A similar series of vesicle aggregation experiments was carried out using plasma membrane vesicles prepared from embryonic and adult chicken brains (31). These experiments also indicated that the rate of N-CAM-mediated adhesion is inversely related to the sialic acid content of the N-CAM molecules and is highly dependent on the N-CAM concentration within the vesicles. The fact that both plasma membrane vesicles and artificial N-CAM vesicles gave similar results indicates that the results obtained are not artifacts due to the use of artificial vesicles or the absence of proteins other than N-CAM from these vesicles.

The adhesive specificity of N-CAM appears to have been highly conserved during evolution. Using plasma membrane vesicles from frog, chicken, and mouse brains, it was found that cross-species N-CAM-mediated adhesion occurs as readily as intra-species adhesion (32). In control experiments, chicken brain and liver plasma membrane vesicles failed to co-aggregate.

The most striking feature of L-CAM-mediated adhesion is that it has an absolute requirement for Ca^{++} (7,23); this is reflected in the fact that liver cells do not aggregate in the absence of Ca^{++}. L-CAM-mediated adhesion appears to be homophilic because purified L-CAM can completely neutralize the ability of antibodies prepared against liver cells to inhibit liver cell adhesion. So far, however, no direct evidence that L-CAM is a homophilic ligand exists.

Ng-CAM is found on neurons and not on glial cells (27); inasmuch as this CAM mediates the heterotypic interaction of these cells, the molecular mechanism of Ng-CAM-mediated neuron-glia adhesion must be heterophilic (i.e., the receptor on glia for Ng-CAM is a different molecule). Antibodies to Ng-CAM also inhibit neuron-neuron adhesion among cerebellar cells but not retinal cells (28). It is possible, therefore, that Ng-CAM is also a ligand in neuronal adhesion. On the other hand, anti-Ng-CAM antibodies may be indirectly affecting neuron-neuron adhesion. For example, the observation that N-CAM and Ng-CAM are co-precipitated by monoclonal antibodies that directly recognize only Ng-CAM (27) is consistent with the possibility that Ng-CAM modulates neuronal adhesion by means of a <u>cis</u> interaction with N-CAM molecules on the same cell. This possibility is currently under investigation.

CAMS IN NEURAL DEVELOPMENT

How are the adhesive properties of the cell adhesion molecules used during embryogenesis and histogenesis to form an organ? We began to approach this problem by studying the expression of N-CAM, L-CAM, and Ng-CAM during embryonic development with immunofluorescent methods. N-CAM and L-CAM are defined as primary CAMs because they appear in the very early stage (blastoderm) of chick embryogenesis; they are present in more than one germ layer, and are expressed on mitotic cells (33). In contrast, Ng-CAM is defined as a secondary CAM because it appears later in neural histogenesis (4-day embryo); it is present only on neuroectoderm derivatives and is expressed only on postmitotic cells (34,35). Two systems, the nervous system and the feather, will be used to exemplify the expression and modulation of these CAMs.

The formation of the nervous system can be described in terms of two major stages: 1. During morphogenesis, induction and segregation of the neural plate occur with the formation of the neural axis, followed by neurulation or neural tube formation; 2. During histogenesis, detailed neural specializations occur, establishing the connectional structures upon which later functional activity depends (36,37). In morphogenesis, variation in the amounts of N-CAM and L-CAM reflect the development of the segregated borders of the neural plate (33,38,39). In histogenesis, Ng-CAM appears, whereas L-CAM disappears (34,35). N-CAM persists through histogenesis and in the adult nervous system but goes through E-A conversion (an example of chemical modulation) particularly in the perinatal period (11,40).

N-CAM and L-CAM are coexpressed in the blastoderm cells (Fig. 5A,B and ref. 33) by stage 2 of Vakaet (41). At the tenth somite stage, when neurulation occurs, N-CAM becomes focused in the neural plate regions and is gradually lost in the lateral ectodermal regions (Fig. 5D and ref. 33). Later, it is reexpressed in localized regions of lateral ectoderm such as lens and pharyngeal placodes (38). In contrast, L-CAM is lost in the neural plate regions but remains on the lateral ectoderm (Fig. 5E and refs. 33,39).

The neural plate invaginates and wraps up to form the neural tube consisting of neuroepithelial cells. These precursors of both neuronal and glial cells are N-CAM positive but L-CAM and Ng-CAM negative. Ng-CAM begins to appear in the ventral region of the neural tube at the 31 somite stage (34). At this time, Ng-CAM is confined to the developing neurites that are to become the white matter of the spinal cord,

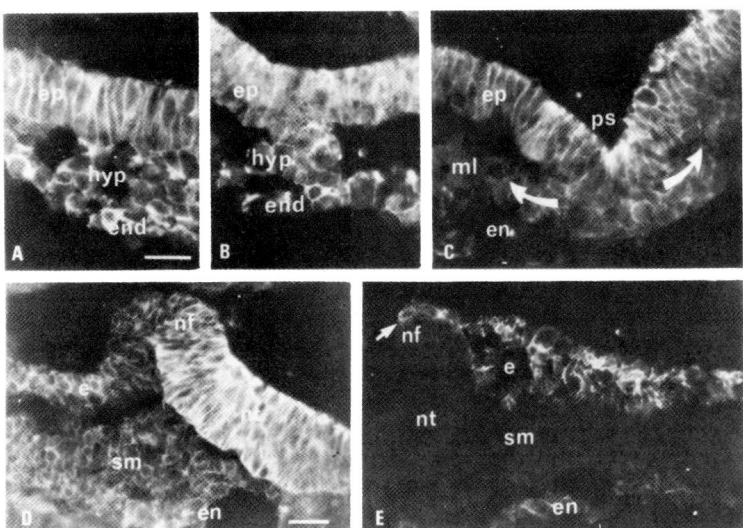

Figure 5. Early appearance of L-CAM and N-CAM during the formation of three primary germ layers. (A) Full primitive streak stage [stage 7 of Vakaet (6)]: laterad to the streak, aggregates of cells released from the epiblast (ep) progressively replace the cells of the endophyll (end). In addition to the epiblast, both the endophyll and the presumptive hypoblast (hyp) are stained with anti-N-CAM. (B) Similar stage, same region: cells from the same aggregates are also stained with anti-L-CAM. (C) Stage 9: head-fold primitive streak (ps) level. In the upper level, the epiblast is labeled by anti-L-CAM antibodies; middle layer cells (ml) that have just been released from the upper layer are also stained. Migrating cells (arrows) are not stained. en, Definitive endoderm. (D) Ten-somite stage; neurulation and ectoderm formation: slightly below the last-formed somite, N-CAM was found in most tissues but the staining intensity was increased dramatically in the neural tube. nf, Neural fold; nt, neural tube; e, ectoderm; sm, somitic mesenchyme; en, endoderm. (E) Same stage, same level; L-CAM is found in all ectodermal cells. The last cells stained (arrow) are in the neural fold; note that the neural tube and the somitic mesenchyme (sm) do not stain with anti-L-CAM and that the endoderm is weakly stained. (Bars = 30 μm).

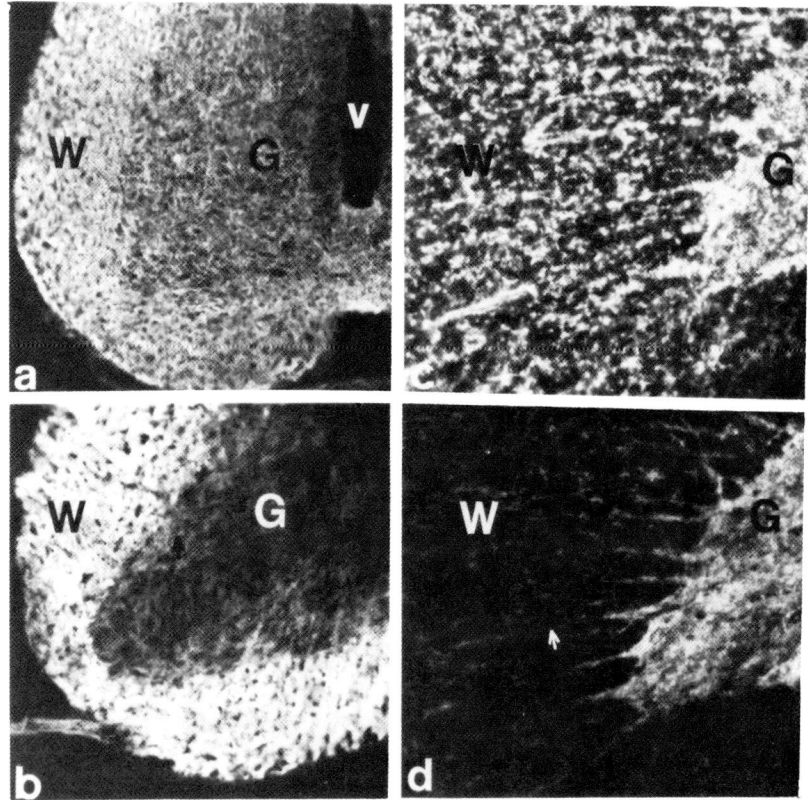

Figure 6. Distribution of Ng-CAM and N-CAM in embryonic and adult spinal cord. Cryostat sections of spinal cords from 8 day embryonic (a,b) and adult chicken (c,d) are stained with anti-N-CAM (a,c) or anti-Ng-CAM (b,d). In the embryo, N-CAM is on the presumptive white (W) and gray (G) matter, and the ventricular layer (V); Ng-CAM is highly enriched in the developing tracts within the presumptive white matter (W), is present in a lower level in the migrating cell bodies and growing nerve fibers (arrow) within the presumptive gray matter, but is totally negative in the ventricular layer. In the adult, N-CAM is present in both white (W) and gray (G) matter; the Ng-CAM staining pattern is reversed and is now enriched in the unmyelinated fibers within the gray matter but is absent in the myelinated fibers in the white matter. Arrows point to nerve fibers.

but is absent from neuronal cell bodies, reflecting a form of polarity modulation. As neuronal layers and clusters begin to form and send out neurites to connect with each other, Ng-CAM is expressed on growing neurites in various neural regions (Fig. 6 and ref. 35). However, when the neurite connections are stabilized and myelinated, Ng-CAM disappears from these myelinated tracts (Fig. 6 and ref. 35) suggesting that oligodendroglia might play a role in the down regulation of Ng-CAM. In the adult CNS, Ng-CAM remains only in unmyelinated tracts such as the olfactory tract and local circuits such as the cerebellar molecular layer (35).

Although the polarity modulation of Ng-CAM in the CNS is striking, Ng-CAM is also transiently present on neuronal cell bodies at the time of cell migration (28). This is seen in the presumptive gray matter of the spinal cord (35) and the granule cells of the cerebellum (28,35). In the latter case, external granule cells in the proliferative zone are positive for N-CAM but negative for Ng-CAM. These cells first express a large amount of Ng-CAM in the premigratory zone, and the level of Ng-CAM remains high as the cells migrate through the molecular layer. When the cells finally reach the internal granule layer, they lose Ng-CAM on their cell bodies. It is known that this migration involves radial glial fibers (42). The direct involvement of Ng-CAM in this neuron-glia interaction has been suggested further by the fact that anti-Ng-CAM Fab' fragments inhibit the movement of granule cells in cerebellar explants (Chuong & Edelman, unpublished observations).

In contrast, the distribution of N-CAM in the nervous system as judged by immunocytochemical studies is rather more uniform. N-CAM is present on all neuronal cell bodies and neurites (Fig. 6). Nonetheless, at different developmental times there are quantitative (prevalence modulation) and qualitative (chemical modulation) differences in N-CAMs from different neural regions (40). In each region of the developing mouse brain, the amount of N-CAM, as judged by ELISA, peaks in perinatal periods and falls to the adult level (about 1/3 of the peak value) by 3 weeks of age. E-A conversion of N-CAM occurs gradually between 1 to 3 weeks after birth. The rate of E-A conversion is different in various nervous tissues, which can be listed in order of decreasing rates of conversion as follows: retina, cerebellum, spinal cord, cerebral cortex, hippocampus, olfactory bulb, followed by diencephalon and tectum (40, see Fig. 7 as an example). The possible significance of these regional differences may be best understood in terms of vesicle aggregation experiments (31) indicating that a two-fold increase in CAM concentration can

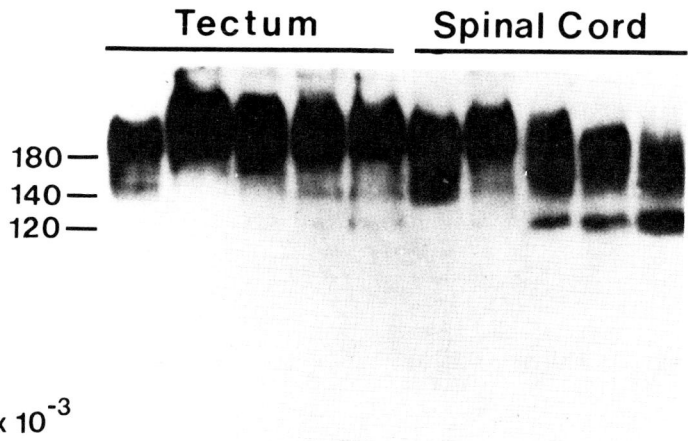

Figure 7. E-A conversion of N-CAM in tectum and spinal cord. Mouse brain lysates of different ages (arrows denote birth, embryonic day toward the left, postnatal days toward the right) were immunoblotted with anti-N-CAM. In the tectum, N-CAM appears as a diffuse zone from Mr 140-200K at E16 then appears as a diffuse zone of Mr 180-250K from P1 to P21 without much change. In contrast, in the spinal cord, N-CAM appears as the diffuse E-form at E16 and P1, but the A-form (discrete bands) gradually appears from P7 to P21, indicating that E-A conversion occurs faster in the spinal cord than the tectum.

result in a greater than 30-fold increase in aggregation rate, and that A-form N-CAM binds much faster than E-form N-CAM. These observations suggest that there is differential N-CAM-dependent adhesivity in the various brain regions during development. It is attractive to hypothesize that, during formation of neural connections, the adhesivity is relatively weak but sufficient to promote fasciculation and contact; later, at the time of functional stabilization, N-CAMs are converted to the highly adhesive A-form to fix the connections. The delay of E-A conversion seen in the <u>staggerer</u> mouse mutant, in which granule cells contact Purkinje cells but do not form mature synapses, is consistent with this hypothesis (12).

CAMS AND HISTOGENESIS: FEATHER AS AN EXAMPLE

In our survey of the presence of N-CAM in embryogenesis, we found that N-CAM is not only expressed in the nervous system, but is also expressed in a dynamic way at key sites of morphogenesis such as the lens placode, pharyngeal placode, limb bud, and somites (38). We choose feather histogenesis as an example to discuss in detail here because the system contains hundreds of induction sites (43) and the end product has a distinct branching pattern (44). This allows us to correlate the expression of CAMs at each stage of histogenesis and cytodifferentiation.

The stages of feather morphogenesis are generally classified as: feather placode formation, feather bud formation, feather follicle development, and feather filament formation (Fig. 8 and ref. 43). At the feather placode stage, induction occurs between dermal condensations and placode epithelia, and at the feather follicle stage they occur between the dermal papilla and papillar ectoderm. After each induction, the epithelial cells such as those in the collar undergo active cell proliferation to generate feather structures. From the collar toward the feather tip, the stratified epithelia gradually segregate into many ridges (barb ridges) that will form the presumptive rami and barbules (Fig. 8 and ref. 44). A layer of cells (basilar layer) between the barb ridges will form marginal plates and later disintegrate to ensure the separation of barbs. Another layer of cells (axial plate) appears at the midline of the barb ridge and will also disintegrate later to ensure the separation of barbules (Fig. 8 and ref. 44). While the marginal and axial plate cells die and leave behind empty spaces, the rest of the epithelial cells keratinize to form the permanent structure of the feather.

In the feather placode stage, N-CAM first appears (45) on a thin homogeneous layer of dermal cells. These cells then form a periodic array of dermal condensations which stain brightly for N-CAM. L-CAM is present on the peridermal cells,

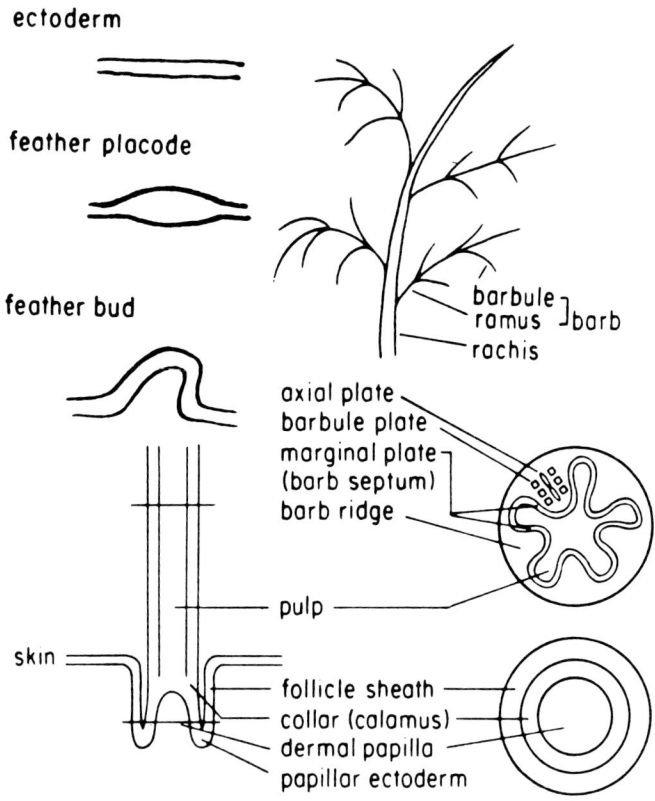

Figure 8. Schematic drawing of stages in feather histogenesis. Mesenchyme-epithelial induction first occurs in the feather placode, resulting in the growth of the feather bud. Induction occurs again between the dermal papilla and the papillar ectoderm, resulting in the proliferating collar that generates the feather filament. Cross sections of the base and tip of the feather filament are shown at the right. As the feather develops, the smooth stratified epithelial cylinder forms many barb ridges. The marginal and axial plates in between and within the barb ridges later degenerate to ensure the separation of the keratinized rami and barbules. The endproduct is a three level branched structure with barbules (tertiary branch) inserted on the rami (secondary branch), and rami inserted on the rachi (primary branch).

basal cells, and placode cells. In the placode, the staining is more on the apical and basal cell surfaces than on the lateral cell surfaces (an example of polarity modulation). In the later placode stage, placode cells express N-CAM transiently. In the feather bud stage, N-CAM is localized in the dorsal-basal regions of the buds (Fig. 9) which later will turn into the N-CAM-rich dermal papilla. The feather epithelia are L-CAM positive and after the induction of the dermal papilla, the collar epithelial cells also become transiently N-CAM positive. Therefore, the general scheme is that induction occurs between groups of N-CAM positive cells and groups of L-CAM positive cells (45). These "CAM couples" occur in many strategic points of the morphogenetic process (Table 2 and ref. 45,46).

During the formation of the feather filament, the stratified epithelia and their basilar layer are L-CAM positive (47). Later, the epithelia begin to form barb ridges, and cells within the ridges are organized into two rectangular rows of L-CAM positive barbule cells. N-CAM appears (47) in a single basilar cell in the valley between two ridges and N-CAM expression then propagates bilaterally cell by cell toward the tips of the ridges. As cells on both sides of the valley express N-CAM, the basilar layer "zips in" to become the marginal plate which is highly N-CAM positive (Fig. 9). N-CAM also appears periodically in the axial plates (rows of spindle-shaped cells present between any two rows of barbule cells (Fig. 9)). The cells in both the marginal and axial plates are epithelial in origin and are at first L-CAM positive and N-CAM negative. When they begin to express N-CAM, the level of L-CAM in these cells decreases and, instead of keratinizing, they die leaving a space. In a process similar to the "lost wax " method in sculpture, the epithelia thus complete the morphological transformation from a smooth cylinder into a branching pattern (47).

The most striking features of this process of feather histogenesis are the periodic appearance of CAM couples at each site of induction, and the correlation of specific cytodifferentiation events with cells carrying one or the other CAM. The regulation of CAM gene expression must be coordinated with that of other genes such as those of keratin in a quite strict fashion.

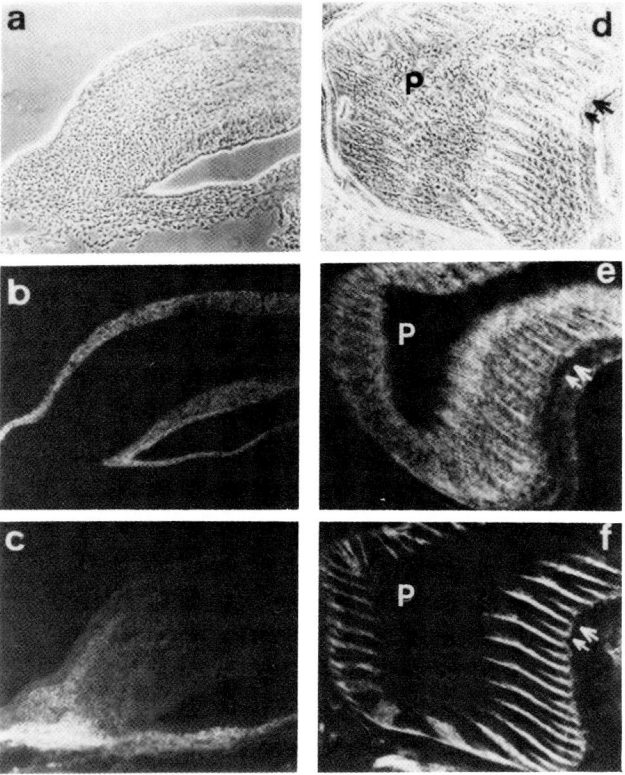

Figure 9. Distribution of L-CAM and N-CAM in the feather bud and the feather filament. Cryostat sections of skin from 12 day embryonic (a,b,c) and newborn chicken (d,e,f) are shown in phase (a,d) and stained with anti-L-CAM (b,e) or anti-N-CAM (c,f). In the longitudinal sections of the feather bud (a,b,c), L-CAM is present (b) on both the feather epithelium and the interplumar epithelium; N-CAM is localized (c) in the basal and dorsal regions of the feather bud, but is not on the feather epithelium, nor on the major inside portion of the bud that is to become the pulp. The stain in the dermis diminishes toward the interplumar region. In the cross section of the feather filament (d,e,f), L-CAM is present (e) on all the barb ridge epithelia; N-CAM appears (f) periodically and alternatingly in the marginal plate epithelia (big arrow) and axial plate epithelia (small arrows). These N-CAM positive cells will later die and leave spaces behind. P, pulp. x126.

Table 2.   CAM Couples

Tissue	L-CAM	N-CAM
neural plate	lateral ectoderm	neural plate and notochord
lens placode	placode ectoderm	placode ectoderm and optic vesicles
limb bud	apical ectodermal ridge and somatic ectoderm	apical ectodermal ridge and underlying mesenchyme
feather bud	placode ectoderm	placode ectoderm and dermal condensation
feather filament	barbule and ramus epithelia	marginal and axial plates
gut	endoderm	endoderm and underlying mesenchyme
kidney	Wolffian duct and late mesonephric tubules	Wolffian duct and early mesonephric tubules

CONCLUSIONS AND SUMMARY

In the above descriptions of the development of the nervous system and the feather, various modulations of CAM expression were observed that could lead to differential adhesivity on different portions of the cellular surface. In brain development, N-CAM is present from neural induction to the formation of mature neuronal networks but goes through a change in adhesivity (chemical modulation, E-A conversion). L-CAM disappears early after the neural induction; later, Ng-CAM appears and disappears in a very dynamic manner (prevalence modulation) at just the times of movement of cell bodies and extension of neurites. When present on neurites of non-moving cells, Ng-CAM is excluded from the cell body (polarity modulation).

During feather development, L-CAM is present from the earliest ectoderm to the final keratinization, and it shows polarity modulation in feather placodes and barbule cells. In contrast, N-CAM appears periodically in a very dynamic way (prevalence modulation) both at sites of induction and in morphogenetic formations of barbs and barbules.

Regulation of the modulation of these CAMs thus may provide the differential adhesivity needed in histogenesis. It is clear that cell adhesion is a fundamental process at the cellular level, but its significance in embryogenesis can differ in various histological milieu. In induction, it brings cell groups together for interaction. In detailed histogenetic processes such as those of the feather filament, it may couple cell adhesion with cell differentiation and cell death. In the nervous system, it is used for fasciculation and neural connections. Further work on the biochemical properties of CAMs and control of their gene expression, as well as additional exploration of their function in vivo, should lead to better understanding of the roles of cell adhesion in the determination of animal forms.

REFERENCES

1. Saxen L, Ekblom P, Thesleff I (1980). In Johnson MH (ed): "Mechanisms of Morphogenetic Cell Interactions, Development in Mammals," Amsterdam: Elsevier, p 161.
2. Edelman GM (1983). Cell adhesion molecules. Science 219:450.
3. Edelman GM (1984). Cell adhesion and morphogenesis: The regulator hypothesis. Proc Natl Acad Sci USA 81:1460.
4. Edelman GM (1976). Surface modulation in cell recognition and cell growth. Science 192:218.
5. Brackenbury R, Thiery J-P, Rutishauser U, Edelman GM (1977). Adhesion among neural cells of the chick embryo. I. An immunological assay for molecules involved in cell-cell binding. J Biol Chem 252:6835.
6. Thiery J-P, Brackenbury R, Rutishauser U, Edelman GM (1977). Adhesion among neural cells of the chick embryo. II. Purification and characterization of a cell adhesion molecule from neural retina. J Biol Chem 252:6841.
7. Bertolotti R, Rutishauser U, Edelman GM (1980). A cell surface molecule involved in aggregation of embryonic liver cells. Proc Natl Acad Sci USA 77:4831.
8. Grumet M, Edelman GM (1984). Heterotypic binding between neuronal membrane vesicles and glial cells is mediated by

a specific neuron-glial cell adhesion molecule. J Cell Biol 98:1746.
9. Hoffman S, Sorkin BC, White PC, Brackenbury R, Mailhammer R, Rutishauser U, Cunningham BA, Edelman GM (1982). Chemical characterization of a neural cell adhesion molecule purified from embryonic brain membranes. J Biol Chem 257:7720.
10. Finne J, Finne U, Deagostini-Bazin H, Goridis C (1983). Occurrence of α 2-8 linked polysialosyl units in a neural cell adhesion molecule. Biochem Biophys Res Commun 112:482.
11. Rothbard JB, Brackenbury R, Cunningham BA, Edelman GM (1982). Differences in the carbohydrate structures of neural cell-adhesion molecules from adult and embryonic chicken brains. J Biol Chem 257:11064.
12. Edelman GM, Chuong, C-M (1982). Embryonic to adult conversion of neural cell adhesion molecules in normal and staggerer mice. Proc Natl Acad Sci USA 79:7036.
13. Rougon G, Deagostini-Bazin H, Hirn M, Goridis C (1982). Tissue and developmental stage-specific forms of a neural cell surface antigen linked to differences in glycosylation of a common polypeptide. EMBO J 1:1239.
14. Cunningham BA, Hoffman S, Rutishauser U, Hemperly JJ, Edelman GM (1983). Molecular topography of N-CAM: Surface orientation and the location of sialic acid-rich and binding regions. Proc Natl Acad Sci USA 80:3116.
15. Friedlander DR, Brackenbury R, Edelman GM (1985). E to A conversion of N-CAM in vitro results from de novo synthesis of adult forms. J Cell Biol, in press.
16. Sorkin BC, Hoffman S, Edelman GM, Cunningham BA (1984). Sulfation and phosphorylation of the neural cell adhesion molecule N-CAM. Science 225:1476.
17. Crossin KL, Edelman GM, Cunningham BA (1984). Mapping of three carbohydrate attachment sites in embryonic and adult forms of the neural cell adhesion molecule (N-CAM). J Cell Biol 99:1848.
18. Murray BA, Hemperly JJ, Gallin WJ, MacGregor JS, Edelman GM, Cunningham BA (1984). Isolation of cDNA clones for the chicken neural cell adhesion molecule (N-CAM). Proc Natl Acad Sci USA 81:5584.
19. Kornblihtt AR, Vibe-Pedersen K, Baralle FE (1983). Isolation and characterization of cDNA clones for human and bovine fibronectin. Proc Natl Acad Sci USA 80:3218.
20. Gallin WJ, Edelman GM, Cunningham BA (1983). Characterization of L-CAM, a major cell adhesion molecule from embryonic liver cells. Proc Natl Acad Sci USA 80:1038.

21. Cunningham BA, Leutzinger Y, Gallin WJ, Sorkin BC, Edelman GM (1984). Linear organization of the liver cell adhesion molecule L-CAM. Proc Natl Acad Sci USA 81:5787.
22. Hyafil F, Morello D, Babinet C, Jacob F (1980). A cell surface glycoprotein involved in the compaction of embryonal carcinoma cells and cleavage stage embryos. Cell 21:927.
23. Yoshida C, Takeichi M (1982). Teratocarcinoma cell adhesion: Identification of a cell-surface protein involved in calcium-dependent cell aggregation. Cell 28:217.
24. Damsky CH, Richa J, Solter D, Knudsen K, Buck CA (1983). Identification and purification of a cell surface glycoprotein mediating intercellular adhesion in embryonic and adult tissues. Cell 34:455.
25. Gallin WJ, Prediger EA, Edelman GM, Cunningham BA (1985). Isolation of a cDNA clone for the liver cell adhesion molecule (L-CAM). Proc Natl Acad Sci USA 82:2809.
26. Grumet M, Rutishauser U, Edelman GM (1983). Neuron-glia adhesion is inhibited by antibodies to neural determinants. Science 222:60.
27. Grumet M, Hoffman S, Edelman GM (1984). Two antigenically related neuronal CAMs of different specificities mediate neuron-neuron and neuron-glia adhesion. Proc Natl Acad Sci USA 81:267.
28. Grumet M, Hoffman S, Chuong C-M, Edelman GM (1984). Polypeptide components and binding functions of neuron-glia cell adhesion molecules. Proc Natl Acad Sci USA 81:7989.
29. Quarles RH, Everly JL, Brady RO (1972). Demonstration of a glycoprotein which is associated with a purified myelin fraction from rat brain. Biochem Biophys Res Commun 47:491.
30. Rutishauser U, Hoffman S, Edelman GM (1982). Binding properties of a cell adhesion molecule from neural tissue. Proc Natl Acad Sci USA 79:685.
31. Hoffman S, Edelman GM (1983). Kinetics of homophilic binding by E and A forms of the neural cell adhesion molecule. Proc Natl Acad Sci USA 80:5762.
32. Hoffman S, Chuong C-M, Edelman GM (1984). Evolutionary conservation of key structures and binding functions of neural cell adhesion molecules. Proc Natl Acad Sci USA 81:6881.
33. Edelman GM, Gallin WJ, Delouvée A, Cunningham BA, Thiery J-P (1983). Early epochal maps of two different cell adhesion molecules. Proc Natl Acad Sci USA 80:4384.
34. Thiery J-P, Delouvée A, Grumet M, Edelman GM (1985).

Initial appearance and regional distribution of the neuron-glia cell adhesion molecule (Ng-CAM) in the chick embryo. J Cell Biol 100:442.
35. Daniloff JK, Chuong C-M, Levi G, Edelman GM (1985). Differential distribution of cell adhesion molecules during histogenesis of the chick nervous system. Submitted for publication.
36. Rakic P (1981). Developmental events leading to laminar and areal organization of the neocortex. In Schmitt FO, Worden FG, Adelman G, Dennis SG (eds): "The Organization of the Cerebral Cortex," Cambridge: MIT Press, p 592.
37. Fraser SE, Hunt RK (1980). Retinotectal specificity: Models and experiments in search of a mapping function. Ann Rev Neurosci 3:319.
38. Thiery J-P, Duband J-L, Rutishauser U, Edelman GM (1982). Cell adhesion molecules in early chick embryogenesis. Proc Natl Acad Sci USA 79:6737.
39. Thiery J-P, Delouvée A, Gallin WJ, Cunningham BA, Edelman GM (1984). Ontogenetic expression of cell adhesion molecules: L-CAM is found in epithelia derived from the three primary germ layers. Dev Biol 102:61.
40. Chuong C-M, Edelman GM (1984). Alterations in neural cell adhesion molecules during development of different regions of the nervous system. J Neurosci 4:2354.
41. Vakaet L (1962). Some new data concerning the formation of the definitive endoblast in the chick embryo. J. Embryol Exp Morph 10:38.
42. Rakic P (1972). Neuron-glia relationship during granule cell migration in developing cerebellar cortex. A Golgi and electronmicroscopic study in Macacus rhesus. J Comp Neurol 141:283.
43. Sengel P (1976). "Morphogenesis of Skin." New York: Cambridge University Press.
44. Lucas AM, Stettenheim PR (1972). Avian anatomy. Integument. Part I and Part II. In: "Agriculture Handbook. 362," Washington, D.C.: Agricultural Research Service, US Dept of Agriculture.
45. Chuong C-M, Edelman GM (1985). Expression of cell adhesion molecules in embryonic induction: I. Morphogenesis of nestling feathers. J Cell Biol, in press.
46. Crossin KL, Chuong C-M, Edelman (1985). Expression sequences of cell adhesion molecules. Proc Natl Acad Sci USA, in press.
47. Chuong C-M, Edelman GM (1985). Expression of cell adhesion molecules in embryonic induction: II. Morphogenesis of adult feathers. J Cell Biol, in press.

SELECTIVE CELL ADHESION MECHANISM: ROLE OF THE CALCIUM-DEPENDENT CELL ADHESION SYSTEM[1]

Masatoshi Takeichi, Kohei Hatta, and Akira Nagafuchi

Department of Biophysics, Faculty of Science, Kyoto University, Sakyo-ku, Kyoto 606, Japan

ABSTRACT Animal cells show preferential adhesiveness to their own type. In elucidating molecular mechanisms of the selective cell adhesion, we studied cell-type-specific property of the Ca^{2+}-dependent cell-cell adhesion system (CDS). We obtained two kinds of monoclonal antibodies which block the function of CDS, ECCD-1 and NCD-1. These antibodies recognize CDS present in distinct cell types: ECCD-1 reacts with epithelial cells and NCD-1 reacts with cells of nervous tissue, muscle cells and lens cells. These results suggested that CDS can be divided into subclasses of different immunological specificity. Aggregation experiments showed that, when heterotypic cells with different subclasses of CDS were mixed, each cell type tended to aggregate separately. All these results suggested that CDS is the machinery responsible for selective adhesion of animal cells.

INTRODUCTION

Selective adhesiveness of animal cells has frequently been demonstrated by in vitro cell aggregation experiments. Cells of animal tissues tend to preferentially adhere to their own type when artificially mixed with different cell types. Such property in the adhesiveness of cells is thought to be essential for histogenetic processes in

---

[1]This work was supported by research grants from the Ministry of Education, Science and Culture of Japan.

animal development, such as the segregation of tissues (1, 2), homing of migratory cells (3) and selective connections of neurons (4).

A prerequisite to elucidating the mechanism of selective adhesion is clarification of the general mechanisms of cell-cell adhesion. Recently several classes of cell surface proteins (or glycoproteins) have been identified as the cell-cell binding molecule, and their biological and biochemical properties have been studied extensively (see 5 for review). Among these molecules the one involved in the Ca^{2+}-dependent cell-cell adhesion system (CDS) is of particular interest in relating cell adhesion molecules in selective adhesion mechanisms. CDS is detected in every cell forming solid tissues, indicating its general importance for cell-cell adhesion (6). The most interesting property of CDS is its cell-type specificity. For example, CDS of teratocarcinoma cells is not cross-reactive with that of fibroblasts or nervous cells; these heterotypic cells tend to segregate in mixed cultures when CDS is active (7). Such phenomenon suggests that CDS is involved in selective adhesion mechanisms.

We recently obtained two kinds of monoclonal antibodies which can block specifically the function of CDS in mouse cells, ECCD-1 (8) and NCD-1 (9). ECCD-1 was obtained as the antibody blocking teratocarcinoma cell-cell adhesion mediated by CDS. It reacts with a cell surface protein with a molecular weight of 124,000, termed E-cadherin. The same or similar molecules have also been identified by other laboratories (10, 11, 12, 13). NCD-1 was obtained as the antibody disrupting brain cell-cell adhesion mediated by CDS (9). The target molecule reacting with NCD-1 has not been identified, although the hypothetical target was termed N-cadherin. Using these antibodies we have studied distribution of E-cadherin and N-cadherin in mouse embryo and fetus, and analyzed the adhesive specificity of these adhesion molecules to explain the role of CDS in animal morphogenetic processes.

METHODS

Monoclonal antibodies to calcium-dependent cell adhesion molecules. Monoclonal antibodies, ECCD-1 and NCD-1, have been obtained as described (8, 9). Methods for assaying the effect of these antibodies on cell-cell adhesion and their complement-dependent cytotoxicity have

TABLE 1
SENSITIVITY OF VARIOUS MOUSE CELLS TO NCD-1 AND ECCD-1[a]

	NCD-1[b]	ECCD-1[c]
Lens epithelium	+	−
Neurohypophysis	+	−
Cardiac muscle	+	−
Skeletal muscle		
Myotube	+	NT
Myoblast	±	−
Epithelium of		
Mammary gland	−	+
Liver	−	+
Pancreas	−	+
Stomach	−	+
Salivary gland	−	+
Thyroid	−	+
Adenohyophysis	−	+
Epidermis	−	+
Otic vesicle	±	+
Fibroblast[d]	−	−
Endothelium of artery	−	−
Adrenal tumor Y1	+	−
Melanoma B16	+	−
Teratocarcinoma AT805	−	+
Neuroblastoma Neuro 2a	−	−
PSA5-E	−	−

[a] See 9 for the age of embryos used to obtain cells.
[b] Complement-dependent cytotoxicity of NCD-1 was assayed. Monolayer cell cultures were incubated in the presence of NCD-1 and complement for 4 hrs. Cells giving the cytotoxicity index (see 9 for definition) higher than 90 were scored with + and cells giving the index approximately 0 were scored with −. Myobloasts and otic vesicle scored with ± gave the index 37 and 15, respectively. Some examples of cells lysed by NCD-1 are shown in Fig. 1.
[c] Effect of ECCD-1 in inducing disruption of cell-cell contact in monolyer cultures was assayed. Cells reacting with ECCD-1 within a few hours were scored with +, and cells not reacting with the antibody even after 24 hr incubation were scored with −. NT, not tested.
[d] Fibroblasts in muscle, skin, endodermal organs.

been described elsewhere (9).

Labelling of cells with fluorescent beads. Suspension of Fluoresbrite carboxylate microspheres (Polyscience Inc., cat. # 15700) was added to cultures of cells at a concentration of 0.2 % when they were transferred, and the cultures were then maintained for one or two days. These cells were harvested and used for aggregation experiments. Nearly 100% of them were labelled with fluorescence by phagocytosis of the beads after this treatment.

RESULTS

Cell Type Specificity of E-cadherin and N-cadherin

ECCD-1 actively disrupts cell-cell adhesion of E-cadherin-positive cells. In Table 1, cells whose cell-cell adhesion is sensitive to ECCD-1 are summarized. E-cadherin was detected in very early embryos, as revealed by the observation that ECCD-1 inhibits "compaction" of embryos occurring at the 8- to 16-cell stage (14). This adhesion molecule is probably expressed even in one cell-stage embryos, since they show CDS activity (15). In the fetal stage, epithelial cells of various organs are sensitive to ECCD-1, but other types of cells such as neuron and fibroblast did not react with the antibody. This pattern of distribution of E-cadherin is maintained to the adult stage.

When ECCD-1 was added to cultures of the sensitive epithelial cells, the "epithelial" sheet was broken and cells converted their morphology into "fibroblastic" form. These epithelial cells reacting with ECCD-1 lose their tight cell-cell adhesiveness.

NCD-1 reacts with CDS of cells of various tissues as shown in Table 1. Activity of NCD-1 in disrupting cell-cell adhesion of the sensitive cells was not as high as that of ECCD-1. Therefore, to detect cells reacting with NCD-1, we employed the complement-dependent cytotoxicity assay which was found to be the most sensitive method for this purpose.

Examples of cells reacting with NCD-1 were neuron and glia of embryonic nervous tissues (e.g., brain, spinal cord, and dorsal root ganglion) (Fig. 1a, b), lens epithelial cells (Fig. 1c, d), cardiac muscle cells (Fig. 1e, f) and myotubes of skeletal muscle. The majority of

Selective Adhesion Mechanism 227

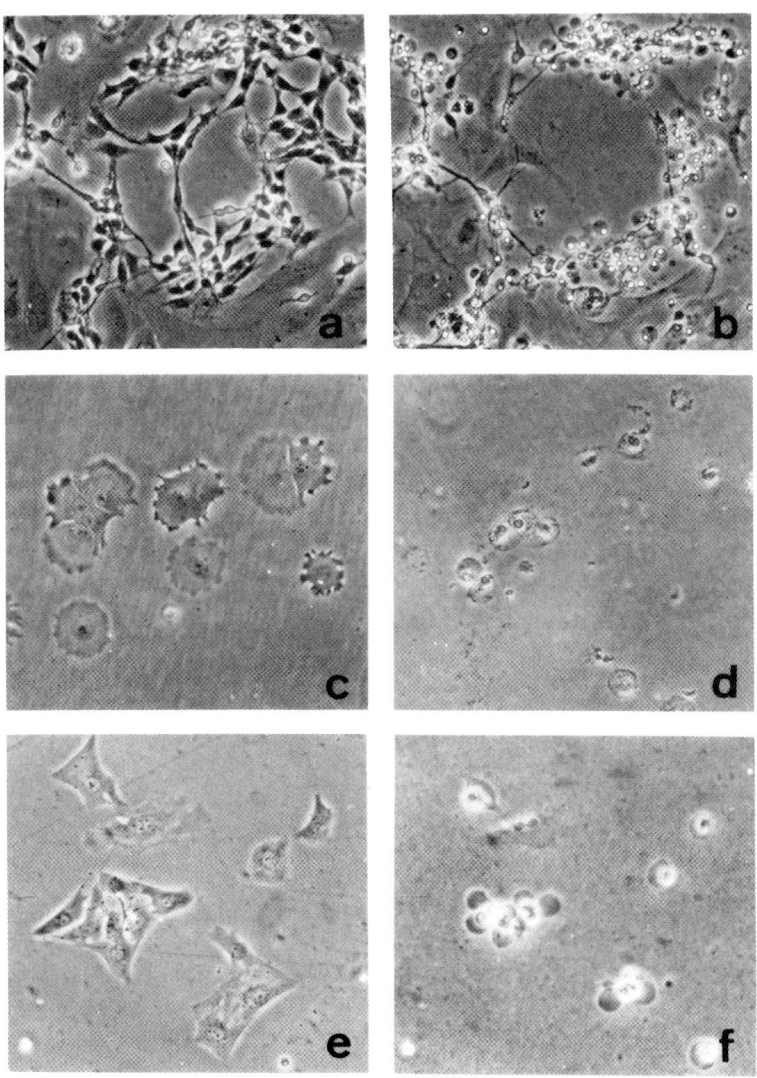

FIGURE 1. Complement-dependent cytotoxic effect of NCD-1 on cells of brain (a, b), lens epithelium (c, d) and cardiac muscle (e, f). Left lane, before addition of the antibody and complement; right lane, 4 hrs after addition of them.

myoblasts, fibroblasts, epithelial cells, vascular endothelial cells and compacted 8-cell stage embryos did not react with NCD-1, and none of E-cadherin-positive cells reacted with it. Therefore, it is possible that E-cadherin and N-cadherin are not simultaneously expressed in a single cell type. A previous study also suggested that expression of E-cadherin and fibroblast-type CDS is mutually exclusive in hybrid cells of teratocarcinoma and fibroblast (16).

Segregation of E-Cadherin- and N-Cadherin-Positive Cells

As indicated from the above experiments, E-cadherin and N-cadherin are immunologically distinct. We then asked whether they are also distinct in their functional specificity: in other words, do these two adhesion molecules cross-react with each other? To answer this, we performed the following aggregation experiment.

We prepared two cell lines, PCC3 and G26-20 which express E-cadherin and N-cadherin, respectively. These cells were dissociated by trypsin-Ca^{2+}-treatment leaving only CDS intact (17) and then mixed. Cells of either line were prelabelled with the fluorescent beads to distinguish them from the other. They were allowed to aggregate for 30 min in a suspension culture, and the distribution of the fluorescence-labelled and unlabelled cells in aggregates formed were examined by fluorescent microscopy. Figure 2 shows that PCC3 and G26-20 cells aggregated independently, indicating that E-cadhrin and N-cadherin are not cross-reactive to each other at least in this cell type combination.

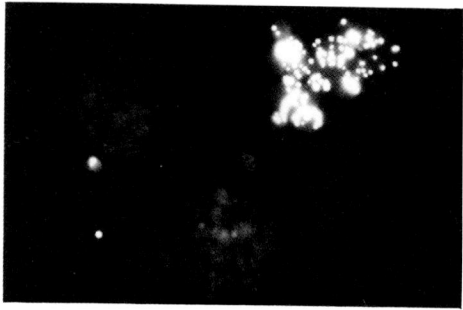

FIGURE 2. Segregation of PCC3 and G26-20 cells. The latters are labelled with fluorescent beads.

DISCUSSION

The Ca^{2+}-dependent cell adhesion system is probably essential for cell-cell binding in the majority of vertebrate tissues, since its activity can be detected in all cells forming solid tissues except non-cohesive cells like blood cells. Strong adhesion-blocking effect of antibodies to CDS supports the importance of this adhesion system for the maintenance of multicellular structures. The finding that different cell types express CDS with different immunological specificity suggests that this adhesion system is also important for selective cell adhesion mechanisms.

So far we have detected at least three subclasses of CDS, E-cadherin type, N-cadherin type and the others which do not react with either ECCD-1 or NCD-1. Fibroblasts, vascular endothelial cells and PSA5-E cells belong to the last group. We have succeeded in the molecular identification of E-cadherin (8), but not of other subclasses yet. As to CDS of fibroblasts and neural retina, candidates for their molecular components have been reported. All these molecules show similar molecular weight and similar protease sensitivity. The total number of subclasses of CDS present in the animal body is not known, but it may not be large because E-cadherin- and N-cadherin-positive cells already account for a large population of cells constituting the body, although micro-heterogeneity may exist in each subclass of CDS.

Results of the previous (7) and present experiments of heterotypic cell aggregation suggested that immunologically distinct subclasses of CDS do not cross-react with each other. Teratocarcinoma cells with E-cadherin cannot stably adhere to G26-20 glioma cells with N-cadherin or to fibroblasts with another subclass of CDS, when their aggregation depends upon CDS. The early observation by Roth and Weston (18) that cells of neural retina and liver adhere preferentially to their own type when mixed can be explained by difference in the specificity of their CDS; it should be noted that liver parenchymal cells express E-cadherin and neural retina cells presumably express N-cadherin. These strongly suggest that CDS is the machinery responsible for segregation of heterotypic cells in various cell combinations.

Distribution of cells expressing different subclasses of CDS in tissues is interesting in respect to the above possible role of CDS. Most tissues are comprised of

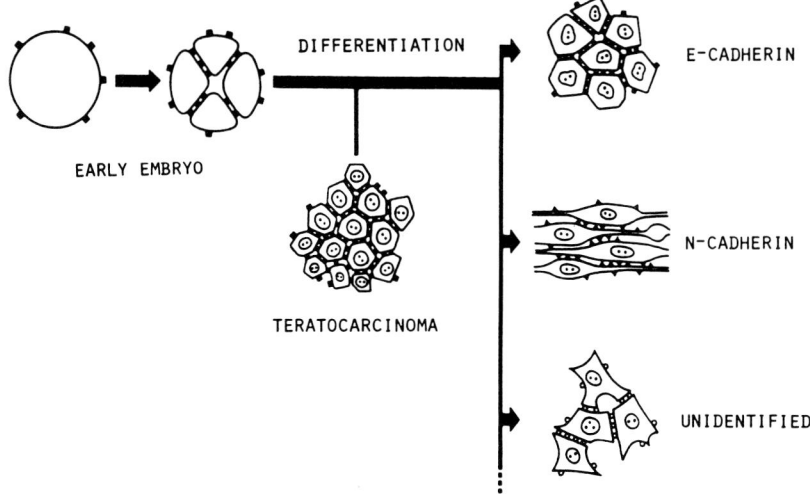

FIGURE 3. Developmental change in expression of various subclasses of CDS.

heterotypic cells such as epithelial and fibroblastic cells. These cell groups are usually separated by clear boundaries and do not form direct cell-cell contact with each other. Such segregation pattern of heterotypic cells in tissues is apparently correlated with the distribution of different subclasses of CDS.

The question is then raised how CDS is involved in the morphogenetic processes occurring in embryonic development. In the earliest stage of development, cells seem to express only E-cadherin since they do not react with NCD-1 and do not adhere to fibroblasts. Other subclasses of CDS are, therefore, expressed in some later developmental stages (see Fig. 3). It is interesting to know at which stage and in which part of the embryo expression of these CDS begins during development. If a fraction of cells in some cell population starts expressing a new subclass of CDS, it may cause segregation of this cell group from the others. Actually there is a good example of a correlation of tissue separation with de novo expression of a new subclass of CDS. Neural tube and lens are formed by pinching off of a group of cells from ectoderm. Thiery et al. (19) demonstrated that L-CAM, which is thought to be the same molecule as E-cadherin (20), disappears from neural tube and lens vesicle during their invagination although overlying ectoderm continues expressing L-CAM. Since lens and neuronal tissues derived from neural tube express N-cadherin, it may be possible that L-CAM is replaced by N-cadherin during the process of invagination causing separation of these tissues from overlying ectoderm.

This kind of model of tissue separation has already been proposed by Townes and Holtfreter (1) to explain the mechanism of segregation of amphibian germ layers. Immunohistological studies to determine the appearance and localization of various classes of CDS in developing embryos may provide this tissue separation model with molecular evidence.

On the other hand, there seems to be another type of mechanism for tissue separation. Otic vesicle, Rathke's pouch and thyroid are also formed by detachment of their primodia from ectoderm or endoderm. In this case, no replacement of cadherin molecules occurs after these organs detach, and they continue expression of E-cadherin. It should be noted that the pattern of separation of these organs from the parental cell layer is quite distinct from that of lens and neural tube. In the latter cases, the tissues are detached clearly and rapidly from ectoderm as soon as their invaginated vesicular or tubular structures close. In the E-cadherin-positive organs, however, even after closure of the invaginated vesicles, they remain connected to the surface epithelium by a stalk-like structure, such as thyroglossal duct, for a certain period: their severance from the origin is completed by degeneration of these interconnecting tissues rather than by detachment. Thus, there are probably two distinct mechanisms for tissue separation, the replacement of specific adhesion molecules followed by detachment, and the degeneration of interconnecting epithelial structures.

It should be finally noted that the pattern of tissue distribution of N-cadherin is similar to that of N-CAM, one of the Ca^{2+}-independent cell adhesion molecules (21). This may suggest that these two functionally independent cell adhesion molecules are expressed coordinately. Thus, the specificity of cell adhesiveness seems to arise from the presence of multiple cell-type-specific adhesion molecules. Such coordinate expression of specific cell adhesion molecules must play some unknown but key role in the cell adhesive behavior involved in animal morphogenesis.

REFERENCES

1. Townes PL, Holtfreter J (1955). Directed movements and selective adhesion of embryonic amphibian cells. J Exp Zool 128: 53.
2. Moscona A (1956). Development of heterotypic combinations of dissociated embryonic chick cells. Proc Soc Exptl Biol Med 92: 410.
3. Gallatain WM, Weissman IL, Butcher EC (1983). A cell-surface molecule involved in organ-specific homing of lymphocytes. Nature 394: 30.
4. Goodman CS, Bastiani MJ, Doe CQ, Lac S, Helfand SL, Kuwada JY, Thomas JB (1984). Cell recognition during neuronal development. Science 225: 1271.
5. Damsky CH, Knudsen KA, Buck CA (1984). Integral membrane glycoprotein in cell-cell and cell-substratum adhesion. In Ivatt RJ (ed): "The Biology of Glycoproteins," Plenum Publishing Corporation, p 1.
6. Takeichi M, Atsumi T, Yoshida C, Ogou S (1982). Molecular approaches to cell-cell recognition mechanisms in mammalian embryos. In Muramatsu T, Gachelin G, Moscona AA, Ikawa Y (eds): "Teratocarcinoma and Embryonic Cellular Interactions," Tokyo: Academic Press, p 283.
7. Takeichi M, Atsumi T, Yoshida C, Uno K, Okada TS (1981). Selective adhesion of embryonal carcinoma cells and differentiated cells by Ca^{2+}-dependent sites. Develop Biol 87: 340.
8. Yoshida-Noro C, Suzuki N, Takeichi M (1984). Molecular nature of the calcium-dependent cell-cell adhesion system in mouse teratocarcinoma and embryonic cells studied with a monoclonal antibody. Develop Biol 101: 19.
9. Hatta K, Okada TS, Takeichi, M (1985). A monoclonal antibody disrupting calcium-dependent cell-cell adhesion of brain tissues: Possible role of its target antigen in animal pattern formation. Proc Natl Acad Sci USA in press.
10. Gallin WJ, Edelman GM, Cunningham BA (1983). Characterization of L-CAM, a major cell adhesion molecule from embryonic liver cells. Proc Natl Acad Sci USA 80: 1038.
11. Damsky CH, Richa C, Solter D, Knudsen K, Buck CA (1983). Identification and purification of a cell surface glycoprotein mediating intercellular adhesion in embryonic and adult tissues. Cell 34: 455.

12. Peyrieras N, Hyafil F, Louvard D, Ploegh HL, Jacob F (1983). Uvomorulin: A nonintegral membrane protein of early mouse embryo. Proc Natl Acad Sci USA 80: 6274.
13. Vestweber D, Kemler R (1984). Rabbit antiserum against a purified surface glycoprotein decompacts mouse preimplantation embryos and react with specific adult tissues. Exptl Cell Res 152: 169.
14. Shirayoshi Y, Okada TS, Takeichi M (1983). The calcium-dependent cell-cell adhesion system regulates inner cell mass formation and cell surface polarization in early mouse development. Cell 35: 631.
15. Ogou S, Okada TS, Takeichi M (1982). Cleavage stage mouse embryos share a common cell adhesion system with teratocarcinoma cells. Develop Biol 92: 521.
16. Atsumi T, Takeichi M, Okada TS (1983). Selective expression of cell type specific cell-cell adhesion molecules in mouse hybrid cells. Differentiation 24: 140.
17. Takeichi M, Ozaki HS, Tokunaga K, Okada TS (1979). Experimental manipulation of cell surface to affect cellular recognition mechanisms. Develop Biol 70: 195.
18. Roth SA, Weston JA (1967). The measurement of intercellular adhesion. Proc Natl Acad Sci USA 58: 974.
19. Thiery JP, Delouvee A, Gallin WJ, Cunningham BA, Edelman GM (1984). Ontogenetic expression of cell adhesion molecules: L-CAM is found in epithelia derived from the three primary germ layers. Develop Biol 102: 61.
20. Ogou S, Yoshida-Noro C, Takeichi M (1983). Calcium-dependnt cell-cell adhesion molecules common to hepatocytes and teratocarcinoma stem cells. J Cell Biol 97: 944.
21. Edelman GM, Gallin WJ, Delouvee A, Cunningham BA, Thiery JP (1983). Early epochal maps of two different cell adhesion molecules. Proc Natl Acad Sci USA 80: 4384.

Molecular Determinants of Animal Form, pages 235–252
© 1985 Alan R. Liss, Inc.

TWO CELL ADHESION MOLECULES: CHARACTERIZATION AND ROLE IN EARLY MOUSE EMBRYO DEVELOPMENT[1]

Caroline H. Damsky,[2,4,5] Margaret J. Wheelock,[2] Ivan Damjanov,[3] Clayton Buck[2]

The Wistar Institute, Philadelphia, PA 19104
Hahnemann Medical College, Dept. of Pathology, Phila., PA

ABSTRACT The cell-cell adhesion glycorprotein, cell-CAM 120/80 and the 140 kd cell-substratum attachment glycoprotein complex (CSATag) are involved in adhesive interactions in early mouse development and in adult tissue. Localization studies and adhesion perturbation studies using specific antibodies indicate that cell-CAM 120/80 participates in the dominant cell-cell adhesion mechanism of early mouse embryos. As additional cell adhesion mechanisms become expressed, the relative contribution to adhesion of cell-CAM 120/80 appears to be lessend although the antigen is still present. The 140kd cell substratum attachment glycoproteins are expressed as early as the blastocyst stage and are later present on cells of virtually all differentiated tissue types. They contribute to differing degrees to the cell adhesion strategies of various cell types, participating in the dominant mechanism of myogenic but not of several other cell types in which they are expressed. We hypothesize that these two sets of adhesion molecules are early participants in the progressive display of

---

[1] This work was supported by CA27909, CA32311, CA07572 from NIH.
[2] Wistar Institute, Philadelphia, PA 19104
[3] Hahnemann Medical College, Department of Pathology, Philadelphia, PA 19102
[4] Department of Stomatology and Anatomy, Schools of Dentistry and Medicine, University of California, San Francisco, CA 94143
[5] To whom correspondence should be directed at present address

cell adhesion molecules which accompanies differentiation and that in this role, they participate in key cell-cell and cell-matrix recognition events in early morphogenesis.

INTRODUCTION

The processes of cell recognition and adhesion are intimately related. Neighboring cells make a decision to adhere to one another and that signal is transmitted across the surface membrane causing changes in the organization of the cytoskeleton, alterations in cell shape and a stabilization of the initial encounter (1). The extent to which the molecule(s) responsible for the initial recognition event participates in the later stabilizing events is poorly understood. We have identified and isolated a cell-cell glycoprotein from cultured mammary epithelial cells (cell-CAM 120/80). A monospecific antiserum against this molecule, anti GP80, disrupts cell-cell adhesion in some epithelial tumor cell lines and in mouse embryos at well defined time points in preimplantation and periimplantation development. It has no effect on cell-cell adhesion at other well defined time points in early mouse development, or on such interactions in most other cell lines and mature epithelial tissue even though the antigen is present as judged by immunofluorescent localization (2-4). We have also identified a complex of three cell surface glycoproteins that are involved in cell-matrix adhesion (5-10). Antibodies that recognize this complex, disrupt cell-matrix adhesion in a wide variety of adherent tissue cells in culture and are effective in early mouse embryos as well (4). However, just as is the case with cell-CAM 120/80, these glycoproteins do not contribute to the same extent to the overall adhesive stability of all cell types.

This article will discuss the data we have accumulated over the past several years from the point of view of trying to understand how these two sets of adhesion molecules fit in functionally to the overall adhesion mechanisms developed by tissue cells in complex organisms.

RESULTS AND DISCUSSION

<u>Identification of the cell-cell adhesion molecule cell-CAM 120/80.</u> This glycoprotein was identified and

purified by monitoring its ability to block the cell-cell adhesion disrupting effects of a polyspecific antiserum (anti-SFM II; 2, 6) prepared against serum-free medium (SFM) conditioned by the human mammary carcinoma cell line, MCF-7. Conditioned SFM was used as the starting material for purification and an 80kd soluble glycopeptide was isolated that blocked disruption of cell-cell adhesion by anti SFM II (the purification scheme and the antibody blocking assay are described in (2)). An antiserum was prepared against this 80kd fragment (anti GP80). Anti-GP80 recognizes a single band at 80kd in unfractionated SFM and a band at 120kd in unfractionated NP40 extracts of epithelial cells (2). Further biochemical characterization of the 120kd form of this molecule has shown that it is most likely an integral membrane glycoprotein, since it cannot be extracted in the absence of detergent, and that it is very protease-sensitive, breaking down to molecules of 105kd and 92kd unless protease inhibitors are included during extraction. The molecule is, however, resistant to total degradation by trypsin in the presence of calcium. It shares this as well as other biochemical and biological properties with uvomorulin (11, 12) and cadherin (13), isolated from murine teratocarcinoma cells and L-CAM isolated from embryonic chick liver (14, 15).

Effect of anti-GP80 on early mouse development. A feature that makes the study of this molecule so compelling is that it is clearly involved in interactions of early mammalian embryos (2, 4, 12, 16, 17). Anti-GP80 inhibits the first major adhesive event in mammalian development: compaction of the 8-cell embryo. The decompacted embryo continues to divide, but cannot organize a blastocyst, as it is unable to form the junctional complexes required for directional pumping of blastocoelic fluid. Exposure of embryos to anti-GP80 will result in decompaction up to the 32-cell stage after which the antibody has no effect on the intact embryos (4). Our explanation for this transition is that sufficient tight junctions have formed so that when embryos are exposed to antibody, either the antibody does not have access to its antigen, or the other junctional elements become strong enough to maintain intercellular adhesion even when the contribution of cell-CAM 120/80 is prevented. A similar scenario exists slightly later in development. Following formation of the fluid filled blastocyst, the trophectoderm layer can be removed by immunosurgery, releasing

the compact ball of inner cell mass (ICM) cells. If left
in culture for 48 hours, these ICM will segregate a layer
of primative endoderm around them that is analagous to the
layer of primative endoderm formed at the interface of ICM
and the blastocoel cavity. Before an endoderm layer has
formed, the ball of ICM cells decompacts in response to
anti-GP80 much like the 8-16-cell embryo. However,
following endoderm formation, the antibody is no longer
effective (4). Determining the distribution at high resolution of cell-CAM 120/80 and its relationship to developing junctional elements before and after compaction of
8-16-cell embryo and before and after endoderm segregation
from ICM, should contribute to an understanding of its
function in adhesion and recognition. This point is discussed more extensively below.

Localization of cell-CAM 120/80 and its relationship
to intercellular junctions

In localization studies using frozen cryostat sections of adult or differentiated fetal tissues, anti-GP80
has been found on almost all types of epithelial cells,
but not on other types of cells (endothelium, muscle, connective tissue, brain). There are a few interesting exceptions to this universality: In the adult mouse kidney, for example, cell-CAM 120/80 is undetectable by immunofluorescence in both the glomerulus and in a subset of
kidney tubules. By contrast, other tubular elements of the
nephron and the collecting ducts display strong staining
for cell-CAM 120/80 (Damjanov and Damsky, in preparation;
Fig. 1; A,B). Small amounts of cell-CAM 120/80 may be
present in the regions of the kidney which display no
fluorescence, but there are clearly large differences in
the level of expression of cell-CAM 120/80 in the different regions of adult kidney. The Sertoli cells of the
male germinal epithelium which interact closely with one
another by tight junctions also show no staining with
anti-GP80. Cell-CAM 120/80 is found on all cultured
epithelia tested, and on mouse and human teratocarcinoma
stem cell lines but not on fibroblastic or PYS teratocarcinoma lines. Anti-GP80 staining is restricted to the
lateral borders of epithelial cells that are in contact
with their neighbors. In many cases (e.g., adult human
lung alveolae, skin, placental cytotrophoblast, liver,
carcinomas from lung and breast and 17d mouse embryo

intestine) the distribution appears diffuse without
obvious enrichment at the apical end of the cells where
junctional complexes reside (2, 3). More recent studies
of adult mouse kidney, however, suggest that the antigen
is enriched at the apical ends of the epithelial cells of
distal tubules and collecting ducts (Fig. 1; C,D).
Studies by Boller et al. (18) using antibody to uvomorulin
and immunoelectron microscopy, have shown quite convincingly that in the adult mouse small intestine, uvomorulin
is enriched in the region of the zonula adherens
(intermediate junction).

In speculating about the functional relationship between cell-CAM 120/80 (uvomorulin) and a particular junctional element, a brief discussion of junctional elements
in different cell types and of what is known about their
molecular composition seems warranted. Well established
cell-cell interactions are a fundamental requirement for
normal epithelial tissue architecture. Transporting and
secretory epithelia and epithelia which maintain a high
transepithelial resistance possess a well characterized
junctional complex, consisting of tight junctions to prevent intercellular leakage of ions and small metabolites
and adherens junctions, both zonular and macular (intermediate and desmosomal junctions, respectively) which are
predominantly adhesive (19, 20). Stratified epithelia
such as skin and esophagus emphasize their adhesive junctions, in particular desmosomes, in order to resist abrasion and stretching. Keratinzing epithelia do not possess
tight junctions at all, relying instead on a lipid-containing surface coat secreted by the stratum corneum to
provide the permeability barrier for the skin (21). The
Sertoli cells of the germinal epithelium on the other
hand, interact with one another almost exclusively by
tight junctions which enable them to sequester later stages of the male germ cell line from the remainder of the
animal. These observations suggest that epithelia with
different physiological roles will emphasize particular
junctional elements in accordance with their functions.

Cell types other than epithelia display analagous
junctional elements. For example, the intercalated discs
between heart muscle cells consist of well developed
desmosome and intermediate junction-like structures.
Endothelial cells also display structures resembling
intermediate and focal tight junctions. Most cell types,
including all of those discussed above as well as connective tissue cells, can also interact via gap junctions who

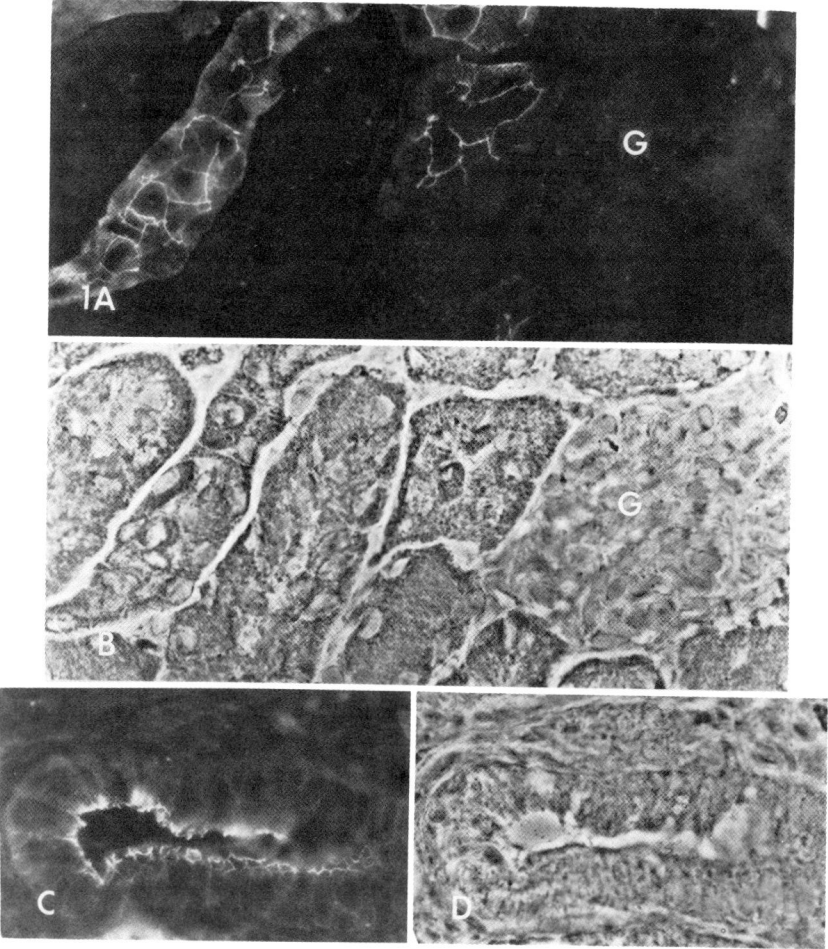

Figure 1. Distribution of cell-CAM 120/80 in adult mouse kidney. Frozen 5 um sections of kidney were fixed in -20°C methanol and stained with anti-GP80 and rhodamine conjugated goat anti-rabbit. The glomerulus (G) and a subset of kidney tubules do not react with anti-GP80 while other tubular structures stain strongly. In positive tubules, the staining is restricted to a tight band near the apical borders of the cells. A,C) anti-GP80, B,D) phase of same fields. Bar = 10 um

role is primarily intercellular communication rather than adhesion.

The integral membrane glycoproteins involved in mediating adhesion at desmosomes have been identified in bovine muzzle epidermis using monoclonal antibodies. There appear to be 3 families of these glycoproteins called desmogleins (22, 23). These antigens are found in a wide variety of other epithelial tissues and in heart muscle as well (24). Unanswered questions remain as to when the membrane glycoproteins mediating adhesion at desmosomes first appear during development. Recognizable desmosomes are present in the chick as early as the blastoderm (28). It is not known whether any of the membrane glycoproteins (desmogleins) are present prior to the appearance of discrete desmosomes, perhaps in a diffuse distribution, or whether the desmosome is assembled as a unit with all elements present in a focal pattern from the outset.

The molecular composition of the zonula adherens junctions has also come under recent scrutiny. It has been known for some time that a marginal band of microfilaments is associated with the zonula adherens (19,20). The microfilament associated proteins α-actinin and vinculin are also closely associated with these filaments at their sites of attachment to the plasma membrane (25). Recently, Volk and Geiger have identified a 135kd integral membrane glycoprotein that is specifically associated with the zonula adherens. It is present in both heart and epithelial tissue (26). As yet, there is no information on when this protein appears during early development or during the differentiation process, and whether it is ever present in a diffuse distribution.

How does what one knows about cell-CAM 120/80 fit in with the information described above. Cell-CAM 120/80 is clearly involved in cell-cell adhesion, since antibodies against it disrupt adhesive events in early development and in some cultured tumor epithelial cell lines. Localization studies indicate that it is present in a diffuse distribution over the surface of blastomeres of the early cleavage and precompaction 8-cell mouse embryo (40) although no recognizable junctional elements are present at that time and the first significant cell-adhesion event, compaction, does not take place until the late 8-cell stage. Gap junctions form at the 8-cell stage (27) and focal tight junctions begin to appear at the 8-16-cell stage. However, most of the area of contact between blastomeres just after compaction reveals extensive close apposition of the plasma membranes of neighboring cells without distinctive specializations. As mentioned previously, cell-CAM 120/80 is restricted to epithelial tissue in the adult organism, being absent from other tissue types rich in junctional elements analagous to those found in epithelia. One must consider as well that even though cell-CAM 120/80 is present in most differentiated normal epithelia, antibodies to cell-CAM 120/80 do not disrupt cell-cell adhesion in these differentiated systems. This suggests that although cell-CAM 120/80 appears to play a dominant role in cell-cell interactions in early development, its relative contribution is lessened by addition of other adhesive complexes which develop as differentiation proceeds. In considering the information discussed here, one might hypothesize that cell-CAM 120/80 is involved primarily in a cell-cell recognition event which results in a relatively weak cell-cell adhesive interaction. Following this event, firmer more highly stabilized junctions are formed which provide the major adhesive strength for the differentiated cell.

If this is the case, however, why is cell-CAM 120/80 so carefully conserved in the differentiated epithelia of the adult after the need for such recognition events is presumably past? Is it necessary for epithelial cells in differentiated tissue to continue to wear a label identifying them as such? In epithelial cells, does cell-CAM 120/80 become associated with a surface membrane molecule that is junction-specific in a larger set of cells?

This latter suggestion is supported by the localization data from Boller et al. (18) that uvomorulin is enriched at the zonula adherens of intestinal epithelium. Our preliminary data in the adult kidney (Fig. 1)also support a junction associated location for this antigen. Thus, one might speculate that concommitant with or following its function in an initial recognition process, cell-CAM 120/80 (uvomorulin) may interact with and become co-sequestered with a zonula adherens-specific molecule that is found not only in epithelium but in heart and endothelium as well. This raises the prospect that these other tissues have their own specific recognition molecules that then become associated with the structural proteins of their junctional assemblies. Further systematic examination of the relationship between cell-CAM 120/80 and junctional complexes in epithelia with different functions and in epithelial tissues at different stages of differentiation, should be illuminating in resolving these interesting questions.

Regulation of expression of cell-CAM 120/80. Cell-CAM 120/80, like the neural cell adhesion molecule N-CAM (29, 30) becomes restricted in its expression as development proceeds. We have determined by fluorescence and can infer by the antibody disruption experiments, that Cell-CAM 120/80 is present on all blastomeres of the 8-cell embryo. It has been detected as early as the 2-cell stage and is present on all cells of the expanded blastocyst. As implantation proceeds, the trophectoderm of the blastocyst becomes the trophoblast of the implant and the primative endoderm differentiates further to produce the epithelial-like visceral endoderm, surrounding the embryonic ectoderm, and the migratory parietal endoderm. Subsequently, at the primative streak stage, mesenchymal cells appear between the embryonic ectoderm and endoderm. These early differentiation events produce a limited number of distinct cell types which can be examined for the presence of cell-CAM 120/80. A summary of results obtained for the 7-day mouse embryo is given in Table 1 (Damjanov and Damsky, in preparation). The two most striking observations are that the parietal endoderm cells which have migrated out from the visceral endoderm to colonize the roof of the blastocoel, no longer display cell-CAM 120/80. Similarly, the mesenchymal cells which come to lie between the embryonic ectoderm and endoderm also lose cell-CAM 120/80. Loss of expression by mesenchyme is reversible, however, and cell-CAM 120/80 becomes re-expressed in mesothelial cell sheets (e.g., Fig. 2) and during the formation of kidney tubule epithelum from mesenchyme under the influence of the ureteric bud (31). An easily manipulable in vitro cell system for turning on and off the expression of cell-CAM 120/80 should be very instructive in understanding how its expression is regulated in vivo.

Table 1
DISTRIBUTION OF CELL-CAM 120/80 IN THE 7 DAY MOUSE EMBRYO

Ectoplacental Cone	+	Chorion	+
Trophoblast Giant Cells	-	Ectoderm	+
Visceral (Proximal) Endoderm	+	Mesoderm	-
Parietal Endoderm	-		

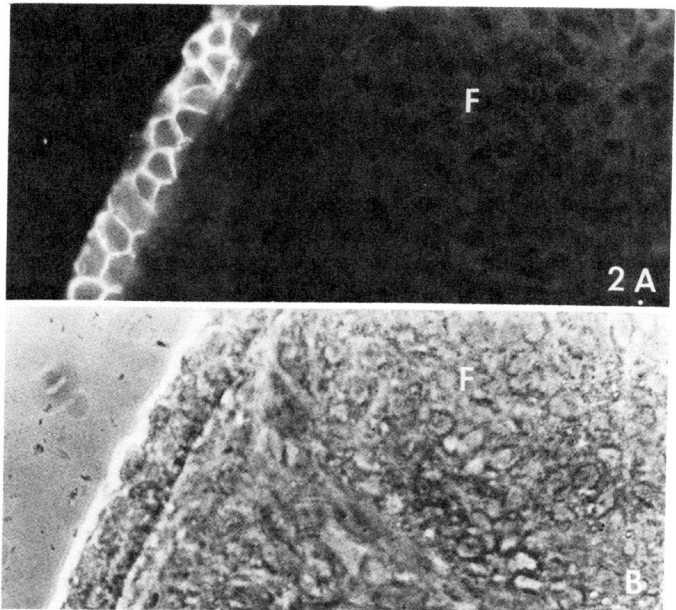

Figure 2. Frozen section of mouse ovary prepared as in Figure 1. The mesothelia sheet covering the ovary is strongly positive for cell-CAM 120/80. The stromal tissue and developing follicle (F) do not react. a) anti-GP80, b) phase of same field.

GP140 Glycoproteins and Cell-Substratum Adhesion in Mammalian Cells

Cell-matrix interactions are also crucial for normal development and the maintenance of appropriate tissue architecture in the adult. A complex of three integral membrane glycoproteins of 120-160kd has been found to contribute significantly to this process in virtually all adherent tissue cells. Antibodies which recognize this complex from mammalian cells, (anti-GP140; 5, 6), disrupt the adhesion of cultured fibroblasts and epithelial tumor cell lines as well as cells in primary cultures of heart, liver and aortic endothelium. The correlation of the presence of the 140kd glycoproteins and adhesion is clearly demonstrated in studies using adherent and non-adherent hamster melanoma cell lines. The adherent line expresses the 140kd glycoproteins at normal levels while in the nonadherent line, expression is reduced 20-fold. The nonadherent cells can be induced to adhere in the present of BrdU, at which time, they re-express normal levels of the 140kd glycoproteins (32, 33).

The 140kd glycoproteins like cell-CAM 120/80 are expressed early in development. The first major cell-matrix adhesion event in mammalian development involves the attachment of the blastocyst to the uterine wall. Blastocysts will also attach to many kinds of substrates in vitro. Anti-GP140 will inhibit this attachment. However, if the antibody is first preincubated with the purified 140kd glycoproteins, its disruptive effects are inhibited (4). This experiment although not a definitive model for determining the role of the 140kd glycoproteins in implantation, does indicate that the 140kd glycoproteins are expressed very early in development. Cell-matrix adhesion should also be important in the attachment of newly formed primative endoderm to the thin layer of extracellular matrix (ECM) which is laid down between it and the underlying ICM, as well as for the migratory activity of parietal endoderm cells as they move out from the periphery of the primative endoderm to colonize the roof of the blastocoel. Antibody disruption experiments at these and other defined periods during early development should shed light on the extent to which this complex is involved in key early events in cell-matrix attachment.

The CSAT antigen complex in avian cells.

More specific questions about the role of the 120-160 kd cell-matrix adhesion glycoproteins have been addressed using a monoclonal antibody, raised against chick myoblast surface membranes, CSAT Mab (Cell Substratum Attachment; 8, 9, 10, 34). In avian cells this antibody recognizes a complex of 3 glycoproteins of M_r 160, 135, 120 when the antigen is analyzed by SDS-PAGE under nonreducing conditions (9). These resemble closely the glycoproteins purified biochemically from mammalian fibroblasts and epithelial cells which block the effects of anti-GP140. When added to the medium of 11 day chick embryo myoblast cultures plated for 48 hours on gelatin in the presence of serum, CSAT Mab selectively rounds and detaches myoblasts leaving the fibroblasts behind. In this complex culture environment, many adhesive factors are available, derived both from the medium and from the cells' synthetic machinery. Despite this, the CSAT Mab can undermine the adhesion of myoblasts. CSAT Mab can also affect the adhesion of fibroblasts, but this is difficult to detect unless the adhesive options for these cells are restricted. If fibroblasts from different embryonic tissues are plated on defined substrates in the absence of serum and the presence of cycloheximide (CHI), interesting and distinctive effects of CSAT Mab are observed (34; Horwitz et al. in preparation). CSAT Mab will inhibit attachment of tendon fibroblasts to laminin (Lm) and fibronectin (Fn) under these conditions. However, it inhibits attachment of embryonic cardiac fibroblasts to Lm but not to Fn. These results suggest that myoblasts and different kinds of fibroblasts have distinct adhesive strategies. The 140kd CSAT antigen (CSATag) complex is present on all these cell types in similar amounts and contributes to their overall adhesion mechanisms, but to different degrees. Myoblasts appear to rely heavily on this complex for adhesion. Tendon fibroblasts, and to a greater extent cardiac fibroblasts have more complex adhesion strategies, the latter cell type having a CSAT Mab-resistant mechanism for adhesion to Fn as well as being able to utilize spreading factors in serum. The CSATag is clearly relevant to cell-matrix interactions <u>in vivo</u>. CSAT Mab will remove myogenic cells from explants of 11 day chick embryonic muscle (8). Furthermore, it will specifically remove myogenic precursor cells from stage 20-22 chick limb buds, leaving chondrogenic precursors behind (35). Interestingly, in both these cases, the CSAT Mab reacts with all of the cell types in the explants, but detaches only the myogenic cells.

Several lines of evidence suggest that the 140kd CSAT antigen glycoproteins are involved in the mechanisms by which cells attach to both Fn and Lm. First, the <u>in vitro</u> plating experiments described above indicate that CSAT Mab can inhibit attachment of at least some kinds of cells to both Fn and Lm. Secondly, localization studies (10) show that the CSAT antigen is present at sites of cell-matrix attachment probably of the close contact variety (see also 36, 37) and co-localizes with Fn at these sites. Thirdly, experiments with RVS transformed chick tendon fibroblasts show that the presence of a defined substrate, either Fn or Lm, can cause the redistribution of CSATag from the diffuse pattern found when these cells are plated on glass in the presence of serum, to a discrete pattern typical of normal tendon fibroblasts (Fig. 3). Finally, experiments designed to determine the contributions of the three in-

dividual glycoproteins to the overall function of the
CSATag complex demonstrate that although cells treated
with trypsin such that band 1 is cleaved, still adhere to
fibronectin, cells treated with trypsin plus dithio-
threitol, such that band 3 as well as band 1 is cleaved,
no longer adhere to fibronectin (Table 2; Damsky et al.
in preparation). Similar results were obtained with mammalian cells (38). Thus, an intact band 3 of the CSATag
complex is required for adhesion to fibronectin.

Although suggesting strongly that CSATag participates
in the adhesion of cells to both Lm and Fn, these data do
not distinguish a direct role for CSATag as a multipurpose
ligand receptor, from a more indirect role in which CSAT
antigen would serve as an organizer of transmembrane cell-
matrix adhesion complexes once initial contacts had been
made between each ligand and its specific cell surface
receptor. Evidence that CSATag can act as a ligand receptor for either Fn or Lm will require demonstration of a
direct binding between the CSATag complex and each of the
ECM ligands. Experiments addressing this point by Pytela
et al (39) have shown that 140kd glycoproteins are bound
by immobilized Fn and are released from the column by the
cell binding peptide of Fn.

Table 2

EFFECT OF CELL SURFACE PROTEOLYSIS ON
THE FUNCTION OF THE CSAT AG

TREATMENT	POLYPEPTIDE COMPOSITION OF CSAT AG	DISTRIBUTION OF CSAT AG	ADHESION ON FN
EDTA	Bands 1,2,3	Focal Pattern on Spread Cells	+
TRYPSIN 1mg/1ml	Bands 2,3	Diffuse Pattern on Spread Cells	+
TRYPSIN 1mg/1ml DTT 50mM 37°C 20'	Band 2	Diffuse Pattern on Round Cells	-

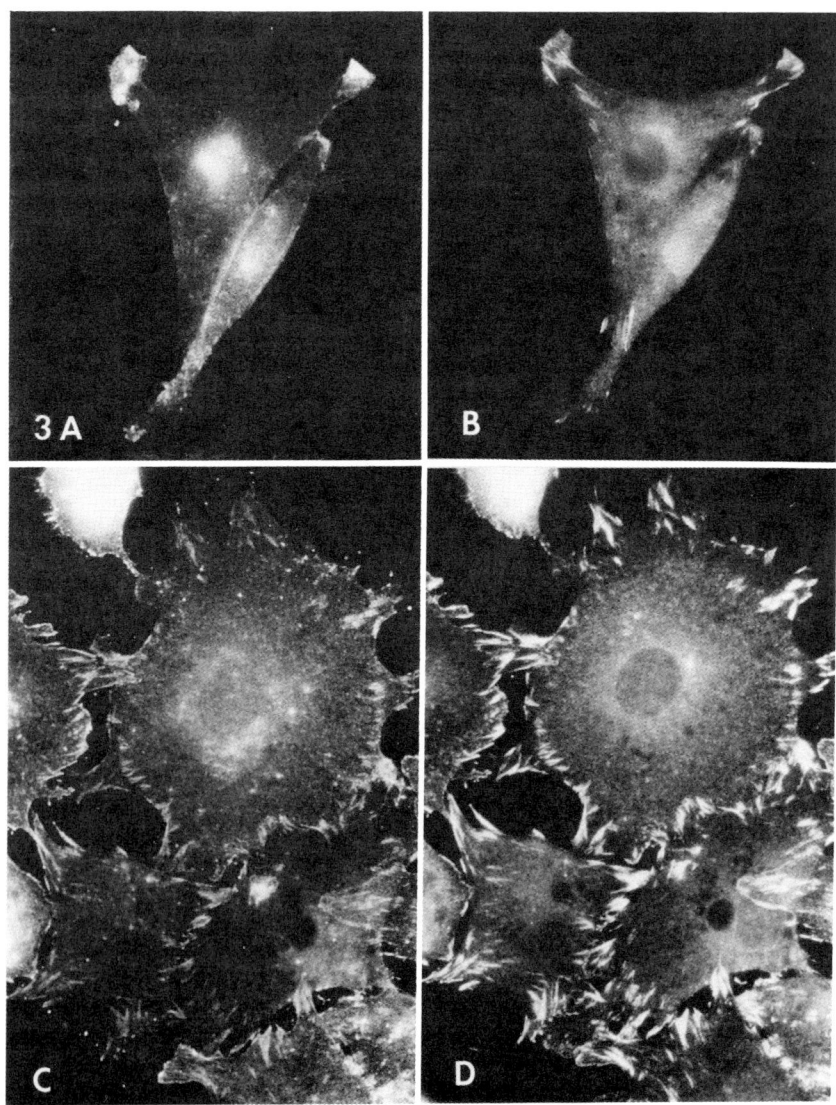

Figure 3.

Figure 3. RVS transformed chick fibroblasts plated on glass coverslips in the presence of serum. Coverslips were fixed in methanol at -20°C and stained with a mixture of CSAT Mab (mouse) and rabbit anti-talin antibody followed by a mixture of fluorescein goat anti-mouse and rhodamine goat anti-rabbit (Cooper Biomedical). (A) CSAT Mab cells are stained with a diffuse pattern with some enrichment in regions of membrane ruffling, (B) anti-talin. Some small talin-rich streaks representing focal contacts are present as well as a diffuse brightness. (C, D) transformed cells plated in serum free medium for 4 hours on glass coverslips coated with Fn and prepared as for A,B. The cells have spread extensively and exhibit a more highly focused distribution of both CSAT ag (C) and talin (D). Thus, the presence of an appropriate extracellular matrix molecule can cause the redistribution of the CSAT antigen.

SUMMARY

The data discussed here demonstrate that cell-CAM 120/80 and the GP140 glycoproteins (CSATag) play important roles in cell-cell and cell-matrix adhesion in vivo. They are both expressed very early in development, and when developing embryos at the appropriate time points are exposed to antibodies specific for these molecules, disruption of adhesion occurs. However, in both cases, the range of expression of the antigens is more extensive than that of the susceptibility of positive cells to antibody-induced disruption of adhesion. Thus, cell-CAM 120/80 is but one of several cell-cell adhesion molecules. Since it is present very early in development, it may precede the expression of other adhesion molecules and therefore play a dominant role in the cell-cell adhesion of the preimplantation mouse embryo. If it becomes sequestered with an element of the apical junctional complex as such junctions appear, as is suggested by the data of Boller et al. (18), and the evidence in Fig. 1, it will be of interest to determine its functional and structural relationships with junction-specific molecules. The GP140 cell-matrix adhesion glycoproteins are clearly the dominant adhesion molecule in myogenic cells since CSAT Mab selectively detaches them in vivo and in vitro when other cell types which express CSATag are left behind. The latter cell types appear to be able to take advantage of a greater range of adhesion factors in their environment and may express more adhesion-related surface molecules. The presence of the GP140 glycoproteins on mouse blastocysts suggests that these molecules may play an important role in the very early cell-matrix adhesion events associated with implantation and cell migration. Thus, both these molecules may represent early players in the orderly and progressive display of adhesion molecules which accompanies differentiation in complex organisms.

ACKNOWLEDGEMENTS

The authors appreciate the gifts of antibody against talin from Dr. Keith Burridge and the CSAT monoclonal antibody from Dr. A.F. Horwitz. The technical assistance of Sena Smith and Eileen Crowley is appreciated as is assistance with manuscript preparation by Marie Lennon.

REFERENCES

1. Damsky CH, Knudsen KA and Buck CA (1984). In The Biology of Glycoproteins. R Ivatt (ed). New York: Plenum Press, pp 1.
2. Damsky CH, Richa J, Solter D, Knudsen KA and Buck CA (1983). Cell 34:455.
3. Damsky CH, Richa J, Wheelock MJ, Damjanov I and Buck CA (1985). In Cell Contact: Adhesions and Junctions as Molecular Determinants. G. Edelman (ed). New York: John Wiley and Sons, in press.
4. Richa J, Damsky CH, Buck CA, Knowles B and Solter D (1985). Dev Biol 108: 513.
5. Knudsen KA, Rao P, Damsky CH and Buck CA (1981). Proc Natl Acad Sci USA 78:6071
6. Damsky CH, Knudsen KA, Dorio R and Buck CA (1981). J Cell Biol 89:173.
8. Neff, NT, Lowrey C, Decker C, Tovar A, Damsky CH, Buck CA and Horwitz AF (1982). J Cell Biol 94:654.
9. Knudsen KA, Horwitz AF and Buck CA. (1985). Exp. Cell Res. 157:218.
10. Knudsen KA, Damsky CH and Buck CA (1982). J Cell Biochem 18:157.
11. Hyafil F, Babinet C and Jacob F (1981). Cell 26:447.
12. Vestweber D and Kemler R (1984). Exp Cell Res 152:169.

13. Yoshida C and Takeichi M (1982). Cell 28:217.
14. Gallin WJ, Edelman G and Cunningham B (1983). Proc Natl Acad Sci USA 80:1038.
15. Cunningham B, Leutzinger Y, Gallin W, Sorkin B and Edelman G. Proc Natl Acad Sci USA 81:5787
16. Peyrieras N, Hyafil F, Louvard D, Ploegh H and Jacob F (1983). Proc Natl Acad Sci USA 80:6274.
17. Shirayoshi Y, Okada TS and Takeichi M (1983). Cell 35:631.
18. Boller K, Vestweber D and Kemler R (1985). J Cell Biol 100:327.
19. Staehelin LA (1974). Int Rev Cytol 39:191.
20. Hull B and Staehelin LA (1979). J Cell Biol 81:67.
21. Elias P, McNutt S and Friend D (1977). Anat Rec 189:577.
22 Gorbsky G and Steinberg MS (1981). J Cell Biol 90:243.
23. Cohen S, Gorbsky G and Steinberg MS (1983). J Biol Chem 258:2621.
24. Cowin P and Garrod D (1983). Nature 302:148.
25. Geiger B, Tokuyasu K, Dutton A and Singer SJ (1980). Proc Natl Acad Sci USA 77:4127.
26. Geiger B. and Volk T (1984). Int. Cong. on Cell Biol.Tokyo: Japan pp. 124a, and as molecular determinants G Edelman (ed) New York, John Wiley, in press.
27. Lo C and Gilula B (1978). Cell 18:399.
28. Overton J. (1962). Dev Biol 4:532.
29. Thiery J-P, Duband J-L, Rutishauser U and Edelman G (1982). Proc Natl Acad Sci USA 79:6737.
30. Edelman G (1983). Science 219: 450.
31. Ekblom P et al. (This volume).
32. Damsky CH, Knudsen KA and Buck CA (1982). J Cell Biochem 18:1.
33. Damsky CH, Knudsen KA, Horwitz AF, Wheelock MJ, Gruber P and Buck CA (1985). In Biochemistry and Molecular Genetics of Metastasis (K. Lapis, L. Liotta and A. Rabson (eds) Martinus Nijhoff, Boston, The Hague, in press.
34. Decker C, Greggs R, Duggan K, Stubbs J and Horwitz A (1984). J Cell Biol 99:1398.
35. Sasse J, Horwitz A and Hoftzer H (1985). J Cell Biol 99:1856.
36. Chen W-T, Greve JM, Gottlieb DI and Singer SJ (1985). J Histochem Cytochem, in press.

37. Chen W-T, Hasegawa E, Hasegawa T, Weinstock C and Yamada K (1985). J Cell Biol 100:1103.
38. Giancotti F, Tarone G, Knudsen KA, Damsky CH and Comoglio P (1985). Exp Cell Res 156:182.
39. Pytela R, Piersbacher M and Ruoslahti, E (1985). Cell 40:191.
40. Hyafil F, Babinet C, Huet C, and Jacob F (1983). In Teratocarcinoma Stem Cells. L.M. Silver, G.R. Martin, and S. Strickland (eds). New York: Cold Spring Harbor Laboratories, pp. 197-207.

Molecular Determinants of Animal Form, pages 253-270
© 1985 Alan R. Liss, Inc.

DETECTION USING MONOCLONAL ANTIBODIES OF A STRUCTURALLY ALTERED FORM OF CELL-CAM 105 ON RAT HEPATOCELLULAR CARCINOMAS.

Douglas C. Hixson and Kerry D. McEntire.

University of Texas System Cancer Center,
Science Park - Research Division,
Smithville, Texas.

Abstract

In an effort to determine the molecular basis for the aberrant behavior of malignant cells, we have used polyclonal and monoclonal antibodies to delineate changes in the expression of cell adhesion molecules during hepatocarcinogenesis. To date we have demonstrated the involvement of two M_r 105,000 acidic glycoproteins (cell-CAM 105) in the initial adhesive interactions of rat hepatocytes. This was suggested by the ability of Fab fragments prepared from anti gp105-2, a heteroantiserum specific for these components, to block reaggregation in suspension and formation of lateral contacts in primary culture. Subsequent immunochemical and immunocytochemical analysis with monoclonal antibodies (MAb) revealed that cell-CAM 105 is phosphorylated and contains a protease sensitive site, two properties it shares in common with L-CAM, the cell adhesion molecule from embryonic chicken liver. Major differences in size, tissue distribution and sensitivity to endoglycosidase digestion suggested, however, that cell-CAM 105 and L-CAM are structurally and perhaps functionally distinct molecules. Results of immunofluorescence analysis of isolated cells and immunoprecipitation analysis with antibodies against a detergent resistant fraction of isolated plasma membranes were consistent with the localization of cell-CAM 105 in intermediate junctions. Immunoprecipitation analysis revealed that on some transplantable hepatocellular

This work was supported by National Institutes of Health Grant CA-31103.

carcinomas, cell-CAM 105 is no longer expressed while on others it is present in a structurally altered form which exhibits a more basic pI than its counterpart on normal hepatocytes. Immunochemical analyses performed to date suggested the altered mobility of cell-CAM 105 on THC cells does not involve major alterations in the peptide moiety.

INTRODUCTION

In the past, investigations into the role of intercellular adhesion in the altered interactions of malignant cells have focused on changes in the frequency or morphology of specialized junctions, the most readily identified elements in the adhesive repertoire of epithelial cells (1 - 3). Results from these predominantly ultrastructural investigations have been inconsistent, with some studies suggesting an association between decreased junctional frequency and malignancy and others concluding that junctional adhesion in malignant tissues was similar to that in normal epithelia and non-invasive tumors (4,5). However, the availability of monoclonal and polyclonal antibodies against isolated junctional components (6-9) has made it possible to delineate recent studies in the structure and distribution of specialized junctions in tumors which are not apparent at the ultrastructural level.

Identification of cell adhesion molecules (CAMs) which mediate the initial and perhaps short term interactions which occur prior to junction formation has proved to be a more difficult problem primarily because these components cannot be isolated as part of a discrete membrane structure. It wasn't until the development of an immunological assay by Gerish (10) that it became possible to definitively identify putative CAMs. In this assay, putative CAMs were identified by their ability to neutralize cell-surface specific antibody fragments which have been previously shown to block the re-aggregation of dissociated cells. As is evident from these proceedings, this approach has been used successfully to identify CAMs on cells from a number of different tissues and species. Interestingly, it appears that a number of these CAMs, namely cell-CAM 120/80 from mammary tumor cells (11), uvomorulin (12) and ECCD-1 antigen (13) from teratocarcinoma cells and L-CAM from embryonic chicken liver (14), are very similar in size and in sensitivity to trypsin digestion and display a similar distribution in adult and embryonic tissues.

Thus, these CAMs appear to be closely related molecules which serve similar functions in their respective tissues.
With the availability of specific anti CAM antibodies, it has now become possible to critically evaluate the role of cell adhesion in determining both normal cellular interactions and the aberrant behavior characteristic of malignant cells. In this regard, we have recently reported that the expression of cell-CAM 105 is decreased or lost on several different transplantable hepatocellular carcinomas (15). In addition, Greenberg et al. have shown that N-CAM is dramatically reduced following transformation of neuronal cells with Rous sarcoma virus (16).

In this paper we extend our previous studies and describe results of immunochemical analysis with a monoclonal antibody (MAb 362.50) specific for cell-CAM 105 which show that while cell-CAM 105 is indeed lost on some THC lines, on others, it is expressed in a structurally altered form. We also discuss the relationship of cell-CAM 105 to other CAMs and present preliminary data which supports the suggestion of Ocklind et al. (17) that cell-CAM 105 may be localized in intermediate junctions.

RESULTS AND DISCUSSION

Identification of a Pair of Acidic Wheat Germ Agglutinin Binding Proteins Missing on THC Cells as CAMs Involved in the Reaggregation of Adult Rat Hepatocytes

In our initial studies, we focused on examining the adhesion inhibitory activity of Fab fragments prepared from heteroantisera raised against subpopulations of cell surface glycoproteins isolated by lectin affinity chromatography from detergent extracts of rat hepatocytes. Antisera showing adhesion inhibitory activity were further characterized by immunoprecipitation analysis to determine their reactivity with detergent extracts of hepatocytes and transplantable hepatocellular carcinomas (THC) surface labeled with ^{125}I. Since many of these THC lines showed altered adhesive interactions in vitro or in vivo, e.g. lack of adherance to hepatocyte monolayers, growth in suspension as single cells or small aggregates or decreased ability to form specialized junctions, we considered components present on normal hepatocytes but missing from THC immunoprecipitates to be the strongest candidates for CAMs.
Results from our initial characterization by SDS-PAGE of ^{125}I-labeled glycoproteins bound to various immobilized

lectins showed that all of the major wheat-germ agglutinin binding components in normal hepatocyte extracts migrated as a single broad band of Mr 105,000 (18). This major band, designated gp105 was also present in the bound fractions eluted from immobilized Ricinus communis agglutinin and Lens culinaris agglutinin. When tested for adhesion inhibitory activity, Fab fragments prepared from rabbit antiserum against gp105 effectively blocked the reaggregation of hepatocytes isolated by collagenase perfusion (Figure 1).

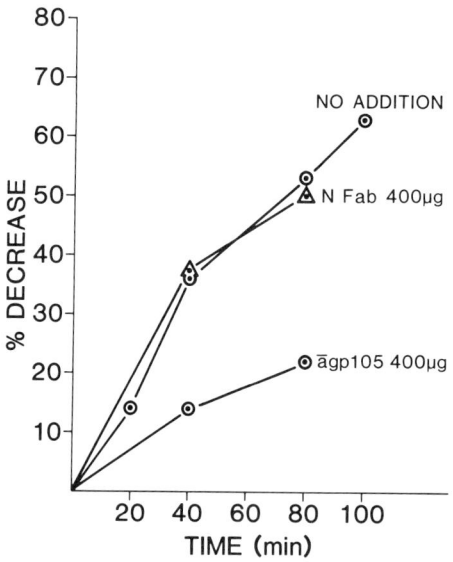

Figure 1. Inhibition of the reaggregaton of isolated hepatocytes by Fab fragments from anti gp105 antiserum. Reaggregation (expressed as the percent decrease in single cells) was inhibited 60% at a concentration of 400 µg/3 x 10^5 cells.

Two dimensional gel analysis of anti gp105 immunoprecipitates revealed however, that what appeared as a single component on one dimensional gels was actually comprised of 9 components with similar molecular weights but differing

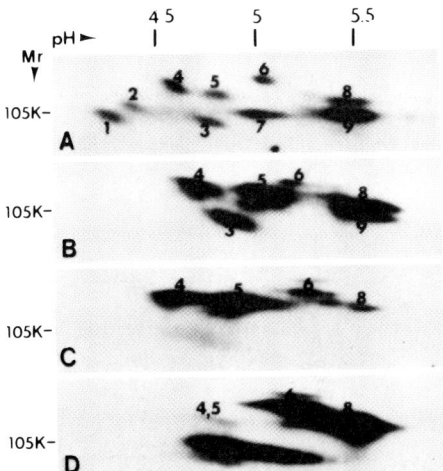

Figure 2. Two dimensional electrophoretic analysis of components immunoprecipitated with anti gp105 from extracts of ^{125}I-labeled normal hepatocytes (A), THC 1677 (B), THC 1682C (C) and THC AS-30D (D). The two most acidic components (components 1 and 2) were missing from two dimensional gel profiles of THC immunoprecipitates.

significantly in pI (15,18)(Figure 2). Immunoprecipitation analysis of THC extracts (Figure 2) revealed both qualitative and quantitative changes in the expression of anti gp105 reactive components with the most consistent change being the apparent loss of a pair of acidic (pI 4.1 - 4.3) glycoproteins (15). Comparison of one dimensional peptide maps of each of the nine components indicated that these two acidic components were closely related in structure but differed significantly from other anti gp105 reactive components (15).

Since the size and lectin binding properties of these acidic glycoproteins were very similar to those reported for cell-CAM 105, the rat hepatocyte CAM isolated by Ocklind and Öbrink (19), we compared by immunodepletion analysis (20) the reactivity of anti gp105 and anti cell-CAM$_2$, an

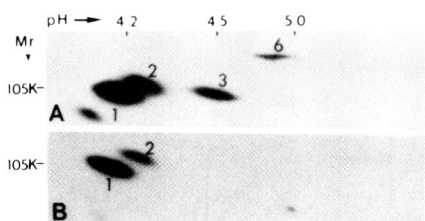

Figure 3. Two dimensional electrophoretic analysis of components immunoprecipitated from extracts of ^{125}I-labeled hepatocytes with anti cell-CAM$_2$ (A) and anti gp105-2 (B) antiserum. Both antisera showed strong reactivity with the two acidic components absent from two dimensional gel profiles of THC immunoprecipitates.

antiserum raised against purified cell-CAM 105 (17,19). Two dimensional gel analysis of anti cell-CAM$_2$ immunoprecipitates (Figure 3) indicated that this antiserum was reactive with 4 (components 1,2,3 and 6) of the 9 components recognized by anti gp105 (15). Of particular interest was the strong reactivity with the two acidic components missing from two dimensional gel profiles of anti gp105 reactive components on THC cells.

Confirmation that the Acidic Components Recognized by Anti gp105 and Anti Cell-CAM$_2$ Antisera are CAMs.

That the two acidic components were adhesion molecules was demonstrated using anti cell-CAM-$_8$ (15) and anti gp105-2, two rabbit antisera specific for cell-CAM 105. In preparing the latter antiserum, advantage was taken of the fact that cell-CAM 105 is the most immunogenic of the components bound to wheat germ agglutinin. Serum (anti gp105-2) was collected after two immunizations with liposomes containing the bound fraction from wheat germ agglutinin which had been immunodepleted with monoclonal antibodies against components 7,8 and 9 (21-23). This serum was very specific and showed no

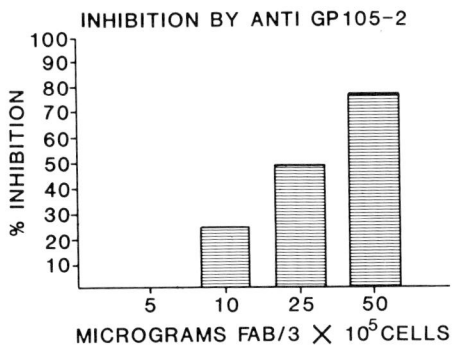

Figure 4. Percentage decrease in hepatocyte reaggregation produced by various concentrations of Fab fragments prepared from anti gp105-2 antiserum.

reactivity with other components in extracts of hepatocytes surface labeled with ^{125}I (Figure 3).
As shown in Figure 4, Fab fragments prepared from anti-gp105 antiserum effectively blocked aggregation of hepatocytes in suspension, giving 77% inhibition at a concentration of 50ug/3 x 10^5 cells. Similar results were also obtained with Fab from anti cell-CAM_8 antiserum (15). In addition, anti gp105-2 Fab fragments were also capable of inhibiting the formation of lateral contacts between hepatocytes attached to collagen gels (24). On the basis of these results, it was thus concluded that the two acidic glycoproteins recognized by anti gp105, anti cell-CAM_2 anti cell-CAM_8 and anti gp105-2 antisera play a role in the initial adhesive interactions of rat hepatocytes.

Examination Using Monoclonal Antibodies Of The Expression Of Cell-CAM 105 By THC Cells

To construct hybridomas secreting antibodies specific for cell-CAM 105, 8643 myeloma cells were fused with spleen cells from Balb/c mice immunized by a protocol similar to that used to prepare anti gp105-2 antiserum. Three hybridoma cultures producing antibodies specific for cell-CAM 105 were identified by a radiometric binding assay and immunoprecipitation analysis. One of these cultures, designated 362.50, was subsequently cloned in soft agar.

Two dimensional gel analysis of components immunoprecipitated from extracts of radioiodinated hepatocytes and THC cells revealed that the mobility of cell-CAM 105 on three THC lines [THC 1677 (15), THC 1682c (15) and THC AS-30D (25)] differed significantly from that on normal hepatocytes with cell-CAM 105 from all three THC showing a more basic pI (4.4 - 4.8) than cell-CAM 105 from normal hepatocytes (Figure 5). In addition, cell-CAM 105 from THC AS-30D displayed greatly increased microheterogeneity and appeared as a family of 14 or more distinct spots spread over a broad range of

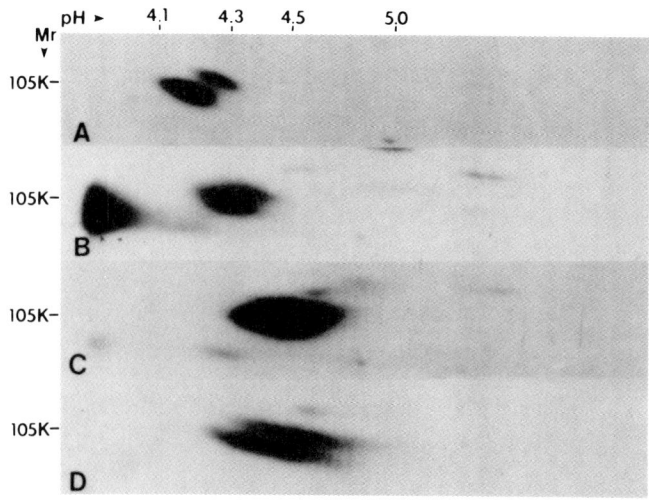

Figure 5. Two dimensional gel analysis of components immunoprecipitated with MAb 362.50 from extracts of ^{125}I-labeled hepatocytes (A), THC 1677 (B) THC 1682c (C) and THC AS-30D (D). Note the more basic pI of cell-CAM 105 from THC cells.

pI's. Thus, it appeared that for these three THC lines, the apparent loss of cell-CAM 105 from two dimensional gel profiles of anti gp105 immunoprecipitates involved a structural alteration in cell-CAM 105 which resulted in its co-migration with component 3 (Figure 2). However, further examination of a number of other THC lines indicated that cell-CAM 105 was indeed absent and could not be detected by immunoprecipitation or immunofluorescence analysis.

Characterization Of Cell-CAM 105 From Normal And Malignant Hepatocytes

A number of different biochemical and immunochemical techniques are currently being employed to determine the structural basis for the altered mobility of cell-CAM 105 on THC cells. As shown in Figure 6, one dimensional peptide maps

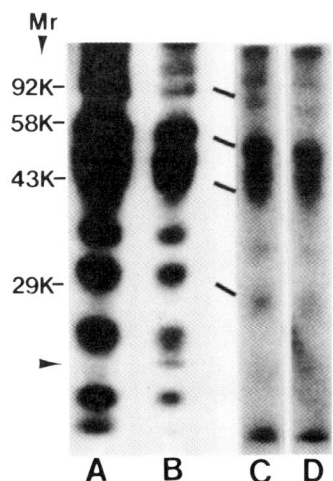

Figure 6. One dimensional V-8 peptide maps prepared by disgestion of cell-CAM 105 immunoprecipitated from extracts of ^{125}I-labeled hepatocytes (A and B), THC 1682c (C) and THC AS-30D (D). Digestion was done as previously described with 0.5 ug of of V-8 protease. Panels A and B show maps of the lower (component 1) and upper (component 2) forms of cell-CAM 105. Note the extra peptide (arrow) in the map from the upper component.

prepared following digestion with V-8 protease are essentially identical, suggesting that the altered mobility does not involve major differences in the peptide moieties. Further, digestion with endoglycosidase H does not alter the mobility on two dimensional gels of either the normal or the THC 1682c form of cell-CAM 105, suggesting that neither form has any unfucosylated high mannose oligosaccharides (26). Similarly, preliminary results indicate that both hepatocyte and tumor cell-CAM 105 are susceptable to neuraminidase. However, cell-CAM 105 from THC 1682c cells still exhibits a more basic pI (4.8 - 5.2) than its hepatocyte counterpart (pI 4.4 - 4.8) even after extended digestion (6 hrs at 37 C with 25 U/ml). This may indicate that there is less sialic acid on THC cell-CAM 105 and that the sialic acid on hepatocytes is more resistant to digestion or alternatively that hepatocyte cell-CAM 105 contains additional negatively charged groups other than sialic acid which are not present or are less numerous on THC cell-CAM 105 e.g. phosphate or sulfate groups.

Comparison Of Cell-CAM 105 To Other CAMs

As was mentioned above, L-CAM, cell-CAM 120/80, uvomorulin and ECCD-1 antigen appear to represent a group of closely related CAMs of which L-CAM is the most extensively characterized. Although cell-CAM 105 shares some characteristics with these other CAMs, it differs from them in a number of important aspects. For example, immunoblot analysis of whole liver cell membranes with monoclonal antibodies specific for L-CAM have revealed reactive components of Mr 124,000, 94,000 and 81,000 (14). Although on occasion a smaller 90 kd component has been observed in MAb 362.50 immunoprecipitates and Ocklind et al (27) have described a 70 kd component which is antigenically related to cell-CAM 105, no molecular weight forms greater than 105 kd have ever been detected even when extractions were done in the presence of protease inhibitors or when hepatocytes were allowed to recover prior to radiolabeling for 12 hours in culture.

Preliminary results from studies in our laboratory also suggest that L-CAM and cell-CAM 105 differ significantly in their sensitivity to digestion with endoglycosidases. Whereas cell-CAM 105 shows no change in either size or pI following digestion with endoglycosidase H (25 mU/ml for 17 hr at 30 C) and thus does not appear to have any high mannose or unfucosylated oligosaccharides of the high mannose type, L-CAM showed a small but significant change in molecular weight consistent with the presence of one high mannose oligosaccharide (28). In contrast, cell-CAM 105 showed a much greater change in size after digestion with peptide: N-glycosidase F (12 U/ml for 17 hrs at 30 C) than that reported for L-CAM following digestion with endoglycosidase F and displayed after digestion an apparent molecular weight of approximately 55,000 as opposed to 110,000 for L-CAM (28). From these preliminary results, it appears therefore that the N-linked oligosaccharides on cell-CAM 105 are either more numerous or are larger in size than those present on L-CAM.

Aside from these structural differences, cell-CAM 105 also shows a more restricted tissue distribution than L-CAM, cell-CAM 120/80, uvomorulin and ECCD-1 antigen. Thus while the L-CAM related CAMs have been found in liver, bile ducts, lung, intestine, skin, kidney and a number of other adult and embryonic tissues (11-13, 29,30), in indirect immunofluorescence assays with MAb 362.50, we have detected cell-CAM 105 only in the liver and kidney (Table 1). Even in the liver,

INDIRECT IMMUNOFLUORESCENCE ANALYSIS OF MAB 362.50
REACTIVITY WITH NORMAL RAT TISSUES

	INTENSITY	LOCALIZATION
LIVER	++++	Canaliculi
TONGUE MUCOSA	-	
TRACHEA	-	
STOMACH	-	
SMALL INTESTINE	-	
LARGE INTESTINE	-	
KIDNEY	++	Tubules
BLADDER	-	
ADRENAL GLAND	-	
THYMUS	-	
SKELETAL MUSCLE	-	
HEART	-	
LUNG	-	
BRAIN	-	
CORNEA	-	
SKIN	-	

Table 1. Localization of cell-CAM 105 in frozen sections of normal rat tissues stained by indirect immunofluorescence with MAb 362.50.

cell-CAM 105 shows a more restricted distribution than L-CAM and is localized in the pericanalicular domain of the hepatocyte plasma membrane (Figure 8). Further, in contrast to L-CAM which is present throughout fetal development, cell-CAM 105 appears to be expressed only in the later stages of fetal development since it is present in 19 day but not in 15 day fetal liver. Similar to other CAMs, however, cell-CAM 105 has never been found in any non-epithelial tissues.

On the other hand, cell-CAM 105, like L-CAM (14,28), is phosphorylated (Fig 7) and appears to have a protease sensitive site.

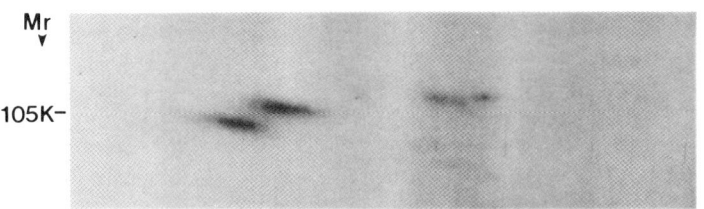

Figure 7. Cell-CAM 105 immunoprecipitated with MAb 362.50 from detergent lysate of 5×10^5 hepatocytes which had been labeled in primary culture for 22 hrs with media containing 1mCi of $^{32}PO_4$.

This latter characteristic is indicated by the fact that the lower molecular weight V-8 peptides from the upper and lower forms of cell-CAM 105 are identical in size (Figure 6), suggesting that a small peptide is rapidly cleaved from the higher molecular weight form producing a large fragment essentially identical in size to the lower form. More importantly, as we have previously discussed (15) the two forms of cell-CAM 105 do not appear to be artefacts of proteolysis during extraction or immunoprecipitation. Indeed, these two forms have been observed in approximately the same relative amounts in extracts prepared from over 30 different preparations of freshly isolated or cultured hepatocytes, an observation which suggests that the conversion of the upper to lower form may be biologically significant and play a necessary role in the function of cell-CAM 105.

Localization in Intermediate Junctions

Since the pericanalicular region of the hepatocyte membrane is rich in specialized junctions, Ocklind et al (17) have suggested that cell-CAM 105 may be a component of intermediate junctions. This concept is consistent with the presence of fluorescent plaques on the surface of isolated hepatocytes stained by indirect immunofluorescence with MAb 362.50 (Figure 8).

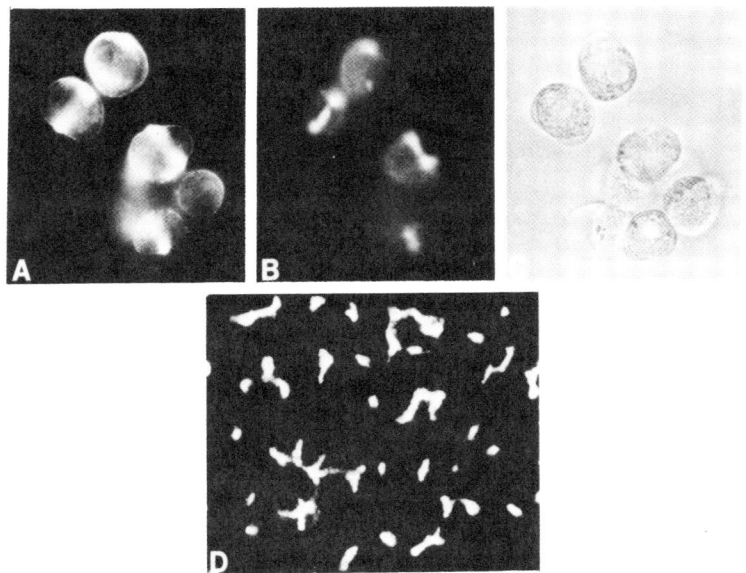

Figure 8. Isolated hepatocytes (A and B) and frozen sections of normal rat liver (D) stained by indirect immunofluorescence with MAb 362.50. Bands and patches of fluorescence were seen in isolated hepatocytes even when all incubations were performed at 4°C in the presence of azide. Panel B shows the upper surface and panel C shows a phase contrast photomicrograph of cells in (A). Fluorescence in frozen sections (D) was localized in the pericanalicular regions of the hepatocyte plasma membrane.

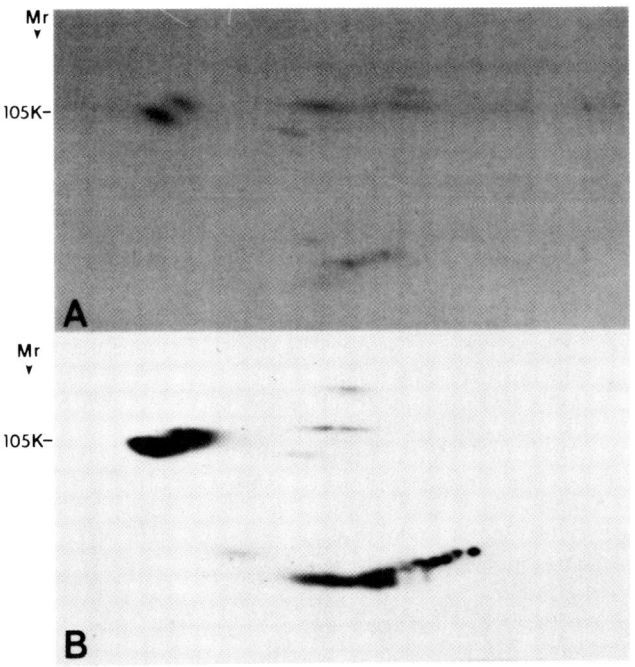

Figure 9. Immunoprecipitation of extracts from ^{125}I-labeled hepatocytes with (A) antisera raised in a rabbit against the detergent resistant fraction of purified hepatocyte plasma membranes; (B) antibodies purified from the antiserum in (A) by affinity chromatography on hepatocyte wheat germ agglutinin binding proteins which had been immobilized on Sepharose. Note the strong reactivity with cell-CAM 105.

Although confirmation will have to wait the results of immunoelectron microscopic analysis currently in progress, it seems possible that these plaques represent previous sites of junctional contact containing stable aggregates of cell-CAM 105 which did not disperse following dissociation into single cells.

Results of immunoprecipitation analysis with antiserum raised against isolated junctions also supports the localization of cell-CAM 105 in intermediate junctions. Antiserum reactive with specialized junctions was raised by immunizing a rabbit with a membrane fraction enriched for intermediate and desmosomal junctions and depleted of detergent-soluble cell-CAM 105 by detergent extraction and sucrose density gradient centrifugation. Antibodies reactive with cell-surface components were isolated by affinity chromatography on Sepharose-bound wheat germ agglutinin binding glycoproteins previously isolated from detergent extracts of hepatocytes on immobilized wheat germ agglutinin. As shown in Figure 9, the major reactivity of both the unpurified antiserum and purified antibodies was against cell-CAM 105, suggesting that at least a portion of cell-CAM 105 on the cell surface is localized in detergent resistant specialized junctions. Although desmosomal junctions are known to contain a 100 kd glycoprotein (7,8), it seems unlikely that this component is cell-CAM 105 since the pattern of fluorescence with MAbs specific for rat liver desmosomes is very different from that observed with MAb 362.50.

Role of Cell-CAM 105 in the Interactions of Normal and Malignant Cells

From the previous discussion, it seems clear that cell-CAM 105 is structurally and perhaps functionally distinct from L-CAM and other L-CAM related cell adhesion molecules. The presence of cell-CAM 105 in 19 but not 15 day fetal liver raises the possiblility that this molecule may be an adult CAM with a more restricted function than L-CAM. Alternatively, cell-CAM 105 may be an accessory CAM [analogous to Ng CAM on neuronal cells (31)] which functions in mediating hepatocyte interactions in concert with the as yet unidentified rat counterpart to L-CAM. Critical analysis of THC lines which have lost the expression of cell-CAM 105 may give insight into the role of cell-CAM 105 in normal cellular interactions. Conversely, detailed information on the biochemical processing of this molecule in normal cells may help to clarify the basis for the structural alterations on THC cell-CAM 105 and the effects these alterations have on function.

ACKNOWLEDGEMENT

We thank Judith Chesner and Jeanette Mowery for their excellent technical assistance. This work was supported by NIH Grant CA 31103.

REFERENCES

1. Weinstein RS (1976). The structure and function of intercellular junctions in cancer. Adv Cancer Res 23:23.
2. Staehelin LA (1974). Structure and function of intercellular junctions. Int Rev Cytol 39:191.
3. McNutt NS, Weinstein RS (1973). Membrane ultrastructure of mammalian intercellular junctions. Prog in Biophys Molec Biol 26:47.
4. Pitelka DR, Jamamoto ST, and Taggart BN (1980). Epithelial cell junctions in primary and metastatic mammary tumors of mice. Cancer Res 40:1588.
5. Pauli UB, Cohen SM, Alroy J, Weinstein RS (1978). Desmosome ultrastructure and the biological behavior of chemical carcinogen-induced urinary bladder carcinomas. Cancer Res 38:3276.
6. Hixson DC, McEntire KD, Donahue KL, Chesner JE (1983). Detection of alteration in adhesive interactions during hepatocarcinogenesis using monoclonal antibodies. J Cell Biol 97:85a.
7. Cowin P, Garrod DR (1983). Antibodies to epithelial desmosomes show wide tissue and species cross-reactivity. Nature 302:148.
8. Cohen SM, Gorbsky G, and Steinberg MS (1983). Immunochemical characterization of related families of glycoproteins in desmosomes. J Biol Chem 258:2621.
9. Meuller H, Franke WW (1983). Biochemical and immunological characterization of desmoplakins I and II, the major polypeptides of the desmosomal plaque. J Mol Biol 163:647.
10. Gerisch G (1977). Univalent antibody fragments as tools for analysis of cell-cell interactions in Dictyostelium. Curr Topics Dev Biol 14:243.
11. Damsky CH, Richa J, Solter D, Knudsen K, Buck CA (1983). Identification and purification of a cell surface glycoprotein mediating intercellular adhesion in embryonic and adult tissue. Cell 34:455.

12. Hyafil F, Morello D, Babinet C, and Jacob F (1980). A cell surface glycoprotein involved in the compaction of embryonal carcinoma cells and cleavage stage embryos. Cell 21:927.
13. Ogou S-I, Yoshida-Noro C, Takeichi M (1983). Calcium-dependent cell-cell adhesion molecules common to hepatocytes and teratocarcinoma stem cells. J Cell Biol 97:944.
14. Gallin WJ, Edelman GM, Cunningham BA (1983). Characterization of L-CAM, a major cell adhesion molecule from embryonic liver cells. Proc Natl Acad Sci USA 80:1038.
15. Hixson DC, McEntire KD, and Obrink B (1985). Alterations in the expression of a hepatocyte cell adhesion molecule (cell-CAM 105) by transplantable hepatocellular carcinomas. Cancer Res, in press.
16. Greenburg ME, Brackenbury R, and Edelman GM (1984). Alteration of neural cell adhesion molecule (N-CAM) after neuronal cell transformation by Rous sarcoma virus. Proc Natl Acad Sci USA 81:969.
17. Ocklind C, Forsum U, and Obrink B (1983). Cell surface localization and tissue distribution of a hepatocyte cell-cell adhesion glycoprotein (cell-CAM 105). J Cell Biol 96:1168.
18. Hixson DC, Allison JP, Chesner JE, Leger MJ, Ridge LL, Walborg EF Jr (1983). Characterization of a family of glycoproteins associated with the bile canalicular membrane of normal hepatocytes but not expressed by two transplantable rat hepatocellular carcinomas. Cancer Res 43:3874.
19. Ocklind C, Obrink B (1982). Intercellular adhesion of rat hepatocytes. Identification of a cell surface glycoprotein involved in the initial adhesion process. J Biol Chem 257:6788.
20. Hixson DC, Allison JP, McEntire KD, Nairn RS, Chesner JE, Walborg EF Jr (1984). Structural differences in envelope glycoproteins associated with rat leukaemia virus produced by Novikoff hepatocellular carcinoma and spontaneously transformed Wistar rat embryo cells. J Gen Virol 65:743.
21. Allison JP, Hixson DC, Lund J, Chesner JE, Ridge LL, Leger MJ, McEntire KD, Walborg EF Jr (1982). Monoclonal antibodies as probes of surface antigenic alterations during experimental carcinogenesis in the rat. In T. Galeotti et al (eds): "Membranes in Tumour Growth," Amsterdam: Elsevier Biomedical Press, p 599.
22. Allison JP, Hixson DC (1983). Monoclonal antibodies as probes of surface alterations associated with hepatocarcinogenesis in the rat. Fed Proc 42:1178.

23. Hixson DC, Ponce MD, Allison JP, Walborg EF Jr (1984). Cell surface expression on adult rat hepatocytes of a noncollagen glycoprotein present in rat biomatrix. Exp Cell Res 752:402.
24. Mowery J, Carter C, Hixson D (1985). Microassay for measuring antibody-mediated inhibition of lateral contact formation between rat hepatocytes. Fed Proc 44:1437.
25. Smith DF, Walborg EF Jr, Chang JP (1970). Establishment of a transplantable ascites variant of a rat hepatoma induced by 3'-methyl-4-dimethylaminoazobenzene. Cancer Res 30:2306.
26. Hughs RC (1983). "Glycoproteins." New York: Chapman and Hall, p 23.
27. Ocklind C, Odin P and Obrink B (1984). Two different cell adhesion molecules - cell-CAM 105 and a calcium dependent protein - occur on the surface of rat hepatocytes. Exp Cell Res 151:29.
28. Cunningham BA, Leutzinger Y, Gallin WJ, Sorkin BC, Edelman GM (1984). Linear organization of the liver cell adhesion molecule L-CAM. Proc Natl Acad Sci USA 81:5787.
29. Edelman GM, Gallin WJ, Delouvee A, Cunningham BA, Thiery J-P (1983). Early Epochal maps of two different cell adhesion molecules. Proc Natl Acad Sci USA 80:4384.
30. Peyrieras N, Hyafil F, Louvard D, Ploegh HL, Jacob F (1983). Uvomorulin: A nonintegral membrane protein of early mouse embryo. Proc Natl Acad Sci USA 80:6274.
31. Hoffman S, these proceedings.

Molecular Determinants of Animal Form, pages 271-292
© 1985 Alan R. Liss, Inc.

FROG GASTRULA CELLS ADHERE TO FIBRONECTIN-SEPHAROSE BEADS[1]

Kurt E. Johnson, Ph.D.

Department of Anatomy
George Washington University Medical Center
Washington, D. C.   20037   U.S.A.

Experiments were performed to examine adhesion of embryonic amphibian cells to beads coated with fibronectin (FN). Blastula cells from three different species of amphibia show poor adhesion to FN-Sepharose beads. Beginning at the early gastrula stage, however, embryonic cells show a progressively increasing tendency to adhere to FN-beads. In one arrested hybrid embryo, cells from all developmental stages lack the ability to adhere to FN-beads.
    Blastula stage cells have the ability to adhere to con A-beads and two kinds of cytodex beads as well but will not adhere to FN-beads. Similarly, cells from arrested hybrid embryos lack the ability to adhere to FN-beads, but will adhere to con A-beads and cytodex beads. Observations in the light and scanning electron microscope show that normal cells form lamellipodia on FN-beads and move about actively on the surface of these beads, much like they do in vivo on surfaces coated by frbrils containing fibronectin.

INTRODUCTION

During gastrulation, prospective mesodermal cells adhere to and migrate across the inner surface of the roof of the blastocoel of early amphibian gastrulae. This cellular substratum is composed of presumptive ectodermal cells and gains an anastomosing network of thin extracellular fibrils during gastrulation (1). In urodeles,

[1]This work was supported by NSF DCB 84 00256.

these fibrils are oriented preferentially along the animalpole-blastopore axis. When deposited on plastic substrata in vivo, these fibrils promote cell attachment and oriented locomotion, suggesting that these oriented fibrils constitute a contact guidance system in vivo, causing oriented mesodermal cell migration (2,3). Indeed, when the orientation of fibrils is changed by production of artificial tension in vitro, and these fibrils are deposited onto artificially conditioned substrata, the orientation of mesodermal cell migration is changed (4).

These fibrils contain fibronectin (FN) in the urodeles Ambystoma mexicanum and Pleurodeles waltlii (5,6) and the anuran Xenopus laevis (7), as judged by their ability to bind labelled antibodies to fibronectin. Antibodies to fibronectin have been injected into the blastocoel of early gastrulae and they prevent gastrulation but not neurulation when they are injected after gastrulation is completed (8). Furthermore, the cell binding peptide of fibronectin prevents gastrulation when injected into the blastocoel of urodele gastrulae while the collagen binding peptide of fibronectin has no effect on gastrulation(9). Fibronectin is synthesized prior to gastrulation and continues to be synthesized during gastrulation (10-12) but does not accumulate on the inner aspect of the roof of the blastocoel as visible, stainable fibrils until after gastrulation begins (13). Presumably, fibronectin is released into the blastocoel fluid and secondarily accumulates preferentially on the inner aspect of the roof of the blastocoel. The present study tests the ability of dissociated cells taken from blastulae and gastrulae of Rana pipiens, Rana sylvatica, and Ambystoma maculatum to adhere in vitro to CNBr-Sepharose beads coupled covalently to fibronectin. I have also tested the ability of cells from arrested interspecific hybrid embryos to adhere to these beads. The results show that blastula cells from normal embryos and cells from all stages of arrested hybrid embryos do not adhere significantly to FN-Sepharose beads. In addition, these results show that early gastrula cells adhere slightly, mid-gastrula cells adhere moderately, and late gastrula cells adhere strongly to Sepharose beads covalently coupled to fibronectin.

MATERIALS AND METHODS

Rana pipiens females, Rana pipiens males, and Rana catesbeiana males were purchased from Nasco(Fort Atkinson, Wisconsin). Normal Rana sylvatica embryos were collected at the two-cell stage in Fairfax County, Virginia. Ambystoma maculatum embryos were collected at early cleavage stages with the kind help of Dr. Albert K. Harris of the University of North Carolina at Chapel Hill around Chapel Hill, North Carolina or by the author in Fairfax County, Virginia. Rana pipiens females were ovulated by pituitary injection (14). Eggs were fertilized in either Rana pipiens (normal) or Rana catesbeiana (hybrid) sperm suspensions made in 10% Wolf-Quimby balanced salt solution (WQBSS) (15) and embryos developed at 22° or 16° C. in 10% WQBSS until the appropriate stages. Embryos were staged according to Shumway (16). Ambystoma maculatum embryos were raised in 10% WQBSS and staged according to Harrison (17). Jelly coats were removed from Rana embryos by incubating them for 15 minutes in 0.7% mercaptoacetic acid in 50% WQBSS with the pH adjusted to 8.7 with 5 N NaOH after addition of mercaptoacetic acid followed by ten washes in 10% WQBSS. Embryos were reared to appropriate stages in 10% WQBSS. Ambystoma embryos were removed from jelly capsules with Dumont #5 watchmaker's forceps. For dissection and dissociation, embryos were transferred to Ca++, Mg++-free WQBSS (CMF) with 2 mM EDTA added, had vitelline membranes removed manually with sharpened Dumont #5 watchmaker's forceps, and were dissected into appropriate fragments, and incubated for 1 hour at 22° C. for dissociation.

CNBr-Sepharose beads for cell adhesion assays were prepared according to manufacturer's instructions (Pharmacia). Bovine plasma fibronectin (Sigma) was dissolved at 1 mg/ml in coupling buffer and 1 ml of this solution was mixed with 1 ml of beads. Blank beads were prepared by incubating washed and swollen beads in 1 M ethanolamine in coupling buffer. Sephadex G-200 beads, cytodex-1, and cytodex-3 beads (Pharmacia) were swollen in WQBSS and washed three times before use in WQBSS containing penicillin G (100 IU/ml) and streptomycin sulfate (1 ug/ml). Sepharose CL-48 beads (Pharmacia) were purchased pre-swollen and were washed three times in WQBSS plus antibiotics. After all coupling and washing steps, beads were allowed to settle out of suspension in graduated centrifuge tubes. Supernatant washes were removed carefully with a pasteur pipette and enough WQBSS plus antibiotics was added to make a 50% v/v suspension of beads.

To set up cultures, Falcon Multiwell tissue culture plates (#3007) with 24 wells in each plate and a capacity of about 2 ml/well received 100 ul aliquots of 50% v/v bead suspensions. This was the correct number of beads to give approximately a single layer of close packed beads on the bottom of the tissue culture plate wells. Next, 1.0 ml of WQBSS + 0.5% BSA + antibiotics was added to each well. These plates were set up while tissue fragments were dissociating in EDTA. After 1 hour of dissociation, cells were gently collected in pasteur pipettes and then expelled into wells containing beads and culture medium. The dissociated fragments were aspirated up and down in the pasteur pipette three times to facilitate dissociation. Then, cells and beads were allowed to settle out of suspension for five minutes and then 0.9 ml of medium was removed along with contaminating EDTA and replaced with a fresh 1.0 ml of WQBSS + 0.5% BSA + antibiotics. Cultures were now transferred to a water bath maintained at 24° C. and equipped with a hood and water trap for creating a 5% CO_2-95% air atmosphere to maintain pH at 7.4. Cultures were incubated for 4 hours and then fixed by the addition of 100 ul of 2.5% glutaraldehyde in 0.05 M PIPES buffer, pH 7.3 with 5 mM $CaCl_2$ added. Cultures for kinetic studies were fixed in the same way after variable incubation times. After fixation, the contents of the well were aspirated into a pasteur pipette and dispersed into a 60 X 15 mm plastic petri dish (Falcon #1007) and then examined in a dissecting microscope at 50X magnification. The beads are large, opalescent, spherical structures that are easily distinguishable from the smaller embryonic cells (FIGURE 3). The cells were sometimes darkly pigmented and always had a granulated appearance due to their yolk platelets. Cells attached to beads were flattened to beads to a variable extent and could not be dislodged from beads by gentle prodding with the tips of forceps. Each culture was scanned and the first 500 beads encountered were scored for number of cells attached. A running total of the number of beads and cells was kept on a differential blood counter so that after 500 beads had been scored, a statistic of number of attached cells/500 beads was recorded. Five replicas were produced for each condition and the average and standard deviation for each determination was calculated.

For scanning electron microscopy, suspensions of beads and cells were transferred to 2.5% glutaraldehyde in 0.05 M PIPES buffer, pH 7.3 with 5 mM $CaCl_2$ added and fixed at 22° C. overnight. Then beads and attached cells were washed in PIPES buffer and then post-fixed in 1% OsO_4 in the same buffer. Following post-fixation, beads were washed again with buffer, transferred to BEEM capsules designed to be used for embedding specimens in plastic for thin sectioning. The cover of the BEEM capsule was cut across the top so that most of the cover was cut away leaving the lower portion of the tight fitting cap as a collar around the outside of the BEEM capsule. A small piece of Nitex screening (45 um squares) was then placed over the top of the BEEM capsule and fixed in place with the collar remaining from the cut top. This allowed dehydration and critical point drying of large numbers of beads without losing any of them during dehydration and critical point drying. After dehydration in a graded ethanol series and critical point drying in liquid CO_2, beads were transferred to SEM stubs coated with double sticky tape, coated with 40 nm of gold-palladium, and viewed in either a JSM-35 or an ISI scanning electron microscope. Dr. John B. Morrill of New College of the University of South Florida kindly helped with the use of his ISI SEM and with the production of stereo pairs of lamellipodia. For light microscopy and time-lapse cinemicrography, standard microscope slides had a thin rectangle of petroleum jelly applied to one surface to serve as a dam for a bead suspension. Suspensions of living cells and beads were then transferred to the slides, excess medium was removed after cells and beads had settled out, and the preparation was sealed with a coverslip. These preparations were viewed while alive under phase contrast optics. Preparations were photographed on Panatomic X film developed with Microdol-X, Ektachrome-ASA 160 color film processed commercially, or in a Nikon CFMA cinemicrography apparatus using 16 mm Plus X-Reversal film processed commercially. Films were analyzed on a Lafayette photooptical data analyzer and tracings of individual frames were made from projected films.

RESULTS

Preliminary experiments were conducted to characterize the adhesion of amphibian embryonic cells to derivatized beads. Normal and hybrid cells adhere very poorly to blank beads, i.e. those coupled to ethanolamine. They also adhere very poorly to BSA-Sepharose beads and plain Sepharose CL-4B beads. There is modest adhesion of cells to Sephadex G-200 beads and to PSM-Sepharose beads.

The kinetics of cell adhesion to FN-Sepharose beads was determined for several different stages of normal and hybrid embryos. When glutaraldehyde was added to cell-bead suspensions immediately after suspensions were prepared, there was no cell adhesion detectable, indicating that cell adhesion observed using longer incubation periods was not non-specific, glutaraldehyde-induced cell adhesion. An incubation period of 4 hours was chosen as a standard assay time because cell adhesion was maximal at this time. Longer incubations were not used because of the possibility of cells differentiating to higher levels of cell adhesiveness during the assay itself. The kinetics of cell attachment are similar for normal and hybrid cells at all developmental stages tested, with maximal cell adhesion being achieved after 4 hours in culture. The absolute level of cell adhesion reached for gastrula stages of normal embryos was invariably greater than that seen for gastrula arrest stages of arrested hybrid embryos (FIGURE 1), indicating that the differences observed between normal and hybrid cells incubated for 4 hours were not merely some trivial result of differences in the kinetics of cell adhesion.

Normal and hybrid cells taken from embryos of different developmental stages were measured for their ability to adhere to FN-Sepharose beads. The results of the experiments show that there is a progressive increase in the ability of normal cells from <u>Rana pipiens</u>, <u>Rana sylvatica</u>, and <u>Ambystoma maculatum</u> to adhere to FN-Sepharose beads as development proceeds. Cells taken from Stage 8 and Stage 9 normal and hybrid blastulae adhere very weakly to FN-Sepharose beads. Cells taken from normal Stage 10 early gastrulae adhere only slightly, but significantly more avidly than comparable cells taken from blastulae. Cells taken from normal Stage 11 mid-gastrulae adhere more often to FN-Sepharose beads and this trend continues in normal Stage 12 late gastrulae (FIGURE 2). In contrast, there is no large increase in cell adhesion with

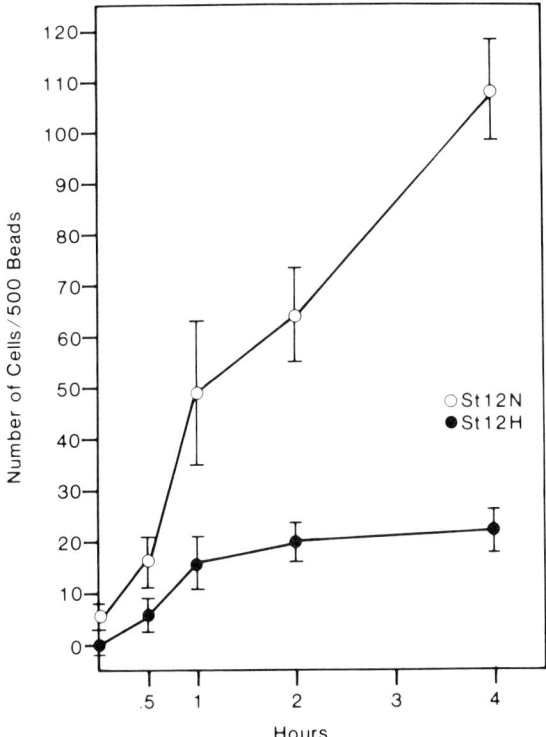

FIGURE 1. The kinetics of cell attachment to FN-Sepharose beads for normal Stage 12 late gastrula cells (open circles) and hybrid Stage 12 late gastrula arrest cells (closed circles).

advancing developmental age in arrested hybrid embryos, although in some instances, there is slightly greater cell adhesion in gastrula arrest stages than in blastula stages (FIGURE 2).

During amphibian development from the early blastula to the late gastrula stage, there is a large increase in the number of cells present in the embryo, due to repeated cell divisions. Thus the differences in number of cells adhering to beads when one compares different developmental stages might simply be a trivial result of different numbers of cell-bead interactions allowed by different numbers of cells being included in cultures. Two observations show that the low level of blastula cell adhesion is not due to a limitation in the number of cell-bead interactions. Standard assay cultures were established using 1 dissociated fragment of the roof of the blastocoel from 1 embryo and 100 ul of 50% v/v suspension of FN-Sepharose beads. There was no significant increase in the number of adherent cells when blastula stage cultures of normal cells were set up using 2 fragments/-

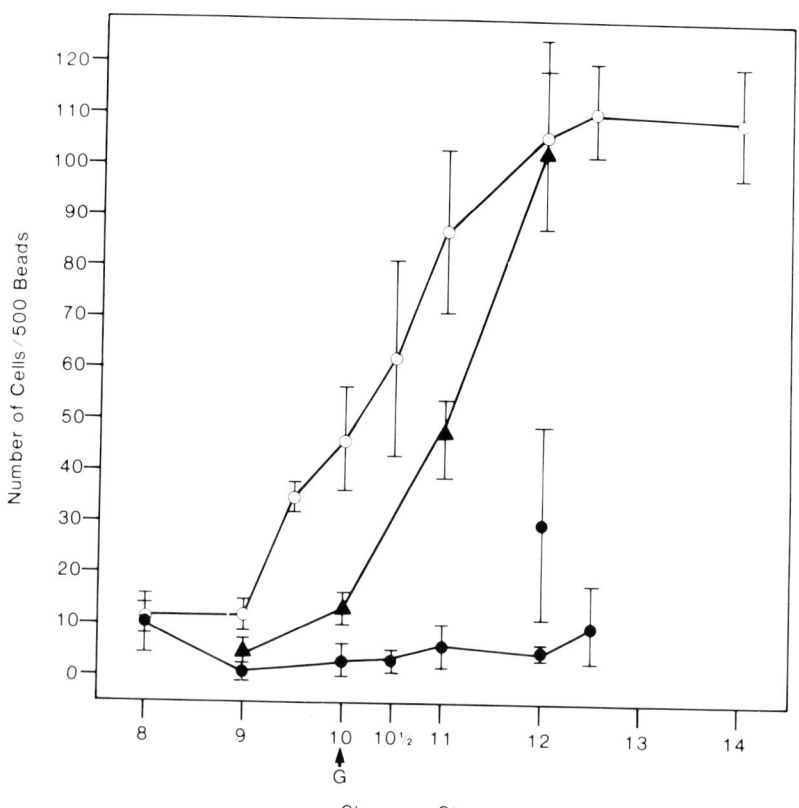

Shumway Stage

Figure 2. Cell attachment to FN-Sepharose beads using four hour incubations. For each time point, the embryos were at the indicated developmental stage at the beginning of the one hour dissociation period. Gastrulation begins at Stage 10 in normal embryos, as indicated by G on the horizontal axis of the graph. Normal Rana pipiens, open circles; normal Ambystoma maculatum, closed triangles; Rana pipiens X Rana catesbeiana arrested hybrid embryos, closed circles. The results obtained with normal Rana sylvatica embryos were similar to those for Rana pipiens and Ambystoma maculatum although the level of cell adhesion was lower.

well, i.e. twice as many cells; and 500 ul of 50% v/v beads, i.e. five times as many beads. Furthermore, when the number of beads was held constant, and the number of cells/well was increased by adding 1, 2, 3, or 4 diss-ociated blastula fragments/well, there was no increase in the number of cells adhering to beads (TABLE 1).

TABLE 1
EFFECT OF INCREASING CELL NUMBER ON
ADHESION OF STAGE 8 BLASTULA CELLS
ON FN-SEPHAROSE BEADS

Number of Tissue Fragments Added to Each Well	Number of Cells Attached/500 Beads Average of 5 Determinations $\pm$ S.D.
1	5 $\pm$ 3
2	5 $\pm$ 1
3	4 $\pm$ 1
4	6 $\pm$ 3

Clearly then, the low number of adherent blastula cells was due to the inability of these cells to adhere to FN-Sepharose beads rather than to limitations in the number of cell-bead interactions.

It may have been that blastula cells failed to adhere to FN-beads because of their complete inability to form cell adhesions. One might imagine, for example, that these cells are too rigid to form the required surface curvatures and extensions needed for cell attachment to beads. This possibility can be eliminated by the argument that blastula cells form adhesions to one another in vivo. Furthermore, by using several other kinds of beads, we have also eliminated this possibility. All developmental stages of all embryos examined show a striking ability to adhere to con A-Sepharose beads, cytodex-1 and cytodex-3 beads (TABLE 2). To be sure, blastula cells do not adhere as avidly to these three kinds of beads as gastrula cells. Nevertheless, their adhesion to these substrata is quite impressive.

The progressive increase in cell-bead adhesion exhibited by developing normal cells is not seen in cells taken from interspecific arrested hybrid embryos formed by fertilizing the eggs of Rana pipiens with the sperm of Rana catesbeiana. In these arrested hybrid embryos, there is a slight increase in cell-bead adhesion as development proceeds, but it is much less striking than the increase seen in normal embryos (FIGURE 2). The rate of cell division in amphibian embryos is maternally determined. Consequently, the number of cells in normal and hybrid embryos are the same. Therefore, the differences in number of cell-bead adhesions can not be due to differences in cell number. Hybrid cells have the ability to adhere to con A-Sepharose beads and also to cytodex-1 and cytodex-3 beads, indicating that hybrid cells have the capability to adhere to some substrata but not to FN-Sepharose beads (FIGURE 3 and TABLE 2).

The fragments taken from different developmental stages of normal embryos had different populations of cells in them. The roof of the blastocoel in Stage 8, 9, and 10 embryos consisted of prospective ectodermal cells and uninvaginated prospective mesodermal cells. As gastrulation proceeded, invaginated prospective mesodermal cells made their way across the inner surface of the roof of the blastocoel. In Stage 11 fragments, there were many invaginated mesodermal cells included with ectodermal cells and uninvaginated mesodermal cells. In Stage 12 fragments, gastrulation had been completed so that the roof of the archenteron was used as a source of cells. In arrested hybrid embryos, gastrulation does not occur. Consequently, when a fragment consisting of the roof of the blastocoel was used, it did not contain the same population of cells as a fragment taken from a normal embryo of comparable age. For example, a fragment taken from a Stage 12 normal embryo contained prospective ectoderm, mesoderm, and endoderm while a fragment taken from an arrested hybrid embryo of the same age as a Stage 12 normal embryo would have only prospective ectoderm and mesoderm in it. Thus, the

TABLE 2
ADHESION OF NORMAL AND HYBRID FROG GASTRULA
EMBRYONIC CELLS TO DIFFERENT KINDS OF BEADS

Frog	Shumway Stage	Description of Stage	Bead Variety	Number of Cells Attached/500 Beads Average of 5 Determinations ± S.D.
Normal	10.5	Mid-Gastrula	Sepharose-4B	1 ± 1
Normal	10.5	Mid-Gastrula	Sephadex G-200	128 ± 25
Normal	9	Blastula	FN-Sepharose	23 ± 5
Normal	9	Blastula	Con A-Sepharose	213 ± 45
Normal	9	Blastula	Cytodex-1	530 ± 84
Normal	9	Blastula	Cytodex-3	526 ± 111
Hybrid	9	Blastula	FN-Sepharose	0 ± 0
Hybrid	9	Blastula	Con A-Sepharose	161 ± 7
Hybrid	9	Blastula	Cytodex-1	412 ± 15
Hybrid	9	Blastula	Cytodex-3	210 ± 9
Normal	12	Late Gastrula	FN-Sepharose	111 ± 26
Normal	12	Late Gastrula	Con A-Sepharose	249 ± 8
Normal	12	Late Gastrula	Cytodex-1	929 ± 58
Normal	12	Late Gastrula	Cytodex-3	1991 ± 211
Hybrid	(12)[1]	Late Arrest	FN-Sepharose	5 ± 1
Hybrid	(12)	Late Arrest	Con A-Sepharose	560 ± 67
Hybrid	(12)	Late Arrest	Cytodex-1	988 ± 139
Hybrid	(12)	Late Arrest	Cytodex-3	1856 ± 128

1. Hybrid embryos arrest at Stage 10. Normal embryos derived from same eggs at Stage 12.

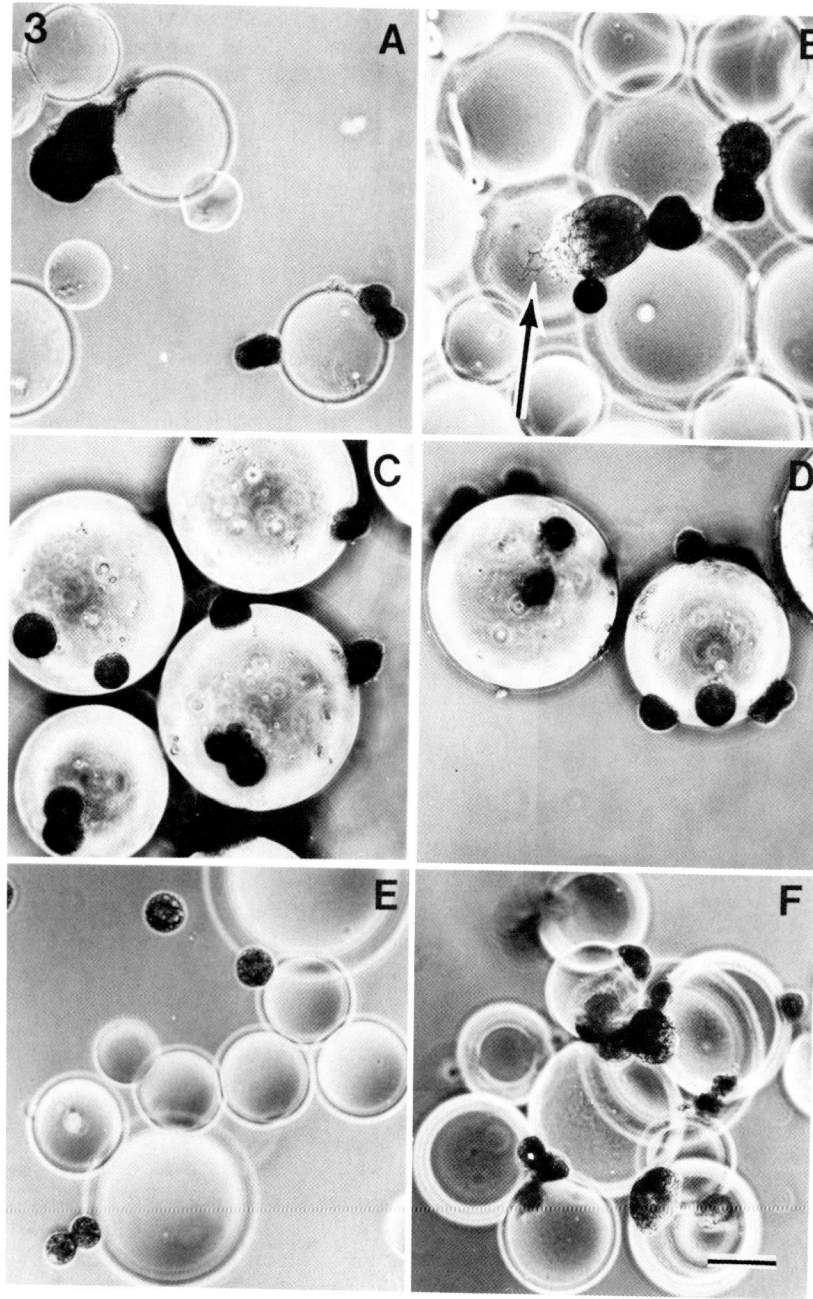

difference between the level of cell-bead adhesion between normal and hybrid embryos at late stages might be the result of using different populations of cells. To examine this possibility, Stage 12 normal embryos and comparable age arrested hybrid embryos were dissected into two frag-ments. For normal embryos, the dorsal fragment consisted of the roof of the archenteron and the other ventral fragment was the floor of the archenteron, i.e. the remainder of the embryo, consisting mostly of the endodermal cell mass surrounded by some ventral mesoderm and ectoderm. These fragments were called dorsal and ventral halves for convenience. For hybrid embryos, one fragment was the roof of the blastocoel and the other fragment was the remainder of the embryo, consisting of uninvaginated mesodermal cells and a large number of yolky endodermal cells. These two different fragments were then used in the standard cell adhesion assay. The results of this experiment, shown in Table 3, indicate that there are large numbers of cells capable of adhering to FN-beads distributed in both halves of a normal embryo with a few more being present in the ventral half than in the dorsal half.

In Hybrid embryos, there are fewer cells capable of adhering to beads in both the dorsal and ventral halves. In this particular experiment, the level of hybrid cell adhesion to FN-beads was a bit higher than that observed in other experiments. This slight but detectable cell adhesion of hybrid cells has been observed using other

Figure 3. Light micrographs of living cells attached to various kinds of beads viewed with phase contrast optics. To be sure that cells were either attached or not attached, the coverslip of the preparation was depressed slightly, creating a fluid stream in the culture. Attached cells moved with beads to which they were attached. Unattached cells remained still while beads moved past them. The arrow indicates a lamellipodium formed on the surface of a bead. A, normal Stage 11 mid-gastrula cells on FN-Sepharose beads; B, normal Stage 12 late gastrula cells on FN-Sepharose beads; C, normal Stage 11 mid-gastrula cells on cytodex-3 beads; D, normal stage 11 mid-gastrula cells on con A-Sepharose beads; E, hybrid Stage 12 late gastrula arrest cells on FN-Sepharose beads (cells shown to be lying close to beads but not attached when coverslip depressed); F, hybrid Stage 12 late gastrula arrest cells on con A-Sepharose beads. Bar represents 50 μm, X200.

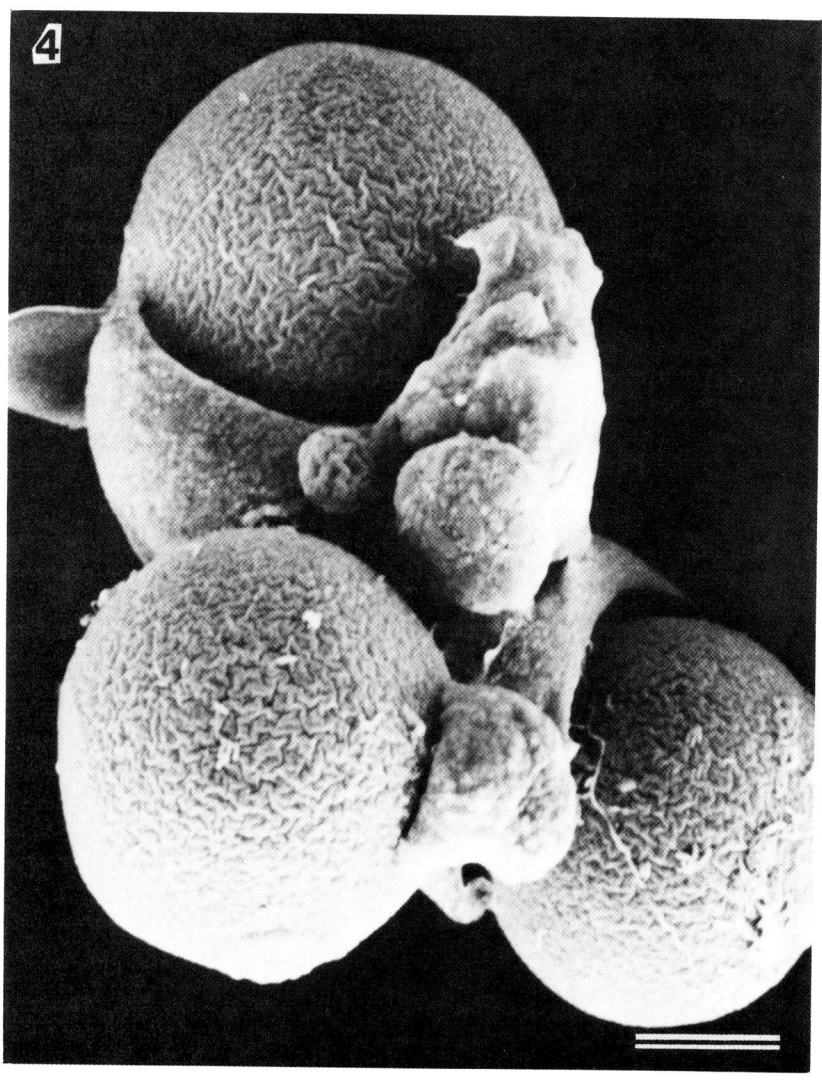

assay systems in the past (Nakatsuji and Johnson, 1984a). Cell-bead aggregates were observed in the light microscope with phase contrast optics, in the scanning electron microscope after fixation, and by time-lapse cinemicrography with phase contrast optics. Cells may be semicircular or crescent-shaped in profile as they partially engulf beads (FIGURES 3, 4, and 6). Cells attach to FN-beads by way of impressive fan-like marginal

TABLE 3
LATE GASTRULA NORMAL AND LATE GASTRULA
ARREST HYBRID CELL ADHESION TO
FN-SEPHAROSE BEADS (STAGE 12)

Fragment	Description	Number of Cells Attached/500 Beads Average of 5 Determinations ± S.D.
Normal Dorsal	Roof of Archenteron	106 ± 23
Normal Ventral	Floor of Archenteron	97 ± 14
Hybrid Dorsal	Roof of Blastocoel	30 ± 19
Hybrid Ventral	Floor of Blastocoel	29 ± 8

extensions that look like lamellipodia in both the light and scanning electron microscope (FIGURE 5). Examination of fibronectin-beads with attached cells at high magnification in the SEM did not reveal the presence of fibronectin-containing fibrils, suggesting that the bound fibronectin is distributed homogeneously over the surface of the beads. Time-lapse films of cells on FN-beads reveal that these marginal extensions are indeed locomotory lamellipodia that move cells over the surface of beads (FIGURE 6). In several instances, pairs of cells attached to two different beads were observed and sometimes, the contractile and locomotory activity of attached cells was sufficient to cause striking and sudden displacement of beads (FIGURE 6). Cells also flatten against con A and cytodex beads and adhere to them, but don't form lamellipodia on these beads.

Figure 4. Scanning electron micrograph of several normal Stage 12 late gastrula cells attached to and spreading upon FN-Sepharose beads. Bar represents 20 μm, X 1,140.

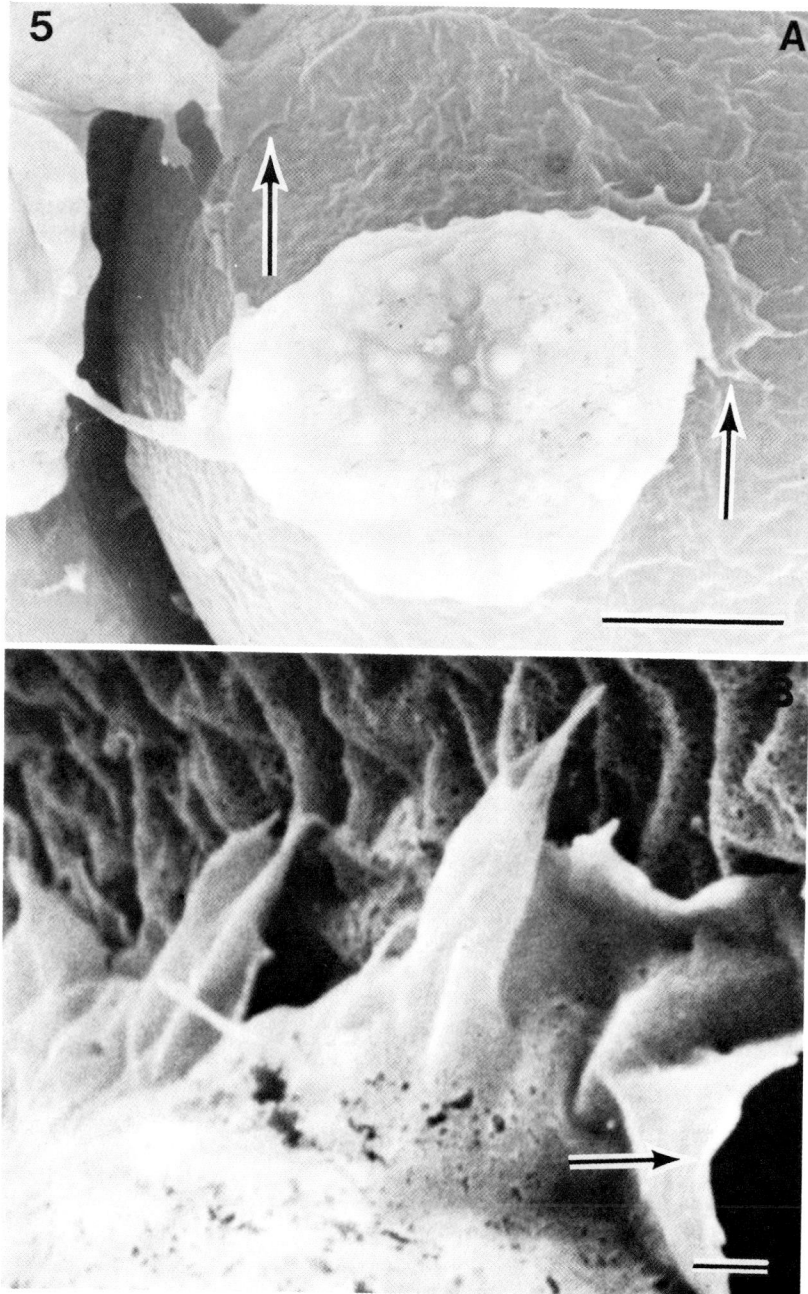

Figure 5. Scanning electron micrographs of normal Stage 12 late gastrula cells attached to and spreading upon FN-Sepharose beads. In A, the arrows mark two different lamelliapodia formed by two different cells on the surface of one bead, bar represents 10 μm, X 2,880. In B, the arrow marks a marginal uplift of the lamellipodium, bar represents 1 μm, X 11,100. The lamellipodium shown in Figure 5B was photographed as a stereo pair with 60° tilt and when the stereo pair was viewed in a stereo viewer, this marginal uplift was clearly visible projecting upwards out of the plane of the micrographs.

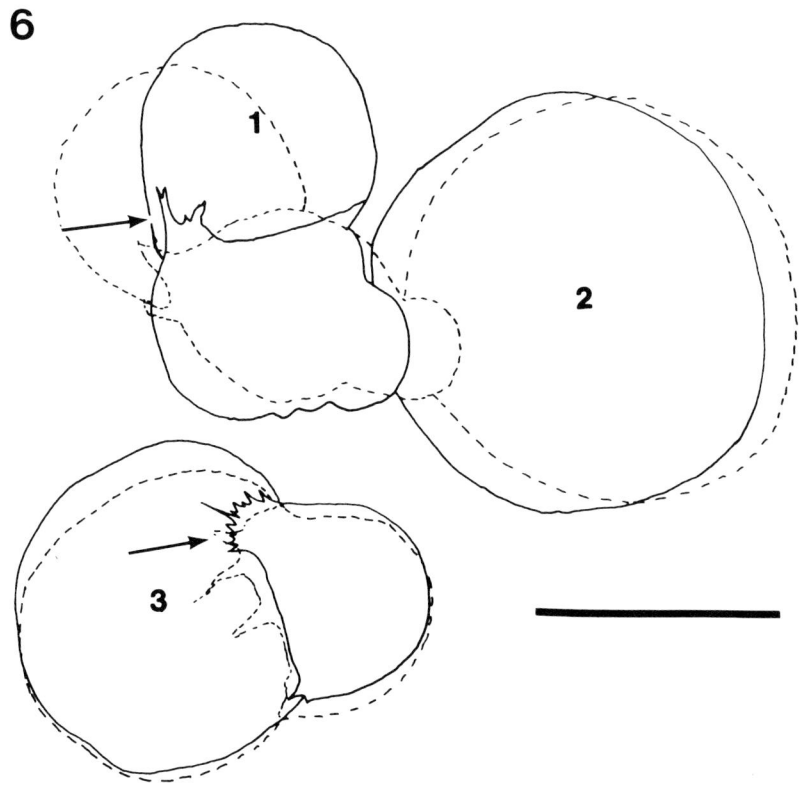

FIGURE 6. Legend appears on following page.

FIGURE 6. Tracings of individual frames from time-lapse movies of normal Stage 12 late gastrula cells attached to and spreading upon FN-Sepharose beads. Beads 1 and 2 have a pair of cells attached, one member of the cell pair being attached to each bead. These cells and beads were observed in one film. Bead 3 has one cell attached and was observed in another film made from a different culture on a different day. The arrows indicate actively ruffling lamellipodia. For beads 1 and 2, the lines indicate the position of the cells and beads at 0 min (solid lines) and 45 min later (broken lines). Both cells are firmly attached, one to each bead. Bead 1 has a large lamellipodium on it (arrow). When the cells contract, bead 1 is displaced markedly with respect to bead 2. For bead 3, the lines indicate the position of cell and bead at 0 min (solid lines) and 35 min later (broken lines). The cell is attached to this stationary bead by a prominent lamellipodium (arrow). Bar represents 50 μm, X 830.

DISCUSSION

The present results show that the ability to adhere to FN-beads is absent prior to the beginning of gastrulation, appears around the beginning of gastrulation, and increases throughout gastrulation. They also show that this change does not occur in the interspecific arrested hybrid embryo formed by fertilizing the eggs of Rana pipiens with the sperm of Rana catesbeiana. This particular hybrid embryo has been studied extensively. Recently, it was shown that extracellular fibrils, known to contain fibronectin in other species, do not appear on the inner aspect of the roof of the blastocoel of this arrested hybrid embryo as they do in normal Rana pipiens embryos (Nakatsuji and Johnson, 1984a). A fragment of the roof of the blastocoel from normal embryos can condition substrata in vitro by deposition of fibrils on these surfaces. Substrata conditioned by fragments from normal embryos promote attachment and locomotion of mesodermal cells from normal embryos. These same fragments also condition substrata so that migrating mesodermal cells from arrested hybrid embryos will attach and migrate, albeit to a degree that is considerably less than that seen for normal migrating mesodermal cells. In contrast, fragments of the roof of the blastocoel taken from arrested hybrid embryos have no conditioning effects for either normal or hybrid migrating mesodermal cells (18). Reciprocal grafting experiments involving transplantation of the roof of the blastocoel between normal and hybrid early gastrulae have been performed in an attempt to learn more about the role of extracellular fibrils in controlling cell migration. The results of these experiments show that normal fragments of the roof of the blastocoel do not become covered by extracellular fibrils and do not support mesodermal cell migration in vivo when these fragments are transplanted into arrested hybrid embryos. In the reciprocal exchanges, hybrid fragments of the roof of the blastocoel when transplanted into normal hosts do become covered by extracellular fibrils and do support mesodermal cell migration in vivo (19). These results suggest that extracellular fibrils are synthesized in cells other than cells of the roof of the blastocoel and that these fibrils then bind secondarily to the roof of the blastocoel. They also suggest that cells of the roof of the blastocoel of hybrid embryos have the ability to bind fibrils made in normal hosts. The present results show that cells from arrested hybrid embryos have a greatly reduced ability to adhere to beads coated with fibronectin. It would be tempting to speculate that cells in arrested hybrid embryos do not show normal morphogenetic cell movements because they lack both the ability to produce normal fibrillar extracellular matrices and they also lack the ability to adhere to these fibronectin-containing matrices. One goal of the current series of experiments is to identify an arrested hybrid embryo which does not synthesize a fibrillar extracellular matrix on the inner aspect of the roof of the blastocoel but does have cells in the embryo with the ability to recognize and adhere to beads coated with fibronectin. This kind of hybrid would be a likely candidate for experiments to inject fibronectin into the blastocoel cavity in an attempt to promote gastrulation under circumstances where it would not otherwise occur without the addition of exogenous fibronectin.

Fibronectin plays an important role in promoting cell adhesion and cell migration in gastrulation in chick embryos (20) and in neural crest cell migration (21). We know a good deal about the molecular biology of fibronectin (22,23). We know its primary structure and have been able to isolate peptide fragments of the intact molecule that contain the cell-binding domain as well as the collagen--and-heparin- binding domains. It has been shown that antibodies to fibronectin will block gastrulation when injected into the blastocoel of amphibian embryos (8) and also that the cell-binding peptide but not the collagen-binding peptide of fibronectin will inhibit amphibian gastrulation and avian neural crest cell migration (9). It is clear that this ubiquitous glycoprotein is important for allowing cell adhesion to the extracellular matrix but also for directing cell migration and neural crest cell migration. Several interesting approaches have been used recently in experiments to investigate the role of fibronectin in controlling neural crest cell migration. Fibronectin coating of latex microspheres prevents their translocation along the neural crest cell migratory pathways in chick embryos (24), presumably because the fibronectin-coated beads adhere to somite cells rather than being swept along with migrating neural crest cells. Fibronectin-coated latex microspheres might be injected into the blastocoel of amphibian embryos in attempts to perturb gastrulation. Lofberg et al. (25) have recently shown that extracellular matrix deposited in vivo onto small fragments of Nucleopore filters ("microcarriers") can cause local stimulation of neural crest cell migration in axolotl embryos. It may be possible to produce "conditioned" microcarriers using the roof of the blastocoel from normal frog gastrulae and then insert these into arrested hybrid embryos in an attempt to promote cell migration locally. To learn more about the morphogentic role of fibronectin and other constituents of the extracellular matrix, we need to think of ways to disturb the synthesis and distribution of extracellular matrix components. It is hoped that the normal and hybrid amphibian embryo system will offer gamete combinations where natural experiments altering e.g. fibronectin synthesis or distribution in the embryo will yield new insights into the control of morphogenetic cell movements.

ACKNOWLEDGEMENTS

The author wishes to thank Dr. Albert K. Harris for collecting Ambystoma embryos used in these experiments and Dr. John B. Morrill for the use of his scanning electron microscope.

REFERENCES

1. Nakatsuji N, Gould AC, Johnson KE (1982). Movement and guidance of migrating mesodermal cells in Ambystoma maculatum gastrulae. J Cell Sci 56:207.
2. Nakatsuji N, Johnson KE (1983a). Conditioning of a culture substratum by ectodermal layer promotes attachment and oriented locomotion by amphibian gastrula mesodermal cells. J Cell Sci 59:43.
3. Nakatsuji N. (1984). Cell locomotion and contact guidance in amphibian gastrulation. Amer Zool 24:615.
4. Nakatsuji N, Johnson KE (1984b). Experimental manipulation of a contact guidance system in amphibian gastrulation by mechanical tension. Nature 307:453.
5. Boucaut J-C, Darribere T. (1983a). Presence of fibronectin during early embryogenesis in the amphibian Pleurodeles waltilii. Cell Diff 12:77.
6. Boucaut J-C, Darribere T. (1983b). Fibronectin in early amphibian embryos. Migrating mesodermal cells contact fibronectin established prior to gastrulation. Cell Tiss Res 234:135.
7. Nakatsuji N, Smolira MA, Wylie CC (1985). Fibronectin visualized by scanning electron microscopy immunochemistry on the substratum for cell migration in Xenopus laevis gastrulae. Dev Biol 107:264.
8. Boucaut J-C, Darribere T, Boulekbache H, Thiery JP. (1984). Prevention of gastrulation but not neurulation by antibodies to fibronectin in amphibian embryos. Nature 307:364.
9. Boucaut J-C, Darribere T, Poole TJ, Aoyama H, Yamada K, Thiery JP. (1984). Biologically active synthetic peptides as probes of embryonic development: A competitive peptide inhibitor of fibronectin function inhibits gastrulation in amphibian embryos and neural crest cell migration in avian embryos. J Cell Biol 99:1822.
10. Darribere T, Boucher D, Lacroix J-C, Boucaut J-C. (1984). Fibronectin synthesis during oogenesis and early development of the amphibian Pleurodeles waltlii. Cell Diff 14:171.
11. Darribere T, Boulekbache H, Shi DL, Boucaut J-C. (1985). Immunoelectron microscopic study of fibronectin in gastrulating amphibian embryos. Cell Tiss Res, in press.

12. Lee G, Hynes R, Kirschner M. (1984). Temporal and spatial regulation of fibronectin in early Xenopus development. Cell 36:729.
13. Nakatsuji N, Johnson KE (1983b). Comparative study of extracellular fibrils on the ectodermal layer in gastrulae of five amphibian species. J Cell Sci 59:61.
14. Rugh R (1962). "Experimental Embryology", 3rd edn. Minneapolis, Burgess Publishing Company, p. 91.
15. Wolf K, Quimby MC (1964). Amphibian cell culture: a permanent cell line from the bullfrog Rana catesbeiana. Science 144:1578.
16. Shumway W (1940). Stages in the normal development of Rana pipiens. Anat Rec 78:139.
17. Harrison RG. (1969). "Organization and Development of the Embryo". New Haven: Yale University Press, p44.
18. Nakatsuji N, Johnson KE (1984a). Ectodermal fragments from normal frog gastrulae condition substrata to support normal and hybrid mesodermal cell migration in vitro. J Cell Sci 68:49.
19. Johnson KE. (1985). Transplantation studies to investigate mesoderm-ectoderm adhesive cell interactions during gastrulation. J Cell Sci, in press.
20. Duband JL, Thiery JP. (1982a). Distribution of fibronectin in the early phase of avian cephalic neural crest cell migration. Dev Biol 93:308.
21. Duband JL, Thiery JP. (1982b). Appearance and distribution of fibronectin during chick embryo gastrulation and neurulation. Dev Biol 94:337.
22. Yamada KM (1983). Cell surface interactions with the extracellular matrix. Ann Rev Biochem 52:761.
23. Yamada KM, Kennedy DW (1984). Dualistic nature of adhesive protein function: fibronectin and its biologically active peptide fragments can autoinhibit fibronectin function. J Cell Biol 99:29.
24. Bronner-Fraser M. (1985). Effects of different fragments of the fibronectin molecule on latex bead translocation along neural crest migratory pathways. Dev Biol 108:131.
25. Lofberg J, Nynas-McCoy A, Olsson C, Johnson L, Perris R. (1985). Stimulation of initial neural crest cell migration in the axolotl embryo by tissue grafts and extracellular matrix transplanted on microcarriers. Dev Biol 107:442.

EXTRACELLULAR MATRIX, CELL POLARITY AND EPITHELIAL-MESENCHYMAL TRANSFORMATION[1]

Elizabeth D. Hay

Department of Anatomy and Cellular Biology, Harvard Medical School, Boston, Massachusetts 02115

ABSTRACT Epithelia form sheets of contiguous cells residing on extracellular matrix (ECM). The tissue phenotype is defined by a high degree of apical-basal surface polarity and the production of cytokeratin, laminin, and type IV collagen. Cytodifferentiation within the tissue can be extensive. Lens cells, for example, produce crystallin protein, and thyroid cells, thyroglobulin. Mesenchymal cells are bipolar or stellate in shape and migrate through ECM. The tissue phenotype is characterized by production of vimentin, type I collagen, and fibronectin. The basal surface of epithelial cells flattens when the cells are placed on top of ECM, such as collagen gel. However, when isolated epithelia are suspended within collagen gels, the surface facing the collagen fibrils extends pseudopodia after 2-3 days. Then the cells elongate, and by 4-5 days, mesenchyme-like cells are leaving the epithelium and migrating through the gel. Although derived from an epithelium that was producing type IV collagen, the mesenchyme-like cells produce type I collagen. They also stop producing laminin and, in the case of thyroid, they switch from cytokeratin to vimentin intermediate filaments. Mesenchyme-like cells derived from thyroid stop making thyroglobulin, and those from lens, crystallin. These results are discussed in the context of mechanisms controlling expression of the epithelial versus the mesenchymal genetic program.

[1]This work was supported by United States Public Health Service grant HO-00143.

INTRODUCTION

In this chapter, the effect of extracellular matrix (ECM) on the differentiation and polarity of epithelial and mesenchymal tissues will be briefly reviewed. Extracellular matrix is defined as a material consisting of collagens, proteoglycans, and structural glycoproteins (fibronectin, laminin, etc.) on which epithelial cells sit or through which mesenchymal cells migrate. It assumes various morphologies and compositions, including collagen fibrils of connective tissue and the meshworks in basal laminae (basement membranes).

During histogenesis in vertebrate embryos, epithelial tissues are the first formed and they give rise in specific places at specific times to the tissue known as mesenchyme (Fig. 1). We have recently discovered that embryonic and even adult epithelia can be induced to give rise to mesenchyme-like cells by the simple step of suspending the tissue within collagen gels (1, 2). This phenomenon will be discussed in some detail, because it provides an experimentally useful example of the effect of environmental factors (Edelman, this volume) on gene expression, including gene turn-off as well as turn-on.

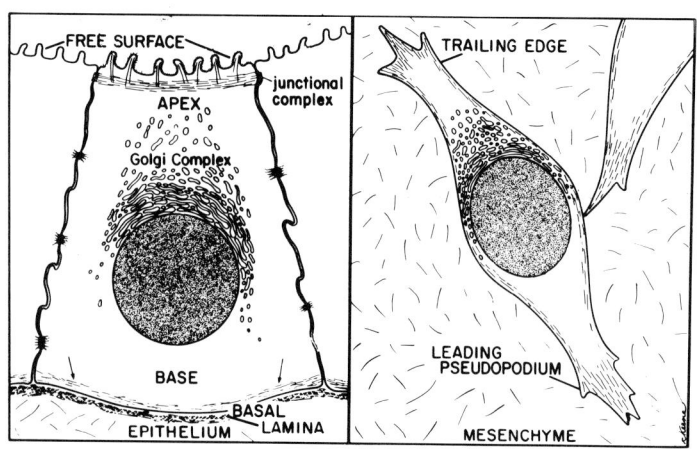

FIGURE 1. Diagram comparing the epithelial and mesenchymal phenotypes. Epithelial cells interact with ECM (basal lamina in most cases) along their basal surface. Mesenchymal cells are surrounded by ECM. From Hay (3).

THE EPITHELIAL AND MESENCHYMAL PHENOTYPES

An epithelium is a sheet of contiguous cells, often cuboidal in shape, residing on ECM. It may be simple or stratified, but the outer surface is always a free surface and is usually covered with microvilli (Fig. 1) and a sialic-acid rich glycoprotein surface coat. Juxtalumenal junctions (zonulae occludentes and adhaerentes) seal the lateral compartment from the free compartment or lumen. Both the cytoplasm and cell surface are highly polarized. The Golgi zone lies in the apical cytoplasm (Fig. 1) and enzymes such as leucine aminopeptitase (4) localize to the apical (free) surface. The lateral surface contains gap junctions and desmosomes and a different set of enzymes, including sodium potassium ATPase. The basal surface interacts with ECM and flattens on contact with ECM, whether or not a basal lamina (basement membrane) is present (5). A highly ordered functional polarity is the result of this careful structural organization (4).

Mesenchymal cells are elongate (Fig. 1) or stellate in shape and they move through ECM with only a limited amount of intercellular contact via gap junctions (6). A migrating mesenchymal cell has a leading and a trailing edge and is bipolar in shape, but the cell surface is not polarized, interacting all around with ECM. When mesenchymal cells are placed on ECM they invade it (Fig. 2), whereas epithelial cells remain on top (7, 8). While epithelia can produce ECM, even striated collagen fibrils on occasion (9), it is the connective tissue cells derived from mesenchyme that produce most of the ECM of the vertebrate body. Mesenchymal cells give rise both to chondrocytes and to osteocytes, cells that produce highly specialized connective products. It is interesting to note that the primitive chordates are primarily composed of epithelial tissues and that the mesenchymal compartment increases enormously in the land vertebrates. The potential of polarized epithelia to give rise to mesenchymal cells that specialize in connective tissue production is used to advantage in the evolution of endoskeletons.

In addition to the morphological and functional characteristics alluded to above, epithelia and mesenchymal cells differ in several biochemical parameters that are easy to measure. Almost all epithelia express L-cam (Edelman, this volume) and have intermediate filaments of the cytokeratin (alpha keratin) type, whereas

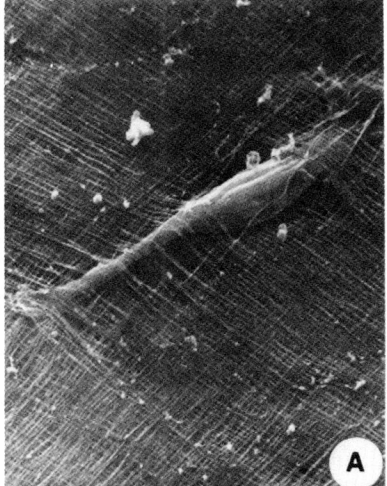 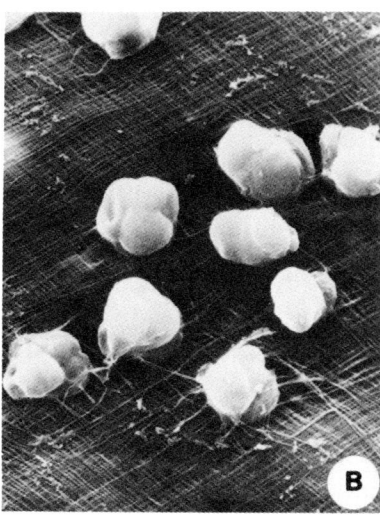

FIGURE 2. Scanning electron micrographs showing the reaction of epidermal cells (A) and a cardiac mesenchymal cell (B) to underlying collagen gel. The mesenchymal cell invades the gel, while the epithelial cell remains on top. Bar, 5 μm. From Overton (7).

mesenchymal cells instead have vimentin intermediate filaments (10, 11) and do not express L-cam. Epithelial cells in general produce type IV collagen, laminin, and proteoglycan (PG), the ECM components that characterize the basal lamina on which they normally reside. With a few exceptions (12), mesenchymal cells do not produce type IV collagen or laminin; they produce type I collagen, PG, and appropriate structural glycoproteins. They are fibroblasts (fiber producing cells). When fibroblasts differentiate into cartilage, bone, or muscle, new ECM patterns (e.g. type II collagen) and specific proteins are expressed.

In this chapter, we shall use the term "phenotype" to refer to tissue type (epithelium or mesenchyme). Within a given tissue, cells synthesize specific proteins (e.g. muscle myosin, crystallin protein, thyroglobulin) that set them apart from other cells of the same tissue phenotype. We shall refer to the acquisition of these cell specific characteristics as "cytodifferentiation."

INTERACTION OF EPITHELIAL CELLS WITH ECM

Cells of the epithelial phenotype respond favorably to contact with ECM molecules along their basal surface (for review, see 3, 13). Two of the most intensively studied systems are the avian corneal epithelium, which steps up corneal stroma synthesis in response to ECM (14-16) and the mammalian mammary epithelium, which steps up casein synthesis in response to ECM, especially if the collagenous ECM is floating (17, 18). The positive reaction of differentiated epithelia to ECM could be classified as a permissive induction (Gurdon, this volume), in the sense that no new gene products are turned on. Existing pathways of cytodifferentiation are enhanced, probably through an effect of ECM on cell shape (17) and/or cytoskeleton (3).

The morphological response of the avian corneal epithelium to ECM has been particularly well studied and will be described briefly here as an example of the way in which epithelia interact with underlying matrix. When corneal epithelia are isolated from ECM using EOTA or collagenase-trypsin (Fig. 3), the basal surface immediately begins to bleb (9, 14). If the epithelia are cultured on Millipore filters or other non-collagenous substrata, the basal surface continues to extend and retract the rounded cell processes called blebs (9, 14). However, if corneal epithelia are cultured on lens capsules or other collagenous substrata, the basal surface flattens in a few hours (Fig. 3) and within 18 hours the epithelia are producing 2-3 times as much corneal stroma as epithelia cultured on non-collagenous substrata (14). The stimulatory effect can be mediated by soluble ECM components (Fig. 4), as well as by polymerized ECM (15, 16).

A striking reorganization of epithelial basal cytoskeleton occurs on contact with polymerized or soluble ECM, presumably due to interaction of ECM receptors in basal plasmalemma with ECM molecules. The cortical actin mat becomes disrupted when epithelium is isolated; it reorganizes in the presence of polymerized ECM or soluble ECM molecules (Fig. 5). As stained by heavy meromyosin subfragments, actin filaments in the reorganized cortical mat form an ordered array, whereas actin filaments in the blebs form a disordered meshwork (15). Presumably, the disordered actin in the blebs reorganizes into a cortical mat when the blebs are withdrawn; protein synthesis is not required for the effect (16).

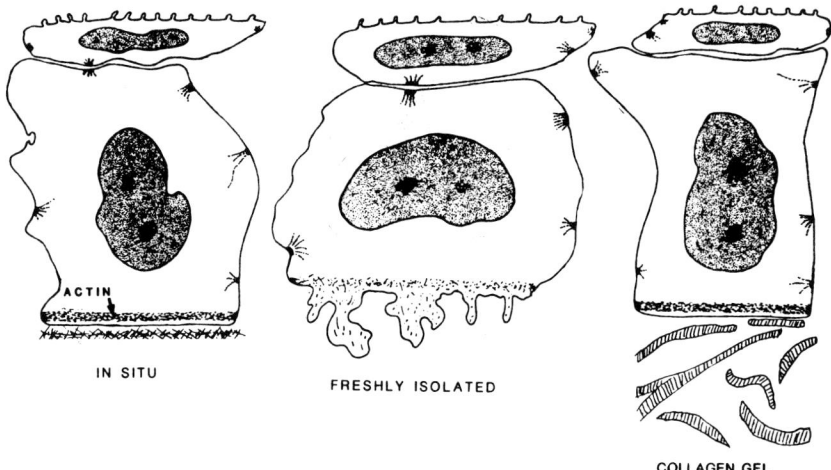

FIGURE 3. Diagrams summarizing the reaction of corneal epithelium to removal and restoration of ECM. In situ, the epithelium resides on a basal lamina and the basal cell surface is flat. The organized actin mat in the basal cortex is disrupted when the cells are isolated and the basal surface blebs (center). When the tissue is placed on a collagen gel, the basal surface flattens and the cortical actin mat reorganizes.

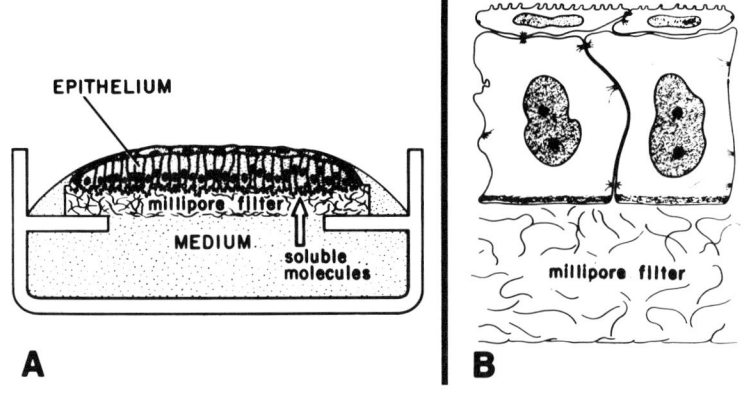

FIGURE 4. (A) Isolated epithelium continues to bleb when cultured on a filter. (B) 6 hours after soluble ECM has been added to medium, basal surface has flattened.

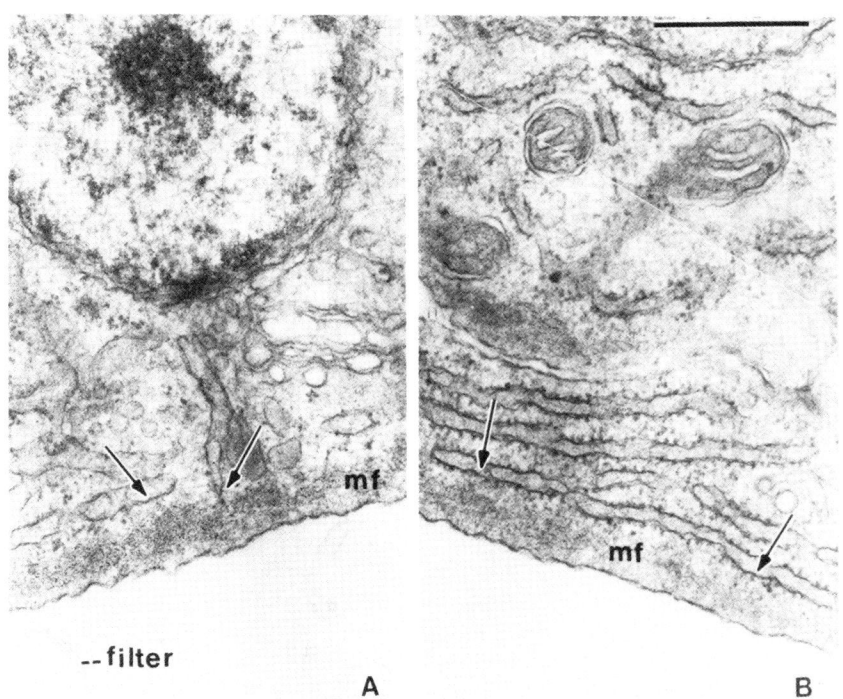

FIGURE 5. Electron micrographs showing reorganized microfilaments (mf) in basal cortex of corneal epithelium treated with soluble type IV collagen (A) and laminin (B). RER (arrows) appears to attach to the actin mat. Bar, 500 nm. From Sugrue and Hay (15).

Evidence that ECM molecules are actually binding to the basal cell surface can be obtained by presenting isolated corneal epithelia with laminin or collagen conjugated to fluorescent covaspheres (19). ECM-coated covaspheres (Covalent Technologies, Inc., Ann Arbor, MI) bind to the basal epithelial surface (Fig. 6). Potential ECM receptors or binding sites in the basal epithelial plasmalemma that have been isolated (19-31) include heparan sulfate proteoglycan (HSPG) and several glycoproteins (Fig. 7). It is tempting to think that ECM acts across the plasmalemma via such receptors to organize the epithelial cytoskeleton and thereby to affect cell shape and cytodifferentiation (3; Bernfield, this volume).

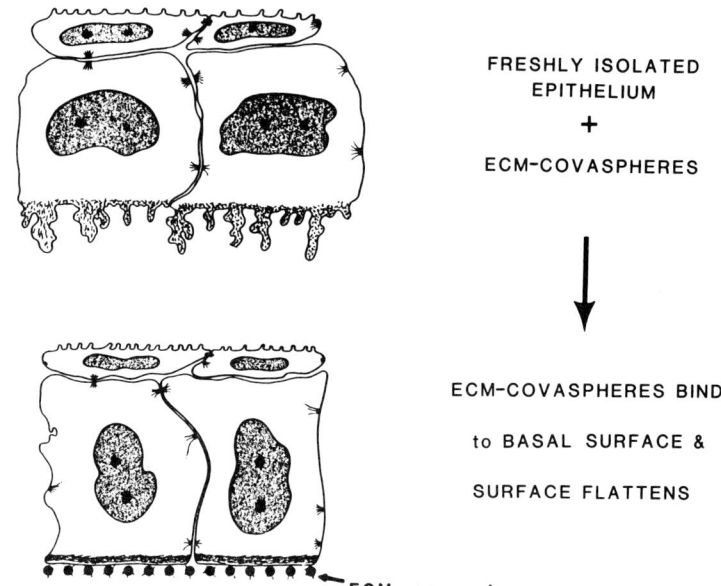

FIGURE 6. Diagram showing attachment of ECM coated covaspheres to basal surface of isolated corneal epithelium.

INTERACTION OF MESENCHYMAL CELLS WITH ECM

Cells of the mesenchymal phenotype also respond to ECM. Matrix molecules such as glycosaminoglycans (GAG) and collagen types I and II stimulate cells determined to form chondrocytes to step up the synthesis of cartilage specific molecules (32, 33). Although factors can be found in ECM that act instructively to induce chondrogenesis (34, 35), the stimulatory effect of "structural" matrix molecules (GAG, collagen) on chondrogenesis can be classified as permissive, in the sense that the chondrocytes affected have already embarked on the chondrogenic pathway. A round cell shape (36-38) promotes cartilage cytodifferentiation and chondrocyte shape can be modified by ECM (36). Drug-induced disruption of the cytoskeleton leads to round cell shape and promotes chondrogenesis (38). However, there is little evidence for a direct effect of ECM on the chondrocyte cytoskeleton as yet.

RECEPTOR	MOLECULE BOUND[1]
31KD	COL
47KD	
65KD	LAM
140KD	FN
HSPG	Complex

1. Active site is often arg-gly-asp.

FIGURE 7. Summary of present status of putative ECM receptors. Collagen binding sites that have been isolated include a 31 kd protein from chondrocytes (21), and a 47 kd from carcinoma (22) and corneal epithelium (19). 65 kd proteins binding laminin have been isolated from carcinoma (23), corneal epithelium (19), muscle (24), and fibrosarcoma (25). Several cell binding sites have been reported for fibronectin, but the most specific one seems to be 140 kd (see 26-28). Most ECM molecules contain arg-gly-asp as the active site interacting with cells (Ruoslahti, this volume; 29). In addition to 47 and 65 kd proteins, larger membrane-intercalated HSPG has been isolated from corneal epithelium (19). Similar putative HSPG receptor for collagen and/or other ECM has been reported in hepatocytes (30) and mammary epithelium (31).

There is good morphological evidence that ECM directly affects the organization of the cytoskeleton of fibroblasts. When the corneal fibroblast is cultured within a hydrated collagen gel (Fig. 8), it elongates, developing actin-rich filopodia on the leading pseudopodium (39). The cytosal contains dispersed myosin, the cortex consists of actin filaments, and microtubules and rough endoplasmic reticulum (RER) course the length of the cell (39). The same cells plated on planar substrata, develop stress fibers and ruffled borders (Fig. 8).

The effect of the collagen gel on assumption of bipolarity by the corneal fibroblast requires an intact actin cytoskeleton; it is cytochalasin-sensitive (40). Subsequent cell elongation requires both intact actin filaments and intact microtubules (40). The actin cortex is dispersed around the entire cell and is in a good position to interact with ECM via putative transmembrane binding sites. Microtubules, however, appear to end in

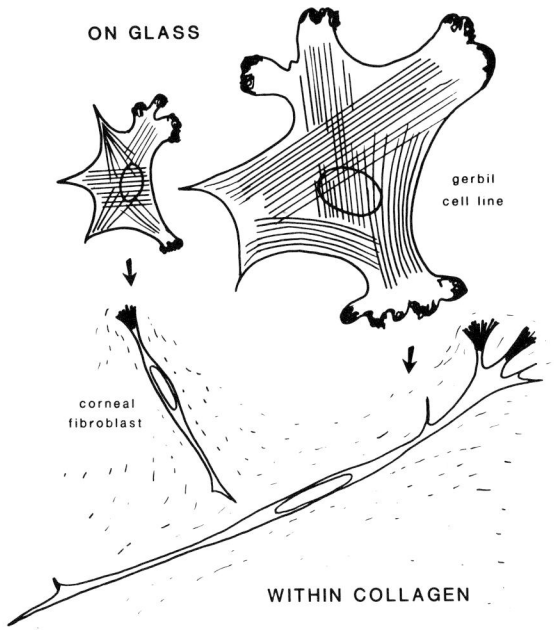

FIGURE 8. Diagram showing flattened fibroblasts with stress fibers and ruffled borders on glass. The same cells elongate when suspended within collagen gels and actin-rich filopodia replace the actin-rich ruffles on the leading end of the cell.

the cortex, not on the plasmalemma (Fig. 9). It is likely that microtubules interact indirectly with ECM, via the actin cortex and putative receptors (40). It is not clear whether or not the metabolism of fibroblasts is stepped up by growing the cells in collagen gels, but there is no doubt that the organization of the cytoskeleton, and possibly of the RER, is dramatically affected by surrounding ECM.

TRANSFORMATION OF EPITHELIA TO MESENCHYME IN COLLAGEN GELS

Hydrated collagen matrices of the type discussed above are very useful for studies of mesenchymal cell

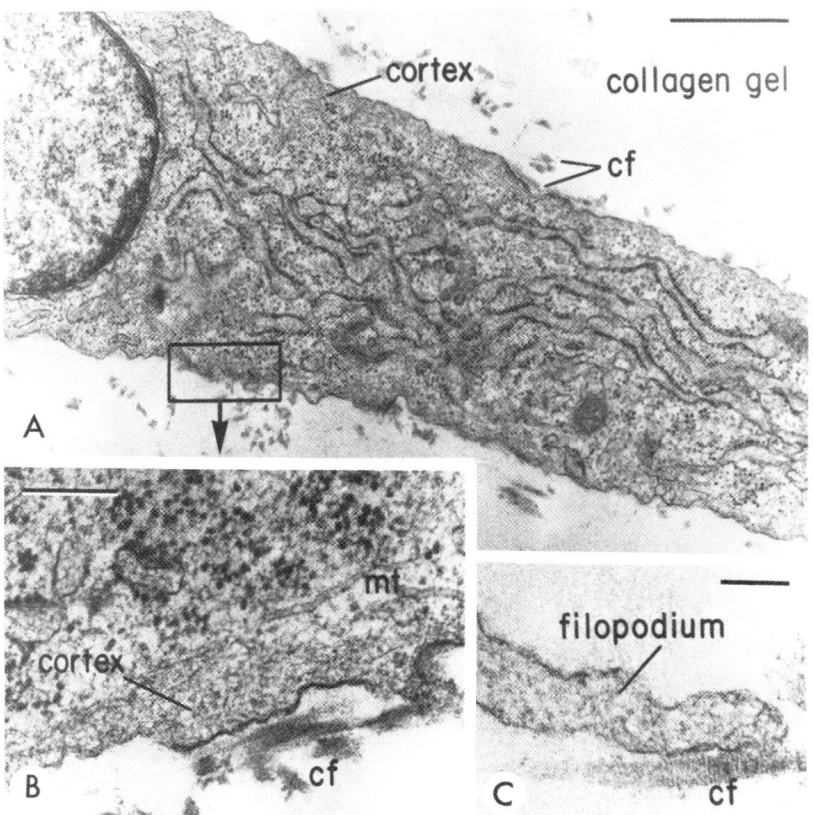

FIGURE 9. Electron micrographs showing the longitudinally arranged organelles (A) and microtubules (mt, B) in fibroblasts cultured within collagen gels. The microfilamentous cell cortex surrounding the cell (A,B) and filopodia (C) have been shown by immunohistochemistry to contain actin. cf, collagen fibril. Bar, 100 nm. From Tomasek et al. (39).

differentiation, because mesenchymal cells are normally embedded in ECM. We reasoned that they would be ideal for conducting experiments in vitro on epithelial-mesenchymal transformation, and we started out to study a normal example of this process, the transformation of embryonic somites to sclerotomal mesenchyme within the 3D gel (1). To our surprise, however, control epithelia that normally do not give rise to mesenchyme in the embryo, did so within collagen gels (Fig. 10 A-C).

Most of the epithelia studied by Greenburg (1) were isolated with trypsin-collagenase and were carefully examined by transmission electron microscopy (TEM) to rule out mesenchymal cell contamination. Greenburg and Hay (2) did a detailed investigation of one epithelium, the anterior lens epithelium, that is not in contact with any mesenchyme. The lens can be readily removed through the anterior eye chamber with its capsule intact and the anterior epithelium can then be dissected from the posterior epithelium without enzyme treatment (41).

The lens epithelial sheet is plunged into a gelling solution of type I collagen, where on the first day of culture it can be seen to have smooth edges (Fig. 10 D). Within 2 days, cell processes begin to protrude into the gel (Fig. 10 E), and by 3 days, cells can be seen migrating away from the explant (Fig. 10 F). By 5 days (Fig. 10 G), numerous elongated, freely migrating cells surround the explant (1, 2). In sections, it can be seen that the original monolayer of anterior lens epithelium has multilayered during the 2-5 day period in the collagen gel (Fig. 11 A, B). In areas where the former apical surface is in close contact with collagen fibrils, cells extend filopodia and move out into the gel (Fig. 11 B).

If, on the other hand, the anterior lens epithelium is placed on top of a collagen gel, the cells that migrate away from the explant give rise to a monolayered epithelium with all of the characteristics (apical tight junctions, new basal lamina formation, etc.) of the original epithelium (1, 2). The cells are cuboidal in shape (Fig. 11 C) and they stain with antibodies to lens crystallin (1). However, the mesenchyme-like cells that migrate away from the explant within a collagen gel do not stain with anticrystallin, nor do they produce delta crystallin as judged by metabolic labeling (1).

Examination of the explant within collagen gel by TEM reveals that mesenchyme-like cells leaving the explant are acquiring RER, prominent nucleoli, and Golgi complexes (2). Cytoplasm loses the fibrillogranular crystallin lens protein deposits (fg, Fig. 12) which characterize lens cells that remain attached to the lens capsule (lower right, Fig. 12). Elongate, mesenchyme-like cells moving away from the explant are rich in RER running longitudinally in the cytoplasm (Fig. 13). Orientation of the RER is similar to that of the normal fibroblast (Fig. 9).

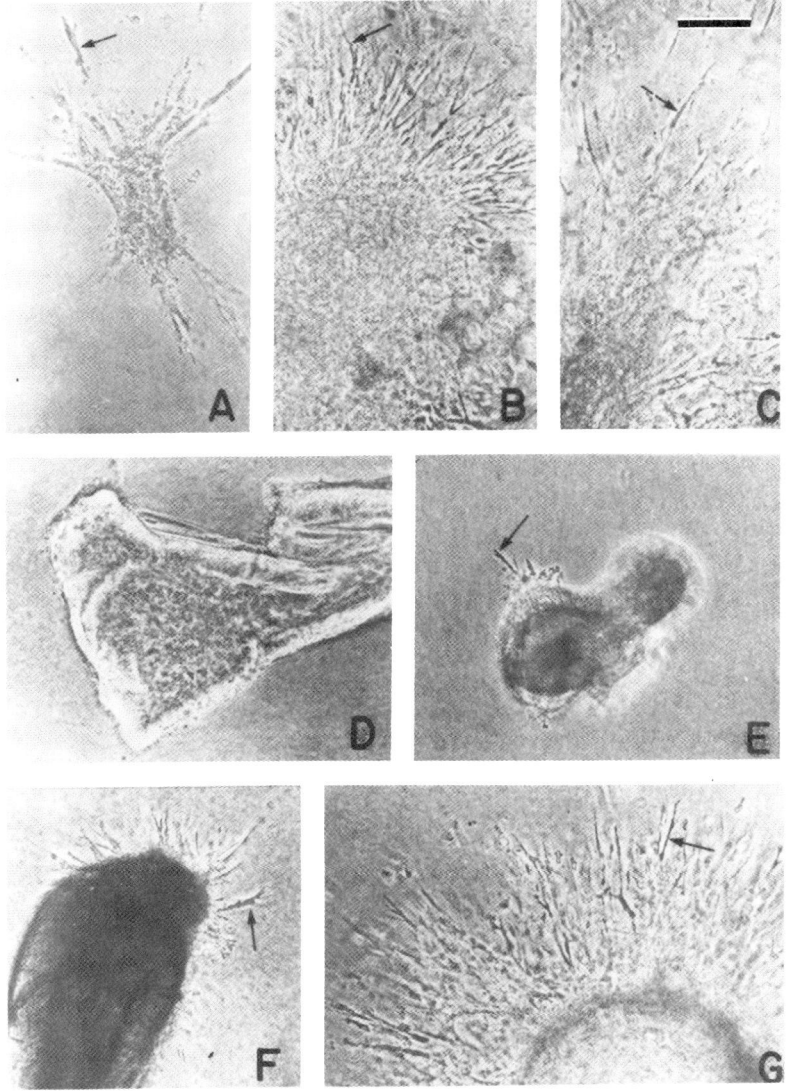

FIGURE 10. Phase contrast light micrographs showing mesenchyme-like cells originating from epithelia cultured in collagen gels for 4 days (A-C, G), 4 hours (D), 2 days (E), and 3 days (F). A. notochord. B. limb ectoderm. C. adult corneal endothelium. D-G. anterior lens epithelium. C, bovine, rest chick. Bar, 100 μm. From Greenburg and Hay (2).

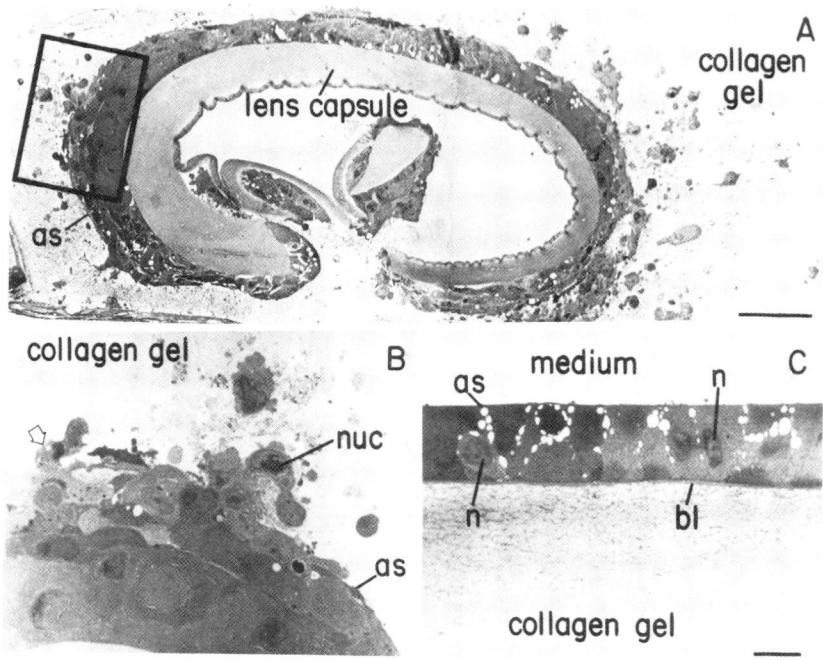

FIGURE 11. (A,B) Light micrographs of sections of a culture similar to that illustrated in figure 10G. Avian anterior lens epithelium isolated with its basal lamina (lens capsule) was grown for 9 days within collagen gel. as, apical surface. (B) is an enlargement of the area in the rectangle in (A). The epithelium is multilayered and the apical cells are sending out pseudopodia (arrow, B) into the gel. nuc, nucleolus. (C) Section of lens epithelium from an explant grown on top of collagen gel. The epithelium on top of gel is coherent and secretes a basal lamina (bl). n, nucleus. Bars, 50 μm. From Greenburg and Hay (2).

Are these mesenchyme-like cells derived from lens epithelium in collagen gel producing the same products as true fibroblasts? We examined them by immunofluorescence, using antibodies specific to chicken type I collagen and found that the newly formed mesenchyme-like cells are rich in type I collagen (1; illustrated in 5). On the other hand, the epithelial cells that grow out of

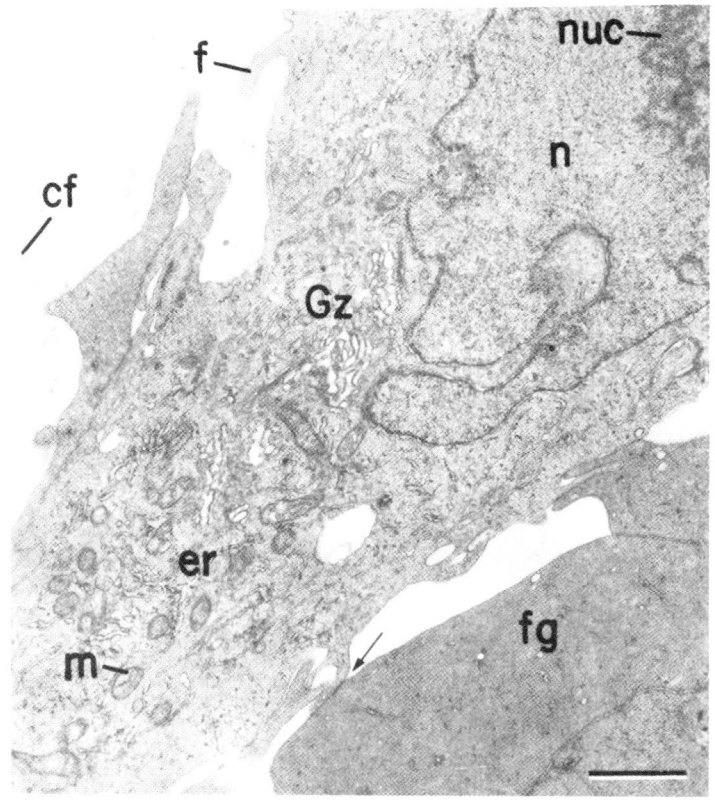

FIGURE 12. Electron micrograph of an elongating apical cell similar to that at the open arrow in figure 11B. It has acquired prominent Golgi zone (Gz), endoplasmic reticulum (er), and mitochondria (m). It has lost the fibrillogranular cytoplasmic material found in underlying lens epithelial cells still containing crystallins (fg). f, filopodium. cf, collagen fibrils of the gel. n, nucleus. nuc, nucleolus. Bar, 1 μm. From Greenburg and Hay (2).

an explant on top of collagen gel (Fig. 11 C), produce a new basal lamina and do not react with antibodies to type I collagen. They react with antibodies to type IV collagen and laminin (1).

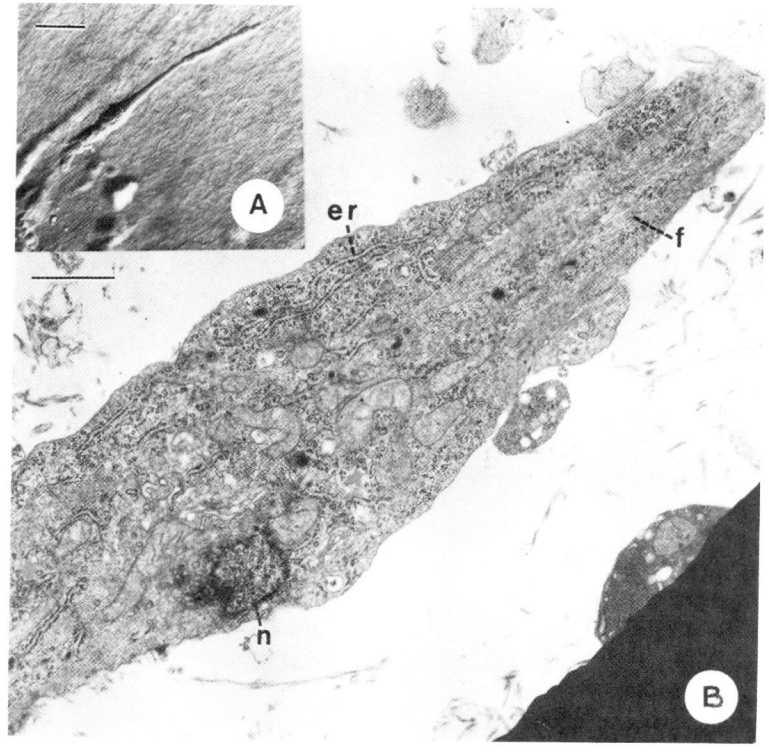

FIGURE 13. (A) Light micrograph and (B) electron micrograph of mesenchyme-like cells derived from anterior lens epithelium in collagen gel. The highly elongate, bipolar cells are rich in granular endoplasmic reticulum (er) and cytoplasmic filaments (f). n, nucleus. Bar, 25 μm (A) and 1 μm (B). From Greenburg and Hay (2).

Thus, the anterior lens epithelium, which is absolutely mesenchyme-free at the beginning of the culture period, gives rise to mesenchyme-like cells when suspended within collagen gel (Fig. 14). Both the ultrastructure and synthetic activities of the new cell type are typical of fibroblasts. It is essential that the epithelium be placed within the solution of collagen while it is gelling so as to develop very close contact with the surrounding collagen fibrils. If a lens or corneal epithelium growing on top of a collagen gel is subsequently overlain with

Cell-Matrix Interaction

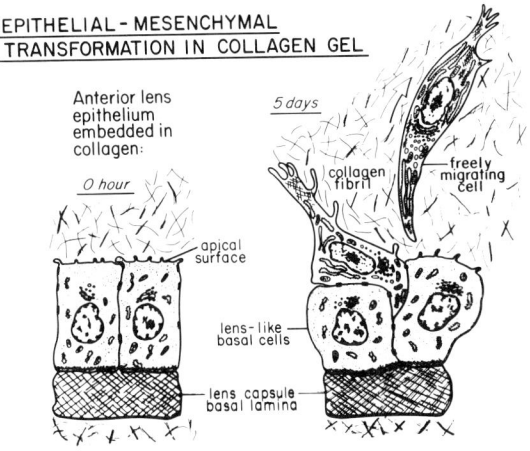

FIGURE 14. Diagram summarizing the behavior of anterior lens epithelium suspended in collagen gel. Based on data of Greenburg and Hay (2).

another collagen gel, it does not give rise to mesenchyme (Greenburg and Hay, unpublished observation). Thus, it is not surprising to find reports in the literature that epithelia tend to form cavities, by multilayering or cell migration, when covered with a collagen gel (42).

We next studied in detail the performance of thyroid follicles suspended within collagen gel. We chose this tissue because there are reports in the literature that the follicles exhibit normal epithelial polarity in collagen gels (43). Greenburg (1) isolated follicles by enzyme treatment and sieving, and suspended them within collagen gels. Within 3-5 days, the majority of follicles formed mesenchyme as described above, except in this case multilayering took place along the basal epithelial surface and then basal cells extended pseudopodia into the collagen, elongated, and migrated away (1).

The difference in our results (1) and those of Chambard et al. (43) on thyroid may lie in differing abilities of the follicles to reconstruct basal laminae, due perhaps to minor differences in isolation procedures. Basal lamina components added to the collagen gel inhibit epithelial-mesenchymal transformation in both lens and thyroid cultures (1). Variation in ability of the isolated epithelium to reconstruct basal lamina along the basal surface of the explants and/or outgrowths could also explain why some gland epithelia grown in collagen gels produce branching tubules (44), while others send isolated cells into the gel (45).

The thyroid follicle provides an excellent example of a cytokeratin epithelial cytoskeleton. The cells are rich in desmosomes and tonofilaments. We next asked whether or not vimentin is acquired during epithelial-mesenchymal transformation. Indeed, cells beginning to elongate from follicles suspended in collagen gel acquire vimentin as they are leaving the follicle. The fully migrating mesenchyme-like cells are very rich in vimentin (1). They still contain cytokeratin at 3-5 days, but after several weeks this rather stable protein gradually disappears. Thus, in all characteristics that we measured, the elongate cells deriving from thyroid are mesenchymal. They stop producing thyroglobulin, acquire vimentin, and ultrastructurally resemble fibroblasts (1).

TRANSFORMATION OF EPITHELIA TO MESENCHYME IN THE EMBRYO

In my concluding remarks, I would like to return to the question of epithelial-mesenchymal transformation *in vivo*. Do our *in vitro* observations shed any light on what might be happening in the real embryo? Well, for one thing, the performance of epithelia suspended within collagen gels shows us that there is nothing special about conducting the transformation on the basal side of the tissue, as occurs *in vivo* (Fig. 15). What is essential is a change in the cell surface, be it apical or basal, from an epithelial cell surface to a mesenchymal one, from a cell surface that behaves in a strictly ordered and polarized fashion, to one that is highly invasive. The common pattern in the *in vitro* transformations described here is that the cell surface exposed to the collagen gel seems to transmit a message to the previously quiescent epithelium to synthesize DNA and divide. The epithelial monolayer acquires one or more additional layers on its apical (as for lens) or basal (as for thyroid) pole and these new cells begin to probe the ECM with filopodia and pseudopodia. It is tempting to think that cell surface proteins typical of epithelia are either replaced or redistributed during cell proliferation under the influence of the adjacent collagen gel.

Cell-Matrix Interaction 311

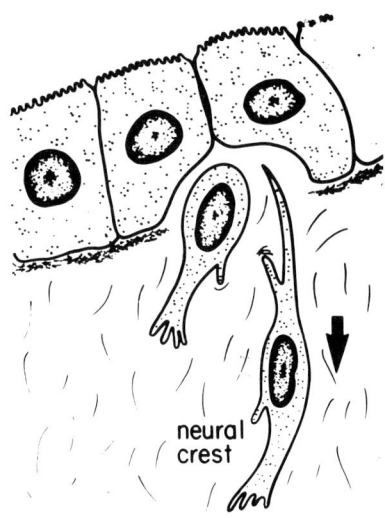

FIGURE 15. Diagram depicting epithelial-mesenchymal transformation in the embryo.

There is evidence from studies of epithelial-mesenchymal transformation in vivo that the cell surface does change its character as cells prepare to leave the epithelium. The presumptive primary mesenchyme of the sea urchin embryo loses affinity for the hyaline layer and outer epithelium, and acquires new antigens and affinity for basal lamina components such as fibronectin (46; 47). The cells then migrate along the inner side of the basal lamina and along strands of blastocoele ECM. In the chick embryo, the sodium pump changes location during epithelial-mesenchymal transformation along the primitive streak (48). Neural crest cells invaginating from the neural tube lose the cell surface glycoprotein, N-cam and develop affinity for fibronectin (Thiery, this volume; 49). Thus, it seems likely that mesenchyme-like cells derived from epithelium suspended in collagen gels turn over epithelia-specific receptors during cell division and acquire ECM receptors and other cell surface proteins that characterize mesenchymal cells.

The exact composition of the cytoskeleton and ECM proteins secreted during epithelial-mesenchymal transformation in the embryo (Fig. 15) has not been studied in any detail. Solursh et al. (50) report that hyaluronic acid appears between transforming cells in somites. Franke et al. (11) observed vimentin in primary mesenchyme either before or after the cells leave the primitive streak.

Our studies of transformation in vitro indicate that vimentin is expressed in some of the cells before they leave the epithelium. At this time, they probably also turn off expression of cytokeratin, although this protein persists in the cytoplasm of the mesenchyme-like cells for a week or more in vitro. Synthesis of the ECM protein characteristic of mesenchyme, type I collagen, begins and the mesenchyme-like cells stop producing type IV collagen and laminin. It is tempting to think that a genetic program for expression of the mesenchymal phenotype is turned on in a synchronous fashion as the cell surface changes its character and the cells lose their epithelial characteristics (Fig. 16). Interestingly, epithelial cytodifferentiation as expressed by lens crystallin or thyroglobulin synthesis is also lost at this time. We are only beginning to think about the way expression of genes for specific protein synthesis (cytodifferentiation) may be linked to expression of the collection of cell surface, cytoskeletal, and ECM genes that seem to comprise the genetic program of the epithelial phenotype (Fig. 16).

The mechanism whereby surrounding collagen gel induces differentiated epithelia to give rise to mesenchyme-like cells is not known. In the embryo, sites of epithelial-mesenchymal transformation are characterized by changes in cytoplasmic staining and loss of basal lamina (51).

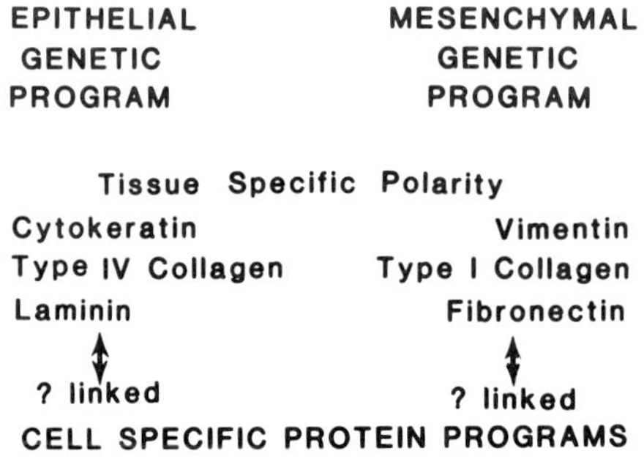

EPITHELIAL
GENETIC
PROGRAM

MESENCHYMAL
GENETIC
PROGRAM

Tissue Specific Polarity

Cytokeratin
Type IV Collagen
Laminin

Vimentin
Type I Collagen
Fibronectin

↕
? linked

↕
? linked

CELL SPECIFIC PROTEIN PROGRAMS

FIGURE 16. Summary of tissue phenotypes.

Basal lamina components placed in the collagen gel inhibit the transformation of lens or thyroid epithelia to mesenchyme-like cells in vitro (1). However, loss of basal lamina is not sufficient to trigger the transformation; there are areas in the embryo lacking in basal lamina (e.g., tips of gland lobules) that do not give rise to mesenchyme. We discussed in the beginning of this chapter the fact that epithelia placed on top of collagenous matrices flatten their basal surface and do not invade the matrix. The only epithelia that have been reported to give rise to mesenchyme from the top of a collagen gel are those, such as cardiac cushion endothelium (52), that seem to be programmed to produce mesenchyme in the embryo.

The idea that particular epithelia are specifically programmed to give rise to mesenchymal cells in the embryo is hard to escape. We can predict with complete accuracy when the primitive streak will produce primary mesenchyme, when this mesenchyme will condense into mesodermal epithelia, and when these epithelia will give rise to so-called secondary (53) mesenchyme. The timing of trunk neural crest formation coincides with that of sclerotome mesenchyme formation. How these programs are set into place and are operated is completely unknown. It could be supposed that all epithelia have an inherent desire to lose surface polarity and transform into cells that can frolic in the underlying ECM. If this is the case, then what is required is an inhibitory mechanism which is turned off at the specific times and places indicated above. The fact that adult epithelia can be induced to give rise to mesenchyme-like cells merely by submerging them in collagen gels suggests that whatever the control mechanism might be, positive or negative in nature, it is a simple one to trigger into operation.

ACKNOWLEDGMENTS

We are indebted to Drs. George Martin, Hynda Kleinman, and other members of the Developmental Biology Laboratory of the National Institute of Dental Research, Bethesda, Maryland, for type IV collagen, laminin, and antilaminin (EHS tumor antigens); to Dr. Charles Little, Department of Anatomy, University of Virginia, Charlottesville, Virginia, for antibodies to several types of collagens, including the antibody against chicken type I collagen that does not interact with rat type I collagen; and to Drs. Joram Piatigorsky and Johan Zwann for anticrystallins.

REFERENCES

1. Greenburg G (1985). "Epithelial-Mesenchymal Transformation In Vitro." Boston: Harvard Medical School, Ph.D. thesis.
2. Greenburg G, Hay ED (1982). Epithelia suspended in collagen gels can lose polarity and express characteristics of migrating mesenchymal cells. J Cell Biol 95:333.
3. Hay ED (ed)(1981). "Cell biology of extracellular matrix." New York: Plenum Publishing.
4. Sabatini DD, Griepp EV, Rodriguez-Boulan EJ, Dolan WJ, Robbins ES, Papadopoulos S, Ivanov IE, Rinder MJ (1983). Biogenesis of epithelial cell polarity. Modern Cell Biology 2:419.
5. Hay ED (1984). Cell-matrix interaction in the embryo: cell shape, cell surface, cell skeletons, and their role in differentiation. In Trelstad RL (ed): "The Role of Extracellular Matrix in Development," New York: Academic Press, p. 55.
6. Hasty DL, Hay ED (1977). Freeze-fracture studies of the developing cell surface. I. The plasmalemma of the corneal fibroblast. J Cell Biol 72:667.
7. Overton J (1977). Response of epithelial and mesenchymal cells to culture on basement lamella observed by scanning microscopy. Exp Cell Res 105:313.
8. Schor SL (1980). Cell proliferation and migration on collagen substrata in vitro. J Cell Sci 41:159.
9. Dodson JW, Hay ED (1974). Secretion of collagen by corneal epithelium. II. Effect of the underlying substratum on secretion and polymerization of epithelial products. J Exp Zool 189:51.
10. Franke WW, Appelhans B, Schmid E, Freudenstein C, Osborn M, Weber K (1979). Identification and characterization of epithelial cells in mammalian tissues by immunofluorescence microscopy using antibodies to prekeratin. Differentiation 15:7.

11. Franke WW, Grund C, Kuhn C, Jackson BW, Illmensee K (1982a). Formation of cytoskeletal elements during mouse embryogenesis. III. Primary mesenchymal cells and the first appearance of vimentin filaments. Differentiation 23:43.
12. Kuhl U, Ocalan M, Timpl R, Mayne R, Hay ED, von der Mark K (1984). Role of muscle fibroblasts in the deposition of type IV collagen in the basal lamina of myotubes. Differentiation 28:164.
13. Trelstad RL (ed) (1984). "The Role of Extracellular Matrix in Development." New York: Alan R. Liss.
14. Meier S, Hay ED (1974). Control of corneal differentiation by extracellular materials. Collagen as a promoter and stabilizer of epithelial stroma production. Dev Biol 38:249.
15. Sugrue SP, Hay ED (1981). Response of basal epithelial cell surface and cytoskeleton to solubilized extracellular matrix molecules. J Cell Biol 91:45.
16. Sugrue SP, Hay ED (1982). Interaction of embryonic corneal epithelium with exogenous collagen, laminin, and fibronectin. Role of endogenous protein synthesis. Devel Biol 92:97.
17. Bissell MJ, Hall HG, Parry G (1982). How does the extracellular matrix direct gene expression? J Theor Biol 99:31.
18. Lee EYH, Parry G, Bissell MJ (1984). Modulation of secreted proteins of mouse mammary epithelial-cells by the collagenous substrata. J Cell Biol 98:146.
19. Sugrue SP (1984). The interaction of extracellular matrix with the basal cell surface of embryonic corneal epithelia. Prog Clin Biol Res 15:77.
20. Chiang TM, Kang AH (1982). Isolation and purification of collagen α 1(I) receptor from human platelet membrane. J Biol Chem 257:7581.
21. Mollenhauer J, von der Mark K (1983). Isolation and characterization of a collagen-binding glycoprotein from chondrocyte membranes. EMBO J 2:45.
22. Kurkinen M, Taylor A, Garrels JI, Hogan BLM (1984). Cell surface-associated proteins which bind native type IV collagen. J Biol Chem 259:5915.
23. Terranova VP, Rao CN, Kalebic T, Margulies IM, Liotta LA (1983). Laminin receptor on human breast carcinoma cells. Proc Natl Acad Sci USA 80:444.

24. Lesot H, Kuhl U, von der Mark K (1983). Isolation of a laminin-binding protein from muscle cell membranes. EMBO J 2:861.
25. Malinoff HL, Wicha MS (1983). Isolation of a cell surface receptor protein for laminin from murine fibrosarcoma cells. J Cell Biol 96:1475.
26. Pytela R, Pierschbacher MD, Ruoslahti E (1985). Identification and isolation of a 140 kd cell surface glycoprotein with properties expected of a fibronectin receptor. Cell 40:191.
27. Damsky CH, Knudsen KA, Bradley D, Buck CA, Horwitz AF (1985). Distribution of the cell substratum attachment (CSAT) antigen on myogenic and fibroblastic cells in culture. J Cell Biol 100:1528.
28. Chen W-T, Hasegawa E, Hasegawa T, Weinstock C, Yamada KM (1985). Development of cell surface linkage complexes in cultured fibroblasts. J Cell Biol 100:1103.
29. Pierschbacher MD, Ruoslahti E (1984). Cell attachment activity of fibronectin can be duplicated by small synthetic fragments of the molecule. Nature 309:30.
30. Kjellen L, Oldberg A, Hook M (1980). Cell-surface heparan sulfate. J Biol Chem 255:10407.
31. Rapraeger AC, Bernfield M (1983). Heparan sulfate proteoglycans from mouse mammary epithelial cells: A putative membrane proteoglycan associates quantitatively with lipid vesicles. J Biol Chem 258:3632.
32. Nevo Z, Dorfman A (1972). Stimulation of chondromucoprotein synthesis in chondrocytes by extracellular chondromucoprotein. Proc Natl Acad Sci USA 69:2069.
33. Kosher RA, Church RL (1975). Stimulation of in vitro chondrogenesis by procollagen and collagen. Nature 258:327.
34. Urist MR, Lietze A, Mizutani H, Takagi K, Triffitt JT, Amstutz J, DeLange R, Termine J, Finerman GAM (1982). A bovine low molecular weight bone morphogenetic protein (BMP) fraction. Clin Orthop 162:219.
35. Sampath TK, Nathanson MA, Reddi AH (1984). In vitro transformation of mesenchymal cells derived from embryonic muscle into cartilage in response to extracellular matrix components of bone. Proc Natl Acad Sci USA 81:3419.

36. von der Mark K (1980). Immunological studies on collagen type transition in chondrogenesis. In Friedlander M (ed) "Immunological Approaches to Embryonic Development and Differentiation, Part II." New York: Academic Press, p.199.
37. Glowacki J, Trepman E, Folkman J (1983). Cell shape and phenotypic expression in chondrocytes. Proc Soc Exp Biol Med 172:93.
38. Zaneti NC, Solursh M (1984). Induction of chondrogenesis in limb mesenchymal cultures by disruption of the actin cytoskeleton. J Cell Biol 98:115.
39. Tomasek JJ, Hay ED, Fujiwara K (1982). Collagen modulates cell shape and cytoskeleton of embryonic corneal and fibroma fibroblasts: Distribution of actin, a-actinin, and myosin. Dev Biol 92:107.
40. Tomasek JJ, Hay ED (1984). Analysis of the role of microfilaments and microtubules in acquisition of bipolarity and elongation of fibroblasts in hydrated collagen gels. J Cell Biol 99:536.
41. Piatigorsky J, Rothschild SS, Milstone LM (1973). Differentiation of lens fibers in explanted embryonic chick lens epithelia. Dev Biol 34:334.
42. Hall HG, Farson DA, Bissell MJ (1982). Lumen formation by epithelial cell lines in response to collagen overlay: A morphogenetic model in culture. Proc Natl Acad Sci USA 79:4672.
43. Chambard M, Gabrion J, Mauchamp J (1981). Influence of collagen gel on the orientation of epithelial cell polarity: Follicle formation from isolated thyroid cells and from preformed monolayers. J Cell Biol 91:157.
44. Yang J, Richards J, Bowman P, Guzman R, Enami J, McCormick K, Hamamoto S, Pitelka D, Nandi S (1979). Sustained growth and three-dimensional organization of primary mammary tumor epithelial cells embedded in collagen gels. Proc Natl Acad Sci USA 76:3401.
45. Dulbecco R, Henahan M, Bowman M, Okada S, Battifora H, Unger M (1981). Generation of fibroblast-like cells from cloned epithelial mammary cells in vitro: A possible new cell type. Proc Natl Acad Sci USA 78:2345.
46. Fink RD, McClay DR (1985). Three cell recognition changes accompany the ingression of sea urchin primary mesenchyme cells. Devel Biol 107:66.

47. Katow H, Yamada KM, Solursh M (1982). Occurrence of fibronectin on the primary mesenchymal cell surface during migration in the sea urchin embryo. Differentiation 22:120.
48. Stern CD (1982). Localization of the sodium pump in the epiblast of the early chick embryo. J Anat 134:606.
49. Thiery JP, Delouvee A, Grumet M, Edelman GM (1985). Initial appearance and regional distribution of the neuron-glia cell adhesion molecule in the chick embryo. J Cell Biol 101:442.
50. Solursh M, Fisher M, Meier S, Singley CT (1979). The role of the extracellular matrix in the formation of the sclerotome. J Embryol Exp Morphol 54:75.
51. Nichols DA (1981). Neural crest formation in the head of the mouse embryo as observed using a new histological technique. J Embryol Exp Morphol 64:105.
52. Bernanke DH, Markwald RR (1982). Migratory behavior of cardiac cushion tissue cells in a collagen-lattice culture system. Dev Biol 91:235.
53. Hay ED (1968). Organization and fine structure of epithelium and mesenchyme in the developing chick embryo. In Fleischmajer R, Billingham RE (eds): "Epithelial-Mesenchymal Interactions," Baltimore: Williams & Wilkins Co.

EXTRACELLULAR MATRIX IN SKIN MORPHOGENESIS[1]

Philippe Sengel, Annick Mauger, Joelle Robert, and Madeleine Kieny

Unité Associée au CNRS n° 682, Laboratoire de Biologie animale, Université scientifique & médicale de Grenoble, 38402 Saint Martin d'Hères, France

ABSTRACT Histochemical detection of various extracellular matrix components in the skin of birds and mammals during the development of cutaneous appendages reveals that some components are evenly distributed (type IV collagen, laminin, and basement membrane proteoglycan), while others exhibit a microheterogeneous distribution pattern which changes with age and site (types I and III collagen, fibronectin, bullous pemphigoid antigen, glycosaminoglycans). This suggests that the latter components might constitute part of the morphogenetic message that the dermis transmits to the epidermis during skin organogenesis. Cultures of embryonic chick dermal cells on two-dimensional substrates or in three-dimensional collagen gels are undertaken to analyze the effect of several extracellular matrix components. Preliminary results show that indeed type I collagen and fibronectin can, under certain circumstances, influence several parameters of cell behavior.

INTRODUCTION

It has become clear in recent years that the extracellular matrix plays an important role in the morphogenesis of various organs during embryonic development (6,12,13,28). Many so-called heterogeneous

[1]This work was supported by the Centre National de la Recherche Scientifique (UA 682 and RCP 533) and by the Ministère de la Recherche et de la Technologie (grants no. 81.E.1082 and no. 83.V.0099).

organs with an epithelial-mesenchymal constitution result from precisely-timed and precisely-located tissue interactions, during which meaningful morphogenetic messages are exchanged between epithelium and mesenchyme (29,37). While the involvement of extracellular matrix components in morphogenesis has been suggested in many developing organs, such as skin (15,17,19), tooth (25,36,38), salivary gland (6), cornea (12), kidney (7,28), the mechanisms whereby morphogenetic messages are generated and transmitted from one tissue to the other are still largely unknown. It appears however that the changing microheterogeneous distribution of various extracellular matrix components, and notably those of the epithelial-mesenchymal junction, may constitute at least part of the language used by the cells.

During skin development in amniote embryos, the mesenchymal component of the organ rudiment, the dermis, plays a predominant role. It controls the size, the shape, the distribution pattern, the growth rate and the region-specific epidermal differentiation of cutaneous appendages (29,30).

Histological studies using indirect immunofluorescence and other histochemical methods were performed on embryonic chick (14,17,18,19,33) and mouse (3 and unpublished results) appendage-forming and glabrous skin in order to localize some of the major constituents of the extracellular matrix during skin morphogenesis and to determine whether their distribution is meaningfully related to developmental events. These studies have revealed that indeed some extracellular matrix constituents, but not others, change their distribution pattern during the development of cutaneous appendages. This led to the idea that the three-dimensional organization of the extracellular matrix might influence not only the behavior of mesenchymal cells, but also that of the overlying epithelial cells, and thus constitute a means for dermal cells to transfer their morphogenetic messages to the epidermis.

In order to further analyze how extracellular matrix macromolecules might influence the behavior of skin cells, the system has been transposed *in vitro* . For the time being, studies were performed on dermal cells only, either in cultures of dissociated cells, or in explant cultures, on two-dimensional substrates (31,34,35) or in three-dimensional hydrated collagen gels (unpublished results). Various parameters of cell behavior were analyzed, such as size, shape, locomotion and patterning. These still preliminary studies have led to the notion that indeed dermal cells are influenced by their extracellular matrix environment in a way which is seemingly related to their morphogenetic performance.

DISTRIBUTION OF EXTRACELLULAR MATRIX COMPONENTS

Various extracellular matrix components were detected in embryonic chick or mouse skin by indirect immunofluorescence on frozen sections using monospecific antibodies, or by other histochemical methods. According to these observations, extracellular matrix macromolecules can be classified into two categories. The first one comprises **laminin**, **type IV collagen** and **basement membrane proteoglycan**, which appear to be evenly distributed along the dermal-epidermal junction throughout the development of feathers, avian foot scales, or hairs, in appendage-forming as well as in glabrous skin regions. These components are probably not directly involved in the morphogenesis of cutaneous appendages. The second category includes **type I and type III collagen**, **fibronectin**, **bullous pemphigoid antigen** and several **glycosaminoglycans**. These components exhibit a heterogeneous distribution, which changes in space and time during the development of appendages. Their appearance and disappearance at specific phases of organogenesis thus appear to be related to skin morphogenesis and to the formation of cutaneous appendages. The controlled synthesis, deposition, and degradation of these macromolecules might thus play an important role in the transmission of morphogenetic messages from dermis to epidermis.

In the chick embryo, before the onset of appendage formation, the distribution of interstitial collagens, of fibronectin, and of sulfated glycosaminoglycans is uniform, and the density of these components is at first sparse. They are mainly located along the dermal-epidermal junction, but later become dispersed throughout the thickness of the dermis, while their density increases. In the dermal condensation of the outbulging feather or scale bud, or in the dermal sheath of the ingrowing feather follicle, where morphogenetic movements are being performed, interstitial collagens and sulfated glycosaminoglycans become sparse or disappear, while fibronectin is accumulated. Conversely, at the base of feather and scale buds, in interappendage skin, and in glabrous regions, which are early stabilized regions of the integument, interstitial collagens accumulate in increasing density with age, while the density of fibronectin decreases. Likewise, hyaluronate (detected by Streptomyces hyaluronidase sensitive Alcian blue stain) is at first uniformly distributed along the dermal-epidermal junction. When the feather rudiments are formed, hyaluronate along the dermal-epidermal junction becomes denser inside the rudiment than in interappendage skin. When the bud grows out, hyaluronate label becomes more intense at the apex of the bud and along its posterior slope (Fig. 11). Later, when the bud elongates, hyaluronate density is higher in the dermal-epidermal junction of the cranial than of the caudal slope of the bud (Fig. 12). Interstitial collagens, in outgrowing buds, also exhibit an asymmetric distribution, with type I collagen being preferentially accumulated under the cranial slope, and type III collagen under the caudal slope and the posterior hinge region.

During later stages of feather development, the density of interstitial collagens decreases along the ingrowing feather follicle, while it continues to increase in interplumar skin. By contrast fibronectin accumulates at the base, inside and around the feather follicle, whereas it almost completely disappears from interfollicular skin regions, except around blood vessels. Alcian blue stained glycosaminoglycans also exhibit a heterogeneous distribution during feather follicle formation: they are laid down in higher amounts underneath the cranial than under the caudal dermal-epidermal junction of the base of the feather follicle.

Recent unpublished observations using an antibody against chick cellular fibronectin (a gift from R.H. Sawyer and K.M. Yamada) reveal a distribution pattern of fibronectin which is superposable to the one we visualized before with an antibody directed against human plasma fibronectin, namely a high density in zones of high morphogenetic activity, and sparseness in histologically stabilized zones (Fig. 15 and 16).

The detection of extracellular matrix components in glabrous skin regions confirms the view that collagen is accumulated in and fibronectin is removed from zones with low or no morphogenetic activity. This is true in normal appendageless regions, such as the midventral apterium,

FIGURES 1-8. Immunofluorescent labeling of extracellular matrix components in frozen sections of mouse skin during hair development. FIGURES 1 and 3. With anti-mouse type I collagen antibody, the dermal-epidermal junction (J) of interfollicular skin, of upper part of hair follicle, and the lower dermis are conspicuously labeled. Note absence of label around hair bud (H), around hair bulb (B), along the dermal-epidermal junction of the dermal papilla and within the dermal papilla itself (P). FIGURES 2 and 4. Phase contrast micrographs of FIG. 1 and 3. FIGURES 5 and 6. With bullous pemphigoid antiserum, the dermal-epidermal junction (J) of interfollicular skin and of upper part of hair follicle is labeled by a continuous line. Note attenuation or absence of label around lower part of hair bud (H) and of hair follicle (F), as well as around dermal papilla (P). FIGURES 7 and 8. With anti-laminin (FIG. 7) and anti-type IV collagen antibody (FIG. 8), the dermal-epidermal junction is labeled by a continuous fluorescent line. Scale bars = 50 μm.

Extracellular Matrix in Skin Morphogenesis   323

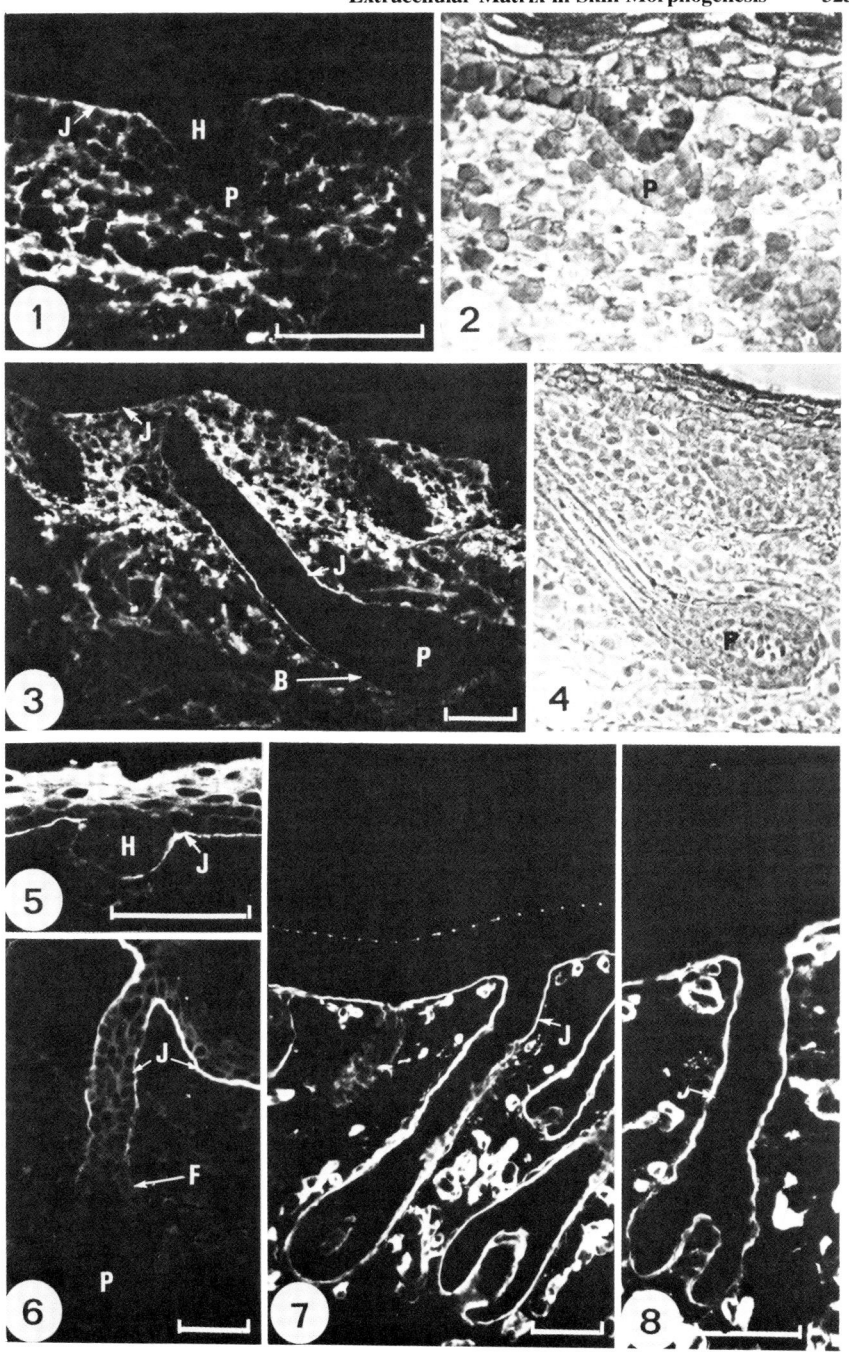

the interappendage skin, as well as in experimentally or genetically induced appendagelessness, such as hydrocortisone treated chick embryos, or *scaleless* mutant chick embryos.

Regarding hair development in the mouse embryo, similar observations have been made on the distribution of types I and III collagen (unpublished results: collagens were extracted and purified from mouse skin by D. Herbage and his group at the Centre d'Etudes et de Recherche de Dermobiochimie in Lyon; the corresponding antibodies were raised in rabbits by D. Hartmann and his group at the Institut

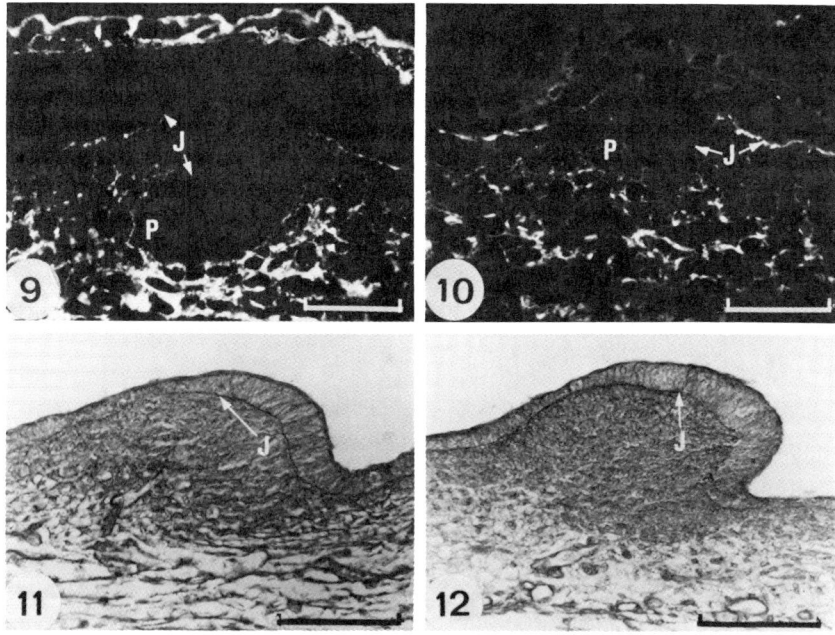

FIGURES 9 and 10. Comparison of anti-mouse type I collagen (FIG. 9) and anti-mouse type III collagen labeling (FIG. 10) of frozen sections of embryonic mouse skin at hair bud stage. Note absence of label in dermal papilla (P) and in a layer of dermal cells around hair bud. Labeling of the dermal-epidermal junction (J) is attenuated or absent around lower part of buds. (Fluorescence on top of epidermis in FIG. 9 is in the amnion). Scale bars = 50 µm.

FIGURES 11 and 12. Alcian blue labeling of glycosaminoglycans in embryonic chick skin, at early (FIG. 11) and late feather bud (FIG. 12) stages. Note changing pattern of staining of dermal-epidermal junction (J). (Photographs by C. Jahoda). Scale bars = 100 µm.

Pasteur in Lyon, France), of fibronectin (32) and bullous pemphigoid antigen (BPA) (3) (Fig. 1 - 14).

In undifferentiated skin, before the onset of hair development, interstitial collagens and fibronectin are present in the dermis at a moderate density. Collagens are at first evenly distributed throughout the depth of the dermis, but later become more abundant in the lower dermis than in the upper dermis. In addition, anti-collagen antibodies strongly underline the dermal-epidermal junction with a continuous fluorescent line. Fibronectin is uniformly distributed in the whole dermis. BPA is

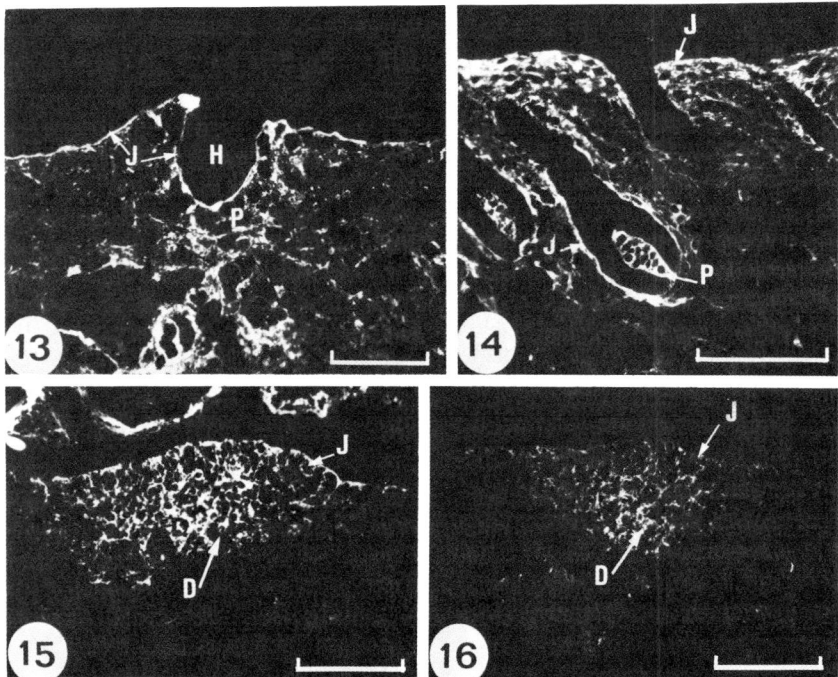

FIGURES 13 - 16. Immunofluorescent labeling of fibronectin in frozen sections of embryonic mouse (FIG. 13 and 14) and chick (FIG. 15 and 16) skin. With anti-human plasma fibronectin antibody, label is conspicuous around hair bud (H), along the dermal-epidermal junction (J), in the dermis underneath the bud, and in the dermal papilla (P). In the outgrowing feather rudiment, label is localized within the dermal condensation (D), similarly with anti-chick cellular fibronectin (FIG.15) or with anti-human plasma fibronectin antibody (FIG. 16). Scale bars: FIG. 13: 50 μm; FIG. 14 - 16: 100 μm.

localized at the dermal-epidermal junction. By the time the hair placodes, and later the hair buds, form, interstitial collagens disappear from the dermal-epidermal junction inside the hair rudiment, but are still detectable outside the rudiments along the dermal-epidermal junction and throughout the dermis with a maximal density in its lower part (Fig. 1, 2 and 10). The small clusters of cells which give rise to the hair papillae, however, remain devoid of anti-collagen label. Conversely, fibronectin accumulates in higher density around the ingrowing hair peg, notably along the dermal-epidermal junction, and also underneath the ingrowing hair bud in the dermal area into which the hair bud is going to penetrate (Fig. 13). Anti-BPA label becomes attenuated and almost completely disappears from the dermal-epidermal junction around the lower part of the hair nodules and buds (Fig. 5).

At later stages of hair development, when the hair follicle is forming, types I and III collagens remain absent from the dermal papilla, and from the dermal-epidermal junction around the hair bulb and around the dermal papilla (Fig. 3 and 4). Elsewhere, namely along the dermal-epidermal junction of the upper two-thirds of the follicles, in the dermis and along the dermal-epidermal junction of interfollicular skin, type I collagen increases in density with age. The distribution of type III collagen is similar to that of type I collagen, except for its overall lower density and its absence from the dermal-epidermal junction around the hair follicle. Anti-fibronectin label remains strong inside the dermal papilla, along the dermal-epidermal junction of the dermal papilla and around the whole follicle, particularly at the posterior part of upper follicle, where it merges into the surrounding epidermis (Fig. 14). BPA is present at the dermal-epidermal junction of interfollicular skin and along the upper two-thirds of the hair follicles. It remains attenuated or absent from around the dermal papilla and the hair bulb (Fig. 6).

These observations are in agreement with those made earlier on feather and scale development in the chick embryo. They again reveal that interstitial types I and III collagens are removed from zones of morphogenetic activity, and that fibronectin is preferentially accumulated in these areas. In addition they indicate that BPA in mammalian embryonic skin might also play a significant role in dermal-epidermal communication during the formation of hairs. They thus strongly confirm the view that the microheterogeneous distribution of interstitial collagens, of certain glycosaminoglycans, of fibronectin and also of BPA might constitute part of the morphogenetic messages that the dermis is known to transmit to the epidermis during the morphogenesis of skin and cutaneous appendages.

According to this concept, interstitial collagens and still unidentified glycosaminoglycans might offer a firm or semi-solid framework on which to found cutaneous appendages, while fibronectin might allow and stimulate cell and tissue motility. In interappendage and in glabrous skin, high amounts of interstitial collagens and low density of fibronectin might cause an early stabilization of skin tissues. Contrariwise, inside developing feather, scale, or hair rudiments, the absence, sparsity or asymmetric distribution of collagens and of certain glycosaminoglycans, and the abundance of fibronectin might cause and facilitate morphogenetic movements (20,21). In embryonic mouse skin, in addition, BPA might constitute a barrier opposed to the transmission of morphogenetic messages in stabilized regions, such as interfollicular skin and the upper part of hair follicles; where BPA is absent, and remains absent, throughout the development of hairs, namely around the dermal papilla and around the hair bulb, the possibility of dermal-epidermal communication is likely to be maintained.

INFLUENCE OF EXTRACELLULAR MATRIX COMPONENTS ON CULTURED DERMAL CELLS

Since it appears from the above-mentioned histochemical observations that extracellular matrix components might constitute one of the means by which skin tissues communicate during embryonic development, several series of *in vitro* culture experiments were set up to investigate in which manner the extracellular matrix macromolecules influence the behavior of skin cells. For the time being, these experiments have been performed on dermal cells only, obtained from the back skin of 7-day chick embryos. The idea is to expose these cells to various homogeneous or heterogeneous environments, either in **two-dimensional** or in **three-dimensional** condition. Two kinds of culture systems have been used: a) **dissociated dermal cells** obtained by the complete disaggregation of pieces of dermis (after removal of the epidermis) and seeded at densities between 10^5 and 5×10^5 per 35 mm diameter plastic tissue culture dish; b) **explants** of small pieces of dermis (approximately 0.5 x 0.5 mm, or less).

Photographic records of selected areas of the cultures were made at 15 min intervals over a period of 9 to 19 hours, during days 1 to 3 of culture, when a majority of cells are still isolated, can thus be easily followed individually and are not disturbed by contacts with encountered cells. Using an image analysis computer program, several parameters of cell shape and behavior were analyzed.

Extracellular matrix macromolecules to which the cultures have been exposed to date comprise bovine type I collagen and human or bovine plasma fibronectin. The behavior of the cells was quantitated using various parameters such as rate of cell patterning, cell density, «cell area» (defined as the surface occupied by the cell's projection on the photographic image plane), and cell movement (speed, directionality and angulation).

Two-Dimensional Cultures

In **cultures of dissociated cells**, the rate of cell patterning was evaluated using an arbitrary scale of 10 stages (34), ranking from initially isolated cells (stage 0) to final conglomeration of dermal cells into several-cell-layer-thick aggregates (stage 9) (Table 1). Cells were cultured for 7 to 10 days in MEM supplemented with 5% fetal calf

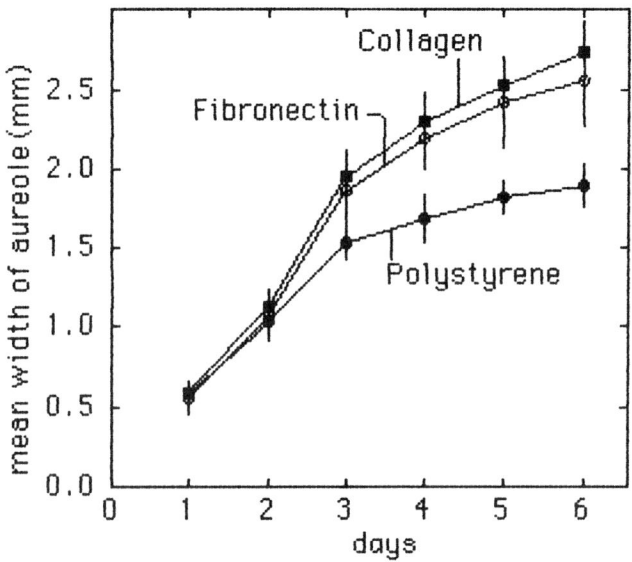

FIGURE 17. Mean width of outgrowth aureole of dermal cells in explant cultures of 7-day chick embryo dermis. The vertical bar represents confidence limits at the 0.05 level.

serum, without change of medium, except on the day following initial seeding. On the ordinary treated polystyrene plastic substrate, which served as control, and with a seeding density of 3.5×10^5 cells/dish, the cultures reached a mean stage 7-8 after 7 days. In explant cultures, using the same medium, the rate of development was evaluated by measuring either the distance travelled by the cells emigrating from the edges of the explants or the surface area of the entire culture together with its expanding aureole of outgrowing cells.

When the bottom of the dishes was coated with fibronectin (using a solution of 40 μg/ml), the rate of cell patterning was unchanged. However, when the dishes were coated with collagen (sprayed on as a methanol solution, collagen being deposited in non-fibrous form as the solvent evaporates), cell patterning was significantly slower than on either plastic or fibronectin, stage 4-5 being reached after 7 days. Fibronectin deposited on top of the collagen coat restored the initial rate of patterning and thus suppressed the inhibitory influence of the collagen substrate. Moreover, when drops of fibrous collagen were allowed to gel on top of the non-fibrous collagen substrate, cells seeded on the gel were further slowed down in the rate of their cell patterning (reaching stage 3-4 after 7 days) as compared to the cells in the same dish which adhered to the non-fibrous collagen substrate.

Table 1
Arbitrary stages of cell patterning and corresponding cell density in cultures of dissociated cells seeded on a two-dimensional substrate

Stages	Salient features	Nb.of cells/mm^2 *
0	isolated cells	180 ± 17
1	initial mutual contacts	237 ± 89
2	early cooperative patterning	279 ± 46
3	loose network	524 ± 96
4	dense network	1049 ± 133
5	subconfluent network	1158 ± 99
6	initial confluence	1556 ± 114
7	parallel crowding	2057 ± 138
8	multilayering	2349 ± 314
9	formation of aggregates	>3000

* ± confidence limits at the 0.05 level

In several other experiments, a heterogeneous substrate was offered to the cells by depositing a roughly circular 5-6-mm diameter drop of fibronectin solution (40 µg/ml) in randomly located spots of the culture dish. The drop was removed by succion after 45 min or left to dry out completely. The wet or dry spots were then rinsed twice locally, so as to avoid dispersal of fibronectin outside coated areas. The position of the spot was carefully recorded using landmarks scratched on the bottom of the dish. This allowed to analyze the behavior of the cells on two kinds of substrates within the same culture dish. Again, there was no difference between the rate of patterning of cells growing on plastic and on the fibronectin spot. However, when the fibronectin spot was deposited on a non-fibrous collagen substrate, the cells outside the spot grew and patterned themselves more slowly than those which grew on the fibronectin substrate. There was no difference between rinsed or dried fibronectin spots.

Another interesting feature was the tendency of cells to align along the circular borderline between the fibronectin spot and the collagen substrate, particularly during the first three days of culture, when they rarely crossed from one substrate to the other.

The cell area of randomly selected cells was found to be highly variable. There was a clear tendency however for the cell area to increase with time, cells on the 7th day of culture being significantly «larger» than on the third or fourth day, on any of the three substrates tested. Moreover, it was found that the cells tended to become «larger» more quickly on collagen (average area: 2700 µm^2) than on plastic (900 µm^2), those on fibronectin being of an intermediate «size» (1800 µm^2), after equal periods of time in culture.

The large «size» of cells on collagen may be due to their high adhesivity to this type of substrate. As it appears that cells attach to collagen molecules through interposed fibronectin (1,16,26), the increased spreading of the cells on collagen might be caused by a collagen

FIGURES 18 - 23. Immunofluorescent labeling of extracellular matrix produced by cultured chick embryo dermal cells and corresponding phase contrast micrographs. FIGURES 18 and 19. With anti-chick type I collagen antibody, after 8 days of culture, nearly all cells are labeled; little if any label is extracellular. FIGURES 20 - 23. With anti-human plasma fibronectin antibody, label of cells cultured on plastic for 5 days (FIG. 20 and 21) is partly intracellular and clustered in thick patches or strands; when cells are cultured on a collagen substrate (FIG. 22 and 23), label after 7 days is mostly extracellular and patterned in delicate parallel fibrils. Scale bars = 50 µm.

Extracellular Matrix in Skin Morphogenesis 331

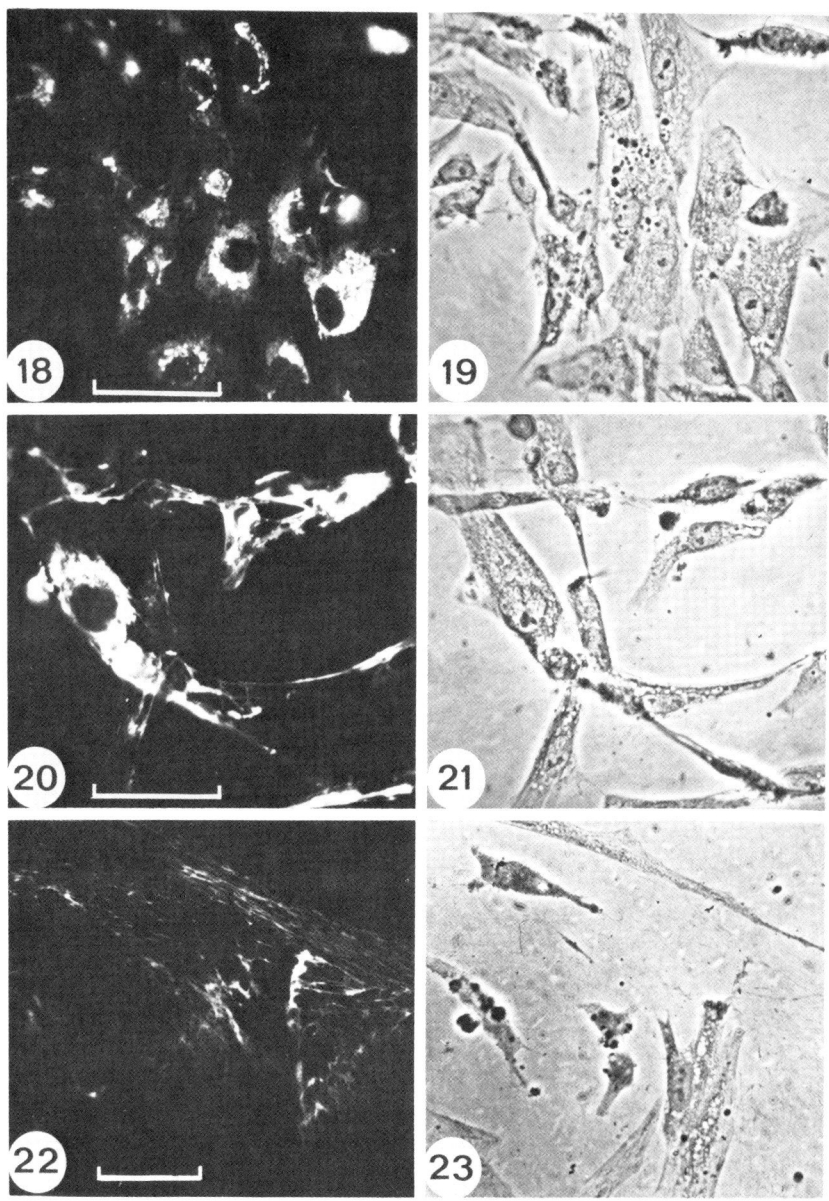

stimulated production of fibronectin. The strong adhesivity of cells to collagen is also in agreement with the view that this extracellular matrix component plays a stabilizing role in organogenesis.

Locomotion in cultures of dissociated and randomly seeded cells is not a salient feature of cell behavior. Indeed randomly seeded cells, being neither attracted nor repelled by exogenous factors, tend to remain on the same spot, without extensive displacements. They rather fidget about by sending out and pulling in cell processes, in an apparently random fashion. Most of them however assume an elongated shape, with two long extensions at their distal ends. This kind of «sur place» movement was analyzed in a number of randomly chosen and isolated cells, by recording the position of their geometric center of gravity in a rectangular coordinate system, at 15 min intervals. Three parameters of movement were considered: **directionality**, which is expressed as the ratio of the distance between starting and end positions (as the crow flies) to the total length of the track; **angulation**, which is the absolute value of the mean angle (in degrees) between two successive track portions; and the **speed** (in μm/h). The two former parameters did not differ significantly on any of the three substrates. Directionality was highly variable from cell to cell and tended to be slightly higher on collagen than on fibronectin, and on fibronectin than on plastic. Angulation was near 90° on all three substrates, indicating that cell fidgeting is caused by the summation of randomly acting minor stimuli. Finally speed was found to be highest on collagen (around 46 μm/h) and lowest on plastic or fibronectin (around 28 or 30 μm/h). Inasmuch as collagen is supposed to play a stabilizing role in morphogenesis by immobilizing cells, it is somewhat surprising that it should support maximal speed. However, one has to remember that the kind of movement measured in these cell cultures is not translocation toward a specific goal or away from a particular spot. The fact that collagen promotes the kind of fidgeting recorded in these experiments does not mean that it should also favor directional cell locomotion. The influence that various extracellular matrix macromolecules might exert on cell movement needs further investigation. The results reported here are to be considered as preliminary and no strong conclusions should yet be drawn from them.

Several recent findings in other systems indicate that indeed fibronectin can promote or facilitate cell migration. This is the case for epithelial epiboly in rabbit cornea wound healing (23) and for chick neural crest cell emigration out of the neural tube (8,24), for instance. However, it has also been shown that chick embryo somitic cells or neural crest cells, when given the choice between fibronectin and laminin substrates, show no preference for either, and migrate at the same rate on any of these two substrates, as well as on matrix deposited on the substrate by cultured mesenchymal or epithelial cells (22).

Explant cultures are characterized by the formation of an aureole of cells growing out from the initial piece of dermal tissue onto the substrate in radial fashion. By measuring the distance between the original position of the explant's edge and the location of the distalmost cells in the aureole at daily intervals, it is seen that collagen, as well as fibronectin, promotes cell migration (Fig. 17). Collagen appears to stimulate locomotion on the two-dimensional substrate slightly more than fibronectin, although the difference between the two curves is not statistically significant. From these data the mean progression rate of the outgrowth can be estimated at around 19 µm/h on collagen, 17 µm/h on fibronectin, and 12 µm/h on plastic, over a period of 6 days. When the migration rate of individual cells was recorded at 15 min intervals over a period of 135 min during the third day of culture, no statistically significant differences were found between cells moving on plastic (11 cells: 45.7 ± 8.2 µm/h [confidence limits at the 0.05 level]), on collagen (9 cells: 39.9 ± 9.0 µm/h), or on fibronectin substrates (18 cells: 39.9 ± 8.6 µm/h). Directionality was also the same (around 0.6) on the three kinds of substrate.

Using immunofluorescent labeling of cultures of dissociated cells or of explant cultures, it is observed that almost all dermal cells synthesize type I collagen (Fig. 18 and 19), while a small minority only produce type III collagen. Little of these collagens is excreted into the extracellular space. Most of it appears to accumulate inside the cells. The majority of cells also produce and excrete fibronectin, which is deposited onto the substrate in large amounts. The distribution pattern and fibrillar organization of the excreted fibronectin is strikingly different according to whether the cells are cultured on polystyrene plastic or on bovine type I collagen (Fig. 20 - 23). On plastic, fibronectin appears to cluster into thick strands and patches, while on collagen the excreted fibronectin network is delicate, consisting of thin, often regularly spaced parallel fibrils.

Three-Dimensional Cultures

Hydrated collagen gels containing MEM supplemented with glutamine, non-essential amino acids, sodium pyruvate, and 10% fetal calf serum, is prepared according to Bell *et al.* (4) using a 5 mg/ml solution of acid soluble type I collagen extracted and purified from calf skin by the CERAD (Lyon). The final concentration of collagen in the gel is approximately 1 mg/ml. For cultures of dissociated cells, cells are seeded into the collagen solution prior to gelling at a density of 5 x 10^5/ml, and 1, 1.5 or 2 ml of the collagen solution with the cell suspension are poured into 35 mm plastic culture dishes. For explant cultures, one or two small pieces of dermis (0.5 x 0.5 mm or less) are placed into the collagen solution prior to gelling, either directly on the bottom of the 35 mm plastic culture dish or on a substrate consisting of 1 ml of pre-gelled collagen medium. In the latter configuration, the explant is so to speak sandwiched between two gels of collagen. In some collagen solution preparations, human plasma fibronectin was added at a concentration of 45 µg/ml.

For the time being no difference was observed between the cultures with or without added fibronectin, either in cultures of dissociated cells or in explant cultures. Morphology, patterning of the cells, and various parameters of cell behavior as described below were the same in either medium. This result might be due either to the overwhelming concentration of collagen in the gel as opposed to that of fibronectin, or to the fact that the medium contains 10% fetal calf serum, which is itself a

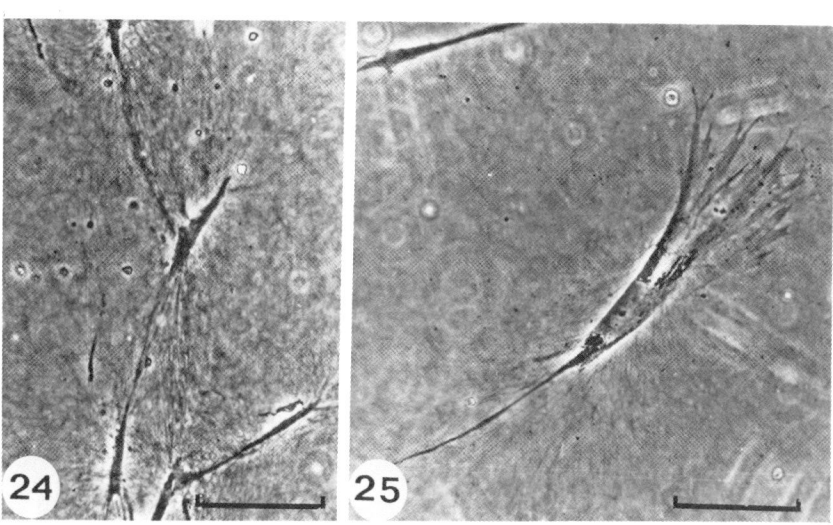

FIGURES 24 and 25. Different morphology of cultured embryonic chick dermal cells within a collagen matrix (FIG. 24) and on the underface of the collagen gel (FIG. 25). Note radially oriented collagen fibers around and between cells in both types of culture. Scale bars = 100 µm.

Extracellular Matrix in Skin Morphogenesis 335

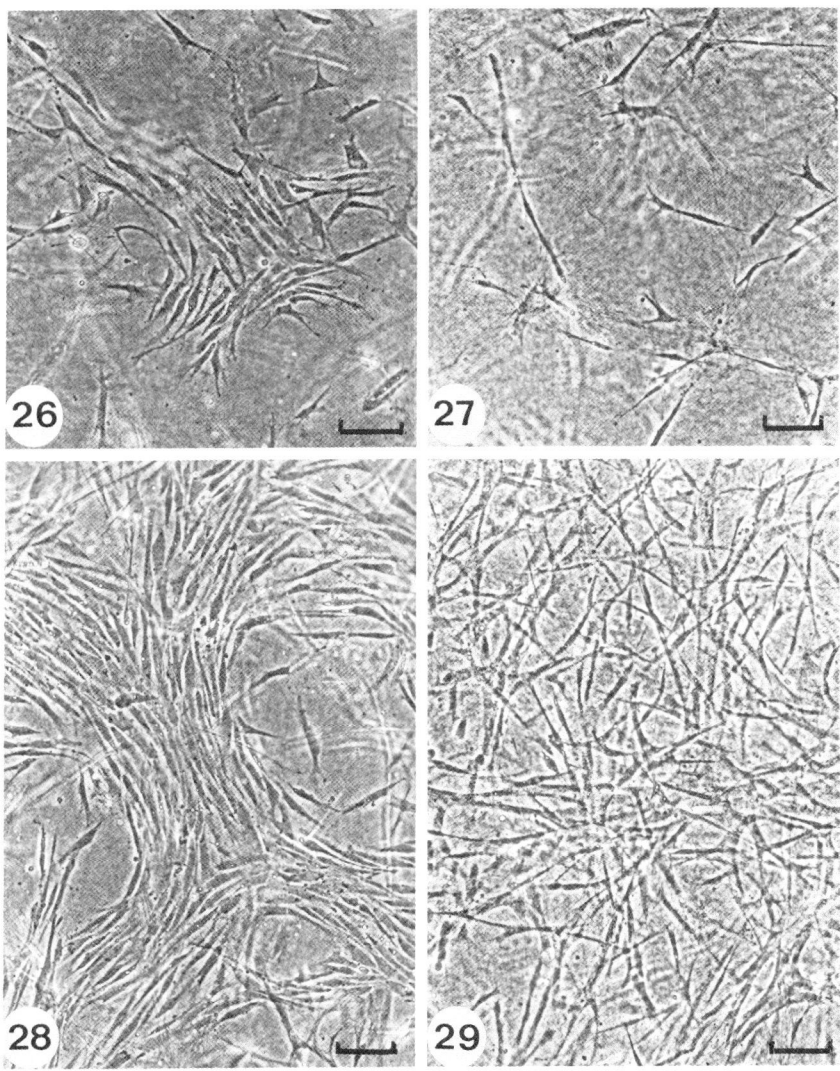

FIGURES 26 - 29. Different patterning of embryonic chick dermal cells cultured on the underface of (FIG. 26 and 28) and within (FIG. 27 and 29) a collagen gel, after 5 (top) and 9 (bottom) days. FIG. 26 and 28 are framed on the same region as FIG. 27 and 29, respectively. Scale bars = 100 μm.

source of fibronectin. No attempts have yet been made to culture the cells in serum-free defined medium gels.

When **dissociated dermal cells** are cultured in a collagen gel, some cells remain in the gel while others reach the underface of the gel and flatten out between the gel and the bottom of the dish. The morphology of these two populations is strikingly different, although in both situations the cell size tends to increase with the duration of the culture.

In the gel, in early cultures, the cells are shaped at random with irregular processes extending from the cell body. At later stages of the culture, they usually become bipolarly elongated with a clear long axis. From the earliest stages of the culture, the cells appear to be surrounded by radially oriented fibrillar structures, extending a long way from the cell body. When two or more cells are sufficiently close to each other, these fibrillar structures take on the appearance of iron filings between the poles of a magnet (Fig. 24). They probably represent aligned collagen fibers (cf. 5). Ultrastructural observation reveals that these oriented fibrillar structures contain few if any cellular processes. While in early cultures, cells in the gel appear to be distributed and oriented at random in the three dimensions of space (Fig. 27), in later cultures they cooperate for the construction of a three-dimensional network of irregularly branched cords (Fig. 29). The density of this network increases with time, and proliferation continues, as attested by numerous mitotic figures, for at least 10 days.

Cells at the interface between the underside of the gel and the bottom of the dish assume a quite different shape, and exhibit a different behavior from those in the gel. They flatten out on the underface of the gel and usually occupy a much larger area on the photographic image plane than do cells inside the gel. They adhere to the gel rather than to the culture dish, as seen after fixation for electron microscopy, when most of these cells remain attached to the gel. When isolated, their morphology resembles that of cells cultured on a two-dimensional substrate. Very often they are fan-shaped and can be observed to locomote with a

FIGURES 30 - 33. Ultrastructure of embryonic chick dermal cells cultured either on the underface of (FIG. 30 - 32) or within (FIG. 33) a collagen gel. Note polarized accumulation of a thick layer of microfilaments in submembranous cytoplasm closest to the bottom of the dish (FIG. 30 and 31), not seen in the cells within the three-dimensional matrix (FIG. 33). Formation of close contacts between epithelioid cells at the underface of the gel (FIG. 32). Note rough endoplasmic reticulum filled with granular or fibrillar material. C, unstained collagen fibers; E, extracellular material seemingly produced by the cultured cells; V, coated vesicles or pits. Scale bars = 0.5 μm.

Extracellular Matrix in Skin Morphogenesis

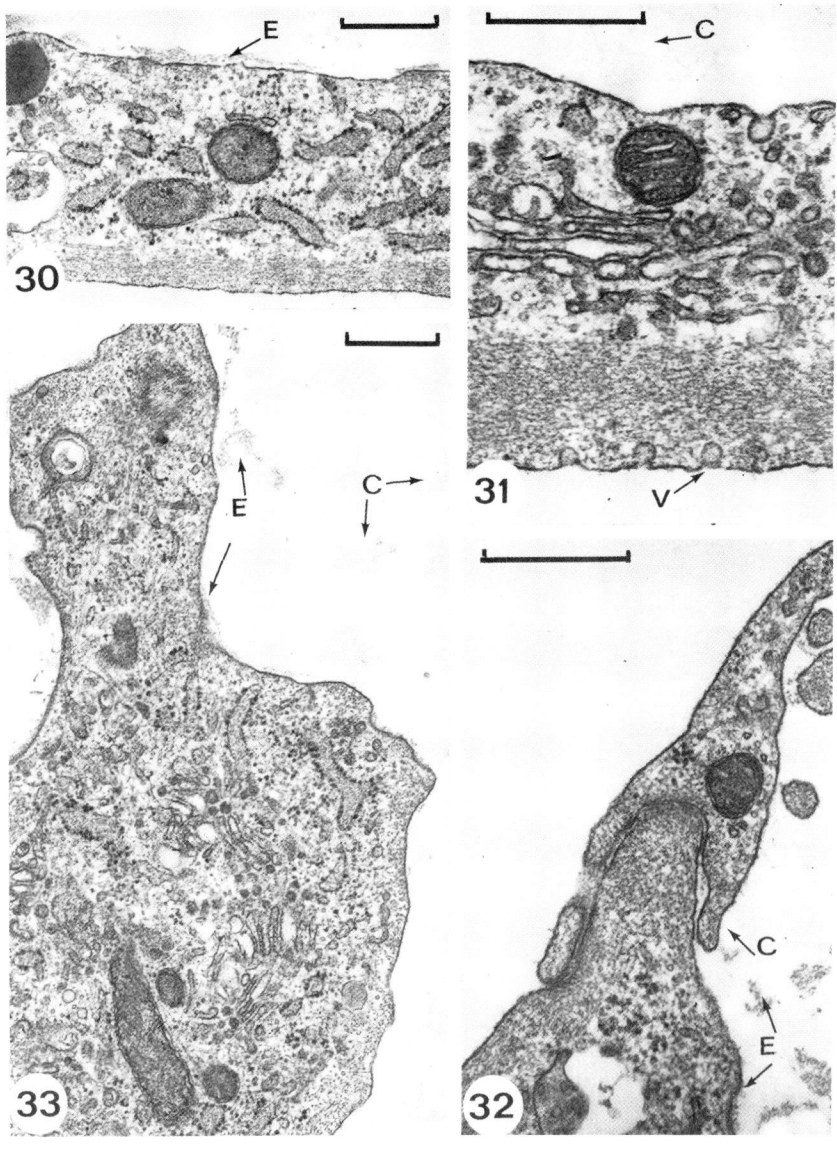

multipolar leading edge and a long thin trailing process (Fig. 25). With time (after 3-4 days of culture), some of them tend to cluster in small aggregates where they align alongside each other in parallel arrays (Fig. 26). Once formed these clusters appear to be stable and to increase in size (Fig. 28). How they arise is as yet unknown: do they derive from the proliferation of one or few parent cell(s) and thus represent small clones or polyclones, or do they result from the congregation and collaboration of several initially unrelated cells? At the ultrastructural level, cells in these clusters establish close contacts between them (Fig. 32) and thus tend to give rise to an epithelioid cell sheet.

Moreover they become strikingly polarized basal-apically: the cytoplasm underneath the apical plasma membrane (the one which is closest to the bottom of the culture dish) is filled with a thick layer of microfilaments (Fig. 30 and 31). The opposite basal side of the cell (the one which is in contact with the gel) contains but little or none of this micro-filamentous network. Cells inside the gel are not polarized and exhibit a rather thin layer of microfilaments all around their peripheral cytoplasm (Fig. 33). Similar submembranous accumulation of microfilaments were also observed in human skin fibroblasts cultured in a collagen gel (2). It is known that microfilaments play an important role in the tensile forces exerted by fibroblasts during the contraction of three-dimensional collagen gels (9). The significance of the polarized apical accumulation of microfilaments in dermal fibroblasts on the underface of the gel is not clear. It might be related to the immobilization of these epithelioid cells. On the other hand, fibroblasts cultured on the surface of a gel are able to contract collagen matrices to the same extent as fibroblasts growing within the gel (11), and therefore the accumulation of apical submembranous microfilament bundles might serve the mechanical contracting function of these cells.

Common features of both «underface» cells and «inside» cells are coated pits in the plasma membrane (Fig. 31) and coated vesicles in the cytoplasm (cf. 11). Both types also deposit finely granular extracellular material(Fig. 30 and 33). This material sometimes takes on the appearance of an electron dense sheet resembling the *lamina densa* of a basement membrane. Otherwise, the cytoplasm is richly provided with often multiple Golgi apparatus, with well-developed rough endoplasmic reticulum the cisternae of which are filled with fibrillar material, signs of biosynthetic activity (Fig. 30 and 33).

In cultures of dissociated cells, seeded at a density of approximately 5×10^5 /ml, contraction of the collagen gel is observed from day 4 of culture. It is characterized by the progressive detachment of the outer edge of the gel from the culture dish and the formation of a peripheral shrinking circular fold. During this process, the flat central part of the gel remains uncontracted and attached to the bottom of the dish.

Explant cultures of pieces of dermis placed inside a three-dimensional collagen gel, just like those placed on a two-dimensional substrate, become progressively surrounded by an increasing aureole of outgrowing cells. However, in the case of the gel, the aureole is three-dimensional and extends in all directions of space, its growth in height being limited by the thinness of the gel (about one millimeter to start with, and approximately 1/10 of a millimeter after several days of culture; the explant itself is about half a millimeter thick at explantation). In the aureole, two kinds of cells can easily be distinguished, according to whether they grow inside the gel (Fig. 34) or on the underside of the gel, at the interface between the gel and the bottom of the dish (Fig. 35). Collagen gels of explant cultures do not contract.

Even in explants which are sandwiched between two layers of collagen, many cells soon reach the underside of the lower gel and there build up an aureole essentially similar to that obtained on two-dimensional substrates: the cells are usually fan-shaped and the majority exhibit a proximo-distal polarity with their leading multipolar edge directed away from the explant (Fig. 34).

Cells building up the aureole inside the gel have a quite different morphology. They are highly refringent in phase contrast microscopy, often spindle-shaped and extend out of the explant like spines on a sea urchin (Fig. 35).

A striking feature of these aureoles is the difference in the rate of expansion between the «underface» aureole and the «inside» one. The former grows quicker and its outer edge is always farther away from the initial explant than that of the «inside» aureole after a given time in culture (Fig. 34). Even in sandwiched explants, where the «inside» cells start emigrating before others reach the underface of the gel, the cells which eventually establish an aureole there soon catch up the «inside» cells, and travel longer distances away from the explant than those which remain in the three-dimensional condition. Apparently dermal cells move more easily on a two-dimensional collagen substrate than inside a meshwork of gelled collagen fibers.

When the individual locomotory behavior of cells is recorded, most are seen to follow an oriented track away from the explant, with a directionality between 0.7 and 1 (Fig. 36A). However, they may also occasionally move back toward the explant for periods of several hours before resuming a general centrifugal direction. By contrast, dissociated cells in the collagen gel move about back and forth in random directions,

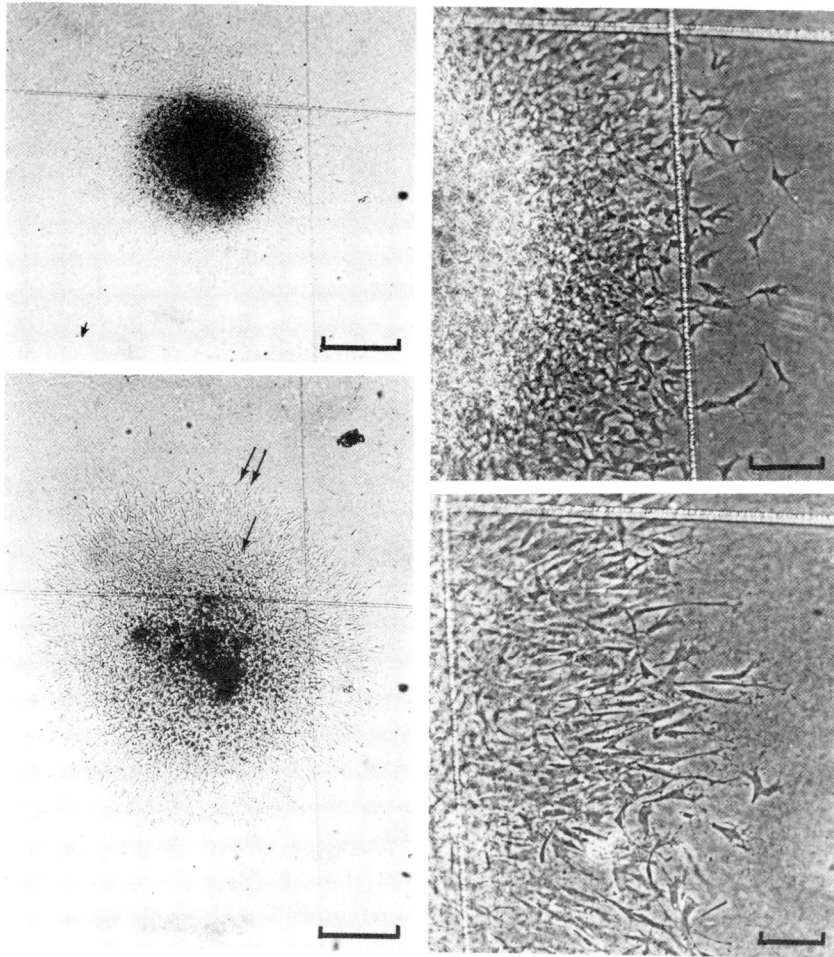

FIGURE 34. Low (left, scale bars = 0.5 mm) and high power (right, scale bars = 100 µm) views of the same explant culture of embryonic chick dermis after 20 (top) and 44 h (bottom) of culture within a three-dimensional collagen gel. The outgrowing cells on which the photographs are focused migrate on the underface of the collagen gel. In the lower left figure, the difference of cell migration rate in the two-dimensional (double arrow) and the three-dimensional aureole (single arrow) is visualized by the distance between the edges of the two aureoles. Straight lines are landmarks scratched on bottom of dish. Compare with cells in three-dimensional aureole (FIG. 35).

Extracellular Matrix in Skin Morphogenesis 341

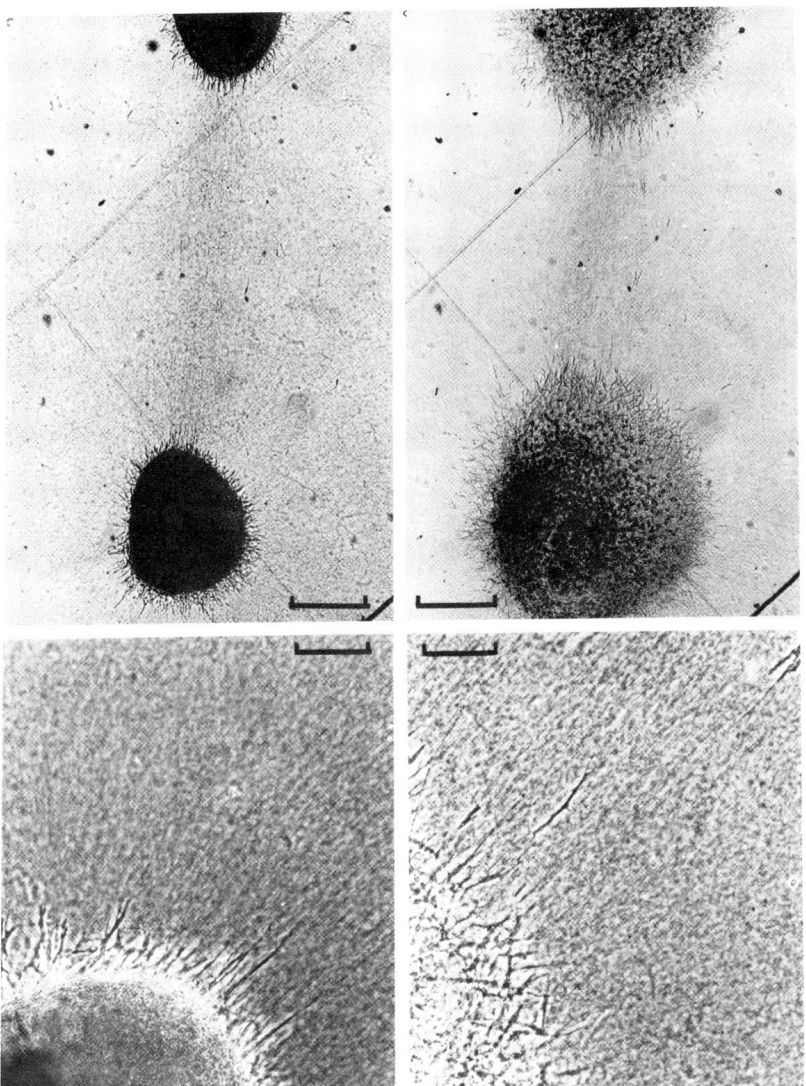

FIGURE 35. Low (top, scale bars = 0.5 mm) and high power (bottom, scale bars = 100 µm) views of the same explant culture of embryonic chick dermis after 20 (left) and 44 h (right) within a three-dimensional collagen gel. Note establishment of oriented collagen fibers between neighboring explants. «Spiny» cells migrate out of the explant and stretch out along collagen fibers. The high power views represent the northern aspect of the lower explant shown in the two top figures. Compare with two-dimensional aureole (FIG. 34).

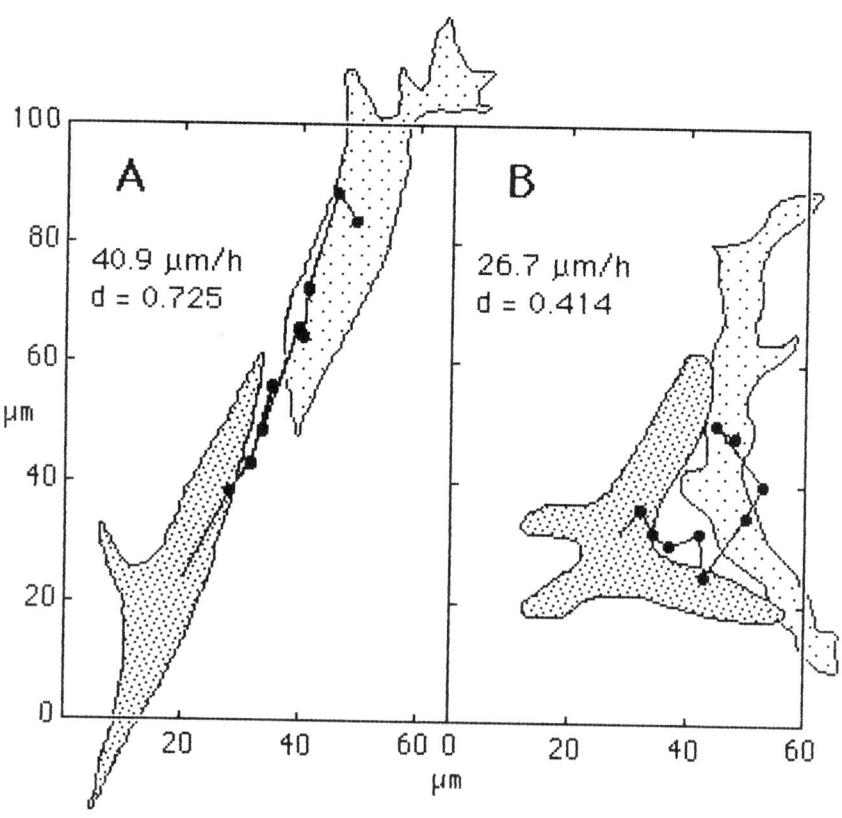

FIGURE 36. Tracks and outlines of embryonic chick dermal cells cultured within a three-dimensional collagen gel, on day 3 of culture. Position of the cells is defined by the coordinates of their geometric center of gravity. Dense stipple, starting position and outline; sparse stipple, end position and outline after 135 min. Mean speed (μm/h) and directionality (d) refer to that period of 135 min. Dots on the tracks represent successive position of the cells at 15 min intervals. **A**, typical cell in an explant culture moving away from the explant in northeastern direction. **B**, typical cell in a culture of dissociated cells, with non-oriented random movement.

a behavior characterized by a low directionality (below 0.5) (Fig. 36B). Moreover, in three-dimensional condition, isolated cells at the outer edge of explant aureoles move more rapidly than dissociated cells (43.8 ± 5.7 µm/h vs. 31.3 ± 4.2 µm/h).

Too few cells have yet been followed by computer-assisted morphometry to yield significant values for the mean speed of randomly chosen isolated cells at the outer edge of «inside» and «underface» aureoles. Although differences are not statistically significant, the mean speed of 7 cells inside the gel was measured to be 40.0 µm/h (± 5.6 µm/h confidence limits at the 0.05 level), while that of 4 cells at the gel underface was 50.4 µm/h (± 14.2 µm/h). When the gel contained 45 µg/ml fibronectin, the values were slightly lower: 9 cells inside the gel travelled at a mean speed of 35.6 µm/h (± 8.4 µm/h), while 4 cells at the underface of the gel migrated at a mean speed of 44.9 µm/h (± 12.4 µm/h). In general these values are rather low, compared to those reported for other cells in collagen matrices, such as human neutrophils, which were observed to migrate at rates between 300 and 480 µm/h (10). In another system using the ability of chick embryo cells to grow out from mesodermal explants into a three-dimensional collagen matrix, it was shown that added fibronectin had no influence on the rate of emigration, but that removal of fibronectin from the serum in the medium significantly inhibited cell migration (27). More measurements are needed to establish whether cells indeed migrate more rapidly on a two-dimensional substrate than in a three-dimensional gel, and what kind of effect fibronectin exerts on embryonic dermal cells.

CONCLUSIONS

Histochemical detection of extracellular matrix macromolecules in the dermis of embryonic chick and mouse skin clearly demonstrates that types I and III collagen, fibronectin, sulfated and other glycosaminoglycans are unevenly distributed in the skin during morphogenesis of cutaneous appendages. These components, the microheterogeneous distribution of which is clearly related to the development of feathers, scales and hairs, may therefore constitute part of the mechanism by which skin tissues communicate. The dermis transmits to the overlying epidermis morphogenetic messages, which are probably expressed as changes in the distribution pattern of interstitial collagens, fibronectin, various glycosaminoglycans, bullous pemphigoid antigen, and other as yet unknown extracellular matrix constituents. The epidermis appears to be able to sense the microheterogeneous texture of its substrate and to react accordingly, either by undergoing morphogenetic movements leading to the formation of appendages or by becoming immobilized in interappendage and glabrous areas.

The attempts to analyze the way in which extracellular matrix components, such as interstitial collagen and fibronectin for instance, might influence the behavior of skin cells, by using cell culture systems, are still preliminary, and the results are to be interpreted with caution. On two-dimensional substrates, type I collagen appears to slow down the rate of patterning of dissociated cells, while fibronectin is able to counterbalance this inhibitory effect. In explant cultures, however, both type I collagen and fibronectin enhance the expansion rate of the cell outgrowth, as compared to bare polystyrene. Also, the extracellular matrix that the cultured cells themselves produce is in part controlled by the nature of their substrate: the architecture of the fibronectin network excreted by the cells in their environment differs according to whether they are cultured on plastic or on collagen.

In three-dimensional collagen gels, dissociated cells or cells emigrating from dermal explants are not influenced by the presence of fibronectin. Their behavior appears to be strongly influenced by the collagen matrix itself, as their shape and ultrastructure differ considerably according to whether they grow inside the gel or on its underface.

Cell cultures in two- or three-dimensional configuration thus offer a powerful means to investigate the relationship between the environment and the morphogenetic performances of the cells. Notably they provide the opportunity to expose the cells to either homogeneous or heterogeneous media, and to quantify their responses. Many more experiments are needed to begin to understand how extracellular matrix modulates the cells' coordinated activities during organogenetic processes.

REFERENCES

1. Akiyama SK, Yamada KM, Hayashi M (1981). The structure of fibronectin and its role in cellular adhesion. J supramol Struct cell Biochem 16:345.
2. Allen TD, Schor SL (1983). The contraction of collagen matrices by dermal fibroblasts. J Ultrastr Res 83:205.
3. Bard S, Micouin C, Thivolet J, Sengel P (1981). Heterogeneous distribution of bullous pemphigoid antigen during hair development in the mouse. Arch Anat micr Morphol exp 70:141.

4. Bell E, Ivarsson B, Merrill C (1979). Production of a tissue-like structure by contraction of collagen lattices by human fibroblasts of different proliferative potential in vitro. Proc natl Acad Sci USA 76: 1274.
5. Bellows CG, Melcher AH, Aubin JE (1981). Contraction and organization of collagen gels by cells cultured from periodontal ligament, gingiva and bone suggest functional differences between cell types. J Cell Sci 50: 299.
6. Bernfield MR (1981). Organization and remodeling of the extracellular matrix in morphogenesis. In Connelly TG, Brinckley LL, Carlson BM (eds): "Morphogenesis and Pattern Formation," New York: Raven Press, p 139.
7. Ekblom P, Saxén L, Timpl R (1982). The extracellular matrix and kidney differentiation. In Hoffman JE, Giebisch GH, Bolis L (eds): "Membranes in Growth and Development," New York: Alan R Liss, p 429.
8. Erickson CA, Turley EA (1983). Substrata formed by combinations of extracellular matrix components alter neural crest cell motility in vitro. J Cell Sci 61:299.
9. Farsi JMA, Aubin JE (1984). Microfilament rearrangements during fibroblast-induced contraction of three-dimensional hydrated collagen gels. Cell Motility 4:29.
10. Grinnell F (1982). Migration of human neutrophils in hydrated collagen lattices. J Cell Sci 58:95.
11. Grinnell F, Lamke CR (1984). Reorganization of hydrated collagen lattices by human skin fibroblasts. J Cell Sci 66:51.
12. Hay ED (1981). Collagen and embryonic development. In Hay ED (ed): "Cell Biology of Extracellular Matrix," New York: Academic Press, p 379.
13. Hay ED (1983). Cell and extracellular matrix: Their organization and mutual dependence. Modern Cell Biology 2:509.
14. Jahoda C, Mauger A, Sengel P (1985). Distribution of glycosaminoglycans in the developing feather. In preparation.
15. Kitamura K (1981). Distribution of endogenous ß-galactoside specific lectin, fibronectin and type I and III collagens during dermal condensation in chick embryos. J Embryol exp Morphol 65:41.
16. Kleinman HK, Klebe RJ, Martin GR (1981). Role of collagen matrices in the adhesion and growth of cells. J Cell Biol 88:473.
17. Mauger A, Démarchez M, Georges D, Herbage D, Grimaud JA, Druguet M, Hartmann DJ, Sengel P (1982a). Répartition du collagène, de la fibronectine et de la laminine au cours de la

morphogenèse de la peau et des phanères chez l'embryon de poulet. C r Acad Sci Paris Série III 294:475.
18. Mauger A, Démarchez M, Herbage D, Grimaud JA, Druguet M, Hartmann DJ, Sengel P (1982b). Immunofluorescent localization of collagen types I and III, and of fibronectin during feather morphogenesis in the chick embryo. Devel Biol 94:93.
19. Mauger A, Démarchez M, Herbage D, Grimaud JA, Druguet M, Hartmann DJ, Foidart JM, Sengel P (1983a). Immunofluorescent localization of collagen types I, III, IV, fibronectin and laminin during morphogenesis of scales and scaleless skin in the chick embryo. Wilhelm Roux's Arch devel Biol 192:205.
20. Mauger A, Démarchez M, Sengel P (1983b). Matrice extracellulaire et morphogenèse de la peau. J Med esthet Chirurg dermatol 10:193.
21. Mauger A, Démarchez M, Sengel P (1984). Role of extracellular matrix and of dermal-epidermal junction architecture in skin development. Progr clin biol Res 151:115, In Kemp JR, Hinchliffe JR (eds): "Matrices and Cell Differentiation," New York:Alan R Liss p 115.
22. Newgreen D (1984). Spreading of explants and embryonic chick mesenchymes and epithelia on fibronectin and laminin. Cell Tissue Res 236:265.
23. Nishida T, Nakagawa S, Awata T, Ohashi Y, Watanabe K, Manabe R (1983). Fibronectin promotes epithelial migration of cultured rabbit cornea in situ. J Cell Biol 97:1653.
24. Rovasio RA, Delouvée A, Yamada KM, Timpl R, Thiéry JP (1983). Neural crest cell migration: requirements for exogenous fibronectin and high cell density. J Cell Biol 96:462.
25. Ruch JV, Lesot H, Karcher-Djuricic V, Meyer JM, Mark M (1983). Epithelial-mesenchymal interactions in tooth germs: mechanisms of differentiation. J Biol buccale 11:173.
26. Ruoslahti E, Engvall E, Hayman EG (1981) Fibronectin: current concepts of its structure and functions. Collagen Res 1:95.
27. Sanders EJ, Prasad S (1983). The culture of chick embryo mesoderm cells in hydrated collagen gels. J exp Zool 226:81.
28. Saxén L, Ekblom P, Thesleff I (1982). Cell-matrix interaction in organogenesis. In Kuehn K, Schoene H, Timpl R (eds): "New Trends in Basement Membrane Research," New York: Raven Press, p 257.
29. Sengel P (1976). Morphogenesis of skin. In Abercrombie M, Newth DR, Torrey JG (eds): "Developmental and Cell Biology Series," Cambridge London New York Melbourne: Cambridge University Press, 277 p.

30. Sengel P (1983). Epidermal-dermal interactions during formation of skin and cutaneous appendages. In Goldsmith LA (ed): "Biochemistry and Physiology of the Skin, Vol I," New York Oxford: Oxford University Press, p 102.
31. Sengel P (1985a). Role of extracellular matrix in the development of skin and cutaneous appendages. Progr clin biol Res 171:123. In Lash JW, Saxén L (eds): "Developmental Mechanisms: Normal and Abnormal," New York: Alan R Liss, p 123.
32. Sengel P (1985b). Epidermal-dermal interactions. In Bereiter-Hahn J, Matoltsy AG, Richards KS (eds): "Biology of the Integument, Vertebrates," Berlin: Springer, in press.
33. Sengel P, Bescol-Liversac J, Guillam C (1962). Les mucopolysaccharides-sulfates au cours de la morphogenèse des germes plumaires de l'embryon de poulet. Devel Biol 4:274.
34. Sengel P, Kieny M (1984). Influence of collagen and fibronectin substrates on the behaviour of cultured embryonic dermal cells. Brit J Dermatol 111 Suppl 27:88.
35. Sengel P, Kieny M (1985). Role of extracellular matrix in skin morphogenesis, analysed by dermal cell cultures. In Marks R (ed): "Skin Models," in press.
36. Slavkin HC, Brownell AG, Bringas P, MacDougall M, Bessem C (1983). Basal lamina persistence during epithelial-mesenchymal interactions in murine tooth development in vitro. J craniofac Genet devel Biol 3:387.
37. Slavkin HC, Greulich RC (1975). Extracellular matrix influences on gene expression. New York: Academic Press 833 p.
38. Thesleff I, Barrach HJ, Foidart JM, Vaheri A, Pratt RM, Martin GR (1980). Changes in the distribution of type IV collagen, laminin, proteoglycan and fibronectin during mouse tooth development. Devel Biol 81:182.

Molecular Determinants of Animal Form, pages 349–363
© 1985 Alan R. Liss, Inc.

CURRENT CONCEPTS OF KIDNEY MORPHOGENESIS

Peter Ekblom and Hannu Sariola

Friedrich-Miescher-Laboratory,
Max-Planck-Society,
Spemannstrasse 37-39
D-7400 Tübingen, West-Germany

ABSTRACT The development of the kidney requires an interaction between three cell lineages, the nephrogenic mesenchyme, the ureter bud, and the endothelium. The trigger for differentiation is an inductive interaction between the mesenchyme and the ureter bud. The induction of the mesenchyme increases the adhesiveness of the mesenchymal cells. They start to express adhesion proteins such as laminin and uvomorulin. At the same time, their proliferation is enhanced, and they become responsive to a serum mitogen, transferrin. The continued growth and differentiation thus requires a supply of serum. This is accomplished by attracting blood vessels close to the developing areas. The stimulation of blood vessels by the induced mesenchyme suggests that the cells in response to induction start to secrete growth factors for endothelial cells.

INTRODUCTION

Although the advances in molecular biology and gene technology during the past ten years have been spectacular, we still do not know very well how genes are activated during embryogenesis. The development of the three-dimensional histoarchitecture is particularly poorly understood. Differentiated tissues are composed of many different cell types, and they are remarkably well organized in relation to each other. This organization is gradually established during embryogenesis. The cell diversification that occurs during embryogenesis must be a result of a differential activation of genes, but it has remained rather unclear how this is brought about.

A number of classical embryological studies suggest that cell differentiation can be influenced by external factors. In the embryo, there is a constant communication between different cell compartments and such interactions thus provide important driving forces for differentiation. Since the studies by Spemann (1), it has been clear that histogenetic processes are initiated by interactions between different cell types. Most tissues form as a result of such inductive interactions (reviews: 2,3,4). These interactions thus provide one important driving force for tissue differentiation, and it would seem of value to analyze the molecular changes that these interactions lead to. It has, however, proven difficult to study these phenomena, and progress in the field has been rather slow. The issues have been comprehensively reviewed by several investigators (2-5). To analyze these problems, several approaches have been used. In this review, we will limit ourself to a short discussion of a few approaches that at present seem to be promising ways to obtain more knowledge on embryonic induction. Since we in our own studies largely have been studying the developing kidney, the examples will be taken almost exclusively from this system.

The Kidney Model System.

Since the differentiation of cells is so intimately related to the development of the three-dimensional histoarchitecture, differentiation can be difficult to analyze merely by biochemical means. A thorough knowledge of the intricate anatomical relationships between the different cell lineages is therefore essential. The differentiation of the kidney was long thought to be a result of an interaction between two cell lineages, the nephrogenic mesenchyme, and the epithelial ureter bud (reviews: 6,7). Detailed morphological studies suggested, however, that some cells of the kidney do not develop from these two cell populations (8,9). Immunohistological studies on basement membrane formation lead us to suggest that the endothelial cells of the kidney are derived from outside vasculature (8,9). Direct experimental evidence supporting this hypothesis is now available (10). Thus, it is now thought that kidney differentiation requires an interaction between three cell lineages (mesenchyme, epithelium, endothelium) rather than only two. The crucial trigger for differentiation is nevertheless the inductive interaction between the nephrogenic mesenchyme and the ureter bud, and these two rudiments can develop without the presence of endothelial cells.

When the nephrogenic mesenchyme comes in contact with the ureter bud, it is still composed of cells with morphological characteristics of normal fibroblasts. In spite of this it is at this stage already predetermined for kidney tubulogenesis, and is converted to epithelial cells if it is properly induced (Fig. 1). In vivo, the ureter bud acts as the inducer, but Grobstein (11,12) could show that a number of other tissues can act as inducers. The reader is referred to previous reviews for a more comprehensive treatment of the anatomical details of the interaction (6,13). The conversion of the mesenchyme to an epithelium can be induced in vitro. In the in vitro cultures it is possible to experimentally manipulate the system to some extent, and the various morphological steps occur very reproducibly, and a good timing of events can be performed. Some disadvantages of the model system should be pointed out. So far, it has not been possible to alter the developmental pathway to any great extent. The mesenchyme can be induced by various embryonic tissues, but the response is always formation of kidney tubules. Some technical problems may also be mentioned. Since the early embryonic kidney tissue is very small (10 000 - 20 000 cells in an 11 day old mouse embryo) the microsurgery is tedious and requires special skills. The amount of material obtained each day is rather limited, and this complicates many biochemical studies. Immunohistology is therefore especially good as a tool, since molecules of potential morphogenetic importance can be studies also in the small tissue samples.

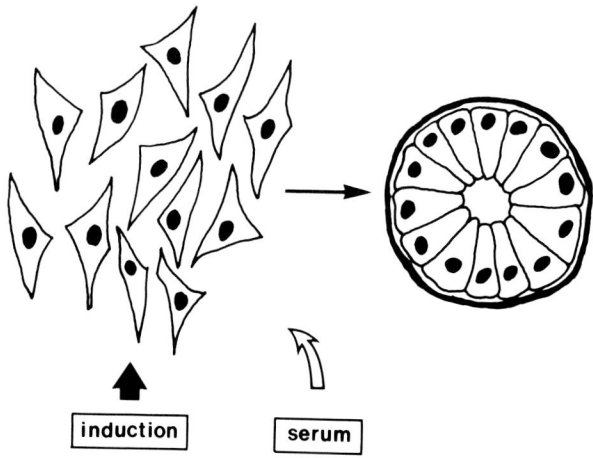

FIGURE 1. Stimulation of the nephrogenic mesenchyme by embryonic inducers and by serum components leads to a conversion of the mesenchyme to epithelium. From (13).

RESULTS

The Interphase Between Mesenchyme and Inducer.

Inductive tissue interaction only occur at short distance, and therefore many investigators have felt it important to analyze the interphase between the inducer tissue and the responder tissue. It was thought by Grobstein (2) who developed the transfilter culture system for the kidney mesenchyme, that the induction of kidney tubules occurs without cell contacts between the mesenchyme and inducer, but this was later shown to be incorrect (14,15). When the mesenchyme is separated by a filter from the inducer tissue, cell processes traverse through the filter, and these processes seem to be of importance for the transmission of the inducing signals (15). In the in vitro cultures, the cell processes from the inducer tissue make close cell contacts with the mesenchyme, and it is thought that these cell contacts are required for induction. It is still not known with certainty that similar contacts are required in vivo. The velocity of the ingrowth of the cell processes into the filters in the transfilter cultures has been studied, and some reports suggested that this ingrowth occurs slowly (7,16). In the model system, induction is completed within the first day of culture, before overt morphogenesis (2), and the cell processes were reported to grow gradually towards the mesenchyme during this 24 hour period. This led Lehtonen (7) to postulate that the time required for induction (24 hours) is used for the transmission of the signals rather than for the response of the mesenchyme. A number of previous and more recent studies speak against this proposal. First, it was soon shown that the ingrowth of the cell processes from the inducer tissue through the filter can occur rather rapidly and it was clearly shown also that a precultivation of the tissues does not shorten the minimum induction time (17). This and other studies (18) make it likely that the time consumed for completion of induction (approximately 24 hours) is required for the response of the mesenchyme to induction. This means that it would be important to search for molecular differences between induced and uninduced cells.

In conclusion, the analysis of the interphase between the mesenchyme and the inducer suggests that induction of the nephrogenic mesenchyme in the in vitro system requires close cell contacts between the inducer and the mesenchyme (15). The hypothesis that a slow ingrowth of the processes explains why induction of the mesenchyme takes 24 hours (7) is most likely not correct. In some other systems, close cell contacts are not required for induction (5), but there is no reason to believe that the inductive interactions should operate in the same fashion in all systems.

The Response to Induction.

For a long time, it was unclear how the mesenchyme at the molecular level responds to induction, as there were no markers that would distinguish between the induced and the uninduced state (6). During recent years, a number of molecular changes have been described to occur during the induction period (first 24 hours), all of which seem to be of biological significance. It is likely that several others will soon be found. All the so far described changes seem to occur at the same time, but it is not easy to judge whether these changes for each individual cell occur in a cascade, or whether induction leads to rather global simultaneous activation of several biosynthetic pathways.

The induction-related changes so far detected could be grouped arbitrarily into the following major categories:
1) changes related to cell adhesion (extracellular matrix and cell surface)
2) changes related to cell proliferation
3) changes leading to an attraction of blood vessels

It is known that kidney-specific antigens also can be detected in the in vitro cultures, but all evidence to date suggests that these antigens appear rather late, several days after completion of induction. They therefore mark terminal differentiation (13,19).

Cell Adhesion.

When the cells are induced, they apparently change the composition of their extracellular matrix. They initially express an interstitial matrix, but this is converted to an epithelial type of matrix. All these switches seem to take place during induction (8,13). The evidence for this comes from morphological studies using immunocytochemistry, and biochemical studies have not yet been performed. Since the epithelial matrix proteins (laminin, type IV collagen) are known to have adhesive properties (20), it is natural to assume that their presence in the induced mesenchyme increases the adhesiveness of the cells.

Several groups have convincingly shown that cell adhesion is not merely a function of the extracellular matrix but is also dependent on cell adhesion molecules found at

the cell surface (21). It is therefore not unexpected that
such cell adhesion molecules appear in the kidney mesenchyme when it is induced. One of these, the N-CAM, seems to
appear during induction, but another, the L-CAM, can be detected slightly after induction (22,23). Our studies are in
line with these proposals, since we have recently found that
a cell adhesion molecule thought to be identical to L-CAM,
uvomorulin, appears in the induced mesenchyme approximately
at 36 hours of in vitro culture, 12 hours after induction
(24). It is noteworthy that we could not disrupt morphogenesis of the kidney tubules with anti-uvomorulin antibodies, although they are functionally active in a number
of other systems. This suggests that other adhesives (laminin, N-CAM) are more important in the adhesion of the induced cells. However, it will be necessary to test this as
soon as functionally active antibodies against these proteins become available.

It can be concluded that induction leads to an early
activation of the expression of cell adhesion molecules.
When the cells start to aggregate, several adhesive proteins are already present. The appearance of adhesive proteins during and after induction was suggested already a
long time ago (25) but the nature of the molecules was not
known. The cell adhesion molecules (matrix proteins, cell
surface proteins) appear well before overt morphogenesis,
and their presence can explain why the cells start to condense and adhere to each other after induction. The early
loss of the interstitial collagens should further facilitate
the condensation process (13).

The appearance of cell adhesion molecules well before
the appearance of kidney-specific antigens suggest a role
for these adhesion proteins in tissue histogenesis. Apparently, the cells must first produce essential building
blocs for morphogenesis (adhesive determinants), and only
later do they produce specialized, tissue-specific products. Since laminin and other basement membrane components
rather soon become deposited in a polar fashion, it is possible that these proteins somehow are involved in the polarization of the cells. The cell-adhesion molecules at the
cell surface do not show a similar polar expression.

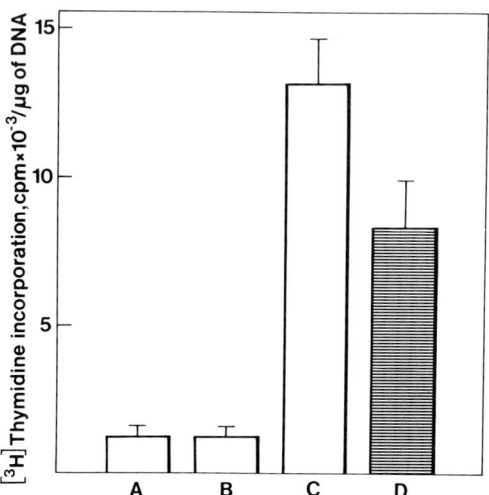

FIGURE 2. Transferrin-dependent cell proliferation on day two of in vitro culture in induced or uninduced cells. From (26).
(A) Non-induced cells do not respond to transferrin.
(B) When induction is prevented by low-pore size filters, cells do not respond to transferrin.
(C) Induced cells respond to transferrin.
(D) Induced cells respond to serum. Bars: SE of triplicates.

Cell Proliferation.

The results on cell adhesion molecules and the extracellular matrix already showed that the mesenchyme respond to induction in a complex fashion. In addition, induction of the mesenchyme also stimulates the cells to proliferate. This was first noted by Vainio et al. (18) but was not further studied until recently (26,27). The proliferation rate, measured by incorporation of radioactive thymidine, reaches a peak by the end of the induction period (at 24 hours). This increase in proliferation is apparently obtained only when the mesenchyme is in contact with inducer tissues (27).

Studies with chemically defined media revealed another very intriguing aspect. The cells respond to the inducer tissue by proliferation, but proliferation drops rapidly after induction unless transferrin is present in the cultures. Detailed studies on the kinetics of cell proliferation (26,28) led us to suggest that the responsiveness to transferrin is acquired during induction (Fig. 2).

For some other tissues, it has previously been shown convincingly that tissue interactions act by changing the responsiveness to serum factors (29). The studies with transferrin in the kidney model system suggest that similar changes in the responsiveness to hormones and growth factors occur in a number of inductive interactions.

The studies on transferrin metabolism do not clarify how the inducer tissue stimulates the first, transferrin--independent proliferation. It is unlikely that the changes in cell adhesiveness could account for this. One should therefore try to search for metabolic changes that are required for initiation of cell proliferation.

Angiogenesis.

Since certain morphological observations suggested that the differentiating kidney mesenchyme starts to attract blood vessels (8), we decided to analyze the vascularization of the embryonic kidney in more detail. It was first necessary to study the origin of the vascular cells of the kidney. Previous morphological studies did not clarify whether the blood vessels of the kidney are derived from the nephrogenic mesenchyme, or whether they develop as a result of an ingrowth of cells derived from outside vasculature (30). Some investigators favored the idea that the blood vessels are derived from outside vasculature (31,32) but more direct experimental data could not be given. Furthermore, in vitro experiments with isolated nephrogenic mesenchymes suggested that blood vessels originate from the nephrogenic mesenchyme (33). However, although some cells from blood vessels occasionally may be present in the in vitro cultures, they are not necessarily derived from the mesenchyme. Observations of in vivo development suggest that the blood vessels are located very close to the mesenchyme already in an 11-day old embryo, and it may be difficult to remove these by microsurgery (8,35). Thus, the results of Emura and Tanaka (32) should be interpreted with caution. They found endothelial cells in the in vitro cultures of the mesenchyme and suggested that the embryonic liver can support neovascularization in the nephrogenic mesenchyme (32).

Because of these uncertainties, we decided to analyze the origin of the kidney blood vessels by transplanting the developing kidney into the chorio-allantoic membrane of the quail. Since quail cells can be distinguished in histology from mouse cells (34) it was possible to demonstrate that quail endothelial cells from the chorio-allantoic membrane invade the developing kidney (Fig. 3). Moreover, the quail endothelial cells organize themselves histiotypically and the end result is a hybrid glomerulus which contains quail endothelium and mouse epithelium (9,35). The endothelial cells produce their own basement membrane and their migration is therefore most likely not very dependent on the extracellular matrix produced by the kidney mesenchyme for epithelium (36). Taken together, all our evidence suggest that the blood vessels of the kidney are not derived from the nephrogenic mesenchyme but from outside vasculature. It should be noted that hatching of the quail egg prevents long-term cultivation on the chorio-allantoic membrane. The muscular walls of the blood vessels apparently develop late, and therefore the origin of the smooth muscle cells of the vasculature is still unknown.

The attraction of the blood vessels into the developing kidney could be due to the secretion of an angiogenesis factor. The secretion of such factors may be differentiation-dependent. This proposal is supported by transplantation experiments. Undifferentiated nephrogenic mesenchymes do not attract blood vessels, whereas experimentally induced mesenchymes elicit a vascular response (35). At present, we are investigating whether the differentiating kidney epithelium secretes diffusible angiogenesis factors. It has recently been shown that growth factors that stimulate endothelial cells can be purified by heparin affinity chromatography (37). So far, such factors have been detected only from tumor cells or adult neural tissue (38). With this approach, it should be possible to detect embryonic angiogenesis factors as well. The developing kidney could be a good source (39).

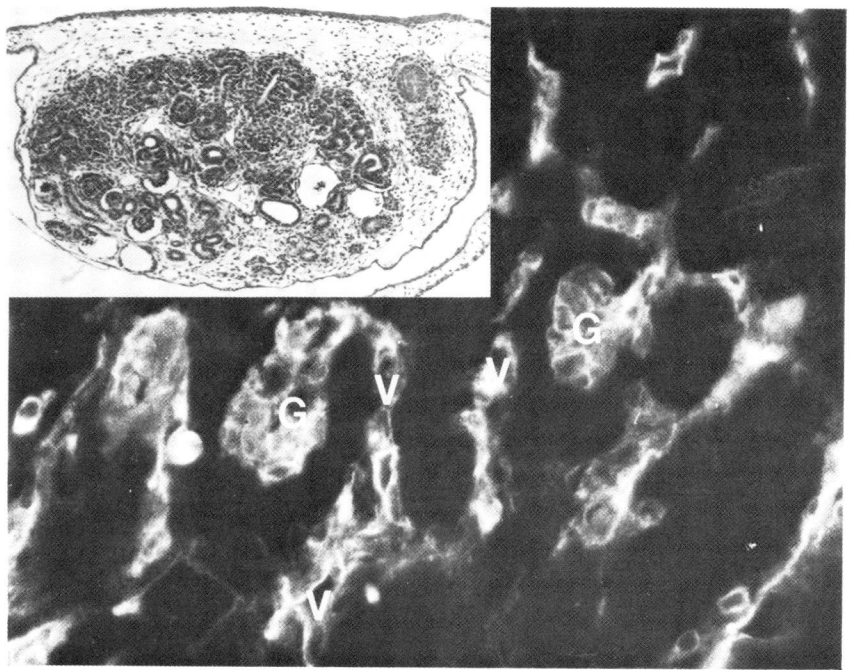

FIGURE 3. The mouse kidneys transplanted on quail chorio-allantoic membrane (insert) were vascularized by quail vessels. The vessels not only express the typical quail nucleolar structure but can also be stained in indirect immunofluorescence with antibodies against quail endothelial cells. (Antibody donated by N. LeDouarin and F. Diaterlen-Lievre). Details, see ref. (39).
g = glomerulus, v = vessel.

In conclusion, it seems well established that the blood vessels of the kidney represent a histogenetically separate cell lineage that is determined early during embryonic life. During organogenesis, the vessels form by a migration and multiplication of predetermined stem cells. This event is obviously controlled by a number of factors, including the surrounding tissue, soluble serum factors, and the vascular wall itself. The process is most likely very similar to neovascularization during inflammation, tissue repair and tumor progression.

DISCUSSION

The studies which have focused on the analysis of the response to induction have revealed a number of important principles for morphogenesis. One consequence of the induction is an increase cell adhesiveness, and molecules that could be responsible for this have been identified. The studies on cell proliferation have identified a serum factor required for kidney organogenesis. It is also now rather clear that the inducer tissue excerts a direct effect on the proliferation of the mesenchyme cells. These proliferating cells, in turn, start to stimulate the proliferation of the third cell lineage, the endothelial cells. This suggests that the induced mesenchyme cells produce angiogenesis-stimulating factors. Thus, when different cell lineages interact with each other during kidney morphogenesis, they often act by stimulating the proliferation of each other. The embryonic kidney should therefore be a good source for embryonic growth factors.

REFERENCES

1. Spemann H (1938). Experimentelle Beiträge zu einer Theorie der Entwicklung. Springer, Berlin, pp. 1-296.
2. Grobstein C (1967). Mechanisms of organogenetic tissue interaction. Natl Canc Inst Monographs 26:279.
3. Kratochwil K (1972). Tissue interactions during embryonic development. In Tarin D (ed). "Tissue Interactions in carcinogenesis", London: Academic Press, pp. 1-47.
4. Saxen L, Lehtonen E, Jääskeläinen M, Nordling S, Wartiovaara J (1976). Inductive tissue interactions. In Poste G, Nicolson G (eds). "The cell surface in animal embryogenesis and development", Amsterdam: North-Holland, p. 331.
5. Toivonen S (1979). Transmission problem in primary induction. Differentiation 15:177.
6. Saxen L, Koskimies O, Lahti A, Miettinen H, Rapola J, Wartiovaara J (1968). Differentiation of kidney mesenchyme in an experimental model system. Adv Morphog. 7:251.
7. Lehtonen E (1976). Transmission of signals in embryonic induction. Med Biol 54:108.

8. Ekblom P (1981). Formation of basement membranes in the embryonic kidney. J Cell Biol 91:1.
9. Bernstein J, Cheng F, Roszka J (1981). Glomerular differentiation in metanephric culture. Lab Invest 45:183.
10. Ekblom P, Sariola H, Karkinen M, Saxen L (1982). The origin of the glomerular endothelium. Cell Differ 11:35.
11. Grobstein C (1955). Inductive interactions in the mouse metanephros. J Exp Zool 130:319.
12. Grobstein C (1956). Trans-filter induction of tubules in mouse metanephrogenic mesenchyme. Exp Cell Res 10:424.
13. Ekblom P (1984). Basement membrane proteins and growth factors in kidney differentiation. In Trelstad RL (ed). "The role of extracellular matrix in development". New York: Alan Liss, p. 173.
14. Nordling S, Miettinen H, Warteovaara J. Saxen L (1971). Transmission and spread of embryonic induction. J Embryol Exp Morphol 26:231.
15. Wartiovaara J, Nordling S, Lehtonen E, Saxen L (1974). Transfilter induction of kidney tubules: correlation with cytoplasmic penetration into Nucleopore filters. J Embryol Exp Morphol 31:667.
16. Lehtonen E, Wartiovaara J. Nordling S, Saxen L (1975). Demonstration of cytoplasmic processes in Millipore filters permitting kidney tubule induction. J Embryol Exp Morphol 33:187-203.
17. Saxen L, Lehtonen E (1978). Transfilter induction of kidney tubules as a function of the extent and duration of intercellular contacts. J Embryol Exp Morphol 147:97.
18. Vainio T, Jainchill J, Clement K. Saxen L (1965). Studies on kidney tubulogenesis. VI. Survival and nucleic acid metabolism of differentiating mouse metanephrogenic mesenchyme in vitro. J Cell Comp Physiol 66:311.
19. Ekblom P, Miettinen A, Virtanen I. Dawnay A, Wahlström T, Saxen L (1981). In vitro segregation of the metanephric nephron. Dev Biol 84:88.
20. Kleinman H, Klebe R, Martin GR (1981). Role of collagenous matrices in the adhesion and growth of cells. J Cell Biol 88:473.
21. Edelmann GM (1983). Cell adhesion molecules. Science 192:218.

22. Thiery JP, Duband JL, Rutishauser U, Edelman GM (1982). Cell adhesion molecules in early chicken embryogenesis. Proc Natl Acad Sci USA 79:6737.
23. Thiery JP, Delouvee A, Gallin W, Cunnigham BA, Edelman GM (1984). Ontogenetic expression of cell adhesion molecules: L-CAM is found in the epithelia derived from the three primary germ layers. Dev Biol 102:61.
24. Vestweber D, Kemler R, Ekblom P (1985). Cell-adhesion molecule uvomorulin during kidney development. Dev Biol (submitted).
25. Wartiovaara J (1966). Cell contacts in relation to cytodifferentiation in metanephrogenic mesenchyme in vitro. Ann Med Exp Fenn 44:469.
26. Ekblom P, Thesleff I, Saxen L, Miettinen A, Timpl R (1983). Transferrin as a fetal growth factor: acquisition of responsiveness related to embryonic induction. Proc Natl Acad Sci USA 80:2651.
27. Saxen L, Salonen J, Ekblom P, Nordling S (1983). DNA synthesis and cell generation cycle during determination and differentiation of the metanephric mesenchyme. Dev Biol 98:130.
28. Thesleff I, Ekblom P (1984). Role of transferrin in branching morphogenesis, growth and differentiation of the embryonic kidney. J Embryol Exp Morphol 82:147.
29. Kratochwil K. Schwartz P (1976). Tissue interaction in androgen response of embryonic mammary rudiment of mouse: identification of target tissue for testosterone. Proc Natl Acad Sci USA 73:4041.
30. Kazimierzak J (1971). Development of the renal corpuscle and the juxtaglomerular apparatus. A light and electron microscopic study. Acta Pathol Microbiol Scand. A Suppl 218:1.
31. Potter E (1965). Development of the human glomerulus. Arch Pathol 80:241.
32. Osathanondh V, Potter E (1966). Development of human kidney as shown by microdissection. V. Development of vascular pattern of glomerulus. Arch Pathol 82:403.
33. Emura M, Tanaka T (1972). Development of endothelia and erythroid cells in the mouse metanephric mesenchyme cultured with fetal liver. Dev Growth Differ 14:237.
34. LeDouarin N (1973). A biological cell labeling technique and its use in experimental embryology. Dev Biol 30:217.

35. Sariola H, Ekblom P, Lehtonen E, Saxen L (1983). Differentiation and vascularization of the metanephric kidney grafted on the chorio-allantoic membrane. Dev Biol 96:427.
36. Sariola H, Timpl R, v der Mark K, Mayne R, Fitch J, Linsenmayer T, Ekblom P (1984). Dual origin of glomerular basement membrane. Dev Biol 101:86.
37. Shing Y, Folkman J. Sullivan R, Butterfield C, Murray J, Klagsbrun M (1984). Heparin affinity: purification of a tumorderived capillary endothelial cell growth factor. Science 223:1296.
38. Klagsbrun M, Shing Y (1985). Heparin affinity of anionic and cationic capillary endothelial cell growth factors: analysis of hypothalamus-derived growth factors and fibroblast growth factors. Proc Natl Acad Sci USA 82:805.
39. Sariola H (1985). Interspecies chimeras: an experimental approach for studies on embryonic angiogenesis. Med Biol (in press).

Molecular Determinants of Animal Form, pages 365-384
© 1985 Alan R. Liss, Inc.

THE FORMATION OF MICROVILLI

G. F. Oster[1]
J. D. Murray[2]
G. M. Odell[3]

ABSTRACT   Microvilli on cells frequently
display a regular hexagonal packing pattern.
We present here a model for how this regular
pattern is established and how the microvilli
are extruded from the cell. The model is
based on the viscoelastic properties of the
actomyosin gel in the cell cortex.

INTRODUCTION

Micrographs of the cellular surface frequently show populations of microvilli which are arrayed in a regular hexagonal packing (see Fig. 1). In this paper we propose an explanation for the beginning stages of this hexagonal pattern. We show that this kind of pattern can arise from the mechanical properties of the contractile actomyosin gel that constitutes the cellular cortex. If our reasoning is correct, it suggests a possible mechanism for the subsequent development of the microvilli. This same model may apply to the development of stereocilia on hair cells of the inner ear, for they too apear in a hexagonal array.

[1]Department of Biophysics and Entomology, University of California, Berkeley, CA   94720
[2]Centre for Mathematical Biology, Mathematics Institute University of Oxford, Oxford OX1 3LB, ENGLAND
[3]Department of Mathematics, Resselaer Polytechnic Institute, Troy, NY 12181

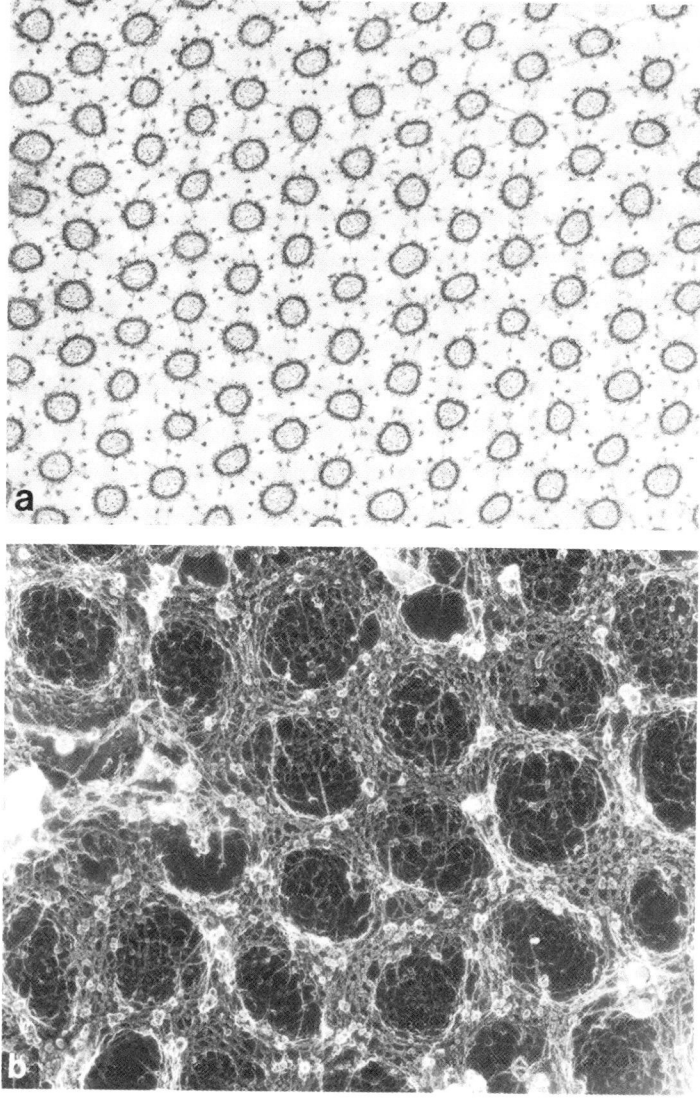

Fig. 1. (a) When microvilli are sheared off from the cell surfaces, the hexagonal arrangement is unmistakable (micrograph courtesy of A. J. Hudspeth). (b) View of a field of microvilli from the cytoplasmic side. Note the bands of aligned actin fibers and the actin sparse regions between the bands (micrograph courtesy of D. Begg).

Our analysis is based on the following sequence of events. First, the actomyosin gel is triggered to contract, presumably by a rise in cytosolic calcium. We show that, because of the mechanochemical properties of actomyosin, a contracting gel cannot remain uniformly distributed, but assembles itself into a hexagonal tension structure. The pattern thus established creates arrays of lacunae which are less dense in actomyosin. Osmotic pressure then expands these regions outward to initiate the microvilli.

In order to implement the above scenario, we must formulate a mathematical model that describes the mechanochemistry of the cytogel cortex. So as to make the paper accessible to the nonmathematical reader we shall present the model as "word equations," and sequester the mathematical treatment in the Appendix.

THE MECHANICS OF ACTIN GELS

In this section we outline the model cytogel upon which our predictions are based. A more complete account can be found in Oster & Odell, (1984a,b); Murray & Oster, (1984a,b); Maini, Murray & Oster, (1984); Oster, (1985); Odell, (1985).
The biological assumptions underlying the model are:
(A) The subcortical region beneath the apical membrane of a cell consists largely of an actin-dense gel.
(B) The gel can contract by a sliding filament mechanism involving myosin cross-bridges linking the actin fibers.

We shall show that these assumptions are sufficient to ensure that an actin sheet will not remain spatially homogeneous, but can form a periodic array of actin fibers. This arrangement of actin fibers could provide the framework for the extrusion of the microvilli by osmotic forces.

Actomyosin Gel is a Fibrous Network

The contractile machinery of the cortical cytogel is formidably complicated, involving not only the fibrous actomyosin and associated proteins but the chemical machinery which regulates contraction as well. Fig. 2 gives a cartoon summary of the principal components of this contractile apparatus. For our purposes, however, we need not deal with the cortex at such a detailed level. By assumption (A) we shall view the cortex as a gel: that is, a fibrous elastic matrix which can contract. The mechanical properties of such a gel are described by macroscopic

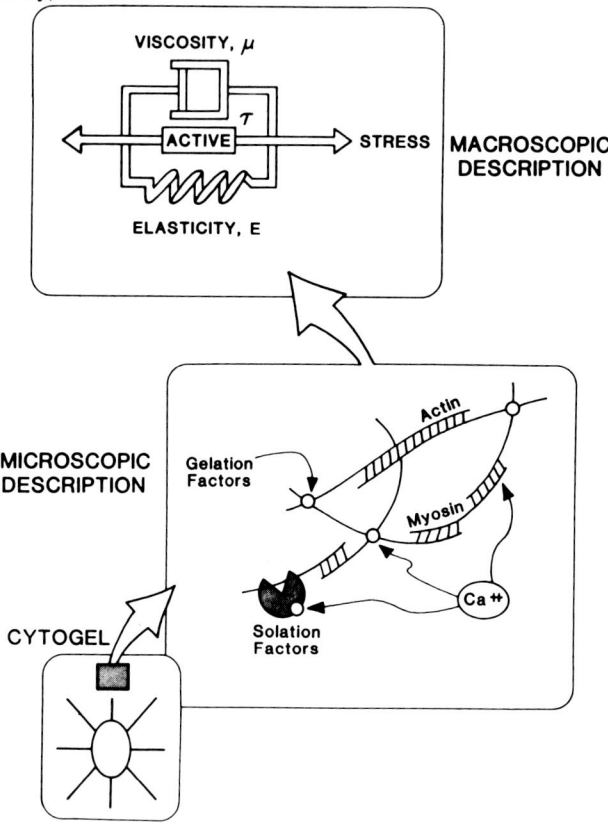

Fig. 2. The cortical cytogel consists of (i) actin and myosin polymers which generate the traction forces (ii) solation and gelation enzymes which control the connectivity of the gel, and thus its viscosity and elasticity, and (iii) a complex of chemical factors (calcium, ATP, cAMP, calmodulin, etc.) which control the activity of the contractile machinery and the solation/gelation actors. However, insofar as the macroscopic properties of viscosity and elasticity are concerned, we can treat the viscoelastic properties of the cortical cytogel as a field of mechanical elements, (e.g. a viscous 'dashpot,' an elastic 'spring,' and an active contractile component). The macroscopic mechanical properties of the cytogel are then contained in the viscosity, μ, the elastic modulus, E, and the active traction per fiber, τ.

viscosity and elastic coefficients. However, in order for the macroscopic description to capture the mechanical consequences of the microscopic mechanism these coefficients must have certain properties, which we discuss below.

The Gel is in Mechanical Equilibrium

Consider a small volume element of cytogel. In order that the element be in mechanical equilibrium with the surrounding cytogel the contraction forces generated within the volume element must just balance the tractions imposed on it by its surroundings.

External Forces = Sum of internally generated forces   (1)

Equation (1) must hold regardless of the nature of the material, since it merely reflects the balance of forces required to maintain mechanical equilibrium. In order to model the cytogel specifically, we must specify how the stress (force/unit area) generated by a volume element of cytogel depends on the strain (deformation) of the volume element.

The Mechanochemical Properties of Actomyosin Gel

Suppose the actin cortex were a passive viscoelastic gel. Then, in a 1-dimensional strip of cytogel, the stress developed by the cytogel could be represented as a "spring" and "dashpot" in parallel, as shown in Fig. 2 (i.e., a Kelvin solid; c.f. Bird, et al., 1977). If such an element is stretched, it responds with a resisting stress (force/ unit area) according to the relation:

Applied stress = - (Viscous stress + Passive elastic stress)

An actomyosin gel differs from this simple mechanical analog in three crucial ways.

1. The sliding filament mechanism underlying the active contraction has the property that as the gel contracts more myosin cross-brides are brought into play, and so the strength of contraction increases. Moreover, since the actomyosin fibers of the gel act in parallel, the strength of contraction increases as the fiber density increases.

Since contraction tends to concentrate fibers, this also increases the strength of contraction as the gel contracts. Consequently, cytogel has the important mechanical characteristic that <u>as it contracts it gets stronger</u>. That is, the force generated by the cytogel strip varies according to the relationship:

Active Traction Force increases with fiber
density, and decreases with strain. (2)

The stimulus for actomyosin contraction involves a calcium pathway; however, as shown in the Appendix, we need not deal with this explicitly in the present discussion.

2. A gel consists of an osmotically swollen polymer network (Tanaka, 1981; Oster, 1985); therefore, the contraction of the gel strip due to active contraction and passive elasticity is opposed by the osmotic swelling pressure. Thus the stress-strain response of a cytogel strip is characterized by the relation (c.f. Fig. 3):

Applied stress = Osmotic pressure - Passive elasticity
  - Active contraction - Viscous stress (3)

The crucial mechanical feature of cytogel is the S-shaped stress-strain curve; Fig. 3a shows graphically how the components of Eq. (3) add to produce this characteristic.

3. An actomyosin gel is a dynamic structure; that is, the actin fibers are constantly being depolymerized and repolymerized. Thus there is a chemical balance between the fibrous (gel) phase and the fragmented, or monomer, phase (sol). If the fiber contracts, the density of gel increases, and the depolymerization (solation) rate increases. Conversely, if the gel is dilated, so its density decreases, the equilibrium is shifted towards gelation. This is illustrated in Fig. 3b.

Cytogel is Spatially Unstable

The simple properties enumerated above turn out to be sufficient to ensure that an initially uniform volume of cytogel, if it commences actively to contract, will not remain uniform, but will form regular spatial structures.

We show in the Appendix that, as the traction of the gel increases, the spatially uniform state becomes unstable

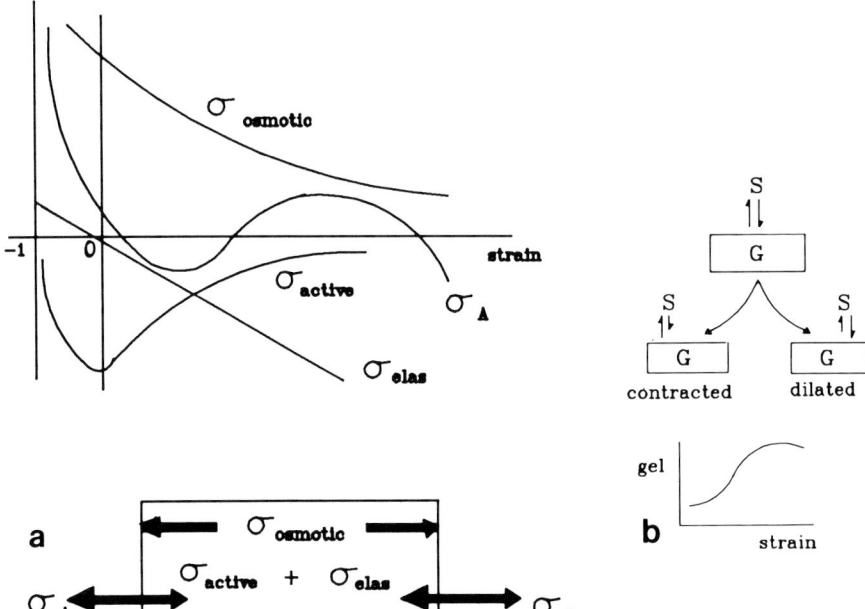

Fig. 3. (a) As stress σ_A (force/unit area) is applied to a strip of cytogel it responds by elongating. The equilibrium strain (fractional elongation, $\varepsilon = (L-L_o)/L_o$) is reached when the stresses developed by the cytogel just balance the applied stress. The internal stresses are composed of the osmotic pressure, $\sigma_{osmotic}$, which tends to dilate the gel, the active stress, σ_{active}, which tends to contract the gel, and the passive elastic stress, σ_{elas}, which is contractile when the strip is in tension ($\varepsilon > 0$) and expensive when the strip is in compression ($\varepsilon < 0$): $\sigma_A = \sigma_{OSMOTIC} + \sigma_{ACTIVE} + \sigma_{ELAS}$. Because of the autocatalytic nature of the active contraction the stress-strain curve, $\sigma_A(\varepsilon)$ is S-shaped. (b) The unstressed gel (G) is in chemical equilibrium with the sol (S). When the gel contracts, the density of gel increases and the chemical equilibrium shifts toward the sol phase; when the gel is dilated the gel density decreases, so that the equilibrium is shifted towards the gel phase. Thus the equilibrium gel fraction is an increasing function of strain.

and the fibers commence to accumulate in periodic foci, as shown in Fig. 4. The spacing and geometry of the foci depend on the specific values of the traction, viscosity and elasticity coefficients and on the boundary conditions. We shall present the full analysis, along with numerical simulations elsewhere. The linear analysis given in the Appendix shows us what to expect.

Intuitively, it is not hard to see why the gel cannot stay uniform once it commences to contract: if a small region initiates contraction it immediately gains a mechanical advantage over neighboring regions, since more actin becomes concentrated there and as the fibers contract they grow stronger. That is, contraction is autocatalytic. Eventually, the swelling pressure and the passive elastic component of the cytogel will equilibrate with the active contraction, but the gel will no longer be uniform.

At first, it might seem surprising that the foci of contraction will not be randomly dispersed throughout the volume, but rather form regular periodic arrays, with hexagonal, square or rhombic spacing. However, it turns out that in this situation, as in numerous other physical systems, hexagonal packing is the expected configuration: anything else would be surprising. The reason is that each focus is surrounded by a circular domain of highly strained gel; within this region contraction is suppressed. Thus the packing of the foci resembles the packing of identical disks in a plane, a process which produces a hexagonal packing array.

Our analysis shows that a uniform field of contractile cytogel will spontaneously form a regular pattern of dense foci. However, if the foci form sequentially from an initiation site then it is easy to show that the pattern of foci will alternate: each new row forms in the space between the previous foci. Such a progressive, or "wave-like" condensation pattern inevitably leads to a hexagonally symmetric pattern (Fig. 4). The mechanisms for generating a wave of development across the cell cortex are discussed in detail in Cheer, et al., (1985).

Osmotic Pressure Initiates the Microvilli

The tension generated by the actomyosin fibers aligns the gel along the directions of stress. Thus the contracting gel forms a "tension structure" consisting of bands of aligned fibers in a hexagonal array (Fig. 1b and 4b). Between the actin dense regions the gel is depleted, and thus

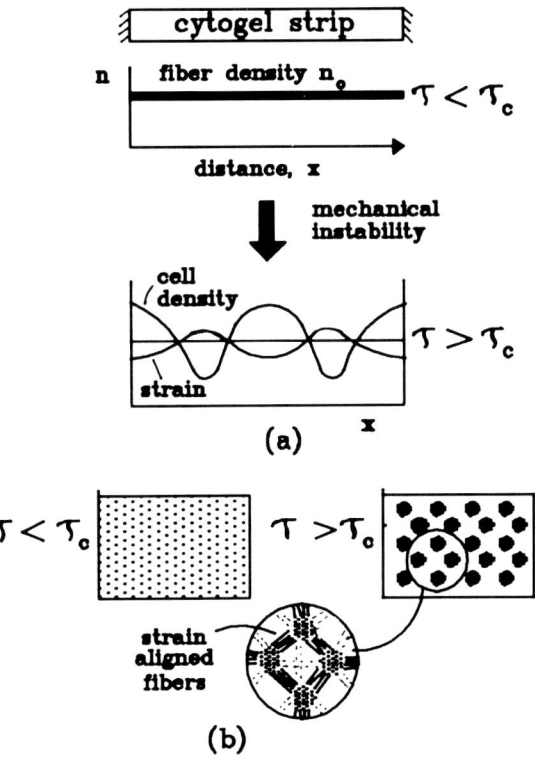

Fig. 4. (a) A strip of cytogel is mechanically unstable: if the fiber traction, τ, and/or the fiber density, n, is increased beyond a certain critical value, τ_c, the uniform distribution of fiber density gives way to stable periodic aggregations.
(b) A planar sheet of cytogel will support hexagonally packed aggregation foci when the traction and/or fiber density exceeds the critical value. The regions between foci are highly strained, thus producing alignment of the actin fibers, which may enhance crosslinking. The interstices between the aggregations are depleted of gel, and thus less able to resist the turgor pressure in the cytogel. So the cell surface will bulge out in these regions (i.e. perpendicular to the plane of the figure), thus initiating the field of microvilli. Note that since the actin-dense foci are hexagonally packed, so will be the packing of the microvilli which form in the regions of low gel density.

less able to resist the osmotic swelling pressure always present in the cell interior. We postulate that this turgor is sufficient to commence to extrude the cortex into incipient microvilli.

Stereocilia May Initiate by the Same Mechanism

Hair cells are the sensory transducers in the inner ear that convert acoustical pressure waves into nerve impulses. The apical surface of each cell is covered by a regular array of rod-like protuberances, the stereocilia. The stereocilia which adorn the mature hair cell are, like the microvilli, packed in a nearly hexagonal array (c.f. Hudspeth, 1983). This array commences as a field of small regularly spaced nodes which later grow into rods, each of which is filled with aligned actin fibers. While the structure of the stereocilia and its role in acoustical sensing has been much studied (c.f. Hudspeth, 1983), the question of what generates the pattern of stereocilia on each cell has hardly been addressed. In light of the above discussion, it is tempting to nominate the same process we have proposed for the initiation and osmotic inflation of microvilli is responsible for laying down the primordia for stereocilia.

CONCLUSIONS

Hexagonal patterns of microvilli and stereocilia arise on the apical surface of hair cells. By analyzing the mechanical properties of the actomyosin gel comprising the cell cortex we have concluded that the hexagonal patterns can arise solely as a consequence of the mechanical instability of a contractile gel.

The cytoskeleton inserts into the membrane at various loci via specialized proteins such as vinculin. However, we have not mentioned the cell membrane in our discussion since it is not mechanically significant in maintaining cell shape. Nor have we discussed the nature of the cross-linking molecules which unite the actin fibers into a gel. However, crosslinking is surely enhanced when fiber density is increased and fibers are aligned in parallel. In some instances crosslinking is strongly influenced by pH. The rise in pH that frequently follows a calcium increase could be the agent which initiates the crosslinking of the actin fibers. However, the model deals only with the inception of the microvilli and their spatial pattern. A more elaborate model is required to investigate these subsequent events.

ACKNOWLEDGMENTS

We would like to thank D. Begg and A. J. Hudspeth for valuable conversations during the course of this research and for providing the photographs in Figure 1. George F. Oster and James D. Murray would like to thank the Center for Nonlinear Studies, Los Alamos National Laboratory, where some of the research was done as an Oppenheimer Visiting Professor and the Stan Ulam Visiting Scholar, respectively. G. F. Oster and G. M. Odell would also like to acknowledge the support of the Science Engineering Research Council of Great Britain (GR/c/63595) during a visit to the Centre for Mathematical Biology, Oxford University. G. F. Oster and G. M. Odell were supported by NSF grants #MCS-8110557 and #MCS-83-01460.

REFERENCES

1. Cheer A., Nuccitelli R., Oster G., Vincent J-P. (1984). Contraction waves on vertebrate eggs: the activation waves. J. Theo.Biol. (submitted).
2. Hudspeth A.J. (1983). The hair cells of the inner ear. Scientific American 248:86-3.
3. Flory P. (1953). Principles of polymer chemistry. Ithaca: Cornell University Press.
4. Jaffe L. (1980). Calcium explosions as triggers for development. Ann. N.Y. Acad. Sci. 339:86-101.
5. Landau L., Lifshitz E. (1959). The Theory of Elasticity, London: Pergamon.
6. Meinhardt H. (1982). Models of biological pattern formation. London: Academic Press.
7. Murray J.D. (1977). Nonlinear differential equation models in biology. Oxford: Clarendon Press.
8. Murray J.D. (1981a). A pre-pattern formation mechanism for animal coat markings. J. Theor. Biol. 88:161-199.
9. Murray J.D. (1981b). On pattern formation mechanisms for lepidopteran wing patterns and mamalian coatmarkings. Phil. Trans. Roy. Soc. (London) B295:473-496.
10. Murray J.D., Oster G. (1984a). Generation of biological pattern and form. IMA J. Math. Appl. in Medic. & Biol. 1:51-75.
11. Murray J.D., Oster G. (1984b). Cell traction models for generating pattern and form. J. Math. Biol. 19: 265-279.
12. Maini P., Murray J.D., Oster G. (1985). A mechanical model for biological pattern formation: a nonlinear

bifurcation analysis. Proc. of conference on 'Ordinary and Partial Differential Equations,' Dundee, July 1984. Heidelberg: Springer Verlag (to appear in proceedings)
13. Odell G. (1985). A mathematical modeled cytogel cortex exhibits periodic Ca^{++}-modulated contraction cycles seen in Physarum shuttle streaming. J. Embryol. exp. morphol. 83:261.
14. Oster G., Odell G. (1984). The mechanics of cytogels I: Plasmodial oscillations in Physarum. Cell Motility 4:469.
15. Oster G. (1985). On the crawling of cells. J. Embryol. exp. morphol. 83:329.
16. Tanaka T. (1981). Gels. Sci. Amer. 244:124-38.

APPENDICES

Model Equations

In this Appendix we briefly describe the mathematical model and the analysis which points to the spatial structures discussed in the text. More complete accounts can be found in the references cited in the text. For simplicity, we shall give only the 1-dimensional analysis; typical 2 and 3-dimensional analyses are given in Murray and Oster (1984a,b).

The model cytogel consists of a 2-component viscoelastic continuum (sol and gel) whose state is regulated by calcium ions. That is, the actomyosin gel consists of the crosslinked fibrous components, and we consider the sol to consist of all of the non-crosslinked components (fibers and monomers). The state of the cytogel is specified by giving the sol, gel and calcium concentration distributions along with the mechanical state of strain of the gel. Thus we define the following field quantities:

$S(x,t)$ = the concentration of sol at position x and time t.
$G(x,t)$ = the concentration of gel at position x and time t.
$C(x,t)$ = the concentration of calcium at position x and time t.
$\varepsilon(x,t)$ = the local strain = $\partial u/\partial x$, where u is the displacement of material points from their original position.

Mass Balance Equations

The conservation equations governing the sol, gel and calcium are (where subscripts denote partial differentation):

Sol: $S_t + (Su_t)_x = D_s S_{xx} - F(S,G,\varepsilon)$ (A1)

Gel: $G_t + (Gu_t)_x = D_g G_{xx} + F(S,G,\varepsilon)$ (A2)

Calcium: $c_t = D_c c_{xx} + B(c,\varepsilon)$ (A3)

The terms $(Su_t)_x$ and $(Gu_t)_x$ are the convections of sol and gel, respectively, by the cytogel moving at velocity u_t. Both sol and gel may diffuse, although the diffusion of gel is much more restricted than the sol by virtue of its crosslinks. The function $F(S,G,\varepsilon)$ is the sol-gel reaction, and is given by the following expression:

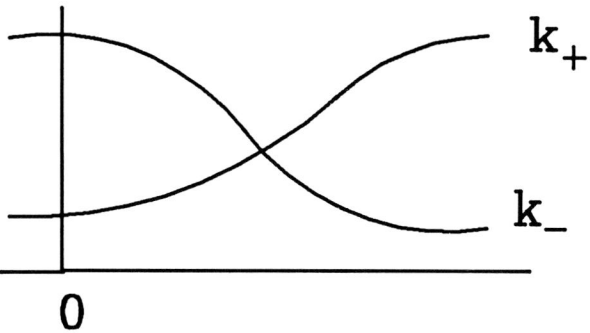

Fig. A1. The gelation rate (k_+) and solation rate (k_-) as functions of strain, ε.

Patterns of Microvilli 379

$$F(S,G,\varepsilon) = k_+(\varepsilon)S - k_-(\varepsilon)G \quad . \qquad (A4)$$

We assume that as the sol is dilated (ε increasing) its density decreases and so the mass action rate of gelation increase. Conversely, when the gel contracts, the gel density increases and the solation rate increases. Hence the gel concentration may increase with increasing strain (c.f. Fig. 3b & Oster & Odell, 1984a,b). Thus the qualitative form of the gelation and solation rate functions are shown in Fig. A1.

Force Balance Equation

The balance of forces within the cytogel involves the following stresses (force/unit area of cytogel) (Oster, 1985):

σ_V = viscous stress due to the hydrodynamic drag between the gel fibers and the fluid componente of the cytogel, as well as the dissipation involved in the actomyosin sliding filament mechanism.

σ_E = the elastic forces generated by the gel fibers. This is largely entropic in origin; i.e. arising from the thermal motion of the polymer strands.

σ_A = the active contraction forces generated by the sliding filament mechanism of the actomyosin fibers.

σ_O = the osmotic swelling pressure of the gel.

Since inertial forces are negligible the force balance for the cytogel is:

$$0 = \nabla \cdot \sigma = \partial/\partial x [\sigma_V + \sigma_E + \sigma_A + \sigma_O] \quad . \qquad (A5)$$

The qualitative shapes of the various stresses are shown in Fig. 3a. For illustrative purposes we shall assume the following functional forms:

$$\sigma = \pi/(1+\varepsilon) - GE(\varepsilon - L\varepsilon_{xx}) - G\tau(c)/(1+\varepsilon^2) - G\mu\varepsilon_t \qquad (A6)$$

    osmotic   elastic   active   viscous

Note that only the osmotic pressure tends to dilate the gel, and that the gel concentration multiples the elastic, active and viscous stresses, since they act through the agent of the gel fibers. The term $L\varepsilon_{xx}$ in the elastic stress arises from the fibrous nature of the gel (c.f. Murray & Oster, op. cit.). The active contraction is triggered by calcium, and so $\tau(c)$ is an increasing function of c.

Thus the model equations consist of the three mass balance equations (A1-3) and the force balance equation (A5), along with the constitutive relations for $F(S,G.\varepsilon)$, $B(C,\varepsilon)$ and σ, and the appropriate boundary and initial conditions.

A Simplified Model System

The model system described above is quite complex, even for numerical simulation. However, the basic pattern forming mechanism can be appreciated by making the following simplifications.

Suppose that the time scale for diffusion of the sol and calcium is much faster than the motion of the gel. Then, from equations (A1, A3) we can take S and c to be essentially constant. Thus S and c appear only as parameters in equations (A2, A5), and we may replace $\tau(c)$ by τ and S by S_o. Thus we can integrate the force balance equation (A5) to obtain, with (A6)

$$G\mu\varepsilon_t = GE\varepsilon_{xx} + H(G,\varepsilon) \qquad (A7)$$

where

$$H(G,\varepsilon) = \pi/(1+\varepsilon) - G\tau/(1+\varepsilon^2) - GE(\varepsilon - L\varepsilon_{xx}) - \sigma_o. \qquad (A8)$$

Here the constant boundary stress, σ_o, is negative, in keeping with our sign convention in (A6).

Note that the form of equation (A7) is of a reaction-diffusion system, where the strain, ε, plays the "reactant", and the kinetics is given by $H(G,\varepsilon)$.

With $S = S_o$, the gel equation (A2) becomes

$$G_t + (Gu_t)_x = k_+(\varepsilon)S_o - k_-(\varepsilon)G + D_G G_{xx} \qquad (A9)$$

Equations (A7 and A9) comprise a coupled system of diffusion-reaction equations similar to those which have been widely studies mathematically, and are known to yield a variety of spatial patterns (c.f. Murray, 1977, 1981a.b; Meinhardt, 1982). Although convection is usually omitted from diffusion-reaction models, its presence here simply enhances the pattern formation potential of the system.

Linear Analysis and Pattern Selection

By nondimensionalizing the equations we both reduce the parameter count and isolate the competing physical processes. We choose the following nondimensionalization:

$G^* = G/S_o$, $\quad x^* = x/L^{\frac{1}{2}}$, $\quad t^* = tE/\mu$, $\quad \pi^* = \pi/S_o E$

$k_+^* = k_+\mu/E$, $\quad k_-^* = k_-\mu/E$, $\quad \sigma_o^* = \sigma_o/S_o E$

$D^* = D_G\mu/LE$, $\quad \tau^* = \tau/E$, $\quad \varepsilon^* = \varepsilon$ \hfill (A10)

Substituting these quantities into (A7) and (A9), and dropping the asterisks for notational simplicity, we obtain the dimensionless equations:

$$G\varepsilon_t = G\varepsilon_{xx} + f(G,\varepsilon)$$

$$G_t + (Gu_t)_x = DG_{xx} + g(G,\varepsilon) \hfill (A11)$$

where

$$f(G,\varepsilon) = -\sigma_o + \pi/(1+\varepsilon) - G\tau/(1+\varepsilon^2) - G\varepsilon \hfill (A12)$$

$$g(G,\varepsilon) = k_+(\varepsilon) - k_-(\varepsilon)G$$

Figure A2 shows the phase plane associated with (A11) and (A12) obtained by suppressing the spatial terms. The nullclines intersect at a stationary state, P, whose location can be placed as shown by adjusting the system parameters appropriately.

We linearize about P and examine the perturbation equations in the usual fashion (see, for example, Murray, 1981a):

$$u_t + G_o v_t = Du_{xx} + g_G u + g_\varepsilon v \hfill (A13)$$

$$G_o v_t = G_o v_{xx} + f_G u + f_\varepsilon v \hfill (A14)$$

where (u,v) are the deviations of (G,ε) from the equilibrium point, P, and subscripts denote the partial derivatives of f and g evaluated at P($\varepsilon=\varepsilon_o$, G=$G_o$).

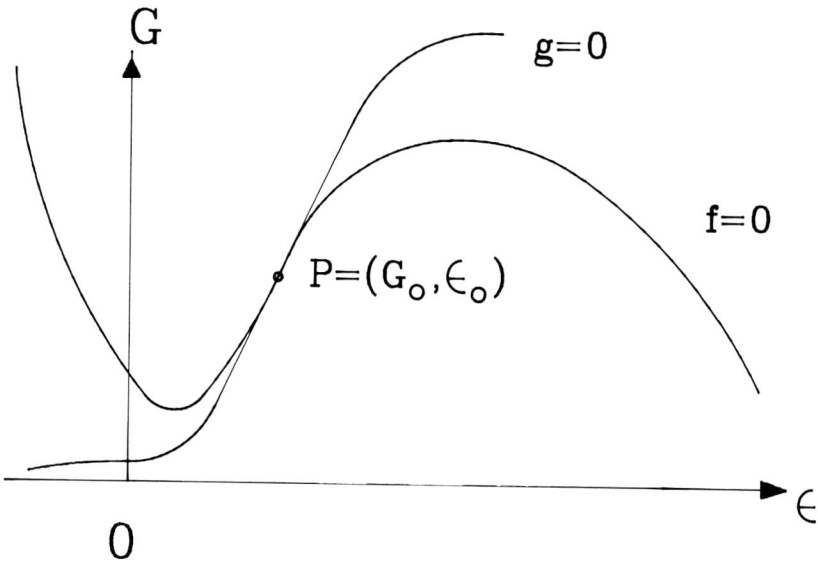

Fig. A2. The (ε, G) phase plane, showing the nullclines $\dot{\varepsilon} = f(\varepsilon, G) = 0$, $\dot{G} = g(\varepsilon, G) = 0$.

The key to diagnosing spatial structures is the dispersion relation obtained by substituting the trial solutions

$$(u,v) \sim e^{\lambda t} e^{ikx} \tag{A15}$$

where λ is the linear growth rate of a mode with wave number k. λ is given as the roots of the quadratic equation

$$G_o \lambda^2 + b(k)\lambda + c(k) = 0 \tag{A16}$$

where

$$b(k) = -(f_\varepsilon + G_o g_G - G_o f_G) + G_o(1+D)K^2$$
$$c(k) = (f_\varepsilon g_G - f_G g_\varepsilon) - (Df_\varepsilon + G_o g_G)k^2 + G_o D k^4 \tag{A17}$$

In order to generate spatial structures, the shape of $\lambda(k)$ must look as shown in Fig. A3. That is, the system must be stable to perturbations which are spatially uniform (k = 0), and unstable for some interval of wavenumbers, k ≠ 0. Homogeneous stability requires that $Re[\lambda(0)] < 0$, which implies that

$$b(0) > 0, \quad c(0) > 0 \tag{A18}$$

i.e.

$$-(F_\varepsilon + G_o g_G - G_o f_G) > 0, \quad f_\varepsilon g_G - f_g g_\varepsilon > 0. \tag{A19}$$

From Fig. A2 we see that

$$f_\varepsilon > 0, \quad f_G < 0, \quad g_\varepsilon > 0, \quad g_G < 0. \tag{A20}$$

Instability requires that $Re[\lambda(k)] < 0$ for some $0 < k < \infty$, which is ensured if

$$b(k) < 0 \text{ and/or } c(k) < 0 \text{ for some } 0 < k < \infty. \tag{A21}$$

The conditions for spatially unstable modes is computed to be

$$D > -G_o g_G / f_\varepsilon$$
$$(Df_\varepsilon - G_o g_G)^2 + G_o Df_G G_\varepsilon > 0$$

These inequalities impose conditions on the parameters, and the forms of the functions f and g are such that these conditions may indeed be met.

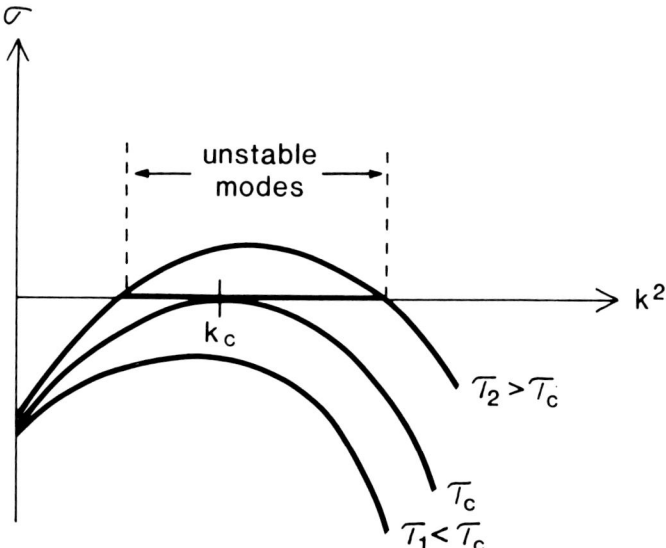

Fig. A3. The dispersion relation (Eq. A16). The system becomes linearly unstable when the traction, τ, and/or the fiber density, G_o, increases to the point where the curve is tangent to the k^2 axis at k_e. The spacing λ_c of the foci are estimated from $2\pi k_c = 2\pi/\lambda_c$.

MONOMOLECULAR INDUCTION OF TWO COMPONENTS OF
THE POSTSYNAPTIC APPARATUS IN MUSCLE

Bruce G. Wallace, Noreen E. Reist, Ralph M. Nitkin,
Justin R. Fallon and U.J. McMahan

Stanford University School of Medicine
Stanford, California 94305

INTRODUCTION

Aggregates of acetylcholine receptors (AChRs) and acetylcholinesterase (AChE) compose part of the postsynaptic apparatus at normal and regenerated neuromuscular junctions. At developing neuromuscular junctions the aggregation of AChRs and AChE on the myofiber's surface is induced by the axon terminal (1, 2). On the other hand, during regeneration of adult muscles the synaptic portion of the myofiber's sheath of basal lamina can direct the formation of these specializations (3, 4, 5). Thus synaptic basal lamina in the adult contains synaptogenic information similar to that provided by the axon in the embryo.
We have undertaken a study aimed at identifying the molecules in the basal lamina that cause the aggregation of AChRs and AChE at regenerating neuromuscular junctions. Our findings to date have led us to the hypothesis that the same molecules direct the formation of both specializations. Here we review some of the evidence that forms the basis for this hypothesis. Detailed accounts of the experiments have been published elsewhere (6, 7, 8, 9).

RESULTS AND DISCUSSION

We began by preparing an insoluble basal lamina-containing fraction from the Torpedo californica electric organ, a tissue which has a far higher concentration of cholinergic synapses than muscle. When added to cultures

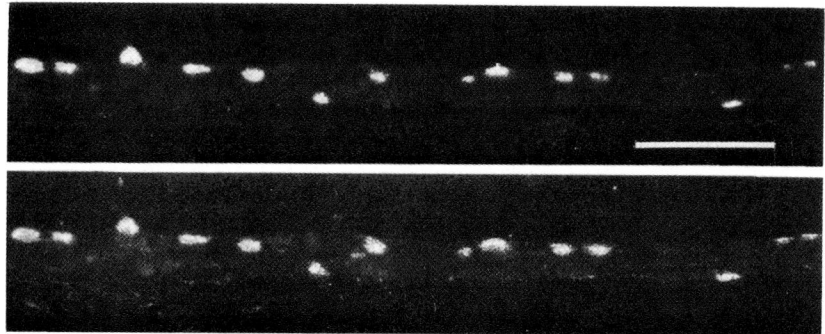

FIGURE 1. Insoluble basal lamina-containing fraction of Torpedo electric organ causes formation of co-aggregates of AChRs and AChE. Fluorescence micrographs of a myotube from a culture of chick myotubes treated for 24 hrs with the insoluble fraction. Top: AChRs, labelled with rhodamine conjugated α-bungarotoxin, are visualized with rhodamine optics. Bottom: AChE, labelled with a monoclonal antibody specific for chick AChE (15) and a fluorescein-conjugated second antibody, is visualized with fluorescein optics. Bar = 50 μm.

of chick myotubes the insoluble material causes the formation of patches of AChRs and AChE on the myotube surface. The aggregates of AChRs overlap those of AChE (Fig. 1) as at neuromuscular junctions (6, 9).

The dose-dependence of the insoluble fraction-induced formation of AChE patches on cultured myofibers was the same as for AChR aggregation, suggesting that a single factor caused both effects. If this were the case, it should be possible to extract the AChE-aggregating activity from the insoluble fraction and purify it by procedures we have developed to solubilize and purify the AChR-aggregating factor several thousand fold (7). Accordingly, we extracted the insoluble fraction with pH 5.5 citrate buffer, fractionated the solubilized material by gel filtration chromatography on Agarose 1.5m, and pooled the column fractions that cause AChR-aggregation. The AChR- and AChE-aggregating specific activities were increased to the same extent by extraction and gel filtration chromatography and the dose dependence for AChR-aggregation was the same as that for the formation of AChE aggregates at each step in the purification (9). Thus AChE- and AChR-aggregating activities copurify.

We also examined whether or not two different monoclonal antibodies, 6D4 and 2F6 (8), directed against the AChR-aggregating factor would recognize the AChE-aggregating factor. When bound to Sepharose beads, 6D4 and 2F6 each removed AChR- and AChE-aggregating activity from the Agarose pool (9). Moreover, when added directly to cultures along with the Agarose pool, 6D4 and 2F6 inhibited the extract-induced formation of both AChR and AChE aggregates (9).

The observations that i) the AChR- and AChE-aggregating activities copurify, ii) the dose dependence for the induction of AChR and AChE aggregates is the same at each stage of the purification, and iii) two different monoclonal antibodies immunoprecipitate and inhibit both activities, lead to the conclusion that a single factor in our extracts of electric organ causes the formation of AChR and AChE patches.

Several lines of evidence indicate that the electric organ AChR/AChE-aggregating factor is similar to the component(s) of the synaptic basal lamina that directs the accumulation of AChRs and AChE on regenerating muscle fibers: as described above, the factor is found in a basal lamina-containing fraction of the electric organ; the corresponding fraction of Torpedo muscle contains an AChR-aggregating activity (6)-(muscle has not been tested for AChE-aggregating activity), antiserum against partially purified factor from electric organ blocks and immunoprecipitates AChR-aggregating activity from muscle and stains muscle basal lamina (7); as shown in Fig. 2, the two monoclonal antibodies (6D4, 2F6) that block and immunoprecipitate the active component in extracts of electric organ and muscle bind in high concentration at the neuromuscular junction; and, by electron microscopy it has been found that at least one of the monoclonal antibodies (6D4) recognizes molecules that are concentrated in or adjacent to synaptic basal lamina (8).

It may be that the electric organ AChR/AChE-aggregating factor is related to a known component of the basal lamina at the neuromuscular junction (10, 11, 12, 13). Several components have been tested for their

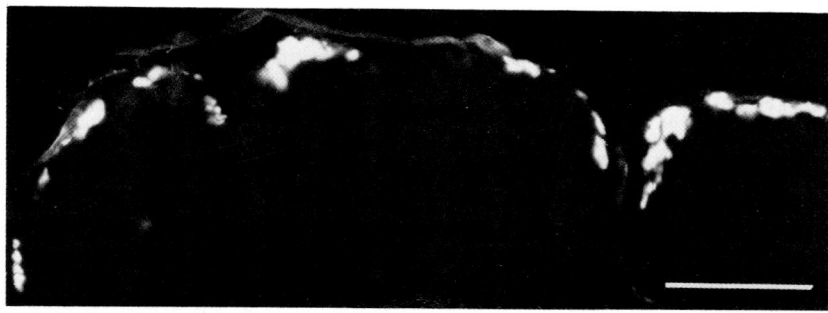
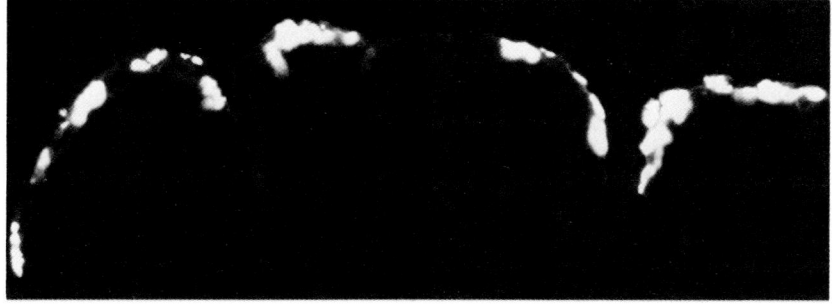

FIGURE 2. 6D4 recognizes molecules concentrated at the neuromuscular junction. Top: Frozen section of Torpedo californica muscle viewed with fluorescein optics to reveal the position of 6D4 binding sites which were labeled with a fluoresceinated second antibody. Bottom: Same field as above but viewed with rhodamine optics to show the location of neuromuscular junctions which were labeled with rhodaminated α-bungarotoxin. 6D4 binds to axons and arteries (not shown) but on muscle fibers it is highly concentrated at the neuromuscular junction. Bar=50 µm.

ability to cause receptor aggregation. Of these, only laminin has been shown to be effective (14). Gel filtration of our solubilized electric organ material did not reveal any AChR-aggregating activity in the size range of laminin (∼880 Kd), most of the activity eluted at a position characteristic of lower molecular

weight proteins (50-100 Kd) (7). Moreover, antiserum
against laminin, which stained basal lamina in our
sections of Torpedo muscle, did not immunoprecipitate
AChR-aggregating activity from electric organ extracts
(8). These observations, coupled with the finding that
6D4 and 2F6 staining co-localized with only a fraction of
anti-laminin staining in muscle (8) indicate that the
electric organ AChR-aggregating factor is not laminin.
However, the fact that two different monoclonal
antibodies against the electric organ AChR-aggregating
factor recognize cell surface antigens in the walls of
arteries and in nerves (see Fig. 2) as well as at
neuromuscular junctions may mean that the factor is
structurally related to one or a family of basal lamina
molecules that has a broad distribution in muscle.

It is now clear that components of the basal lamina
in the synaptic cleft of the neuromuscular junction
direct the accumulation of AChRs and AChE on regenerating
myofibers in vivo (4, 5). Since a single factor in our
basal lamina-containing fraction of electric organ causes
the aggregation of both AChRs and AChE on cultured
myotubes in vitro and since monoclonal antibodies against
the factor recognize molecules concentrated in the
synaptic cleft at the neuromusclar junction, it is only
reasonable to suggest that during regeneration in vivo,
the molecule in the synaptic basal lamina that causes
aggregation of AChRs also causes aggregation of AChE. We
are now performing experiments aimed at testing this
monomolecular induction hypothesis at both regenerating
and developing neuromuscular junctions. We are also
seeking to learn whether the electric organ AChR/AChE-
aggregating factor causes the formation of other
components of the postsynaptic apparatus.

ACKNOWLEDGMENTS

Studies were supported by USPHS grant NS14506 and by
BRSG grant RR5353.

REFERENCES

1. Anderson MJ, Cohen MW, Zorychta E. (1977). Effects of innervation on the distribution of acetylcholine receptors on cultured muscle cells. J Physiol (Lond) 268:731.

2. Dennis MJ (1981). Development of the neuromuscular junction: interactions between cells. Ann Rev Neurosci 4:43.

3. Burden SJ, Sargent PB, McMahan UJ (1979). Acetylcholine receptors in regenerating muscle accumulate at original synaptic sites in the absence of the nerve. J Cell Biol 82:412.

4. McMahan UJ, Slater CR (1984). The influence of basal lamina on the accumulation of acetylcholine receptors at synaptic sites in regenerating muscle. J Cell Biol 98:1453.

5. Anglister L, McMahan UJ (1984). Basal lamina directs the accumulation of acetylcholinesterase at synaptic sites in regeneration muscles. Soc Neurosci Abstr 10:281.

6. Godfrey EW, Nitkin RM, Wallace BG, Rubin LL, McMahan UJ (1984). Components of Torpedo electric organ and muscle that cause aggregation of acetylcholine receptors on cultured muscle cells. J Cell Biol 99:615.

7. Nitkin RM, Wallace BG, Spira ME, Godfrey EW, McMahan UJ (1983). Molecular components of the synaptic basal lamina that direct differentiation of regenerating neuromuscular junctions. Cold Spring Harbor Symp Quant Biol 48:653.

8. Fallon JR, Nitkin RM, Reist NE, Wallace BG, McMahan UJ (In press) Monoclonal antibodies directed against an AChR-aggregating factor from Torpedo electric organ recognize molecules concentrated at neuromuscular junctions. Nature (Lond).

9. Wallace BG, Nitkin RM, Reist NE, Fallon JR, Moayeri NN, McMahan UJ (In press). AChR-aggregating factor from Torpedo electric organ also causes formation of AChE aggregates. Nature (Lond).

10. McMahan UJ, Sanes JR, Marshall LM (1978). Cholinesterase is associated with the basal lamina at the neuromuscular junction. Nature (Lond) 271:172.

11. Anderson MJ, Fambrough DM (1983). Aggregates of acetylcholine receptors are associated with plaques of a basal lamina heparan sulfate proteoglycan on the surface of skeletal muscle fibers. J Cell Biol 97:1396.

12. Sanes JR, Hall Z (1979). Antibodies that bind specifically to synaptic sites on muscle fiber basal lamina. J Cell Biol 83:357.

13. Sanes JR (1982). Laminin, fibronectin and collagen in synaptic and extrasynaptic portions of muscle fiber basement membrane. J Cell Biol 93:442.

14. Vogel Z, Christian CN, Vigny M, Bauer HC, Sonderegger P, Daniels MP (1983). Laminin includes acetylcholine receptor aggregation on cultured myotubes and enhances receptor aggregating activity of aneuronal factor. J Neurosci 3:1058.

15. Rotundo RL (1984). Purification and properties of the membrane-bound form of acetylcholinesterase from chicken brain. J Biol Chem 21:13186.

MORPHOGENESIS OF THE MOUSE MOTOR ENDPLATE[1]

François Rieger

Groupe Biologie et Pathologie NeuromusculairesINSERM
U 153 17 rue du Fer-à-Moulin 75005 Paris, FRANCE

ABSTRACT : The main steps of the development of the neuromuscular system in the mouse are schematically described. Particular emphasis is given to the first events of nerve-muscle contact and to morphological stabilization of the motor endplate. Plasticity of the endplate and the role of activity in its morphogenesis is documented at the cellular and molecular level during normal development and during the pathogenic processes found in two mouse mutants, muscular dysgenesis and motor endplate disease. Muscle inactivity in muscular dysgenesis is directly due to excitation-contraction uncoupling and, in motor endplate disease, to abnormal axonal conduction. Ultraterminal sprouting occurs in both genetic conditions. Motor endplate disease is characterized by an active remodeling of the mutant motor endplate and muscular dysgenesis by a marked morphological immaturity of the neuromuscular contacts. In addition to the abnormalities of motor endplate formation, muscular dysgenesis is characterized by a unique type of multifocal polyinnervation of the mutant myotubes. Thus, not only muscle inactivity but also other abnormal motoneuron-muscle interactions cooperate to produce the complex neurobiological situation found in the muscular dysgenesis embryo.
The neural cell adhesion molecule (N-CAM) is most

[1]This work was partially supported by CNRS, INSERM, the French Ministère de l'Industrie et de la Recherche and the Neuroscience Research Institute at the Rockefeller University.

probably involved in motor endplate morphogenesis. It has been found concentrated at the adult motor endplate, colocalized with the acetylcholine receptor. In developing muscle, N-CAM is found on the surface of the myofibers and also concentrated at the motor endplate. There is a postnatal embryonic to adult conversion of muscle N-CAM, concurrent with the regression of polyinnervation. In adult muscle, denervation causes the decrease of junctional N-CAM and the reappearance of extrajunctional N-CAM. Thus, the main events of motor endplate morphogenesis are accompanied by changes in distribution (polarity modulation) and chemical properties (modification in sialic acid content) as well as prevalence of N-CAM. N-CAM may play a critical role both in the establishement of the initial nerve-muscle contact and in the stabilization and maintenance of the functional motor endplate.

INTRODUCTION

Four different cell types interact during the development of the neuromuscular system in the vertebrate embryo : the motoneurone, the Schwann cell, the muscle cell and the fibroblast. The adult motor endplate is a region of elaborate morphological specialization with postsynaptic folds, postsynaptic densities at the crests of the folds, and synaptic basal lamina and axon terminals containing synaptic vesicles and active zones of neurotransmitter release, enwrapped by the terminal Schwann cell (1). The postsynaptic densities contain a high density of acetylcholine receptor (AChR). Its packing density decreases at least 50 fold a few micrometers away from the axon terminals (2). AChE is essentially concentrated in the synaptic basal lamina. Synaptic basal lamina also contains high levels of a heparan sulfate proteoglycan (3), associated with AChR clusters on developing myotubes in vitro.

I. DEVELOPMENTAL ASPECTS OF THE NEUROMUSCULAR SYSTEM.

The developmental sequence which leads to the ultimate mature motor endplate is schematized in Table 1. No organized basal lamina (BL) is necessary for early synapse formation. There is no visible BL between axon and myotube at day 13-14 in the mouse embryo diaphragm muscle, when the first axon-myotube contacts are observed at the electron microscope level, although synaptic vesicles and postsynaptic densities are already present (Pinçon-Raymond and Rieger, in preparation, and Fig. 1). This is in agreement with recent observations on embryonic rat intercostal muscle by (4).

TABLE 1
DEVELOPMENT OF THE NEUROMUSCULAR SYSTEM IN THE MOUSE

EMBRYONIC OR POSTNATAL AGE	PRESYNAPTIC	POSTYNAPTIC
E12 - 14	Exploratory axons First contacts	Primary myotubes
E14 - 16	Synaptic transmission begins	Deposits of basal lamina (laminin, HSPG, C10 antigen) junctional AChE and AChR clusters; C1 synaptic antigen
E16 - 18	Polyinnervation develops	Extrajunctional AChR decrease
E18 - 19		Secondary myotubes
PN0 - 2 wk	Regression of polyinnervation; myelination and formation of the node of Ranvier	Maturation of the motor endplate : - junctional folds - AChE, AChR clusters - AChR gating changes

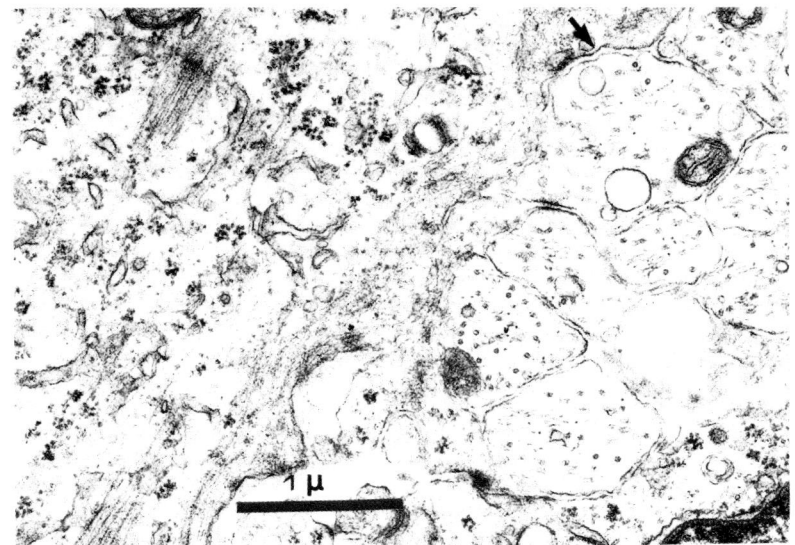

Fig. 1 : Early axon-myotube contacts. Absence of organized muscle synaptic basal lamina. Mouse embryo diaphragm muscle. Embryonic day 13. Electron micrograph obtained by Dr M. Pinçon-Raymond.

The first acetylcholinesterase (AChE ; EC 3-1-1-7) accumulations appear immediately after the first contacts are observed, at day 14-15 (5). The phrenic nerve is stained by the AChE cytochemical reaction as well as focal areas on muscle, corresponding to early nerve-muscle contacts. The first junctional AChR clusters are observed one day later by autoradiography of radioactive ^{125}I α-bungarotoxin (α-BgTX), irreversibly bound to the AChR (6). Junctional basal lamina is induced de novo at the time junctional AChR clusters are detected (7).

Later, at day 18 in the mouse embryo diaphragm muscle, all myotubes are polyinnervated. This polyinnervation decreases during the first two weeks after birth, as it does in rat muscle (8-10), at the time the motor endplate gets a more mature morphology. The terminal nerve arborization can be visualized by silver nitrate impregnation techniques (11). Newborn and 1 week old muscle motor endplates have still simple features. The definitive characteristic pattern of the adult is formed 2-3 weeks after birth (Fig. 2).

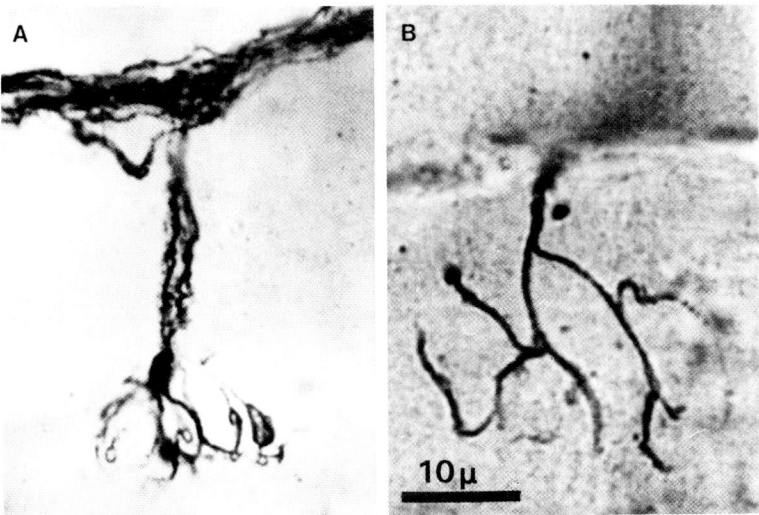

Fig. 2 : Axon terminal arborization in new-born and 3 week old mouse muscle. Diaphragm muscle a) Newborn : polyinnervated motor endplate b) 3 weeks : mature motor endplate - Silver nitrate impregnation technique (11).

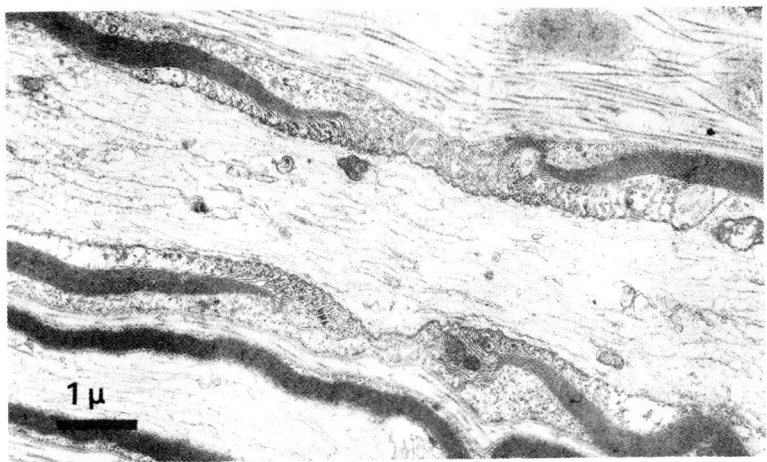

Fig. 3 : Node of Ranvier. Sciatic nerve. Adult mouse.

In the nerve, the main known developmental events are myelination and formation of the nodes of Ranvier. The node of Ranvier is the gap between two adjacent cells (Fig. 3). The morphological maturation of this area, highly specialized for nerve impulse propagation, also occurs during the two first weeks after birth (for a review, see 12).

II. GENETIC ABNORMALITIES OF MOTOR ENDPLATE.

Several studies have already addressed the question of plasticity of the motor endplate and the role of activity in its morphogenesis (13-19). We have used two mouse mutants, motor endplate and muscular dysgenesis, to reveal plasticity and the role of nerve and muscle activity in the developmental maturation of the motor endplate.

Motor endplate disease :

In motor endplate disease (med) a state of partial to total muscle inactivity develops between 1 and 3 weeks after birth (20,21), at the late stages of motor endplate morphogenesis. The homozygote mutant mice (med/med) ; Fig. 4) are characterized by a progressive muscular weakness and die 3 weeks after birth. The main morphological consequence of muscle inactivity is ultraterminal nerve sprouting. The sprouting is the most intense in the soleus muscle (22, Fig. 5). In the mutant muscle, in 3 weeks old med/med mice, the original terminal branches are swollen and small axon sprouts have developed and innervate wider territories and even adjacent myofibers, a phenomenon which is never found in normal muscle.

At the electron microscope level, in the mutant soleus muscle (Fig. 6) the original innervated regions are recognized by the presence of the junctional folds. The original axon terminals are still present without any sign of detachment from the myofiber. The axon sprouts extend over the myofiber surface, in simple appositions. They are filled with neurofilaments, which indicates active axon growth. Small clusters of synaptic vesicles are occasionally seen close to the regions of contact suggesting functional contact.

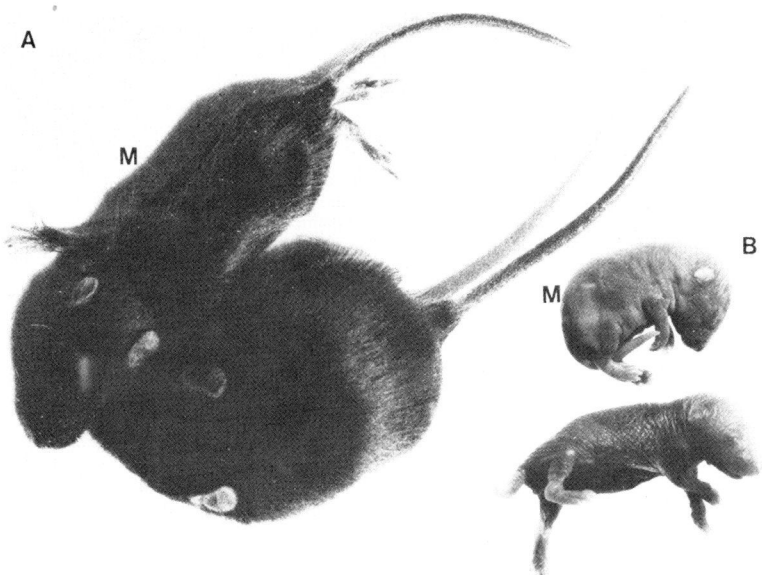

Fig. 4 : <u>Motor endplate disease and muscular dysgenesis</u>. a) normal +/med ? (+) and mutant med/med (m) littermates. 18 day old. Note the curly hair of the normal mouse, due to the expression of the dominant fur marker, caracul (Ca^d). b) normal +/mdg (+) and mutant mdg/mdg (m) littermates at birth.

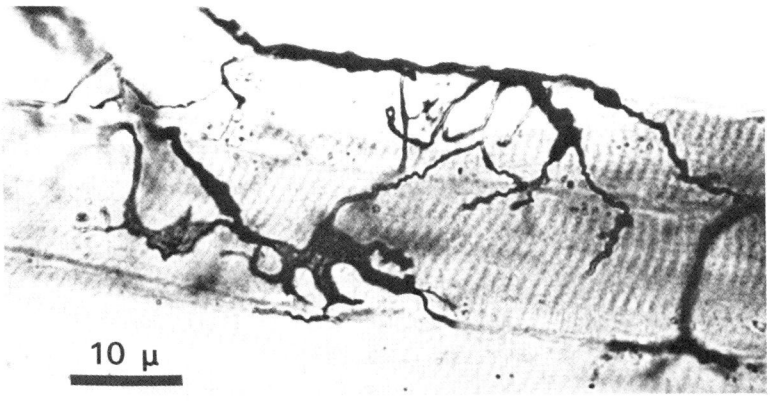

Fig. 5 : <u>Ultraterminal sprouting at med/med soleus motor endplates</u>. Swelling and sprouting. Silver nitrate impregnation.

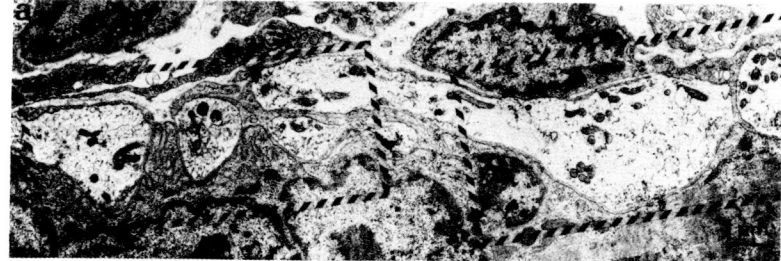

Fig. 6 : Plasticity of med/med motor endplate at the electron microscope level. Areas corresponding to the original innervated region and to the region and to the region of simple axon sprout-myofiber appositions are indicated.

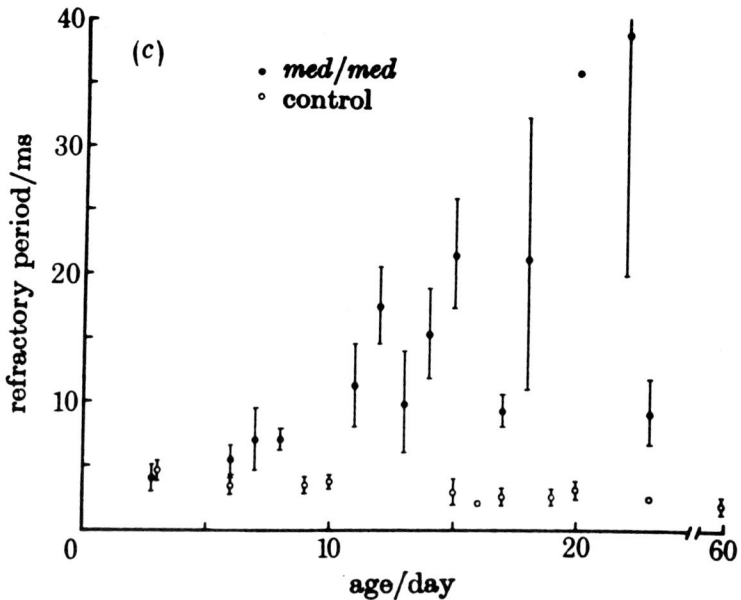

Fig. 7 : Increased refractory period between two successful nerve stimulations. From (24).

This situation of muscle inactivity in motor endplate disease, is caused by early failures of axonal conduction. We have found that the nodes of Ranvier in mutant nerve are widened nearly as early as they are formed (23). At day 9, the average nodal gap length is already significantly wider in the mutant than in the normal nerve. This phenomenon is almost general at postnatal day 18.

The direct physiological consequence of such a widening of the mutant nodes of Ranvier is an increased refractory period between two successful nerve stimulations (24). In the mutant, this period is progressively increased between days 10-20, in contrast to normal, which stays constant or even slightly decreases (Fig. 7). Thus, a state of functional denervation leading to muscle inactivity is responsible for the morphological changes and plasticity of the motor endplate in motor endplate disease. We have found using nerve grafting experiments that the primary abnormality probably does not reside in the Schwann cell but in the motoneurone itself (12).

Muscular dysgenesis :

A second example of genetic abnormality of motor endplate morphogenesis is offered by the mutation muscular dysgenesis (mdg) in the embryo. The homozygous mutant embryo (mdg/mdg ; Fig. 4) is characterized by a state of total muscle inactivity as soon as the myotubes are formed (25-27), due to excitation-contraction uncoupling. The mutation is lethal at birth, because of the lack of respiratory movements.

In all striated muscles, the innervation proliferates. At the electron microscope level, many growth cones can still be observed at day 18 in the mutant diaphragm muscle (Fig. 8), in contrast to normal muscle. The innervation pattern, visualized by silver nitrate impregnation techniques, is simple in the normal diaphragm (28) and restricted to a central region of the muscle, with the main trunk of the phrenic nerve and short secondary and tertiary branches. In the mutant muscle, the innervation pattern is plexiform. The motor endplates in normal muscle are polyinnervated by a few axons with simple terminal arborizations. The regions of synaptic contact in the mutant muscle show a profuse axonal ultraterminal sprouting.

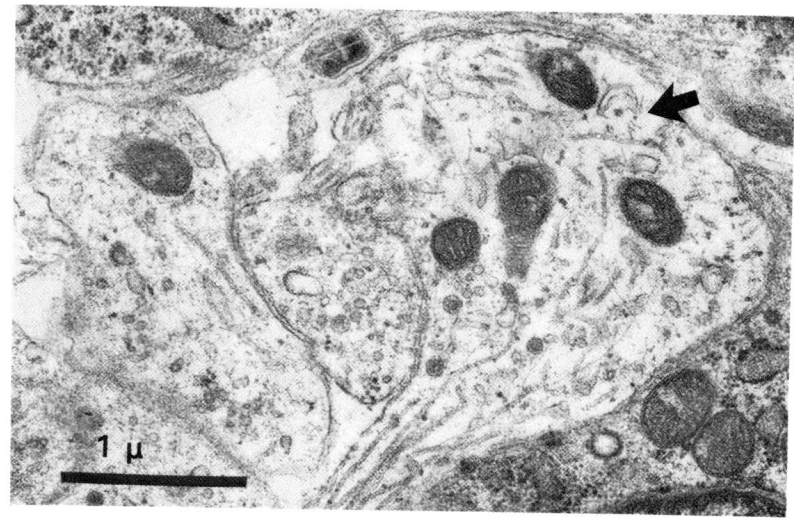

Fig. 8 : <u>Intense axonal growth in mdg/mdg muscle at embryonic day 18</u>. The growth cones are indicated by an arrow. Electron micrograph (Dr M. Pinçon-Raymond).

We have isolated and studied individual muscle fibers. Each normal myotube at E18 is innervated in a single synaptic region (6), demonstrated by autoradiography of ^{125}I α-BgTX bound to the AChR. Outside the synaptic region there is no extrasynaptic AChR. On the mutant myotubes, several junctional, AChR clusters are observed. We have electrophysiological evidence that these areas are sites of ACh release (29,30). Multifocal innervation is not found in any other model of muscle inactivity such as tetrodotoxin-induced inactivity in the rat embryo (31). Thus, muscle inactivity does not explain all the innervation abnormalities found in muscular dysgenesis.

Another important observation in <u>mdg/mdg</u> muscle is that basal lamina does not form on the mutant myotubes as early as it does on normal myotubes and never as well (5). Delayed and incomplete basal lamina formation may also be due to muscle inactivity, as shown <u>in vitro</u> (32). Synaptic basal lamina has been shown to control nerve terminal regeneration in adult muscle (33). Basal lamina may also play a role in earlier nerve-muscle interactions, after ther first contacts are established. Its function may be defective in muscular dysgenesis.

However, the situation may be even more complex. We have evidence that the mutant nerve is directly affected, independently of the muscle, in its capacity to recognize muscle or adhere to it. Recent studies (34) with chimaeric mice, combining normal and mutant tissues in one embryo, have shown that chimaeric diaphragm muscle could be of a rather constant phenotype all along its length, (that is predominantly mutant as determined by glucosephosphate isomerase electrophoresis), and have nevertheless independently, normal, restricted, or the mutant, broad innervation patterns as demonstrated by AChE cytochemistry. Thus, muscle genotype does not strictly control the nerve phenotype.

In conclusion, muscular dysgenesis is probably not a simple model of muscle inactivity alone. We propose that muscle inactivity is the cause of the immature features of the mutant motor endplate and that aberrant nerve-muscle interactions, but whatever they may be involving, lead to the state of multifocal innervation.

III. Is N-CAM THE MOLECULAR DETERMINANT OF MOTOR ENDPLATE MORPHOGENESIS ?

We would like in this section to address the following questions :

What are the molecular determinants of motor endplate morphogenesis ?

What is the relationship between the molecular determinants of shape and the molecular determinants of synaptic function ?

The molecular determinants of motor endplate morphogenesis are probably not organized basal lamina components, AChE nor AChR, because in the embryo, exploratory axons contact the primary myotubes before basal lamina is deposited and before AChE and AChR aggregates appear.

One obvious question is : Does a morphogenetic protein like N-CAM (35) fit into the developmental sequence of motor endplate morphogenesis ? It has been recently demonstrated by G.M. Edelman and his colleagues (36, 37) that some initial events of nerve-muscle contact in vitro require the presence and availability of N-CAM. Table 2 gives a summary of the properties of N-CAM in nervous tissues (see review in 38) and muscle relevant to nerve-muscle interactions. What is the function of N-CAM in the neuromuscular system ? Grumet et al (36) showed that N-CAM is expressed on chick myotubes in vitro. It is possible to obtain nerve-muscle contacts in vitro using co-cultures of spinal cord explants and myotubes, with neurite outgrowth and synapse formation. A perturbation experiment affecting these early interactions was carried out, by introducing into the culture medium anti-N-CAM

Table 2

Properties of N-CAM

NEURONS

Sialic acid-rich, cell surface glycoprotein
Undergoes embryonic to adult conversion with
  - decrease in sialic acid
  - increase in rate of binding
Mediates neuron-neuron adhesion (homophilic)
Implicated in morphogenesis

MUSCLE
Expressed on myotubes in vitro
Mediates adhesion between neurites and myotubes in vitro

Fab' fragments (37). By doing so, neurite fasciculation and adhesion between neurites and myotubes was reported to be perturbed. Time-lapse cinematography showed that neurites grow, but do not remain in contact with the myotubes and often retract giving a general aspect of limited neuritic outgrowth.

Our recent work (39) has been directed towards the biochemical characterization and the localization of N-CAM in developing and adult muscle in vivo.

N-CAM at the adult motor endplate :

We have searched for anti-N-CAM immunoreactive antigens in adult mouse muscle using SDS-PAGE techniques. Fig. 9 shows a 7.5 % polyacrylamide gel which has been loaded with detergent homogenates of sciatic nerve and diaphragm.

Immunoblotting was performed with a specific rabbit anti-N-CAM polyclonal antibody raised after injection of purified mouse brain N-CAM (40). In sciatic nerve, we found two immunoreactive bands, 140 Kd and 120 Kd ; in muscle we found essentially one, 140 Kd. One outcome of

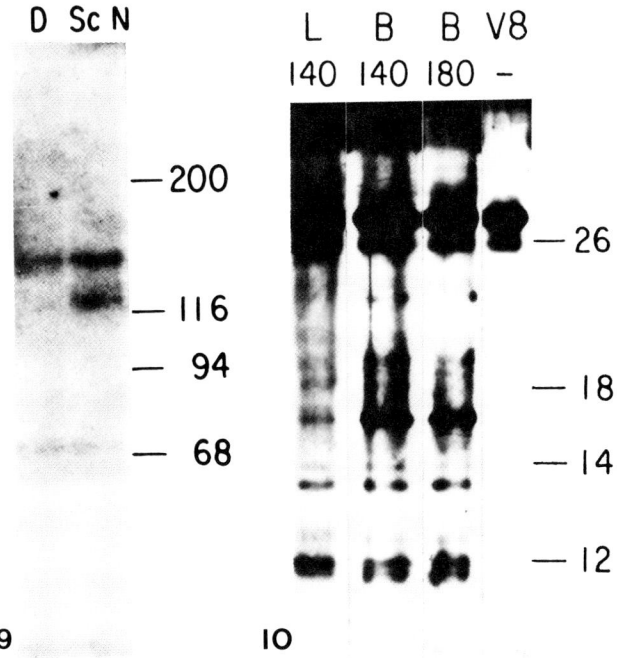

Fig. 9 : Immunoelectrophoretic analysis of N-CAM in adult muscle and nerve. Diaphragm muscle and sciatic nerve (adapted from 39).

Fig. 10 : The immunoreactive 140 Kd band in muscle is N-CAM. Peptide mappings of the muscle 140 Kd component and the brain 140 Kd and 180 Kd components. The fourth lane is the Staphylococcus aureus V8 protease (adapted from 39).

this experiment was that no nerve contamination could be detected in the muscle extract in our experimental conditions. We wanted to verify that the 140 Kd protein present in muscle was N-CAM. We performed a peptide mapping of the 140 Kd protein (41). After proteolytic digestion with the Staphylococcus aureus V8 protease of the limb muscle 140 Kd protein and the brain 140 and 180 Kd N-CAM, we obtained similar peptides (Fig. 10). These observations suggest that the immunoreactive 140 Kd protein in muscle is indeed N-CAM.

We used the polyclonal anti-N-CAM antibody to localize N-CAM in adult diaphragm muscle. Indirect immunofluorescence staining, using the rhodamine labeled second antibody, showed axon staining in axon bundles and N-CAM staining at the motor endplate (Fig. 11).

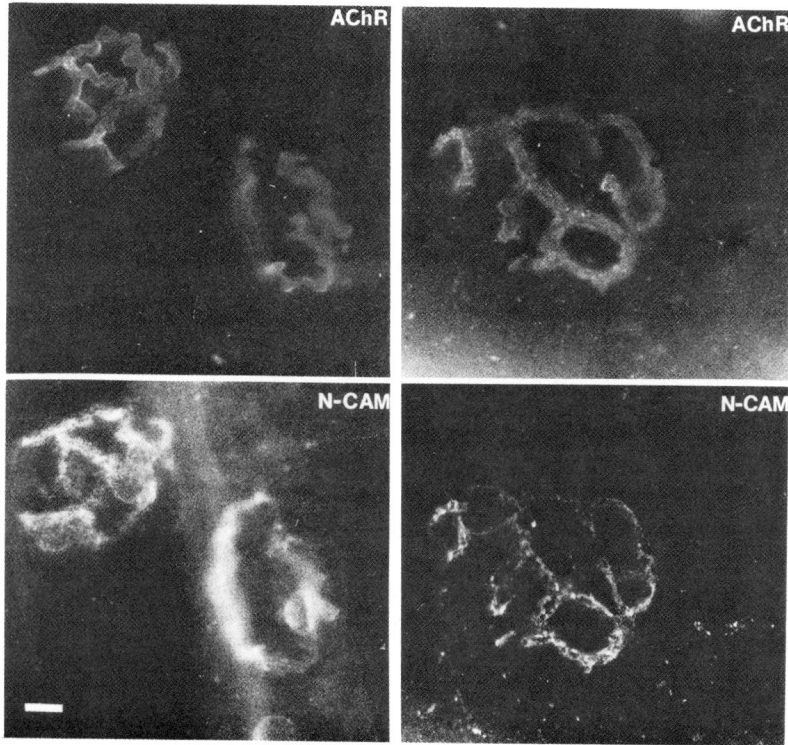

Fig. 11 : Colocalization of AChR and N-CAM at the adult mouse motor endplate. Two examples.

The N-CAM staining presents the characteristic structure of an adult motor endplate. Part of the N-CAM staining seems to correspond to axon terminal branches. It was important to know whether N-CAM and AChR are colocalized at the motor endplate. We performed double staining experiments, using fluorescein-labeled α-BgTX and the anti-N-CAM antibody revealed with rhodamine labeled goat anti-rabbit IgG (Fig. 11). Both stainings appear to be fairly well coextensive. We found that more than 90 % of about 300 MEP observed were double-stained (39 and Table 3). More details, corresponding most probably to the axon terminal branches, are found in the N-CAM stained regions.

Table 3

FITC -BgTX and N-CAM indirect immunofluorescence stainings at adult motor endplates (MEPs)

Double-stained MEPs	α-BgTX staining only	N-CAM staining only*
291 (91 %)	21 (7 %)	8 (2 %)

* staining pattern similar to authentic MEPs

Developmental aspects of N-CAM in muscle :

We analyzed extracts of E14, newborn and adult diaphragm muscles by SDS-PAGE techniques, loading the same amount of protein (100 µg) on each lane of the polyacrylamide gel (Fig. 12). The experiments showed that there is an embryonic increase of N-CAM between E14 and birth by a factor of about 5 and a postnatal decrease by a factor of about 30 (Table 4).

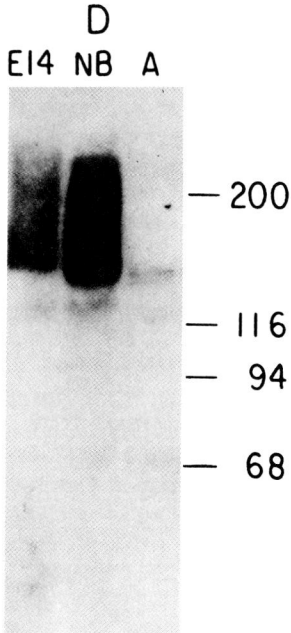

Fig. 12 : Immunoelectrophoretic analysis of developing muscle N-CAM. Embryonic day 14 (E14) newborn and adult (adapted from 39)

Table 4

Quantitation* of polyclonal anti-N-CAM immunoreactivity in muscle and spinal cord during development

	E14/Adult	Newborn/Adult
diaphragm	5.2	31
spinal cord	2.1	4.4

* quantitative autoradiography scanning. Same amounts of protein were loaded on 7.5 % polyacrylamide gels and immunoblotted with polyclonal anti-N-CAM (Summarized from 39).

The same occurs in spinal cord to a somewhat lesser extent, as already reported (42). Moreover, in developing muscle, N-CAM is detected as a polydisperse protein and, as in spinal cord, shows a conversion from an embryonic form, still present at birth, to an adult form. This polydispersity in embryonic muscle is due to the existence of polysialylated forms of N-CAM as previously observed for brain and spinal cord (42). Neuraminidase treatment converts polydisperse N-CAM into a single 140 Kd species (data not shown).

We found N-CAM concentrated at the immature motor endplates in newborn diaphragm muscle (39). Double staining experiments showed a colocalization of AChR and AChE, with coextensive fluorescein-labelled α-BgTX and N-CAM-rhodamine stainings. More than 90 % of over 500 newborn MEPs recognized by their α-BgTX staining were double-stained (Table 5). We also found N-CAM staining on axons and in contrast to adult muscle, on the surface of the myofibers (39).

Table 5

FITC α-BgTX and N-CAM indirect immunofluorescence stainings at newborn motor endplates (MEPs)

Double-stained MEPs	α-BgTX staining only	N-CAM staining* only
549 (92 %)	22 (4 %)	25 (4 %)

* staining pattern similar to authentic MEPs (summarized from 39)

Is expression of muscle N-CAM under the control of the nerve ?

We approached this problem with an experimental perturbation of nerve-muscle interactions : total denervation by section of the adult sciatic nerve. One week after denervation, N-CAM had sharply decreased at the motor endplate. In the gastrocnemius muscle, N-CAM at the motor endplate, recognized by fluorescein-labelled α-BgTX binding, was most of the time below the level of detection by our immunofluorescent technique. A statistical analysis performed on muscle, two weeks after denervation, showed that 14 % of the motor endplates are double-stained, in contrast to more than 80 % in normal, contralateral muscle (33 and Table 6). However, in contrast with this decrease at the motor endplate, N-CAM increased within denervated muscle.

Denervated soleus, extensor digitorum longus, and gastrocnemius muscles showed high levels of 140 Kd N-CAM, in contrast to normal where N-CAM is barely detectable (Fig. 13). There may be up to a 50-100 fold increase in N-CAM in denervated muscle as shown by densitometric scannings. Immunofluorescence studies of muscle cross-sections (39) showed that the increase of N-CAM in denervated muscle essentially occured inside the myofibers, but also on their surface and in the endomysium.

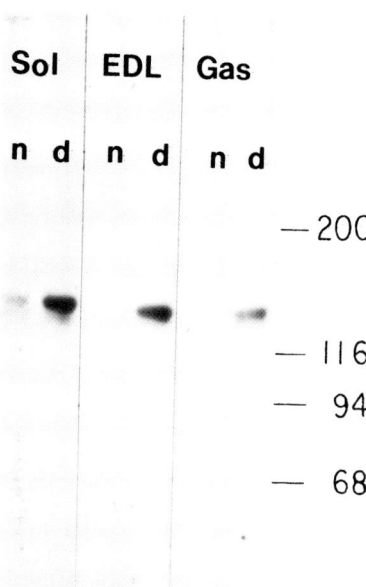

Fig. 13 : Increase in muscle N-CAM after denervation. Soleus (Sol), extensor digitorum longus (EDL) and gastrocnemius (Gas) muscles, one week after section of the sciatic nerve. Adult mouse.

Table 6

FITCα-BgTX and N-CAM indirect immunofluorescence stainings at adult control and denervated* motor endplates

	Double-stained MEPs	α-BgTX staining only	N-CAM staining only
Control (n=22)	18 (82 %)	0 (0 %)	4 (18 %)
Denervated (n=49)	7 (14 %)	38 (78 %)	4 (8 %)

* 2 weeks (taken from 39)

DISCUSSION

Many important issues remain to be elucidated regarding motor endplate morphogenesis. They involve general aspects of the differentiation of the motoneuron, the Schwann cell, the muscle cell and the fibroblast which cooperate and influence each other during the embryonic and postnatal period of acquisition of MEP shape and function. A non-exhaustive list of critical points which are of general interest in the field of nerve-muscle developmental neurobiology is given in Table 7. As regards the first steps of motor endplate morphogenesis, it is clear that the molecular determinants of recognition between the growth cones of exploratory axons and the primary myotubes are not related directly to any organized BL component (4 and this report). It cannot be ruled out, however, that BL constituents not yet part of an electron dense BL act as recognition molecules.

Such a role of BL constituents in early development would be perfectly consistent with the established function of mature synaptic basal lamina in directing nerve terminal regeneration in adult muscle (33). However, we would be led to hypothesize that these BL components can be intimately structurally related to the early myotube plasma membrane or even be an integral membrane component. In this respect, AChE happens to be a good example of such a molecular BL component undergoing biochemical and cell location changes, as muscle differentiation proceeds. In adult mature muscle, AChE is found concentrated at the

Table 7

Central issues in nerve-muscle interactions

1- Neuroblast and myoblast migration and differentiation

- Factors affecting motoneuron differentiation
- Role of extracellular matrix
- Guides for axon growth
- Formation of axon fasciculi.

2- Recognition and adhesion of nerve and muscle at new synaptic sites

- Molecular determinants of initial contact
- Role of muscle basal lamina
- Role of nerve and muscle activity on the formation of the initial contact
- Motoneuron and innervation type.

3- Supramolecular structure of synapse. Maintenance and stabilization

- Relationship between basal lamina, synaptic sarcolemma and cytoskeletal network
- Synthesis and turnover of synaptic site components.

4- Regression of polyinnervation and role of activity. Plasticity of the motor endplate

- Axonal growth factors
- Elimination of polyneuronal innervation
- Survival factors for motoneurons.

MEP, and is a tailed, asymmetric 16 S species mainly extractable under high salt conditions and preferentially associated with the muscle synaptic BL (43). AChE is also present in myoblasts and early myotubes essentially as a hydrophobic, integral protein (44). It is interesting to observe that very early accumulations of AChE appear at the time, or immediately after, the first axon-myotube contacts are formed, even before junctional AChR clusters are detected. It is not known however, whether these AChE spots correspond to specialized areas of plasma membranes or to very early BL patches, although the second possibility is the least probable given the mainly hydrophobic properties of AChE in 12-13 day old embryonic muscle (5 and unpublished observations).

Another important issue in the early steps of synapse formation is relative to the step next to recognition, i.e. adhesion. We have very little knowledge of the adhesion mechanism. Recognition and adhesion may be very related mechanisms, but may also be mediated by different molecular components. Our results suggest a definitive role of N-CAM as a recognition and adhesion molecule at synaptic sites in vivo and in vitro.

Muscular dysgenesis offers a genetic perturbation of the very early nerve muscle interactions in the mouse embryo. There is an obvious muscle inactivity component in the innervation abnormalities of mdg/mdg muscle. A number of the immature features of the muscle and the motor endplate found in the mutant have already been described in other models of muscle inactivity such as botulinum toxin, d-tubocurarine (45) or α-BgTX induced paralysis of the mouse (31) or the chick embryo (46). In all cases, the atrophic muscle presents signs of delayed differentiation and also impairment of the development of the motor endplate. However, it should be mentioned that these experiments were performed after the first contacts between axons and myotubes had occured. For example, in the chick embryo, neurogenic movements are already observed as early as the 4th day of incubation (47,48). This was the time chosen to inject α-BgTX in the work by Giacobini et al (46). Thus, muscular dysgenesis is unique in two respects. First, muscle inactivity is a permanent feature of the myotubes as soon as they are formed causing a natural perturbation of nerve-muscle interactions at the earliest stage of muscle formation. Second, as already mentionned, results obtained with chimaeric mice tend to prove that nerve can be independently affected by the mutation (34). Thus, both early muscle inactivity and defective differentiation of the motoneuron may cause, in a yet undefined proportion, the axonal abnormalities observed in mdg/mdg embryos, namely collateral sprouting, irregular swellings, disorganization of microtubules and neurofilaments and defective Schwann cell-axon interactions (M. Pinçon-Raymond and F. Rieger unpublished observations) and also the terminal multifocal innervation of the myotubes. How both factors cooperate in mdg/mdg embryos to provoke the neurobiological landmarks of muscular dysgenesis in the mutant neuromuscular system is a matter of current research in our laboratory. Our working hypothesis is that the delayed muscle differentiation and the immature features of the motor endplate in mdg/mdg embryos are mainly due to muscle inactivity, caused by excitation-contraction uncoupling (26, 27, 49). Our recent observations on tetrodotoxin treated 12-13 day old mouse embryos show similar impairment of muscle and motor endplate morphogenesis (F. Rieger, M. Pinçon-Raymond, L. Houenou and J. A. Harris ; in preparation) but presynaptic effects are less obvious and in particular multifocal innervation is not observed. Thus, multifocal innervation may be essentially related to a basic defect in nerve-muscle interactions, probably involving defective BL formation.

Muscle inactivity explains most of the morphological, ultrastructural and biochemical changes of the motor endplate observed later during development in motor endplate disease. The abnormalities observed on both sides of the nerve-muscle junction, namely ultraterminal nerve sprouting and atrophic and denervation-like changes of muscle and functional proteins such as AChR and AChE have been found in other studies to be mainly related to muscle inactivity (12). In motor endplate disease, muscle inactivity is a secondary consequence of a primary defect of axonal conduction, with failures of nerve impulse propagation (24) associated with widened nodes of Ranvier and presumably unsuccessful compensatory increases in the amount of voltage-dependent Na^+ Channels (23). The cell primarily affected by the mutation is the motoneurone, as shown by nerve grafting experiments of mutant sciatic nerve into normal recipients leading to regeneration of axons normally myelinated by the grafted med/med Schwann cells (12). The presence of such clusters suggest that active release of ACh may occur at these sites of close axon-myofiber contact. This suggests that there is tentative formation of functional contact in progressively inactive med/med muscle. We can speculate on the ability of the motoneuron to compensate for the progressively less efficient original innervation by extending the innervation territories themselves. These aspects of motor endplate plasticity in motor endplate disease may be suggestive of similar processes during development of normal muscle present in order to match the motor input and the physiological requirements of the developing and

aging muscle.
The fact that N-CAM is concentrated at the vertebrate neuromuscular junction suggests that it is one of the molecular determinants of motor endplate morphogenesis. This is the first known example of a colocalization of a morphogenetic protein, N-CAM and a functional protein, AChR. This colocalization has been observed in developing mouse muscle at birth and at the adult motor endplate. Similar observations have been made at the developing and adult chick MEP, in both fast-twitch (posterior latissimus dorsi, PLD) and slow-tonic (anterior latissimus dorsi, ALD) muscles and also at the adult frog neuromuscular junction (unpublished observations).

More work is required to determine at which precise step of development N-CAM is accumulated at the motor endplate and how it compares with the temporal inductions of AChE, AChR and known elements of synaptic BL. One fascinating possibility is that N-CAM itself is the **organizer** protein for the establishment of stable functional nerve-muscle contacts and further maintenance and maturation. It will be interesting to determine the precise molecular basis for motor endplate shape and explain the great morphological differences between mouse and frog muscle, for example. Precise early localizations of N-CAM agregates are presently underway.

We have studied different aspects of N-CAM localization and biochemistry during development which are summarized in Table 8. We have found that N-CAM distribution on myofibers becomes restricted to motor

Table 8

N-CAM in Muscle

- increases in amount from E14 to birth, then decreases
- is colocalized with AChR at the motor endplate in adult and newborn tissue
- undergoes E-A conversion

endplate during synapse stabilization and that the E to A conversion of N-CAM is concurrent with synapse elimination. Potential functions for N-CAM during development are listed in Table 9 and will not be further discussed. However, it is interesting to stress the fact that the near total disappearance of N-CAM from the surface of the muscle fiber between birth and the adult stage is a good illustration of one of the mechanisms postulated by Edelman (38) to be involved in cell-cell interactions, that is **polarity modulation.**

Table 9

Potential Functions of N-CAM in Muscle

1) Initial binding of motor and sensory neurons to muscle cells early in fetal development
2) Organization of AChR and components of synaptic basal lamina
3) Synapse stabilization by selective strengthening of axon-myofiber bonds
4) Regeneration of neuromuscular connections following injury

Denervation induces increased levels of N-CAM inside the myofibers and return of N-CAM to their surface. Such changes may possibly render the myofiber competent for reinnervation, which would be able to occur at both the old sites of innervation and at other sites on the muscle fiber, as realized in ectopic innervation (50).

Our next efforts will be directed towards answering key questions concerning motor endplate morphogenesis related to biosynthesis and cellular compartmentalization of nerve and muscle N-CAM during normal or abnormal development : Are AChR, AChE and N-CAM coregulated ? Does N-CAM act as an inducer and/or stabilizer of the nerve-muscle contact ? Is N-CAM involved in the pathogenesis of some genetic diseases affecting nerve-muscle interactions ?

ACKNOWLEDGEMENTS

Most of the work on muscle structure and mutants reviewed here was performed in a stimulating and productive collaboration with Dr M. Pinçon- Raymond in Paris. I thank Drs M. Lazdunski, M. Fardeau, R. Bournaud, J. Powell, A. Peterson, R.L. Sidman for many fruitful discussions related to the neuromuscular mutations in the mouse. I am happy to acknowledge their essential contribution. The work on N-CAM was mostly performed in Dr G.M. Edelman's laboratory at the Rockefeller University. I am grateful to Dr M. Grumet with whom the work was carried out and to Dr B. Cunningham for his interest and friendly comments. I thank Dr G. Edelman for having made most of this work possible.

REFERENCES

1. Couteaux R (1981). Structure of the subsynaptic sarcoplasm in the interfolds of the frog neuromuscular junction. J. Neurocytol. 10 : 947.
2. Fertuck H.C. and Salpeter MM (1976). Quantitation of junctional and extrajunctional acetylcholine receptors by electron microscope autoradiography after ^{125}I-alpha-bungarotoxin binding at mouse neuromuscular junctions. J. Cell Biol. 69 : 144.
3. Anderson M.J. and Fambrough DM (1983). Aggregates of acetylcholine receptors are associated with plaques of a basal lamina heparan sulfate proteoglycan on the surface of skeletal muscle fibers. J. Cell Biol. 97 : 1396.
4. Sanes J.R. and Chiu AY (1983). The basal lamina of the neuromuscular junction. Cold Spring Harbor Symposia, 607.
5. Rieger F., Powell J.A. and Pinçon-Raymond M (1984). Extensive nerve overgrowth and paucity of the tailed, asymmetric form (16 S) of acetylcholinesterase in the developing skeletal neuromuscular system of the dysgenic (mdg/mdg) mouse. Abnormal formation of muscle basal lamina. Dev. Biol. 101 : 181.
6. Powell J.A., Rieger F., Blondet B., Dreyfus P. and Pinçon-Raymond M (1984). Distribution and quantification of ACh receptors and innervation in diaphragm muscle of normal and mdg mouse embryos. Dev. Biol. 101 : 168.
7. Anderson M.J., Klier F.G. and Tanguay KE (1984). Acetylcholine receptor aggregation parallels the deposition of a basal lamina proteoglycan during development of the neuromuscular junction. J. Cell Biol. 99 : 1769.
8. Bennett M.R. and Pettigrew A (1974). The formation of synapses in striated muscle during development. J. Physiol. London 241 : 515.
9. Betz W.J., Caldwell J.H. and Ribchester RR (1979). The size of motor units during post-natal development of rat lumbrical muscle. J. Physiol. London 297 : 463.

10. Miyata Y. and Yoshioka K (1980). Selective elimination of motor nerve terminals in the rat soleus muscle during development. J. Physiol. London 309 : 631.
11. Pinçon-Raymond M. and Rieger F (1981). The motor innervation of skeletal muscles in the "motor endplate disease" mutant mouse (med and med^J alleles) 40 : 189.
12. Rieger F. and Pinçon-Raymond M (1984). Normal and defective differentiation of the nodes of Ranvier in neurological mutants of the mouse. In Adv. Cell Neurobiol. 6 : 273.
13. Duchen L.W. and Strich SJ (1968). The effects of botulinum toxin on the pattern of innervation of skeletal muscle in the mouse. Q. J. Exp. Physiol. 53 : 84.
14. Van Essen D. and Janssen JKS (1974). Reinnervation of the rat diaphragm during perfusion with α-bungarotoxin 91 : 571.
15. Bourgeois J.P., Betz H. and Changeux JP (1978). Effets de la paralysie chronique de l'embryon de Poulet par le flaxedil sur le developpement de la jonction neuromusculaire. C R Acad Sci Ser. D. 285 : 773.
16. Brown M.C. and Ironton R (1978). Sprouting and regression of neuromuscular synapses in partially denervated mammalian muscles. J. Physiol. London 278 : 325.
17. Srihari T. and Vrbova G (1978). The role of muscle activity in the differentiation of neuromuscular junctions in slow and fast chick muscles. J. Neurocytol. 7 : 529.
18. Wernig A., Pecot-Dechavassine M. and Stover H (1980). Sprouting and regression of nerve at the frog neuromuscular junction in normal conditions and after prolonged paralysis with curare. J. Neurocytol. 9 : 278.
19. Fishman M.C. and Nelson PG (1981). Depolarization-induced synaptic plasticity at cholinergic synapses in tissue culture. J. Neurosci. 1 : 1043.
20. Duchen L.W., Searle A.G. and Strich SJ (1967). An hereditary motor endplate disease in the mouse. J. Physiol. London 189 : 4P.
21. Duchen LW (1970). Hereditary motor endplate disease in the mouse : light and electron microscopic studies. J. Neurol. Neurosurg. Psychiatry 33 : 239.

22. Pinçon-Raymond M., Ludosky M.A., Cartaud J. and Rieger F (1983). Intense ultraterminal sprouting from motor nerves and ultrastructural aspects of the neuromuscular junction and non-junctional sarcolemma of the soleus (slow-twitch) muscle in motor endplate disease in the mouse. Tissue and Cell 15 (2) : 205.
23. Rieger F., Pinçon-Raymond M., Lombet A., Ponzio G., Lazdunski M. and Sidman RL (1984). Paranodal dysmyelination and increase in tetrodotoxin binding sites in the sciatic nerve of the motor endplate disease (med/med) mouse during postnatal development. Dev. Biol. 101 : 401.
24. Angaut-Petit D., McArdle J.J., Mallart A., Bournaud R., Pinçon-Raymond M. and Rieger F (1982). Electrophysiological and morphological studies of a motor nerve in motor endplate disease of the mouse. Proc. Roy. Soc. London B. Ser. 215 : 117.
25. Gluecksohn-Waelsch S (1963). Lethal genes and analysis of differentiation. Science 142 : 1269.
26. Pai AC (1965a). Developmental genetics of a lethal mutation (mdg) in the mouse. I genetic analysis and gross morphology. Dev. Biol. 11 : 82.
27. Pai AC (1965b). Developmental genetics of a lethal mutation (mdg) in the mouse. II Developmental analysis. Dev. Biol. 11 : 93.
28. Rieger F. and Pinçon-Raymond M (1981). Muscle and nerve in muscular dysgenesis in the mouse at birth : sprouting and multiple innervation. Dev. Biol. 87 : 85.
29. Bournaud R (1980). Electrophysiological studies of neuromuscular transmission in muscular dysgenesis in the mouse (mdg/mdg). Abst. First. Meet. Int. Soc. Develop. Neuroscience. Strasbourg p 190.
30. Rieger F., Bournaud R., Pinçon-Raymond M., Dreyfus P. and Blondet B (1980). Neurobiological defects of muscle and nerve in murine muscular dysgenesis (mdg/mdg). Abst. Am. Meet. Soc. Neuroscience cincinnati, Ohio.
31. Braithwaite A.W. and Harris AJ (1979). Neural influence on acetylcholine receptor clusters in embryonic development of skeletal muscle. Nature (London) 279 : 549.
32. Sanes J.R. and Lawrence JC Jr (1983). Activity-dependent accumulation of basal lamina by cultured rat myotubes. Dev. Biol. 97 : 123.

33. Sanes J.R., Marshall L.M. and McMahan UJ (1978). Reinnervation of muscle fiber basal lamina after removal of myofibers. Differentiation of regenerating axons at original synaptic sites. J. Cell Biol. 78 : 176.
34. Rieger F., Cross D., Peterson A., Pinçon-Raymond M. and Tretjakoff I (1984). Disease expression in $\pm/\pm$ ⟷ mdg/mdg mouse chimaeras : evidence for an extramuscular component in the pathogenesis of both dysgenic abnormal diaphragm innervation and skeletal muscle 16 S acetylcholinesterase deficiency. Dev. Biol. 106 : 296.
35. Edelman G.M. (1983). Cell adhesion molecules. Science 219 : 450.
36. Grumet M., Rutishauser U. and Edelman GM (1982). Neural cell adhesion molecule is on embryonic muscle cells and mediates adhesion to nerve cells in vitro. Nature (London) 295 : 693.
37. Rutishauser U., Grumet M. and Edelman GM (1983). Neural cell adhesion molecule mediates initial interactions between spinal cord neurons and muscle cells in culture. J. Cell Biol. 97 : 145.
38. Edelman GM (1984). Modulation of cell adhesion during induction, histogenesis and perinatal development of the nervous system. Annu. Rev. Neurosci 7 : 339.
39. Rieger F., Grumet M. and Edelman GM (1985). N-CAM at the vertebrate neuromuscular junction. J. Cell Biol. In press.
40. Edelman G.M. and Chuong CM (1982). Embryonic to adult conversion of neural cell adhesion molecules in normal and staggerer mice. Proc. Natl. Acad. Sci. USA 79 : 7036.
41. Cleveland D.W., Fischer S.G., Kirshner M.W. and Laemmli UK (1977). Peptide mapping by limited proteolysis in sodium dodecyl sulfate and analysis by gel electrophoresis. J. Biol. Chem. 252 : 1102.
42. Chuong C.M. and Edelman GM (1984). Alterations in neural cell adhesion molecules during development of different regions of the nervous system. J. Neurosci. 4 : 2354.
43. Dreyfus P.A., Rieger F. and Pinçon-Raymond M (1983). Acetylcholinesterase of mammalian neuromuscular junctions : presence of tailed asymmetric acetylcholinesterase in synaptic basal lamina and sarcolemma. Proc. Natl. Acad. Sci. USA 80 : 6698.

44. Rieger F., Koenig J. and Vigny M (1980). Spontaneous contractile activity and the presence of 16 S form of acetycholinesterase in rat muscle cells in culture : reversible suppressive action of tetrodotoxin. Dev. Biol. 76 : 358.
45. Drachman DB (1967). Is acetylcholine the trophic neuromuscular transmitter ? Arch. Neurol. (Chicago) 17 : 206.
46. Giacobini G., Filogamo G., Weber M., Boquet P. and Changeux JP (1973). Effects of a snake α-neurotoxin on the development of innervated skeletal muscles in chick embryo. Proc. Natl. Acad. Sci. USA 70 : 1708.
47. Alconero BB (1965). The nature of the earliest spontaneous activity of the chick embryo. J. Embryol. Exp. Morphol. 13 : 225.
48. Ripley K. and Provine R (1972). Neural correlates of embryonic motility in the chick. Brain Res. 45 : 127.
49. Klaus M.M., Scordilis S.P., Rapalus J.M., Briggs R.T. and Powell JA (1983). Evidence for dysfunction in the regulation of cytosolic Ca^{2+} in excitation-contraction uncoupled dysgenic muscle 99 : 152.
50. Weinberg C.B. and Hall ZW (1979). Junctional form of acetylcholinesterase restored at nerve-free end plates. Dev. Biol. 68 : 631.

Molecular Determinants of Animal Form, pages 423-433
© 1985 Alan R. Liss, Inc.

POSITIONAL INFORMATION AND PATTERN FORMATION

L.Wolpert
Department of Anatomy and Biology as Applied to Medicine
The Middlesex Hospital Medical School
London, UK

ABSTRACT Pattern formation in chick limb bud
development is considered in terms of models based on
positional information and prepatterning.
Experimental evidence on the effects of the polarizing
region on the humerus suggest that a prepattern
mechanism is involved as the humerus is neither
duplicated nor eliminated when a polarizing region is
placed at various positions along the antero-posterior
axis. Positional information may modify the basic
unit set up by a prepatterning mechanism. Models for
the encoding and interpretation of positional
information are considered. A new class of model
provides a simple way of specifying rectangular
patches. The pigment patterns seen in bird plumage
are briefly discussed in terms of positional
information.

The concept of positional information suggests that
cells have an intrinsic record of their position
that effectively gives them an address (1,2). This
record, the positional value, is used by the cells to
determine cellular behaviour according to the cells'
genetic constitution and developmental history. This is
called the interpretation of positional value. It is as
if the cells were in a coordinate system and the overt
pattern emerges from the process of interpretation. Thus
the pattern need not be isomorphic with a set of graded
positional values. A specified cell type, or boundary

[1]This work was supported by the Medical Research
Council

between cell types, could arise sharply at any point. By
contrast, a prepattern mechanism could provide distribution
of a morphogen which is isomorphic with the overt pattern
of formed structures. For example, a wave-like pattern of
morphogen could give rise to a series of repeated
structures if the structures developed at a concentration
above a threshold, near say, the peak of the waves. In
this case, each structure is equivalent. If a set of
repeating structures were specified by positional
information they would be non-equivalent.

Limb Morphogenesis

A model for chick limb morphogenesis suggests
that positional values along the two main axes, the
proximo-distal and antero-posterior are specified by two
different mechanisms(3). This specification occurs in
the progress zone at the tip of the bud. For the proximo-
distal axis it is suggested that position is specified by
how long cells remain in the progress zone whereas for the
antero-posterior axis it is suggested that there is a
signal, possibly a diffusible morphogen from the posterior
margin of the bud.

While there is quite good evidence for both these
mechanisms it is also held that positional information may
only act on a prepattern(4). The idea would be that the
basic form of the cartilaginous elements, such as the
humerus, radius and ulna and digits are laid down by a
prepattern type of mechanism and that positional
information effectively names the elements and modifies
them. The best evidence that some sort of prepattern may
be present is that when the cells of the early limb bud are
disaggregated, reaggregated and placed in a jacket of limb
bud ectoderm they can form quite good digits. This argues
against the idea that the digits are specified by a signal
from the polarizing region. A prepattern mechanism for
the limb could be provided by a single wave for the
humerus, two for the radius and ulna, and so on.

Most studies on the action of the polarizing region
have concentrated on the digits because they provide good
markers. For example there is very good evidence for a
quantitative signal from the polarizing region specifying
the nature of the digits - that is digit 2, 3, or 4.
Typically, if a polarizing region is grafted to the
anterior margin the pattern of digits is 432234 whereas
that in the normal limb is 234. If the signal is

attenuated, then 32234 or 2234 may result. Similar results have been obtained with local application of decreasing concentrations of retinoic acid (5). There is, however, no evidence that the digits themselves are specified by the polarizing region. Whenever new digits form this is associated with widening of the limb bud: on a general prepattern basis this provides for the formation of another wave. It has been found that for each additional digit the limb bud widens by about 150um from an initial width of about 1000um (Tickle & Stein, personal communication). This suggests that if a prepattern were involved the wave length for a digit would be about 150um.

In order to investigate this problem further we have concentrated on the humerus. If the humerus were specified by the signal from the polarizing region then it should be possible to obtain two humerii if the additional polarizing region is grafted early enough. In terms of a model based on a diffusible morphogen the limb might have to widen substantially for an additional humerus to form. For unless widening occurred the concentration of the morphogen would rise above the threshold for the humerus and no humerus at all, or a very truncated one should form, since the humerus on most fate maps corresponds to an antero-posterior position equivalent to digits 2 and 3. On a prepattern model the wavelength for a humerus would be expected to be about 500um and thus this mechanism also has the requirement for a a substantial widening to fit another wavelength in.

We have grafted a polarizing region to various positions along the anterior posterior axis at very early stages of limb development such as stage 16 when there is no visible sign of a limb bud (Wolpert & Hornbruch, unpublished). When grafted to an anterior position, duplication of the ulna and digits has been obtained but in no case was the humerus duplicated. When the polarizing region was placed in a more central position a duplicated humerus did often develop but this was always associated with the complete splitting of the limb bud into two separate limb buds. The length and width of the humerus was not significantly altered in any of the experiments. The results show that it has not been possible to duplicate the humerus within the domain of a single bud. However if the domain of the limb bud is divided in two, each domain can give a humerus. These results suggest that while a polarizing region is necessary for humerus formation it is

not directly specified by positional information.
We thus suggest, like others, that there is a
prepattern mechanism for specifying the cartilaginous
elements (2). Before considering that mechanism it is
necessary to take into account some results of Solursh (6)
and our own which give a somewhat different picture of limb
bud development than that hitherto adopted. Using
Solursh's micromass culture technique we have found
(Cottrill, unpublished) that the progress zone gives rise
to a cartilaginous sheet, and that only when the cells are
taken from regions where they have left the progress zone
for some 24 to 48 hours, do the cells develop into both
cartilage and connective tissue. This implies that the
formation of the cartilage elements must be thought of in
terms of the inhibition of cartilage formation. Solursh
has provided evidence from in vitro studies that the
ectoderm inhibits cartilage formation and has suggested
this as the mechanism whereby cartilage formation is
confined to the core of the limb bud. Of course this does
not account for the sequence of cartilage elements such as
humerus, radius and ulna, wrist, digits.

Our (Arcuri, Murray, Stein & Wolpert) model for
specifying the cartilage rudiments makes use of a reaction
diffusion mechanism in which the cartilage inhibitor is
produced by the ectoderm and diffuses much faster than the
other reactant, the activator. By altering parameters in
the mechanism it is possible to generate a sequence of
waves with first one, then two, and three peaks – even four
and five peaks. These changes in parameter are thought to
be related to the proximo-distal positional value, which
could alter, for example, one of the diffusion constants.
It is a feature of this model that if the domain of the
early bud is divided in two a humerus could develop in each
domain with rather modest widening, whereas substantial
widening is required to generate two humerii.

Modes of Positional Information
In our model, positional information modifies the
elements laid down by the prepattern. We have drawn a
distinction between two modes of positional information(2).
In the simpler, and probably evolutionarily more
primitive mode, there is an isomorphism between the
positional values and their interpretation. A continuous
gradient in positional information is thus expressed
directly as a continuous gradient in some cellular

property, such as the wave of cell hypertrophy that spreads from the centre of the cartilaginous elements (7). The other mode is that in which a continuous set of positional values is interpreted in a discontinuous manner. This mode of interpretation must involve thresholds so that, at a particular positional value, a particular cellular activity is switched on. This is required for all patterns of any real complexity where sharp changes in the spatial distribution of cell behaviour are required. An example from cartilage morphogenesis is the highly localized changes involved in the morphogenesis of the tibia and fibula in the chick: the distal end of the tibia captures the distal end of the fibula (8).

If positional information is initially specified by the concentration of a chemical substance which could be as simple as hydrogen ions, calcium, cAMP or retinoic acid, two quite separate problems can be distinguished. First, how could these concentrations be converted into a stable record of position, and second how the cell could interpret this record of positional value in order to generate the observed patterns.

The Encoding of Positional Information
It is possible to design models in which a monotonic gradient in a morphogen can switch on, for example, particular genes at a threshold concentration (9), and which remain 'on' even when the morphogen is removed. The 'on' state is present in all those cells whose positional value is above the threshold. In more complex models of this type the gene can only be turned on at a particular concentration of the morphogen (10).

An interesting new possibility is that suggested by the model of Goldbeter & Koshland (reviewed 11) in which they consider the reversible covalent modification, such as phosphorylation, of an enzyme. This model can give particularly sharp thresholds at particular values of the ratio of the rates of action of the phosphorylating and dephosphorylating enzymes, such that all of the protein is phosphorylated. If there is a series of proteins, each with its own pair of enzymes, then it is possible that by increasing the concentration of just one signal substance, the different proteins become switched to the phosphorylated state in succession. First one is switched, then the second, and so on (Goldbeter, personal communication).

Good experimental evidence for some such system is the early insect embryo where it is thought that the gradient in the egg activates successive genes in the bithorax complex and thus defines the positional value of the thoracic and 7 abdominal segments (12). If the positional values are specified cumulatively then markers of all previous positions are specified at successive positions. Thus the first would be 1, the next 12, then 123 and so on. This is an example of a Gray code and represents a qualitative extension of the principle of the continuous mode.

Interpretation of the Positional Values

In principle, patterns of unlimited complexity can be generated with a mechanism based on positional information. All that is required is a specification of position of the cells that is sufficiently fine grained, and more important, a mechanism that provides for the appropriate interpretation of position at every address. Consider, for example, cells in a two dimensional grid in which each cell has a positional value and the formation of pigment in each cell can be turned 'on' or 'off'. This system could form an extremely large number of patterns but would, in principle, require each cell to have a complete table of interpretations for every positional value. For this reason I have tried to find a more efficient mechanism for interpretation, particularly one that did not require every cell to have a complete listing of the interpretation at each position. Such a listing cannot be avoided if patterns were such that cells could be 'on' or 'off' in every conceivable position. But this is not the case, for in most patterned systems cells are on and off in quite large coherent patches. For example the cartilage in early limb buds is in lumps. Similarly, an examination of pigment patterns in the head and neck regions of birds, suggests that the patterns are very often made up of up to 6 patches. The formation of patterns by the combination of a small number of discrete patches offers the possibility of a quite simple mechanism for interpretation which could generate such patterns with much greater economy than would be required by a complete listing of positions and interpretations.

Positional Information and Pattern Formation 429

The basis is the Gray code 1, 12, 123 ..... 12345678910 described above. We may now imagine that position is specified by these genes or marker molecules. Consider now a 10 by 10 grid of 100 cells, in which the horizontal position is specified by markers 1 to 10 and the vertical position by markers 1' to 10'. Each cell then has its position specified ', for example, the cell which is third along the horizontal axis and fifth along the vertical axis will be denoted 3, 5'. It will have the positional marker molecules 1,2,3, 1,' 2,' 3,' 4,' 5'.

The gene for pigment formation is assumed to be turned on or off depending on which of the positional molecules bind to sites adjacent to the gene. These molecules can either activate or inhibit pigment formation. The rules for this activation and inhibition constitute the mechanism whereby the pattern is generated.

Consider first a single binding site for activation. If this site is activated by, say, 3, then all cells whose horizontal positional value is 3 or greater will make pigment since all cells will have a positional marker 3 (Fig.1a). Now consider an inhibitory binding site which is dominant over activation. (I am indebted to Jesse Wolpert for this suggestion). If this inhibitory site is activated by, say, 5, then a vertical black strip will form (Fig.1b). Consider now an activation site that requires simultaneous binding of two positional markers. If this site is activated by the binding of both 3 and 4', then a rectangle will form (Fig.1c). Combining activation with inhibition more localized rectangles can be specified. For example, activation by 3 and 4' with either 5 or 7' inhibiting gives the rectangle bounded by 3, 5, 4', 7' (Fig.1d). Thus this mechanism provides an economical method for specifying rectangular patches and does not require the specification of every cell in the patch. The specification of single cells requires four positional markers, which should be compared with the minimum of two required on any system. In Fig.1e a number of patches are specified which could, with some licence, be thought of as the prepattern for cartilage in a limb. The specification of this pattern is 4' not 4 or 8'; 4 and 3' not 7 or 5', 4 and 7' not 7 or 9'; 9 and 2' not 4'; 9 and 5' not 7'; 9 and 8' not 10'.

The specification of less regular patterns is however more complex. For example, to divide the grid diagonally the specification is (2 and 1', not 2') (3 and 2', not 3') ...... (10 and 9' not 10'). This seems to be a rather

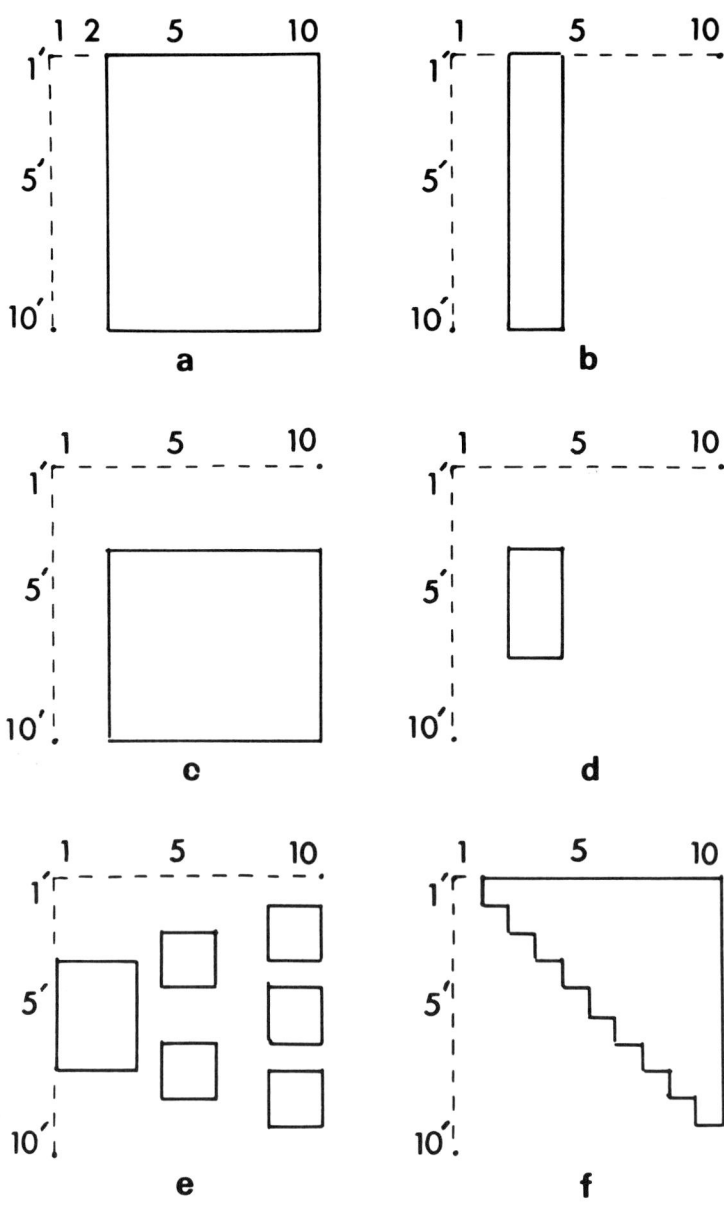

Figure 1

inelegant specification and thus makes an important point. What other ways are there of encoding position so that complex shapes are easily interpreted?

Feather Patterns

We do not know how fine grained the positional grid is nor how locally it can be interpreted. Pigment patterns in birds provide a fruitful area examining such questions. While feathers themselves are quite complex systems, whose initial spacing involves some sort of prepattern, we can examine the system at one level up, that is by considering how precisely each feather is specified. Have all the cells of each feather a unique positional value? Individual feathers can be strikingly different from their neighbours. In the tail of the lyre bird the 16 feathers are in mirror symmetry and the two outer and two inner are quite different from any other. In the Bower bird, the King of Saxony, just one feather on each side of the head becomes extremely long but it is not known whether it is always the same feather. A similar problem arises with the spots of budgerigars: each spot is made by one feather. In the newly hatched quail just three or four feathers posterior to the ear are pigmented. A glance at any book on pterylation shows remarkably detailed patterns characteristic of different birds. It remains to be determined just how precise such patterns are.

It must also be pointed out that the pattern of colouration of birds does not in general arise from changes in colouration of adjacent patches. In fact the pattern is, as it were, stencilled onto the feathers. Thus at the border between a green and black region the green feathers near the boundary will have black tips.

Divergence of Similar Structures

A major feature of many systems is the presence of repeated structures which, while conforming to a basic structure, can have rather different forms. Obvious examples from vertebrates are the fore and hind limbs, vertebrae, and feathers. In insects there is a clear divergence in the behaviour of segments and appendages from a basic form.

It is convenient to think of each of these cases in terms of a basic unit with a particular set of positional values. This set of positional values is the same in each unit but the interpretation of the positional values can be different in each unit because the units have been assigned further positional values in a larger coordinate system. This is particularly clear in insects where each segment may have the same set of positional values but the interpretation differs because of the action of homeotic genes (12). The antenna and leg use the same positional information but the antenna develops if the aristapaedia gene is turned on. In the chick wing and leg a similar situation pertains. The positional signals are the same in both, it is just the interpretation that differs.

It is a central and difficult problem to understand how this difference in interpretation is brought about. There are two main possibilities. In the first, each unit has its own unique set of molecules for interpretation. Thus there could be quite separate binding sites for the positional markers in leg and wing. This seems less likely than the second possibility which is that the postulated interactions are modified by positional molecules that characterise arm and leg. These molecules would need to be able to change the pattern of binding required for activation or inhibition. They might, for example, change an activation by a 3 into an activation by a 4 instead.

We all hope that an understanding of the homeotic genes, and particularly the homeobox segment, will provide answers. Might the products of different homeotic-like genes in wing and leg alter the interpretation of positional information?

ACKNOWLEDGEMENTS

I am grateful to Dr. C. Tickle for her comments and to Miss M.Maloney for preparation of the manuscript.

REFERENCES

1. Wolpert L (1971). Positional information and pattern formation. Curr Top Devel Biol 6:183.
2. Wolpert L, Stein WD (1984). Positional information and pattern formation. In Malacinski GM, Bryant SV (eds): "Pattern Formation. A Primer in Developmental Biology," New York: Macmillan Publishing Company, p 3
3. Wolpert L (1981). Pattern formation in limb morphogenesis. In Sauer HW (ed): "Progress in Developmental Biology" Stuttgart: Fisher,p 141

4. Ede DA (1982). Levels in complexity in limb mesoderm culture systems. In Yeoman MM, Truman DES (eds) "Differentiation in vitro" Cambridge: Cambridge University Press, p 207
5. Tickle C, Lee J, Eichele G (1985). A quantitative analysis of the effect of all-trans-retinoic acid on the pattern of chick wing development. Dev Biol 108 (in press)
6. Solursh M (1984). Ectoderm as a determinant of early tissue pattern in the limb bud. Cell Diff 15: 17
7. Rooney P, Archer C, Wolpert L (1984). Morphogenesis of cartilaginous long bone rudiments. In Trelstad RL (ed) "The Role of Extracellular Matrix in Development" New York: Alan R. Liss, Inc. p 305
8. Archer CW, Hornbruch A, Wolpert L (1983). Growth and morphogenesis of the fibula in the chick embryo. J Embryol exp Morph 75:101
9. Lewis J, Slack JMW, Wolpert L (1977). Thresholds in development. J Theoret Biol 65:579
10. Meinhardt H, (1978). Space-dependent cell determination under the control of a morphogen gradient. J Theoret Biol 74,307
11. Goldbeter A, Koshland Jr, DE (1982). Sensitivity amplification in biochemical systems. Quart Rev Biophys 15,3:555
12. Lawrence PA, Morata G (1983). The elements of the bithorax complex.Cell 35:595

Molecular Determinants of Animal Form, pages 435-454
© 1985 Alan R. Liss, Inc.

SHAPING OF THE BODY COLUMN IN HYDRA: IS A PRE-PATTERN NECESSARY?*

Hans R. Bode, Patricia M. Bode, Lorette C. Javois and Shelly Heimfeld

Developmental Biology Center and the Department
of Developmental and Cell Biology
University of California
Irvine, California 92717

ABSTRACT. Normal hydra will regenerate from excised pieces of tissue. Does the final shape of the body column of the animal arise by morphogenetic mechanisms alone or is a patterning process a necessary prerequisite? The alternatives are discussed. Evidence is presented that the head plays a role in determining the final dimensions of the body column and that a prepattern of the head appears before morphogenesis occurs. This suggests a prepattern is part of the process of shaping a body column.

INTRODUCTION

In trying to elucidate the processes underlying the development of animal form, a useful approach is to determine what general classes of mechanisms are most likely involved. A distinction which particularly concerns us is the following. Does the formation of a structure first require a process which lays down a prepattern to which the cells respond with an appropriate morphogenetic process? Or can the cells carry out morphogenesis based solely on their intrinsic properties and those of the external substrata, without any

---

*This work was supported by an NIH post-doctoral fellowship (GM 08513) to LCJ, and by NIH research grants (GM 29130 and HD 16440) to HRB.

information beyond an asymmetry which says where to start? People concerned with pattern formation have usually assumed the former to be true (e.g., 1, 2). In recent years, attention has focused on the alternate viewpoint due primarily to the efforts of Oster and his colleagues. They have developed mechanical models that render the need for prepatterns unnecessary (e.g., 3, 4). These two views can lead to rather different predictions as to what to expect in terms of the molecular determinants of form.

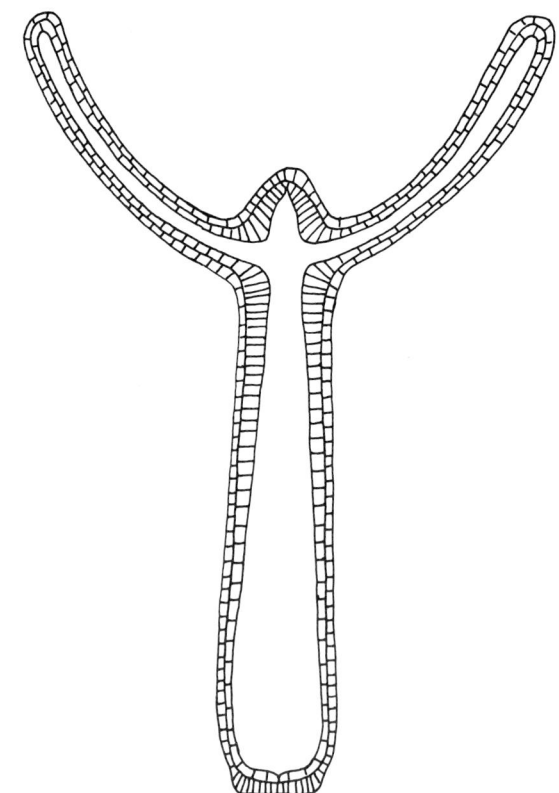

FIGURE 1: Longitudinal cross-section of a hydra indicating the two-layered structure.

Hydra is a useful animal to address this question for several reasons. It has been studied extensively to elucidate patterning events and processes (see 5 for review). It has a simple body plan, implying a smaller number of necessary morphogenetic processes, and it can be manipulated in many ways. As shown in Figure 1, the animal is essentially a tube that consists of two concentric epithelia, the ectoderm and the endoderm, surrounding a gastric cavity. The two layers are separated by the mesoglea, a basement membrane. Apically, the tube ends in a dome or cone, the hypostome, which is surrounded by a whorl of tentacles, usually six in number. This constitutes the head of the animal. At the other end, the tube closes to form the basal disc or foot.

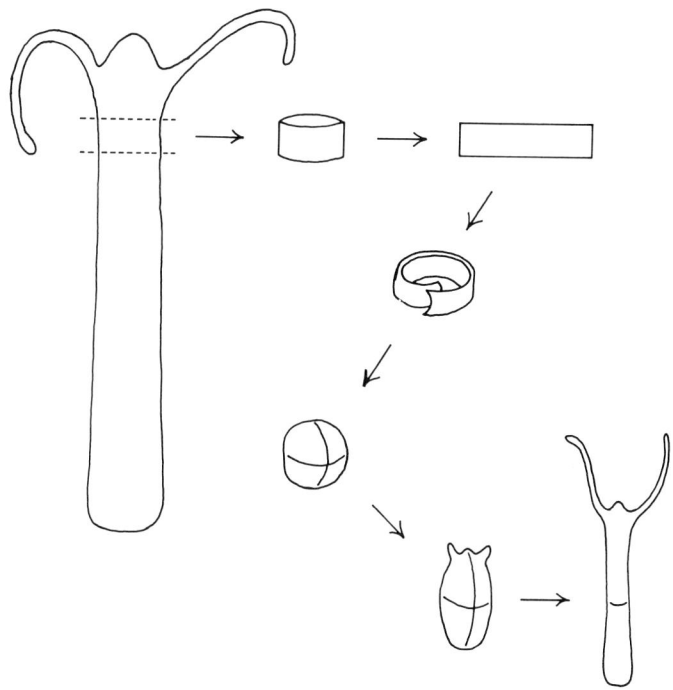

FIGURE 2: Regeneration of a hydra from an excised rectangular piece of tissue. A ring excised near the apical end of the body column is cut open yielding a rectangle. The ends of the piece scroll toward one another, forming a ball which is a hollow sphere after one day. By day two the sphere has elongated and tentacle rudiments are visible. By day six the final form is apparent.

That the number of processes required to develop this body are probably few can be illustrated by taking advantage of hydra's remarkable capacity for regeneration. Excision of a piece of the body column, as shown in Figure 2, leads to the following events. The tissue will first roll up into a ball. Within a few hours all the free edges have healed together, and a day later the tissue is a spheroid consisting of the normal two layers. In the next day or so, the sphere becomes a short stubby cylinder, and the first signs of head formation appear as tentacle rudiments. From days three to six the body column elongates further, reaching the final cylindrical shape. The shape of the hypostome becomes defined and the tentacles evaginate to their final length. At the opposite end the most pronounced effect is the differentiation of the epithelial cells of the ectoderm so that they secrete copious quantities of mucous. This disc of cells is the foot which serves to attach the animal to a substrate. An important aspect of the process is that the head always regenerates at the original apical end of the excised piece, and the foot at the basal end, which implies the body tissue has an intrinsic polarity.

Hence, there are really only two major morphogenetic events involved once wound healing is complete. One is the conversion of the sphere into a cylinder which has the proportions found in the final regenerate, and the other is the evagination of a set of tentacles in a ring below the hypostome. Minor changes occur in the hypostome area giving it the conical or domed form typical of the particular strain or species of hydra. In the following we will concentrate on the first of the two events, conversion of a sphere into a cylinder, and discuss if morphogenesis does or does not require preceding patterning.

DESCRIPTION OF THE FORMATION OF THE BODY COLUMN

To this end, detailed information is needed on how the shape changes procede and what affects the final proportions of the cylinder. Measurements of the length and circumference, in terms of the number of epithelial cells in each dimension, have been made during the process of regeneration (6). In addition, the effects of the shape and the size of the original piece on the final dimensions were examined. There are three general results.

The first is that the initial shape of the piece has no effect on the final shape of the cylinder (6). This result is illustrated in Figure 3. Three pieces of different shapes, each consisting of about 1/10th of the body column, were excised and allowed to regenerate. The pieces were (A) a rectangle derived from a ring of tissue so that the long dimension included the entire circumference, (B) a square, and (C) an axial rectangle whose long dimension included the entire length of the body column. One day later, the circumferential rectangle had formed a spheroid, the square formed a spheroid or an ovoid, while the axial rectangle had become a short cylinder.

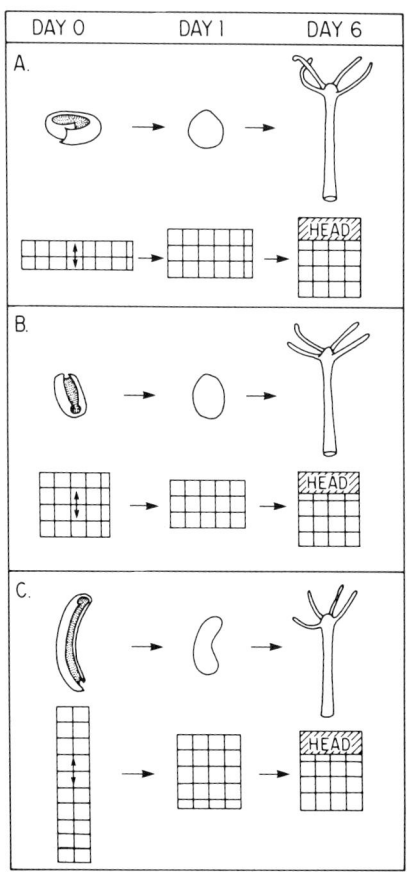

FIGURE 3: Cell rearrangements in the ectoderm during the regeneration of tissue pieces approximately the same size but having different shapes. The tissue is depicted schematically opened up into two dimensions. For simplicity, spheres are displayed as rectangles with the length equal to one-half the circumference. Each block represents 5 x 5 cells. The arrows on Day 0 indicate the long axis of the donor body column.

The main process in these initial shape changes was a rearrangement of the cells. In the circumferential rectangle, cells had shifted so that the number of cells along the axial direction was increased, and the number around the circumference decreased. The reverse occurred in the axial rectangle. Although the arrangement of cells in the square was close to that of the final cylinder, the cells also rearranged initially into a more spherical form. Over the next few days all three underwent further alterations in cell arrangement, until on day six they had cylindrical body columns with similar dimensions.

The second result indicates that the proportions of the final body column cylinder depend on the original size of the excised piece (6). Cylinders formed from small pieces were squatter, while those derived from large pieces were more elongate. Over a 20-fold size range the linear dimension of the length increased sevenfold while the circumference increased less than twice (Fig. 4a). Or phrased another way, the length/circumference of the final cylinder increased 4X over the size range. Measurement of the number of cells along the length and around the circumference indicated that half of the difference in body shape between small and large regenerates was due to cell arrangement. The number of cells along the length of the column relative to the number around the circumference increased by a factor of two from smaller to larger regenerates. The other half of the four-fold difference over the size range was due to changes in the shape of the cells. As shown in Fig. 4b, the epithelial cells in the small regenerates were wider and more squamous, while those in the large cylinders were narrower and more columnar.

Finally, a related result is the relationship of the head size to the body shape. Regardless of animal size, the diameter of the body column remains the same as that of the head (6). This result would be uninteresting if the proportions of the regenerates were wholly unaffected by size. However, the two regions do not constitute the same fraction of the total cells in the animal over the size range. The fraction of tissue that forms the head dome decreases as the size of the regenerating piece increases. The hypostome fraction decreases from 26% in the small animals to 8% in the largest, while the body fraction increases from 60% to 70% (7). The remainder is in the tentacles and basal disc. Despite these differences in proportions over the size range, the diameters of the head and body column are always the same. This suggests that the relationship of the two regions is not independent.

Shaping of the Body Column in Hydra 441

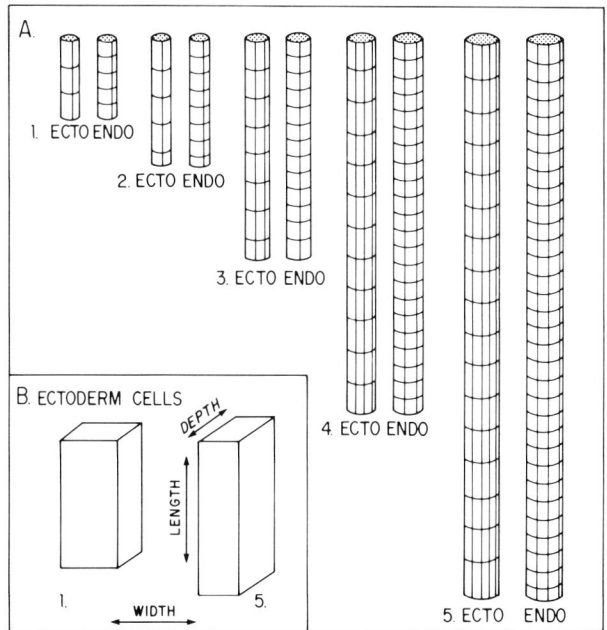

FIGURE 4: A. The three-dimensional shape of the extended body columns, and the shape and arrangement of cells in the ectoderm and endoderm for five different size categories of regenerates. Each block represents 5 x 5 cells, and its shape represents the average shape of an individual cell. Only one-third of the cells around the circumference are visible. B. Comparison of the three-dimensional shapes of ectodermal cells from the smallest (1) and the largest (5) regenerates.

Hence a mechanism for converting a sphere into a cylinder must be able to account for the change in cylinder shape with size, and the constancy of the head-body diameter. We will first consider mechanisms that do not rely on patterning and then examine one that does.

SHAPING THE BODY COLUMN WITHOUT PATTERNING

As more is learned about the cell surface, the mechanics of the cytoskeleton, and the interaction of cells within the extracellular matrix, the need for elaborate prepatterns to instruct the cells lessens. Using known tissue components, models have been developed which involve self-propagating chains of events to generate cell patterns and three-dimensional form directly. Here we consider two possible mechanisms to elongate a sphere into a cylinder.

A mechanism based on the ideas of Campbell (8) that is easily envisioned in hydra involves contraction of cell filapodia on the substratum to drive cell rearrangements. It is analogous to the mechanism described in Calpodes, in which the metamorphic change in cell arrangement of segments was linked to contraction of epidermal "feet" on the basal lamina (9). The recent work of Harris and his colleagues have added to its attraction, as they have demonstrated directly that cells can exert force on the substratum, deforming it to create patterns of cells (10-13).

In hydra, the epithelial cells of the body column have muscle processes which emerge from their bases and extend in opposite directions for perhaps two cell lengths, interdigitating with those of neighboring cells (8). In the ectoderm the processes are oriented parallel to the long axis of the body, while in the endoderm they run circumferentially (14). They run along and articulate with the mesoglea, and their contraction is responsible for the normal contraction and elongation of the animal respectively. Creeping of the processes has been observed, but presumably when the forces in the tissue are balanced no net movement occurs (8).

A sphere of tissue might be converted into a cylinder using the force generated by the muscle processes to rearrange the cells. A critical asymmetry is necessary, however, to produce the elongate shape. The contractile force exerted by the endodermal muscle fibers which run circumferentially must be greater than that of the ectodermal fibers which run axially. That this is likely is demonstrated whenever an excised sheet of tissue rounds up and heals over. Regardless of the shape of the tissue, the edges which meet first are those curving around in what was the circumferential direction in the donor body column.

Presumably the forces within the early spherical regenerate are out of balance. Therefore, as regeneration progresses, continued contractile and creeping movements of the muscle processes would lead to cell rearrangements. Eventually an equilibrium would be achieved between the tendency to elongate due to the stronger endodermal muscle fibers, the tendency to round up due to internal hydrostatic pressure (8), and to tissue strain.

Another intuitively appealing mechanism uses the countervailing forces of differential adhesion and strain. According to Mittenthal and Mazo (15) graded differences in cell-cell adhesion will convert a flat epithelial sheet into a tube. Assume that rings of cells having differing adhesive affinities are arranged in a bull's-eye pattern with the strongest in the center. As cells rearrange to maximize contacts of like cohesive strength within a ring and minimize contacts with cells in adjacent rings, the sheet will be deformed into a cylinder. The greater the adhesive disparity between bands of like cells, the longer and narrower the cylinder will be. This is opposed by the stiffness of the sheet which is assumed to have elastic properties. The greater the mechanical strain of bending the sheet, the shorter and wider the cylinder. The final shape will be a balance between these two opposing forces.

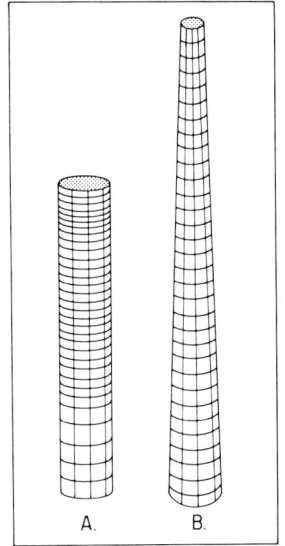

FIGURE 5. A. The variation in cell surface area along the length of the endoderm in large animals (drawn to scale from the average measurements of stained whole mounts). Each block represents 5 x 5 cells and its shape represents the shape of the individual cells. B. The predicted shape of the cells when the body column is extended, resulting in the secondary shape normally seen.

To apply this model to hydra, the cells of the apical end of the sphere where the head will form would have the maximum cohesive value. The apical end would be the high point of a monotonically decreasing gradient of values radiating outward from the presumptive head.

Consistent with such a gradient in the intact animal is the axial distribution of cell shape in the endoderm. The cells at the apical end of the column are columnar with large areas of mutual contact, suggesting that cohesion is indeed strongest there (Fig. 5). Proceeding down the column, the cells become less columnar, reducing their adjacent areas, and at the basal end become quite squamous.

The model predicts that the proportions of cylinders would change with size in a particular way. They should become shorter and wider at small sizes, consistent with an increased strain of bending around a tighter curve. When the various leg segments of <u>Drosophila</u> were compared, they were found to conform approximately to the expectations of the model (15). For hydra, the model fits qualitatively, but the observed diameters for the smaller regenerates are larger than predicted. The discrepancy suggests that an additional factor may be superimposed on the process, thereby obscuring what may be the "natural" diameter arising from the morphogenetic mechanism(s).

MORPHOGENESIS PRECEDED BY PATTERNING

Forming a cylinder from a sphere is not a very complicated example of morphogenesis, and either of the mechanisms described above are probably sufficient to carry it out. However, development of a hydra from a piece of tissue involves head and foot formation as well as shaping of the body column. Since the head and body column have the same diameter even though they make up different proportions of the animal over the size range of regenerates, it seems likely that the shaping of the two structures are causally linked.

Which is dependent on which? If the head diameter is dependent on mechanisms for shaping the body column, then a morphogenetic mechanism is sufficient If the reverse, then the situation is more complicated and probably involves a series of processes. It would include an initial patterning mechanism that defines a region of the sphere to be the presumptive head. Subsequently, the cells of this region would respond by forming a dome. Thereafter, a morphogenetic mechanism would guide the rest of the tissue into the cylindrical shape of the body column under the constraint of the head dome.

This viewpoint leads to two predictions. One would expect to find indications of the head before any sign of morphogenesis has occurred. Secondly, the absence of the head should affect the shape of the final cylinder. There is evidence consistent with both of these predictions.

Indications of a Prepattern for the Head

One reason to believe that the head plays a controlling role in regeneration is that it has the character of an "organizer". Implantation of a piece of hypostome into the wall of the body column induces a secondary axis (16). The nearby host body cells take part in forming the new head and some are realigned into a section of body column (17). This organizing property is dominant in the hypostome and weakly present in the body (17,18). In addition, a secondary body column is only organized if head structures are formed also. Otherwise the implanted tissue is nothing more than a temporary deformation in the host body.

Related to this role, there is a variety of evidence that "headedness" returns very early in the developmental process, suggesting the presence of a prepattern. When a hydra is decapitated, the apical end of the body column regenerates a head, and with it the high capacity for induction. It has been shown that after only three hours an increase in the ability to induce a secondary axis in a host has occurred in the regenerating apical tip, and by seven hours this ability has reached the normal level of a differentiated head (17,18). As the earliest indications of the morphology of the head do not become visible for 30-48 hours, this patterning event clearly precedes morphogenesis.

Another expectation of a prepattern is that it would affect cell differentiation as well as morphogenesis. Thus, it is plausible that one could detect head-specific antigens in the designated head region before there was any overt sign of a head. We have looked for such antigens by generating monoclonal antibodies against hydra cells and have found two that are relevant here.

CP8 is a monoclonal antibody that in the adult binds to vesicles or granules just under the apical surface of the ectodermal epithelial cells of the hypostome and tentacles, but nowhere else (19). The density of the vesicles is very high in the hypostome, and drops steeply to background levels across the ring of tentacles to the top of the body column. Hence, it binds to a head-specific antigen. To determine when the antigen first appeared during regeneration, pieces were stained at various stages of the process.

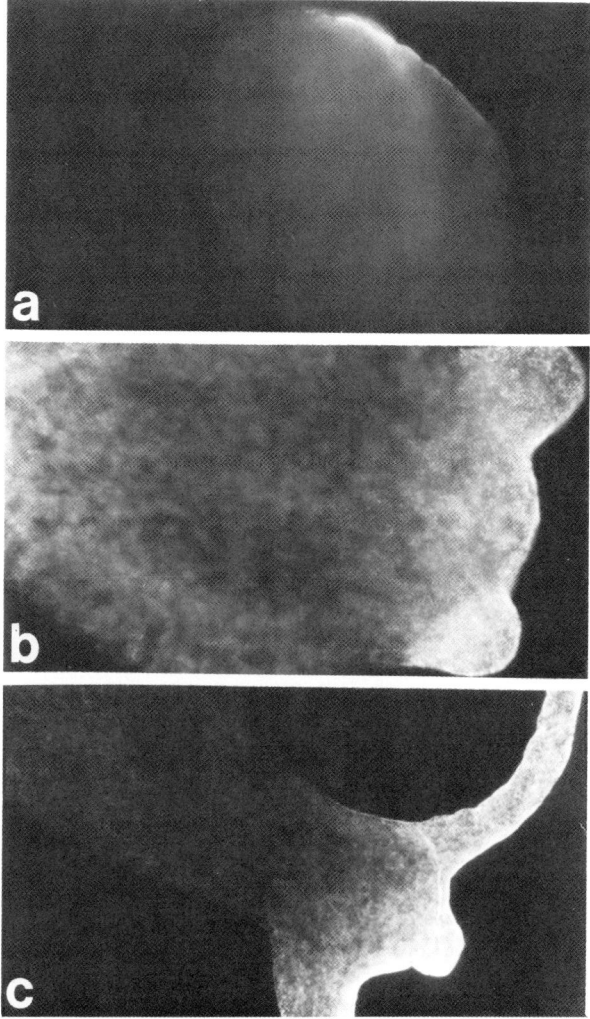

FIGURE 6: Distribution of the CP8 antigen at three stages of development of hydra from excised pieces of the body column. Binding of the antibody was visualized with indirect immunofluorescence. A. Sphere stage. B. Short cylinder with tentacle rudiments. C. Final form of the animal. Magnification: 180X

Pieces of tissue shaped like postage stamps were excised and by 8 hours had rounded up into spheroids.

Within 12 to 16 hours some of them bound CP8 in a localized area. By 24 hours the size of the stained area had grown to cover a domed area. The stain was most

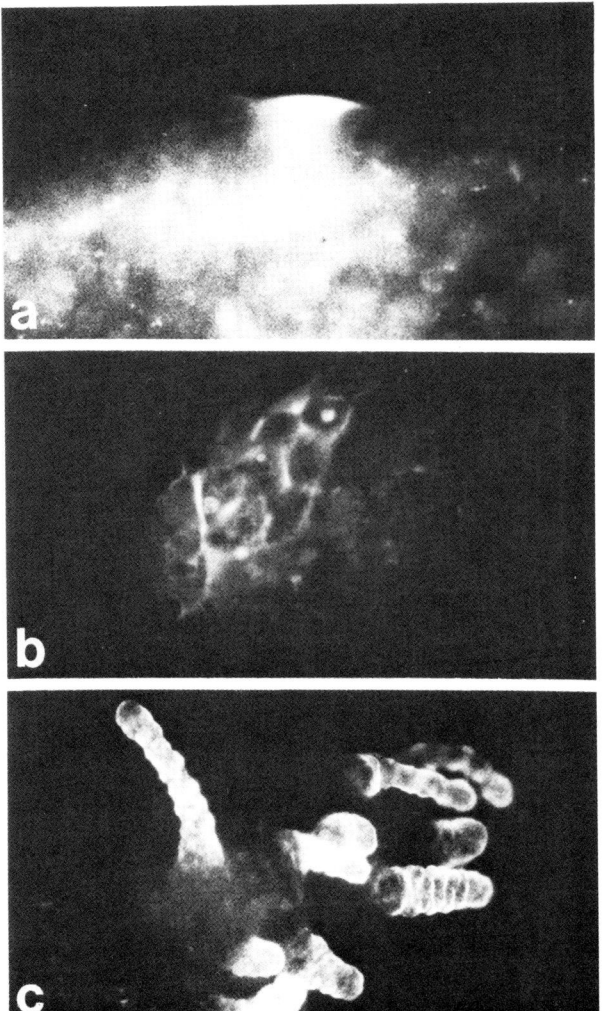

FIGURE 7: Distribution of the TS19 antigen during aggregate development as visualized with indirect immunofluorescence. A. Hollow sphere at 24 hrs. Patch of stained ectodermal cells on upper edge of sphere seen from the side. B. Hollow sphere at 48 hrs. Patch of epithelial cells seen from above exhibiting an intensity of stain comparable to tentacle epithelial cells in the adult. C. 96 hrs: several heads and numerous tentacles have formed. The antigen is confined to the cells of the tentacles. Magnifications: A: 180X; B: 360X; C: 90X.

intense in the middle and decreased outward, suggesting a continuous increase in the antigen (Fig. 6a). By two days the first tentacle rudiments appeared from within the region stained with the antibody (Fig. 8b). By six days, when regeneration was complete, the CP8-staining was restricted to the upper end of the axis (Fig. 8c).

The other monoclonal antibody, TS19, is specific for the epithelial cells of the tentacles. It was used in a regeneration experiment involving aggregates. Hydra can be dissociated into a suspension of viable cells, and these centrifuged into an aggregate which subsequently develops into a normal animal again (20). To approach the question of a head prepattern, the temporal and spatial distribution of the TS19 antigen during development of aggregates made from body tissue was examined.

During the first day of development the solid sphere of cells is converted into a two-layered hollow sphere with the usual ectoderm on the outside and an endoderm as the inner layer. Although there were no signs of tentacle rudiments at this stage, there are patches of epithelial cells that were TS19+ (Figure 7a). The density of immunofluorescent stain was low at 24 hours, yielding a speckled appearance. By two days it had increased to the smooth, uniform pattern typical of adult tentacle cells (Figure 7b). Tentacle rudiments were evident by three days, and complete heads with substantial tentacles were present by four days. By this time the TS19 antigen was confined to epithelial cells of the tentacles (Figure 7c).

Since markers of head differentiation also appear long before any sign of morphogenesis, the formation of a head prepattern very likely occurs, and it is a very early event.

Evidence that the Head Affects the Dimensions of the Body Column

The second prediction dealt with the head, or the presumptive head, playing a causal role in the shaping of the body column. If it did, the absence of a head should affect the shape of the final cylinder. Clear cut evidence of this type is difficult to obtain because the head regenerates in most experimental situations. Also, in those instances where headless animals result, the inability to osmoregulate could have an affect on the

Shaping of the Body Column in Hydra

final shape. Tissue excised from the peduncle region, especially the lower portion, frequently regenerates only basal disc areas (6). Such regenerates remain spheroids and often become enveloped by the basal disc cells. Those which form heads, whether this occurs early in the process or is delayed, gradually develop the appropriate cylindrical shape thereafter.

The second situation which can produce headless regenerates is the removal of the apical portion from an animal with a developing bud. Buds, the asexual mode of reproduction, can inhibit head formation. These regenerates are initially longer and narrower than before, but become increasingly rounded. At the end of the usual regeneration period, both types of headless regenerates have very different overall shapes from normal cylindrical animals (6). These results imply that the presence of a head or a presumptive head may be instrumental in initiating the appropriate morphogenetic mechanisms for column shaping, or stabilizing them if they are already present. Further, the shape of the cylinder formed by tissue mechanics may be different from that when the fully formed head is present since the head appears to impose its diameter on the column below it.

A Model for Shaping the Body Column Involving a Head Prepattern

The following process can be envisioned for shaping the body column if a prepattern for a head is a prerequisite. Since isolated pieces form hydra with the appropriate proportions, a mechanism to allot tissue to each of the structures would exist. One that has been applied to hydra for this purpose (21,22) is a reaction-diffusion model developed by Gierer and Meinhardt (23,24) and extended by MacWilliams (25). Patterns are created by the interaction of two components: activation and lateral inhibition. Activation, an autocatalytic process, sets up a region of tissue to form a structure. Patterns are created by the interaction of two components: activation and lateral inhibition. Activation, an autocatalytic process, sets up a region of tissue to form a structure. Inhibition, presumed to be a diffusible substance whose production is linked to the concentration of activation, diffuses into the surrounding tissue and prevents the formation of the same structure. This same inhibition also functions to delimit the spread of the activated region, thereby allocating a specific fraction of the available tissue to the particular structure.

Applying this model to the conversion of a spherical shell into a cylinder, the apical end of the original piece will undergo the activation process. When the process is complete, a circular region will be allocated to the future head as indicated by the dashed line on the spheres in Figure 8. The remainder will form the body column. (For simplicity we will ignore the foot because its absence has little affect on column shaping. Normally, it will be formed in a similar way at the opposite end since it too has organizing properties [26-28]).

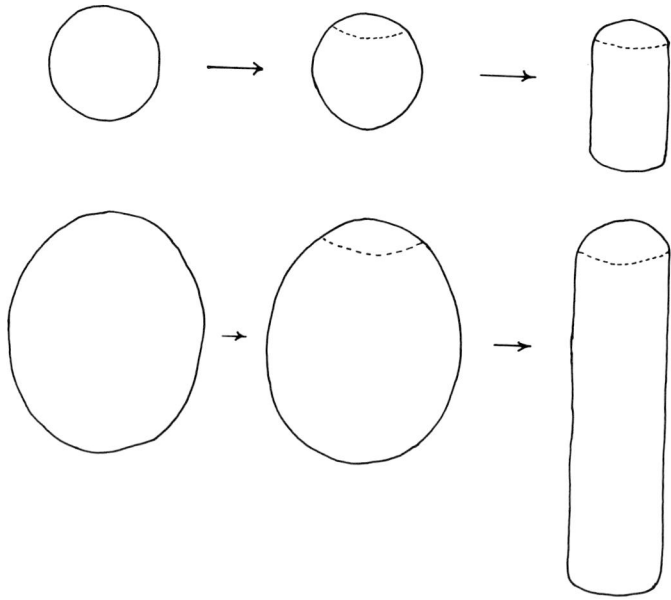

FIGURE 8: Schematic of the development of a small and a large regenerate, assuming a prepattern for the head is the initial event. The dotted line indicates the allotted tissue for the head.

Then, based on a model of tissue evagination developed by Gierer (29), we assume that an early response of the cells in the presumptive head is to undergo a shape change so that a cap with a defined radius of curvature is set up. This would dictate the diameter of the dome that is the future hypostome. To elongate the body, we assume that during patterning or early differentiation the presumptive head in some way signals the remaining tissue chemically or triggers a series of mechanical events. The epithelial cells of the body column might then align themselves to the perimeter of the head so that the muscle processes are in a functional position to exert contractile and traction forces leading to cell rearrangement, and/or a graded pattern of adhesive properties may be initiated in the tissue, with the head at the high point, also facilitating cell rearrangements. The result would be a cylindrical animal with a domed head.

Formally, the model accounts for the observed geometrical relationships described earlier. The initial shape of the piece would be irrelevant. Also, the head and body column diameters would automatically be the same. It would also explain why the final shape of the cylinder is a function of the initial size of the excised piece. As small regenerates have a relatively larger fraction of tissue allocated to the head, the remaining tissue would have to be arranged around a relatively larger diameter. This would result in short, stubby cylinders (Figure 8). Conversely, in larger pieces, a smaller fraction of tissue forms the head which results in a relatively smaller diameter. A larger fraction and amount of tissue remains to be rearranged and would yield an elongate body column.

CONCLUSIONS

The two views, patterning coupled with subsequent morphogenesis, and morphogenesis without any prepattern, lead to very different predictions as to what one should look for in terms of the molecular determinants of form. If patterning is a prerequisite, one must understand the nature of the patterning process, find the morphogens or molecules involved, and determine the cellular response to them to carry out the morphogenesis. In contrast, if morphogenesis proceeds entirely by self-propagating mechanical means, then more than likely there is no need to search for new molecules that specify form. The ones involved are probably already known and are part of the structural machinery of the cell and the extracellular matrix. It will be a matter of devising experiments to demonstrate that these structural molecules can carry out the process.

The few results presented here all indicate that a restricted portion of the regenerating tissue begins head development before there is any sign of morphogenesis.

The absence of the head in some regenerates suggests that the head plays a causal role in shaping the body column. Hence, it seems unlikely that mechanisms of morphogenesis alone would be sufficient to turn a rectangular piece of tissue into the shape of a hydra. Instead, at the moment, the most consistent view is that a patterning process delineates the regions followed by subsequent differentiation and morphogenesis. Thus, seeking molecules involved in these patterning processes and the subsequent morphogenesis will probably be necessary to figure out how hydra gets its form.

ACKNOWLEDGEMENTS

We wish to thank Richard Campbell and Scott Fraser for stimulating discussions.

REFERENCES

1. Wolpert L (1969). Positional information and the spatial pattern of celular differentiation. J. Theoret Biol 25:1
2. Meinhardt H (1982). "Models of Biological Pattern Formation." San Diego: Academic Press.
3. Odell GM, Oster G, Alberch P, Burnside B (1981). The mechanical basis of morphogenesis. I. Epithelial folding and invagination. Develop Biol 85:446.
4. Oster GF, Murray JD, Harris AK (1983). Mechanical aspects of mesenchymal morphogenesis. J Embryol exp Morph 78:83.
5. Bode PM, Bode HR (1984). Patterning in hydra. In Malacinski G, Bryant SV (eds): "Pattern Formation. A Primer in Developmental Biology, Vol 1", New York: Macmillan & Co, p 213.
6. Bode PM, Bode HR (1984). Formation of pattern in regenerating tissue pieces of Hydra attenuata. III. The shaping of the body column. Develop Biol 106:315.
7. Bode PM, Bode HR (1984). Formation of pattern in regenerating tissue pieces of Hydra attenuata. II. Degree of proportion regulation is less in the hypostome and tentacle zone than in the tentacles and basal disc. Develop Biol 103:304.
8. Campbell RD (1980). Role of muscle processes in hydra development. In Tardent P, Tardent R (eds): "Developmental and Cellular Biology of

Coelenterates", Amsterdam: Elsevier/North Holland Biomedical Press, p 421.
9. Locke M, Huie P (1981). Epidermal feet in pupal segment morphogenesis. Tissue & Cell 13:787.
10. Harris AK, Wild P, Stopak D (1980). Silicone rubber substrata: a new wrinkle in the study of cell locomotion. Science 208:177.
11. Harris AK, Stopak D, Wild P (1981). Fibroblast traction as a mechanism for collagen morphogenesis. Nature 290:249.
12. Stopak D, Harris AK (1982). Connective tissue morphogenesis by fibroblast traction. I. Tissue culture observations. Develop Biol 90:383.
13. Harris AK, Stopak D, Warner P (1984). Generation of spatially periodic patterns by a mechanical instability: a mechanical alternative to the Turing model. J Embryol exp Morph 80:1.
14. Mueller JF (1950). Some observations on the structure of hydra, with particular reference to the muscular system. Trans Amer Microscop Soc 69:133.
15. Mittenthal JE, Mazo RM (1983). A model for shape generation by strain and cell-cell adhesion in the epithelium of an arthropod leg segment. J Theor Biol 100:443.
16. Browne EN (1909). The production of new hydranths in hydra by the insertion of small grafts. J Exp Zool 7:1.
17. Webster G (1971). Morphogenesis and pattern formation in hydroids. Biol Rev 46:1.
18. MacWilliams HK (1983). Hydra transplantation phenomena and the mechanism of Hydra head regeneration. I. Properties of the head activation. Develop Biol 96:239.
19. Bode HR, Dunne JD, Heimfeld S, Huang LW, Javois LC, Westerfield J, Yaross MS (1985). Transdifferentiation occurs continuously in hydra. Curr Top Develop Biol (in press).
20. Gierer A, Berking S, Bode H, David C, Flick K, Hansmann G, Schaller H, Trenkner E (1972). Regeneration of hydra from reaggregated cells. Nature New Biol 239:98.
21. Bode PM, Bode HR (1980). Formation of pattern in regenerating pieces of Hydra attenuata. I. Head-body proportion regulation. Develop Biol 78:484.
22. Bode PM, Bode HR (1982). Proportioning a hydra. Amer Zool 22:7.
23. Gierer A, Meinhardt H (1972). A theory of biological pattern formation. Kybernetik 12:30.

24. Meinhardt H, Gierer A (1974). Applications of a theory of biological pattern formation based on lateral inhibition. J Cell Sci 15:321.
25. MacWilliams HK (1982). Numerical simulations of hydra head regeneration using a proportion-regulating version of the Gierer-Meinhardt model. J Theor Biol 99:681.
26. MacWilliams HK, Kafotos FC (1968). Hydra viridis: Inhibition by the basal disk of basal disk differentiation. Science 159:1246.
27. Cohen JE, MacWilliams HK (1975). The control of foot formation in transplantation experiments with Hydra viridis. J Theor Biol 50:87.
28. Hicklin J, Wolpert L (1973). Positional information and pattern regulation in hydra: formation of the foot end. J Embryol exp Morph 30:727.
29. Gierer A (1977). Biological features and physical concepts of pattern formation exemplified by hydra. Curr Top Develop Biol 11:17.

POSITIONAL MAPS AND CELLULAR INTERACTIONS IN
INSECT DEVELOPMENT[1]

Vernon French

Department of Zoology, University of Edinburgh,
Kings Buildings, West Mains Road, Edinburgh EH9 3JT,
Scotland, U.K.

ABSTRACT Pattern formation involves cellular interactions, generating a map of positional value which the cells then individually interpret. Positional maps may be relatively simple and may be repeated through the organism. Hence insects may have equivalent epidermal positional maps on different segments (or parts of segments) but interpret them in various segment-specific ways.
Grafting and similar experiments on many insects suggest that the limb epidermis has a two-dimensional polar map and that its regenerative behaviour results from two rules for cellular interaction: shortest route intercalation between cells from different proximal-distal or circumferential positions, and distalization if newly intercalated cells are identical to nearby pre-existing cells. Most of this work, however, has not addressed the relationships between positional maps on the limb and the rest of the thoracic segment. Grafting and extirpation experiments on the beetle, Tenebrio, demonstrate that a supernumerary leg is regenerated after interaction between regions of thoracic epidermis anterior and posterior to the leg base, and between the leg base and inappropriate regions of surrounding thorax. At present, the results are compatible with the formation of the leg either as the central part of a positional map covering much of the ventral thoracic segment, or at the intersection of "compartment borders" on the thorax.

[1]This work is supported by the British Science and Engineering Research Council.

INTRODUCTION

Zygotes develop into complex organisms consisting of a large number of differentiated cell types arranged in intricate spatial patterns. Most zygotes and early embryos have localised determinants, consisting of special regions formed during oogenesis or as a result of some extrinsic factor such as site of sperm entry. The distribution of determinants by a precise cell lineage is very important in directly setting the developmental fate of most cells in embryos such as those of the nematode or leech. Even in these species, however, the development of every cell is not autonomous or in strict accordance with its allocation of zygotic cytoplasm. The fate of some cells is clearly influenced by their position, by interaction with their neighbours. Cellular interaction is of major importance in the development of echinoderm, vertebrate and insect embryos. Interactions mediating pattern formation can be studied by surgically manipulating cells, removing them or transplanting them elsewhere. Some interactions may involve a specific short-range instruction (induction or inhibition) from an adjacent cell (or underlying cell layer), but experiments on a wide range of embryos have suggested that patterns are specified by a more extensive system of interactions acting within a cell sheet to generate Positional Information (29). A co-ordinate system (or map of positional values) is established relative to initial boundary positions and cells then interpret their positional values, acquiring cellular properties and developmental fates appropriate to their location.

In the early insect embryo, the results of a wide range of ligature, translocation, centrifugation and irradiation experiments (24) suggest that a gradient (or a pair of gradients) forms down the longitudinal axis, perhaps by a source-diffusion or diffusion-reaction mechanism with the posterior (and possibly the anterior pole) of the egg as an important reference point (24,21). The pattern of body segments becomes determined in relation to this simple one-dimensional map. Although segments are demarcated and segment type determined at an early stage, within the embryonic segment further interactions are involved in forming the surface patterns of the epidermis and the intricate internal patterns of neurons and mesodermal tissue. These interactions continue, in many insects, throughout embryonic and larval life and have been intensively studied by grafting and other types of experiment, particularly on the legs and abdominal segments of hemimetabolous insects and the imaginal discs of Drosophila.

One central feature of the concept of Positional Information is that the cells receive geographical information rather than specific instructions. The positional map is not directly related to the pattern that will form, so the same map may be used in different parts of the organism but be interpreted differently (29). Hence in insects, the same cellular interactions may build the same positional map in the different segments or parts of segments but, with the different segmental determinations leading to various modes of interpretation, this will result in the formation of the diversity of different segmental structures.

I shall now discuss one specific proposal about the nature of the positional map on the insect appendage and the evidence suggesting that 2 simple rules are followed if the map is disrupted, promoting interactions between cells which are not normally neighbours. I shall then discuss some preliminary experiments investigating the extent of the map on the thoracic segment.

THE POLAR CO-ORDINATE MODEL AND THE INSECT LEG

The insect leg is a complex three-dimensional jointed structure consisting of a surface cuticle which is secreted by the underlying monolayer of epidermal cells, internal apodemes formed by invagination of epidermis at the joints, muscles attaching to the apodemes and to areas of surface epidermis, nerves and trachea. There is considerable evidence that the important cell layer with respect to pattern formation and regeneration is the epidermis (14). The characteristic cuticular patterns of bristles, spines, pigmentation etc. seem to depend only on interactions within the epidermis, the position of muscles may depend on the location of epidermal apodemes and surface attachment sites, and the pattern of nerves and trachea probably reflect the arrangement of epidermis and muscle.

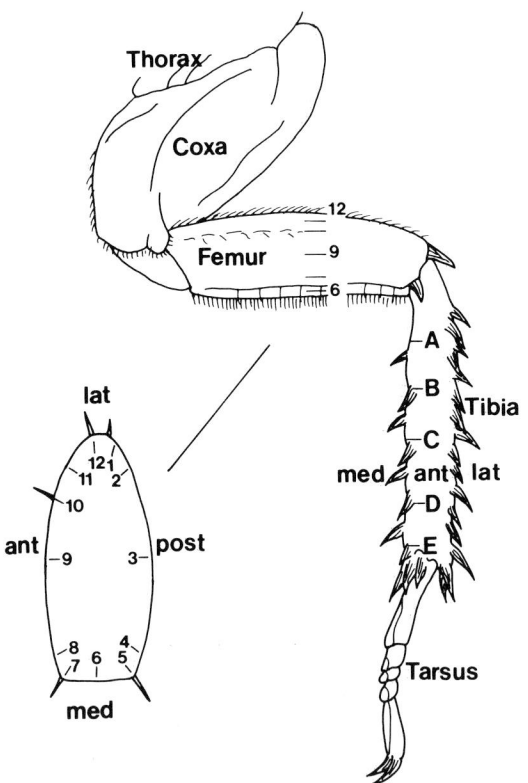

FIGURE 1. The larval cockroach left metathoracic leg shown in anterior view and in transverse section through the femur. Five proximal-distal levels (A - E) are marked down the tibia. Twelve positions (1 - 12) are marked around the leg circumference, as well as the anterior (ant), posterior (post), medial (med) and lateral (lat) faces of the leg.

The Polar Co-ordinate Model (16, 8) proposes that epidermal cells of the leg have 2 major positional values characteristic of (i) their proximal-distal level within a leg segment (sequence A-E in Fig. 1) and (ii) their

Insect Positional Maps 459

circumferential position around the leg (sequence 0-12 in
Fig. 1). These positional values were acquired by interactions earlier in development, they determine which
structures are made when new cuticle is secreted, they may

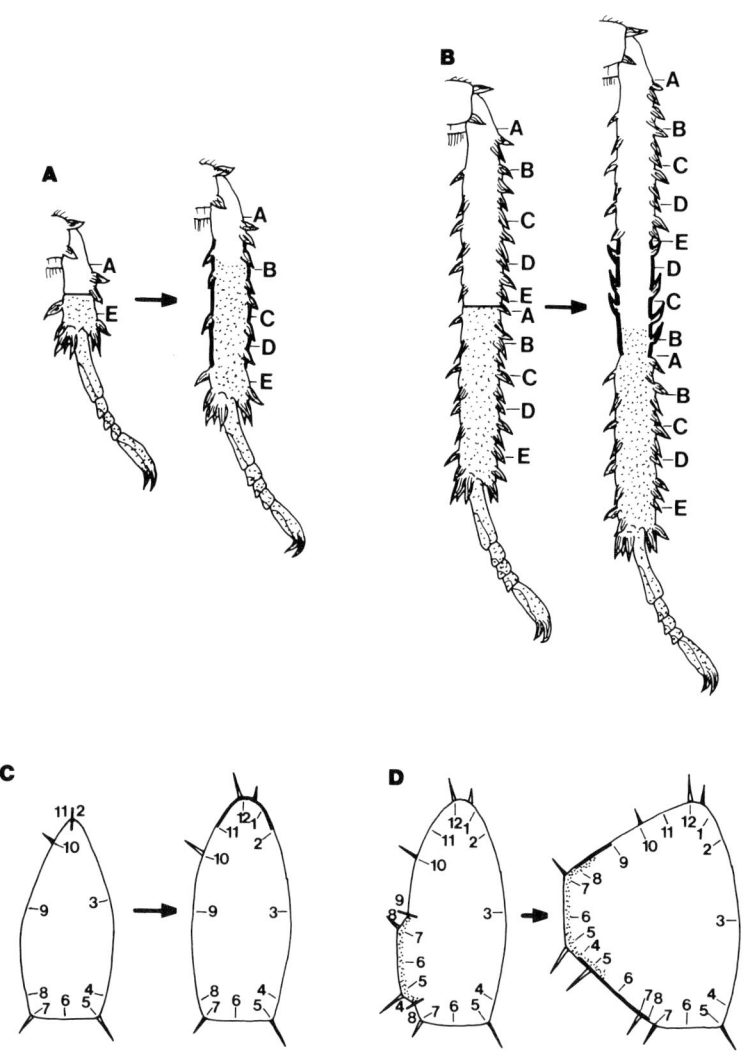

FIGURE 2. Legend appears on following page.

FIGURE 2. Intercalary regeneration.
(A, B) Interaction between proximal (A) and distal (E) levels of the cockroach tibia leads to local growth and the formation of an intercalary regenerate consisting of the intermediate levels (B, C, D).
(C) After removal of a strip of lateral epidermis, cells from circumferential positions 11 and 2 are confronted and local growth leads to the replacement of cells with positional values 12 and 1.
(D) The graft of medial face of left femur to the anterior face of the host left femur results in intercalary regeneration of the shortest section of circumference which normally separates graft and host positions. Graft and graft-derived tissue is stippled.

control the extent to which particular cells divide during larval life (7), and in insects which can regenerate they govern the results of abnormal cellular interactions provoked by damage or experimentation. This proposal rests largely on the results of grafting operations on larval legs and comparable experiments on Drosophila imaginal discs.

Intercalary Regeneration

If part of the length of a larval leg segment is removed and the distal level grafted back onto the proximal level of the stump (Fig. 2A), local growth occurs at the junction and the intervening mid-tibia is formed by intercalary regeneration, restoring the segment to an approximately normal size and pattern (4, 10). If a proximal level is grafted onto a distal level, a similar intercalary regenerate is formed, but now in reversed orientation so that the resulting abnormal segment has no positional discontinuities between adjacent cells.

If a longitudinal strip is removed from the leg circumference, non-adjacent cells heal together and intercalary regeneration occurs to restore the normal pattern (Fig. 2C). Similarly, if a longitudinal strip of epidermis and cuticle is moved around the leg circumference and grafted into a different position, normally non-adjacent cells interact along the graft-host junctions, provoking local cell division (1) and the intercalary regeneration of the intervening section of circumference, by the shortest route (11). Hence, in Fig. 2D, positions 5, 6, 7

(and not 3, 2, 1, 12, 11, 10, 9) are intercalated at the
junction between graft position 4 and host position 8.
In terms of the Polar Co-ordinate Model (or P.C.M.),
these experiments show that interaction between cells with
different proximal-distal or circumferential positional
values provokes local cell division. While cells remote
from the junction retain their original values, the new
cells acquire intermediate values, thereby restoring
continuity in the positional map. This intercalation rule
is followed in a wide range of grafting operations and
there is no sign that particular levels or positions have
special boundary properties.

Grafts between differently pigmented species or
differently structured leg segments show that both graft
and host contribute to the regenerate (Fig. 2), although a
proximal-distal intercalary regenerate is produced mostly
from the more distal level (5). In circumferential intercalary regeneration cells seem unable to cross over
2 positions, approximately mid-medial and mid-lateral (12),
and this may reflect the sub-division of the leg into
2 lineage compartments, as has been directly demonstrated
by clonal analysis of the Drosophila leg disc (25).

Distal Regeneration

If a leg is amputated or autotomised, haemolymph clots
over the wound, the stump epidermis migrates in to become
confluent under the clot, and cell division gradually
forms a distal regenerate beneath the old cuticle of the
stump. A normal leg stump will reliably regenerate a
fair copy of the missing distal structures, but a
surgically-constructed symmetrical double-half stump can
form a tapering distally-incomplete regenerate (8).
These results are explained by the P.C.M. in terms of
interactions occurring during epidermal wound healing at
the amputation site. This will bring together cells from
different circumferential positions and will stimulate
cell division, intercalating cells with intermediate,
circumferential but more distal positional values (Fig. 3).
From a normal circumference this process will continue
until all the more distal parts are regenerated, but the
healing of an abnormal symmetrical circumference will
oppose some cells which do not differ in circumferential
value and the regenerate will taper and may be distally
incomplete (8, 14).

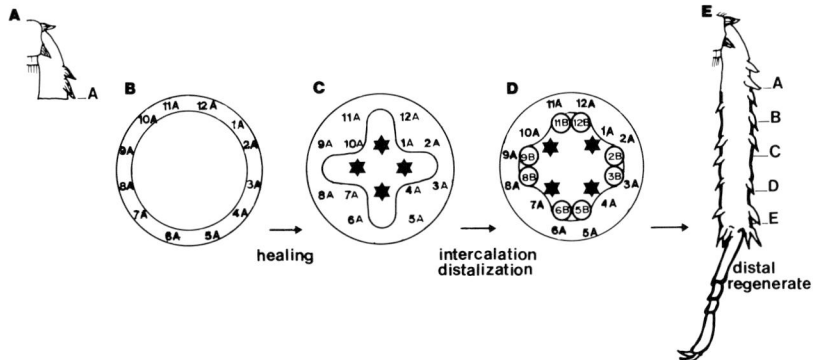

FIGURE 3. After amputation through the proximal tibia (A), distal regeneration occurs (E). B, C and D are schematic distal views showing healing, intercalation and distalization at the amputation site. The proximal epidermis of the stump (level A) forms a complete circumference of cells which will heal together confronting cells with different values. Shortest route intercalation (stars) will produce new cells with values identical to adjacent stump cells, so they adopt more distal radial values, becoming 2B, 3B etc. Subsequent intercalation completes the B level and the process is repeated until all distal levels have been formed. This result is almost independent of the initial healing pattern.

This explanation involves continued intercalary regeneration at the distal tip of the stump and developing blastema, as the regenerate is gradually formed in a proximal-to-distal sequence. However, from recent histological studies (27, 28), it seems that cell division is not limited to the healing tip but spreads back into the stump. If the distalization shown in Fig. 3 occurred in both daughter cells then a positional discontinuity (and resulting cell division) would spread back some way into the stump, but this would not be a large effect (20) and the extent of cell divisions during distal regeneration does suggest that the current P.C.M. explanation is inadequate.

Different Co-ordinate Systems

The P.C.M. proposes that the insect limb epidermis or the Drosophila imaginal disc has a two-dimensional polar map of positional values arranged along its longitudinal and circumferential axes, and that its regenerative behaviour results from 2 rules for cellular interaction : <u>shortest route intercalation</u> between cells with different positional values, and <u>distalization</u> if the intercalated cells have values identical to those of nearby pre-existing cells.

Many of these results can be explained simply in terms of any form of continuous map which can re-establish continuity (20, 23), but without specific rules and a specific co-ordinate system the results of many grafting experiments cannot be predicted. Transformation between co-ordinate systems is always mathematically possible so, in a sense, experimental results cannot <u>prove</u> that the positional map has a particular form. However, different co-ordinate systems require different rules, in addition to the re-establishment of continuity, to explain the results (Fig. 4) and, so far, the PCM provides the simplest and most coherent and predictive model of insect limb regeneration.

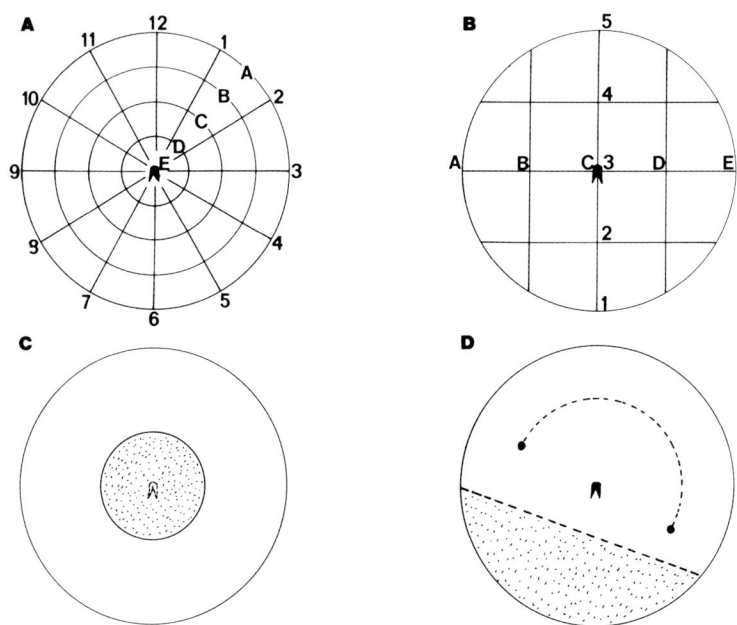

FIGURE 4. Positional maps (A, B) and experimental results (C, D) on insect leg epidermis which is represented as a disc with the distal tip (marked by claws) in the centre.
(A) Polar Co-ordinate Map. Cells have positional values characteristic of their level on a radius (A - E) and their position on a circumference (1 - 12). There is no discontinuity between 12 and 1, so a continuous sequence of values runs around the limb.
(B) A Cartesian Co-ordinate map. Cells have positional values in 2 orthogonal axes, A - E and 1 - 5.
(C) When the distal part of the limb is removed, simple averaging along Cartesian axes will replace the missing tissue, since its positional values (e.g. C3) lie between those of the stump. If the map is Polar a special rule is required since the regenerated structures (e.g. D, E) are in no sense "between" the remaining proximal ones.
(D) If opposite positions on the limb (shown as two spots) are grafted together, averaging along the circumferential

axis of a Polar map produces the observed arc of tissue
(dashed line) but averaging along Cartesian axes would
form distal structures (e.g. C3), so a Cartesian model
requires a special rule to accommodate this result.
Similarly, removal of a small segment of an imaginal disc
(heavy dashed line) produces regeneration of the missing
tissue (15) which has values between those of the stump
on a Polar map (e.g. value A6) but not on a Cartesian map
(e.g. value C1).

Supernumerary Regeneration

After various grafting operations or after extensive
damage, insect limbs can regenerate supernumerary branches.
If the distal part of the right larval leg of a cockroach,
cricket, stick insect or beetle is grafted onto the stump
of the left leg then one transverse axis is reversed
relative to that of the stump. After reversal of the
anterior-posterior (A-P) axis supernumerary distal
regenerates form from the junction in anterior and
posterior positions and, after reversal of the medial-
lateral (M-L) axis they form in medial and lateral
positions (3, 9, 2, 13). The supernumeraries are of host
handedness and orientation.

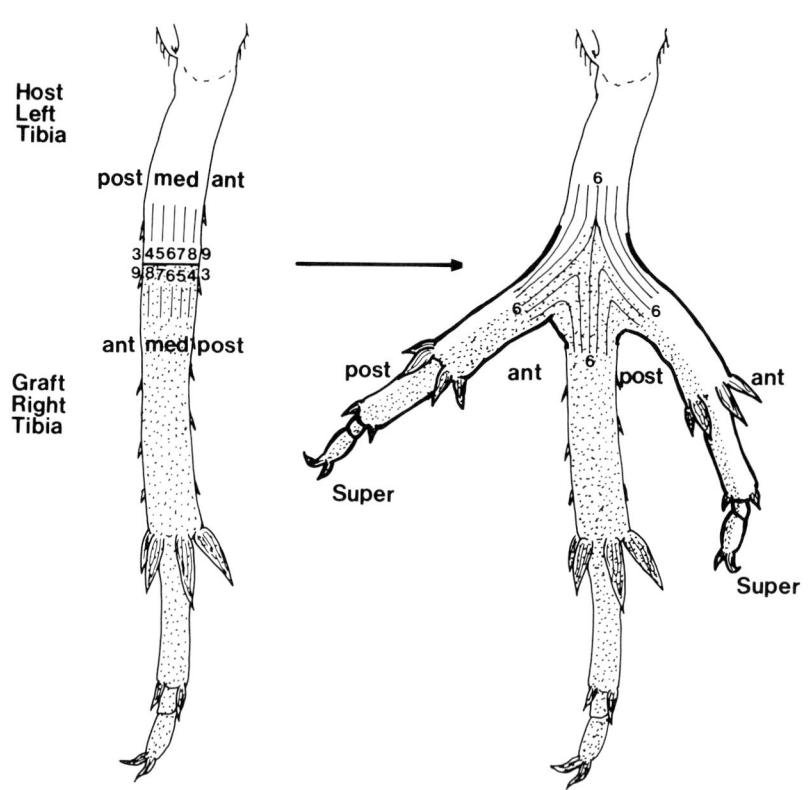

FIGURE 5. Graft at proximal tibia level between cricket left and right metathoracic legs. The medial-lateral (med - lat) axes are in agreement but the anterior-posterior (ant - post) axis of the graft is reversed relative to the host. Circumferential positional values (3-9) are shown on the graft and host tissue confronted at the junction. Shortest route intercalation will occur, except mid medially (6/6) and mid laterally (12/12). Complete circumferences are formed anteriorly and posteriorly (since intercalation will produce values 8, 7, 6, 5, 4 on one side and 10, 11, 12, 1, 2 on the other side of these positions) and from each a supernumerary (super) will regenerate. Graft and graft-derived tissue is stippled.

After A-P axis reversal, cricket leg supernumeraries are reliably derived half from the graft and half from the host, while after M-L axis reversal they are of variable origin (13, 17). In terms of the PCM, reversal of one transverse axis confronts graft and host cells from different circumferential positions and shortest route intercalation should occur to different extents at different points around the junction. This will generate a complete circumference at each of the positions where opposite values are confronted (e.g. anterior and posterior after the A-P reversal - Fig. 5) and each of these will undergo distal regeneration to form a supernumerary of observed handedness and orientation. The origin of these supernumeraries supports the suggestion that anterior-posterior compartment borders are not crossed in intercalary regeneration (13, 17).

If a left graft is placed, 180° rotated, onto a left stump, both transverse axes are reversed and at all points around the junction cells from opposite positions are confronted. The results of this operation are very variable, between animals and between species, but the graft usually de-rotates and one or two supernumeraries may be formed. These may also be explained by patterns of intercalation at the junction (16).

Leg duplications and triplications can be generated in Drosophila by a temperature pulse at particular stages in the larval development of a temperature-sensitive cell-lethal mutant. These structures have been analysed in terms of localised cell death in the imaginal disc, followed by healing and intercalation according to the rules of the PCM (18). This interpretation made novel predictions about the origin of the central and side branches of the legs, and these have subsequently been shown to be correct (19).

THE LEG AND THE THORACIC SEGMENT

Grafting experiments on the larval legs of several species of hemimetabolous insect, and a variety of regeneration experiments on the imaginal discs of Drosophila, suggest that the leg epidermis has a positional map arranged along polar co-ordinates. Most of these studies have not been directly concerned with the relationship between the appendage and the rest of the thoracic segment, which consists of the ventral sclerotised sternites, lateral pleura, the wing primordium (on two of the thoracic segments) and the dorsal sclerotised tergites. The relationship between the leg and its surroundings can be readily studied by grafting and extirpation experiments on the larvae of the beetle, Tenebrio. Like other holometabolous insects, Tenebrio goes through metamorphosis but, unlike Drosophila, it has larval legs which survive, together with the rest of the larval segment epidermis, and make the corresponding adult parts.

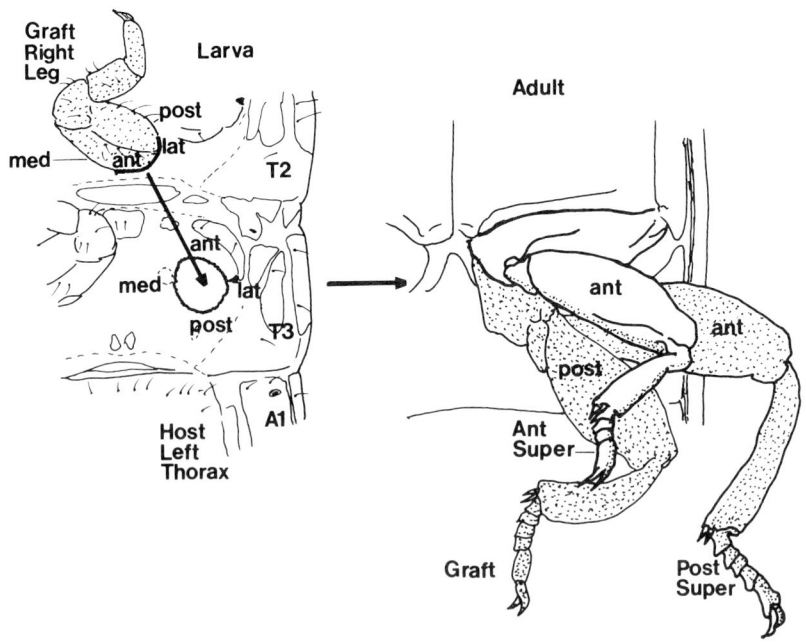

FIGURE 6. Graft of the right larval leg of Tenebrio to the leg site on the left side of the ventral thorax (T_2, T_3 and $A1$ are the mesothoracic, metathoracic and first abdominal segments). Reversal of the anterior-posterior (ant-post) axis of the graft results in the regeneration of anterior and posterior supernumerary legs (Super). The graft is stippled and the supernumerary legs are shown half graft (stippled) and half host in origin (see 15).

If the entire larval right leg is removed at its base and grafted into the site of the left leg, reversing its A-P axis relative to that of the surrounding thorax, then entire supernumerary legs are formed anterior and posterior to the graft leg (Fig. 6). If the operated animal moults to a further larval stage, the supernumeraries form as larval legs, while if it metamorphoses to an adult the graft and supernumeraries form typical adult legs. If the graft is orientated to reverse the M-L axis, the supernumeraries form medially and laterally (15). These results show that interactions between positions around the base of the leg and inappropriate regions of the surrounding thorax result in regeneration of supernumeraries, just as occurs at a more distal level on the leg (Fig. 5), suggesting that the positional map of the leg may extend back onto a ring of surrounding ventral thorax (16).

The relationship between the leg and its surrounds can be further investigated by extirpating and grafting regions of thoracic epidermis. If a transverse band of thorax is removed and remaining areas grafted together, then interaction between a zone posterior to the mesothoracic leg and a zone anterior to the metathoracic leg can result in the regeneration of a supernumerary leg of reversed A-P orientation (Fig. 7A, B).

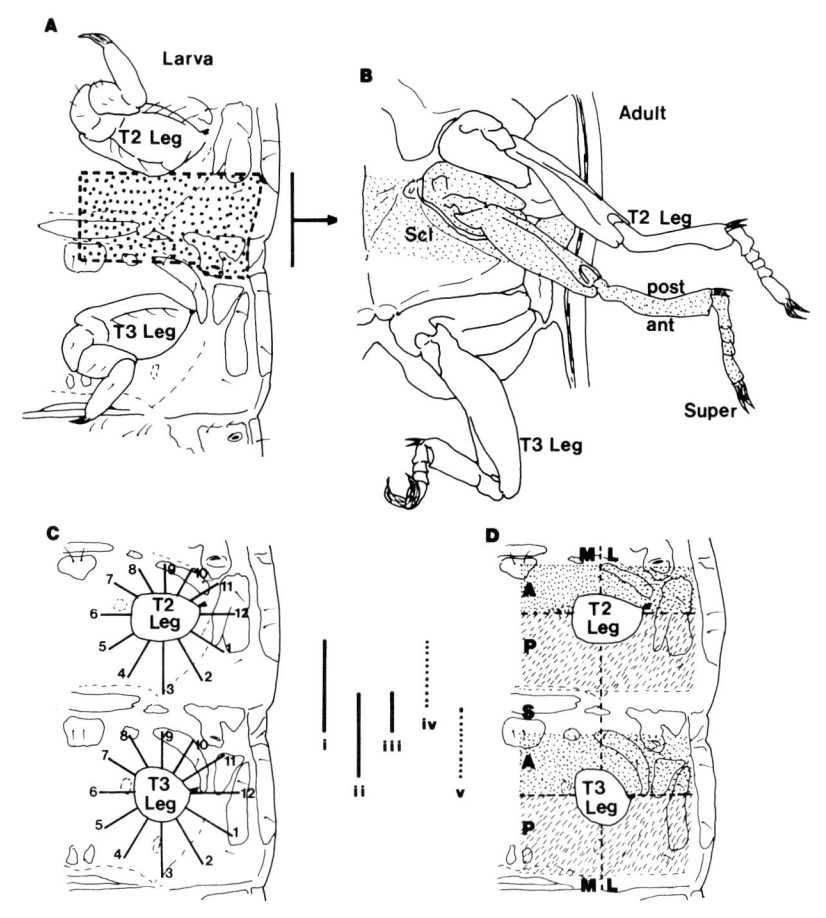

FIGURE 7. Excision of transverse strips of ventral thorax between mesothoracic (T_2) and metathoracic (T_3) larval legs of Tenebrio.
(A, B) Removal of large strip of thorax (dotted) posterior to T_2 leg (A) and resulting regeneration of supernumerary leg (Super) and sclerites (Scl) in reversed anterior-posterior orientation (13).
(C, D). Two possible arrangements of positional information around the bases of T_2 and T_3 legs. Between

C and D are shown by heavy solid lines the smallest strips which can be removed immediately posterior to T_2 (i - as in A), immediately anterior to T_3 (ii) and midway between the legs (iii), and produce regeneration of a supernumerary leg. Heavy dotted lines show 2 examples (iv, v) of excisions which do not give supernumerary regeneration.

(C) shows extension of polar map of leg positional values (1-12) back onto the surrounding thorax (see 16). After excisions i, ii and iii (but not iv and v) interaction between posterior regions (value 3) of the T_2 map and the anterior regions (value 9) of the T_3 map results in intercalation of a complete circumference and regeneration of a leg.

(D) shows subdivision of the segment into intersegmental (S), anterior (A) and posterior (P) transverse compartments, and into medial (M) and lateral (L) longitudinal compartments (see 22, 6) and formation of a leg where the A/P and M/L compartment borders cross. This will occur in an ectopic position after excisions i, ii and iii.

The pattern of our results from extirpations of various widths and from various positions (and the results from previous extirpations and grafts around the base of the cockroach leg (6)) are, at present, compatible with two rather different models of the relationship between leg and thorax.

The thorax may consist of three qualitatively different bands (6), perhaps corresponding to lineage compartments within the segment (22). A leg, with an **independent** positional map, would form where anterior and posterior bands abutted, perhaps where this transverse border intersects a longitudinal border between medial and lateral 'compartments' (23), as shown in Fig. 7D. There is direct evidence from clonal analysis that the Drosophila leg disc is divided into anterior and posterior compartments (25) and indications from limitations of intercalary regeneration that this may be generally true of insect legs (Fig. 5, 6), but there is no evidence that legs are divided into medial and lateral compartments. There is, furthermore, no evidence that the ventral thorax consists of three transverse and two longitudinal lineage compartments.

An alternative interpretation of the extirpation results (Fig. 7C) is that the limb map may also cover the surrounding thorax, with maps in adjacent segments being separated by an intersegmental region. The results of further experiments, particularly grafts between regions of the leg and thorax, may decide between these alternatives. Similar experiments will be used to investigate the dorsal parts of the segment and the wing primordium.

The results of regeneration studies on Drosophila imaginal discs (16, 8) might suggest that the insect thoracic segment consists of just two rather similar positional maps, one centred on the leg and another on the wing. However the two imaginal discs which form the adult fly thoracic segment are derived from only a small part of the embryonic segment, the rest of which forms the larval segment and dies during metamorphosis, so their positional maps will probably not reflect the organisation of the entire thoracic segment in other insects.

CONCLUSION AND PROSPECTS

The mechanisms of cellular interaction underlying pattern formation are a major problem in developmental biology and insects offer exciting possibilities for their investigation because surgical, genetical and molecular techniques can all be used.

Surgical (and, to some extent, genetical) experiments suggest that the normal development and the regeneration of insect limbs depends on a detailed epidermal map of stable positional values, arranged down and around the limb. This model (the P.C.M.) links change in positional value directly to cell division, so intercalary regeneration, and perhaps normal growth (7), is provoked by a positional discontinuity between adjacent cells. The use of these techniques to study the relationship between positional maps in the limb and the rest of the segment, and between different segments, should test the assumption that the same positional map can underlie very different patterns, and could help us to relate the initial events of segmentation of the insect embryo to the subsequent development of those segments.

We need somehow to augment this level of analysis with investigation of the molecular nature of "positional value" and "cellular interaction". The local growth and patterning response to a positional discontinuity suggests that positional value could be a property of the cell surface, rather than the traditional "concentration of a diffusible morphogen". Hence it may be amenable to monoclonal antibody searches for position-specific antigens. An alternative molecular approach in Drosophila could be via investigation of the transcripts and protein products from genes whose mutant alleles show specific abnormalities in pattern or growth control. Some clues may eventually come from the current analysis of "segmentation" and homeotic genes, although at least the latter seem to affect interpretation rather than the formation or nature of positional maps.

ACKNOWLEDGEMENTS

For their collaborative endeavours I wish to thank Hilary Anderson, Tamara Rowlands and Neil Toussaint.

REFERENCES

1. Anderson H, French V (1985). Cell division during intercalary regeneration in the cockroach leg. J Embryol exp Morph (in press).
2. Bart A (1971). Morphogenèse surnuméraire au niveau de la patte du Phasme Carausius morosus Br. Wilhelm Roux Arch 166:331.
3. Bohn H (1965). Analyse der Regenerationsfähigkeit der Insekten-extremität durch Amputations- und Transplatationsversuche an Larven der afrikanschen Schabe Leucophaea maderae Fabr. (Blattaria). II. Mitt. Achsendetermination. Wilhelm Roux Arch 156:449.
4. Bohn H (1970). Interkalare Regeneration und segmentale Gradienten bei den Extremitäten von Leucophaea-Larven (Blattaria). I. Femur und Tibia. Wilhelm Roux Arch 165:303.
5. Bohn H (1971). Interkalare Regeneration und segmentale Gradienten bei den Extremitäten von Leucophaea - Larven (Blattaria) III Die Herkunft des interkalaren Regenerates. Wilhelm Roux Arch. 167:209.
6. Bohn H (1974). Extent and properties of the regeneration field in the larval legs of cockroaches (Leucophaea maderae), I Extirpation experiments. J. Embryol exp Morph. 31:557.
7. Bryant P, Simpson P (1984). Intrinsic and extrinsic control of growth in developing organs. Q Rev Biol. 59:387.

8. Bryant S V, French V, Bryant P J (1981). Distal regeneration and symmetry. Science 212:993.
9. Bullière D (1970). Interprétation des régénérats multiples chez les Insectes. J Embryol exp Morph 23: 337.
10. Bullière D (1971). Utilisation de la régénération intercalaire pour l'étude de la détermination cellulaire au cours de la morphogenèse chez Blabera craniifer (Insecte Dictyoptère). Dev Biol 25:672.
11. French V (1978). Intercalary regeneration around the circumference of the cockroach leg. J Embryol exp Morph 47:53.
12. French V (1980). Positional information around the segments of the cockroach leg. J Embryol exp Morph 59:281.
13. French V (1984). The structure of supernumerary leg regenerates in the cricket. J Embryol exp Morph 81:185.
14. French V (1984). A model of insect limb regeneration. In Malacinski G, Bryant S V (eds): "Pattern Formation: A Primer In Developmental Biology", New York, London: MacMillan, p339.
15. French V (1985). Interaction between the leg and surrounding thorax in the beetle, Tenebrio (in prep).
16. French V, Bryant P J, Bryant S V (1976). Pattern regulation in epimorphic fields. Science 193:969.
17. French V, Toussaint N (1985). The origin of supernumerary leg regenerates after interspecies grafts in the cricket (in prep).
18. Girton J R (1981). Pattern triplications produced by a cell-lethal mutation in Drosophila. Dev Biol 84:164.
19. Girton J (1983). Morphological and somatic clonal analysis of pattern triplications. Dev Biol 99:202.
20. Lewis J (1981). Simpler rules for epimorphic regeneration: the polar co-ordinate model without polar co-ordinates. J theor Biol 88:371.
21. Meinhardt H (1982). "Models of Biological Pattern Formation". London, New York: Academic Press.
22. Meinhardt H (1984). Models for positional signalling, the threefold subdivision of segments and the pigmentation pattern of molluscs. J Embryol exp Morph 83 Suppl: 289.
23. Mittenthal J (1981). The rule of normal neighbours: a hypothesis for morphogenetic pattern regulation. Dev Biol 88:15.

24. Sander K (1976). Specification of the basic body pattern in insect embryogenesis. Adv Insect Physiol 12:125.
25. Steiner E (1976). Establishment of compartments in the developing leg imaginal disc of Drosophila melanogaster. Wilhelm Roux Arch 180:9.
26. Stumpf H (1968). Further studies on gradient-dependent diversification in the pupal cuticle of Galleria mellonella. J exp Biol 49:49.
27. Truby P R (1983). Blastema formation and cell division during cockroach limb regeneration. J Embryol exp Morph 75:151.
28. Truby P R (1985). Separation of wound healing from regeneration in the cockroach leg. J Embryol exp Morph 85:177.
29. Wolpert L (1971). Positional information and pattern formation. Curr Top Dev Biol 6:183.

Molecular Determinants of Animal Form, pages 477–488
© 1985 Alan R. Liss, Inc.

POSITION-SPECIFIC ANTIBODIES AS PROBES FOR THE REGIONAL
IDENTITY OF DROSOPHILA IMAGINAL DISC CELLS[1]

Danny L. Brower

Department of Molecular and Cellular Biology and Department
of Biochemistry, University of Arizona, Tucson, AZ 85721

ABSTRACT Position-specific (PS) antigens show non-uniform distributions on imaginal discs from wild type Drosophila melanogaster larvae. Here, I show that the PS antigen distributions are essentially unchanged in discs from mutants that alter growth properties and in discs from other Drosophila species; thus genetic differences that do not alter the specification of adult pattern also fail to alter PS antigen patterns. These findings are consistent with the proposed role for the PS antigens in morphogenesis, and especially in combination with data from patterning mutants, indicate that the PS antigens can be used as reliable markers for the regional identity of undifferentiated cells in imaginal disc epithelia.

INTRODUCTION

Drosophila imaginal discs provide an excellent system for studying intercellular interactions in undifferentiated tissues. The epithelial cells of each disc are stably determined to make adult cuticle of a given regional character, for example, dorsal mesothorax in the case of the wing disc (1). Moreover, each region within the wing disc is specified to make a particular dorsal mesothoracic structure (2), however this more precise specification is not stably inherited and appears to depend on continuous intercellular signalling (3). Because the disc epithelium is composed of one cell type, cell surface molecules that are expressed with regional specificity within discs are good

[1]Supported by grant GM 34112 from the NIH.

candidates for molecules involved in either the pattern formation process, or processes, such as morphogenesis, that depend on the positional fields that are established. We set out to identify and characterize such molecules, using monoclonal antibodies against Drosophila imaginal discs and embryos.

A number of antibodies have been found that bind to members of a related family of cell surface glycoproteins (4,5). The antigens recognized by these antibodies are expressed non-uniformly on imaginal discs from mature larvae, and the expression of a particular antigen by a given disc region appears to depend only on the position of the region in the undifferentiated epithelium. There is no clear correspondence between the expression of a particular antigen and the type of adult structure that a region is specified to make at metamorphosis; cells that will make identical structures in different locations can display essentially qualitative differences in the antigens on their surfaces. Because of this positional specificity, the 3 related antibody classes were designated PS1-3. PS1 antibodies bind primarily to the dorsal region of late third instar wing discs, and PS2 antibodies primarily to the ventral region. PS3 antibodies recognize both the PS1 and PS2 antigens, plus at least one other related component, the distribution of which remains to be determined.

In a number of instances, the distributions of the PS antigens can be correlated with morphogenetic events in the epithelia of various imaginal discs (6). In particular, discrete morphological entities are often found where there are discontinuities in the distributions of the antigens. For example, distinct grooves on the apical and basal surfaces of the disc epithelium are found at the presumptive wing margin (7,8), and the cells on either side of these grooves show essentially qualitative differences in the amounts of PS1 and PS2 antigens expressed on their surfaces. This and similar observations in other discs and at other times during development have led us to the hypothesis that the PS antigens are involved in cell-cell recognition or adhesion processes during morphogenesis (6).

The above hypothesis predicts that the patterns of PS antigen expression should be similar both in discs from strains carrying mutations that affect some cellular functions without altering pattern (for example, mutations affecting growth rate), and in discs from closely related species that undergo similar morphogenesis. These tests are especially relevant in light of work on the distribution of the enzyme aldehyde oxidase (AO) in Drosophila wing discs. In wild type discs, AO shows non-uniform distributions that appear, in some cases at least, to correlate with patterning processes in the undifferentiated epithelium (9). The AO distributions are very different, however, in slow-growing discs (carrying Minute mutations, which do not detectably alter disc morphology or the specification of adult pattern) and in discs from related species of flies (10), indicating that AO expression is not directly associated with the processes of disc morphogenesis or the specification of adult pattern elements.

Regardless of their function, the nonuniform distributions of the PS antigens in imaginal discs makes the PS antibodies potentially useful as probes for the regional identity of cells in the undifferentiated disc epithelia. Pattern disruptions in discs can result from mutations in numerous genetic loci, but for the most part analysis of these disruptions has depended on examination of the adult cuticle made by the disc cells at metamorphosis. Thus, a dynamic picture of disc patterning is derived only from inferences following experimental manipulation and examination of the finished product. And although the adult cuticle is rich in positional markers, some regions, such as the dorsal and ventral wing blade, still cannot be distinguished from one another in the adult but do show clear PS antigen differences. A demonstration that PS antigen patterns are unaffected by variables such as growth rate is important if the PS antibodies are to be useful as probes for the regional identity of cells in mutant discs, especially since mutations that alter pattern specification may also affect growth rate. Also, establishing the species range for cross-reactivity of the PS antigens is important for determining if the PS antibodies can be used in studying developmental processes in species other than Drosophila melanogaster.

MATERIALS AND METHODS

The antibodies used in this study were CF.5E5 (PS1), CF.2C7 (PS2), and CF.6G11 (PS3). The production of these antibodies and details of the immunofluorescence staining procedure are described in reference 4. Briefly, Drosophila melanogaster wing imaginal discs were dissected from balanced $M(2)c^{33a}$ (11,12) and l(2)gd (13) stocks (homozygous l(2)gd larvae were selected using the orange marker for malpighian tubule color), and incubated in PS antibodies in a nitro blue tetrazolium staining solution. Washes and dilution of fluorescein-conjugated goat anti-mouse IgG (Antibodies Incorporated) were in RPMI 1640 medium (Flow Labs.) with 5-10% bovine serum. After the final wash, the discs were fixed in 2% formaldehyde for 10 min and mounted in 70% glycerol/30% 0.1 M Tris Cl, pH 9, to which was added 2% n-propyl gallate. The discs were viewed under epi-illumination on a Zeiss universal microscope. Photographs were on Kodak Tri-X film developed with Diafine developer.

For species other than <u>Drosophila melanogaster</u>, the nitro blue tetrazolium solution sometimes interfered with the immunofluorescence. Therefore, primary antibody incubations for these discs were in RPMI 1640, with subsequent processing as above. This procedure results in some diffuse background fluorescence and antibody-induced aggregation of the antigens on the cells (4), but does not interfere with the localization of the antigens on the disc epithelium.

RESULTS

The PS antibody binding patterns are particularly striking on wing discs from late third instar larvae, and these patterns have been characterized in more detail than the patterns on other imaginal discs. PS1 antibodies bind primarily to the dorsal region of the disc, and PS2 antibodies to the ventral region (4). (The coordinates of the disc - dorsal, anterior, etc. - used here are defined with respect to the location of the adult structures that a disc region is specified to give rise to.) Although PS3 antibodies bind nonuniformly as well, the PS3 patterns probably result from cross-reaction with the PS1 and PS2 antigens (4,5), and the variations in PS3 binding to different regions of the disc are not nearly so great as for the other antibodies. For these reasons, this study has focused on the PS1 and PS2 immunofluorescence patterns on mature wing discs. In some instances other discs have been examined, and those results are consistent with the findings reported here.

Mutations That Affect Growth.

Minute mutations reduce the growth rate of Drosophila cells, including those of the imaginal discs. I examined the PS antigen distributions in discs carrying the strong Minute, $M(2)c^{33a}$ (12). No differences from the wild type PS antigen patterns were detected (Figures 1 & 2). Not only are the gross dorsoventral asymmetries maintained, but more detailed aspects of the patterns, such as the pattern of disappearance of the PS2 antigen from the most anterior ventral part of the disc, are also similar to the wild type patterns.

Growth control apparently is lost in the imaginal primordia of animals homozygous for the mutation lethal(2) giant discs. For example, if pupariation is delayed by overcrowding, or if discs are cultured in adult hosts, wing discs from these mutants can grow to many times their normal size (13,14). Although extreme overgrowth may be accompanied by pattern disruptions, wing discs dissected from uncrowded, mature-looking l(2)gd larvae typically appear similar to wild type discs, except that the large folds of the epithelium are not as deep as normal (13). This flattening effect is particularly pronounced in the wing pouch.

The growth properties and aberrant disc morphology of l(2)gd larvae make it difficult to assign precise developmental ages to the discs that were examined. It is clear, however, that the PS antigen patterns of mutant discs from larvae that are nearing pupariation are similar to those of late third instar wild type discs (Figures 3 & 4). Not only is the dorsoventral asymmetry preserved, but it appears that detailed aspects of the patterns are also similar. For example, the faintly fluorescent dorsal PS2 antibody-binding patches observed in wild type wings are clearly visible in figure 4. Indeed, because the folds of the l(2)gd disc are not so deep as those in wild type discs, it is possible to see that these two dorsal patches are not continuous at the bottom of the fold, an observation that was not possible in whole mounts of wild type discs (4,6).

Related Species of Drosophila.

The patterns of binding of the PS antibodies to some related species of Drosophila are shown in figures 5-8, and

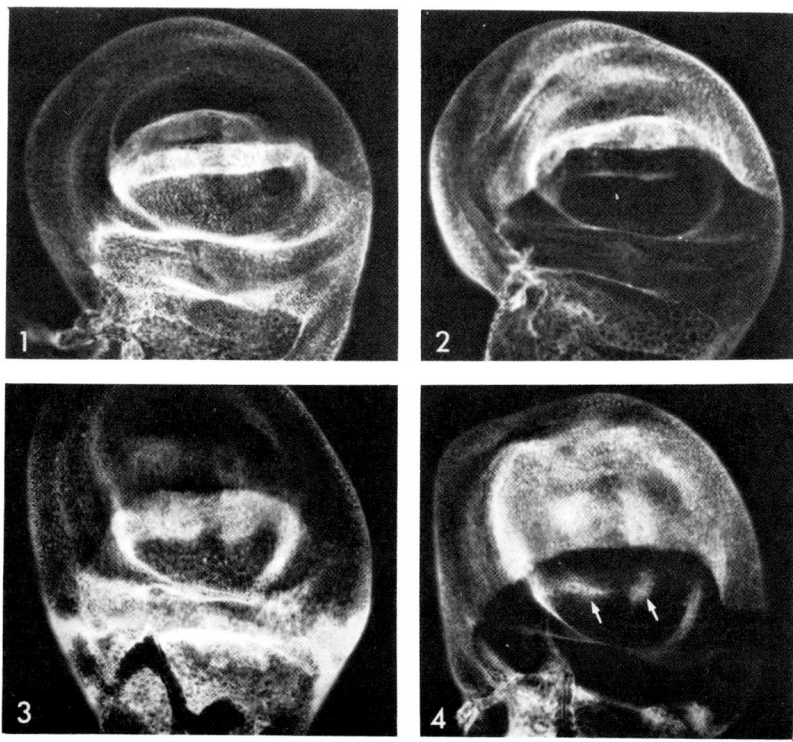

FIGURES 1 & 2. PS1 and PS2 immunofluorescence patterns on wing imaginal discs from late third instar larvae heterozygous for Minute(2) c^{33a}. The patterns are indistinguishable from those of wild type discs, with the PS1 antigen found primarily on dorsal cells (lower part of disc in figure 1) and the PS2 antigen primarily on ventral cells (upper part of disc in figure 2).

FIGURES 3 & 4. PS1 and PS2 immunofluorescence patterns on wing discs from larvae homozygous for lethal(2) giant discs. The disc folds in this mutant are not so pronounced as in wild type discs, so that the wing pouch seems to be extended in the dorsoventral axis (top to bottom of the figure). The antibody binding patterns are very similar to those of the wild type; even the faintly fluorescent PS2 patches normally observed in the wild type dorsal wing pouch can be seen in figure 4 (arrows). All X135.

TABLE 1
PS ANTIBODY BINDING TO DIFFERENT DROSOPHILA SPECIES[a]

	PS1 (CF.5E5)	PS2 (CF.2C7)	PS3[b] (CF.6G11)
D. simulans	+	+	+
D. ananassae	−	+	−
D. willistoni	−	+	−
D. pseudoobscura	+[c]	−	−

[a]These results are consistent with our preliminary affinity purification experiments, with the exception of a failure to purify the PS1 antigen from D. simulans.
[b]Although shown as negatives, very faint binding is often observed with the PS3 antibody on the lower three species; this binding is too faint to tell if it corresponds to the PS3 pattern on D. melanogaster.
[c]Faint binding is sometimes observed in the wing pouch. This may be due to cross-reaction with a non-PS antigen.

summarized in table 1. Drosophila simulans, which is very closely related to Drosophila melanogaster, possesses antigens that cross-react with all three PS antibodies. As can be seen in figures 5 & 6, the major features of the PS1 and PS2 patterns are similar in Drosophila melanogaster and Drosophila simulans.

Although cross-reactivity disappears rather quickly as one moves away from Drosophila melanogaster, when strong antibody binding is observed, the patterns are similar to those seen in melanogaster. For example, the ventral specificity of the PS2 antigen is conserved in both Drosophila ananassae and Drosophila willistoni (Figures 7 & 8). No clear binding was observed with the other PS antibodies in these species or in Drosophila pseudoobscura. In some of these cases where there was no clear binding, extremely faint fluorescence was seen but the levels were almost at the background levels, and it was impossible to see a clear pattern. Moreover, it is quite possible that at least some of this very faint fluorescence is due to weak cross-reactivity with antigens unrelated to the PS antigens.

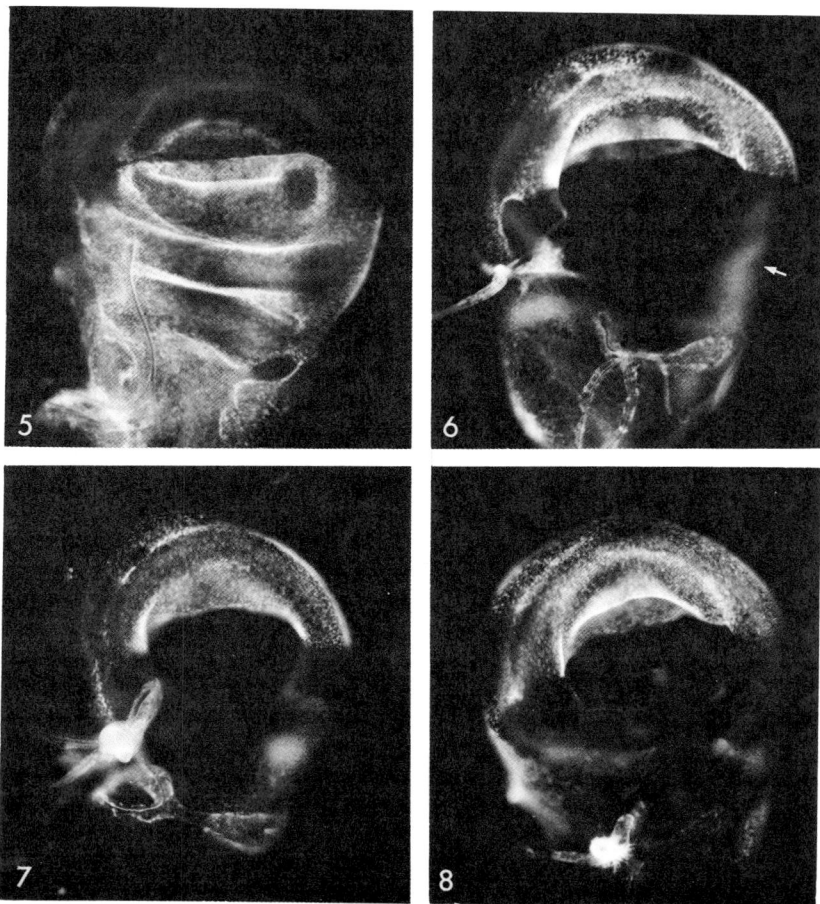

FIGURES 5 & 6. PS1 and PS2 patterns on late third instar wing discs from Drosophila simulans. The patterns are very similar to the Drosophila melanogaster patterns. The blurry patches of fluorescence in figures 5-8 (e.g., arrow in figure 6) are from binding to the reverse side of the disc. X156.

FIGURES 7 & 8. PS2 antibody binding to wing discs from Drosophila ananassae (Fig. 7) and Drosophila willistoni (Fig. 8). The patterns are very similar to those of Drosophila melanogaster. X169.

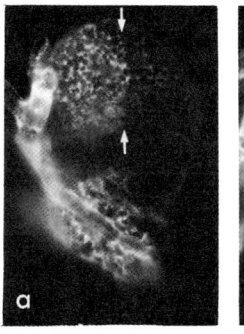
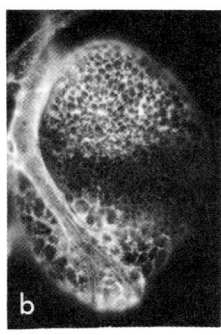

FIGURE 9. PS2 antibody binding to wild type (a) and engrailed[1] (b) second instar wing discs. The smaller cells in the upper part of each disc are part of the epithelium proper; the bright large cells on the lower part of the discs are adepithelial cells, which are muscle precursors. In the wild type, the brightest binding to the disc epithelium is on the anterior ventral quadrant (upper left). The sharp line separating the anterior and posterior halves of the wild type disc (arrows) is missing in the engrailed mutant, where the posterior half appears to be transformed into another anterior half (see ref. 15). X265.

DISCUSSION

The relatively rapid loss of PS antigen cross-reactivity as one moves away from Drosophila melanogaster is unfortunate, as the usefulness of the antibodies to examine developmental events in other species is quite limited. Of course, one may be able to extend the range of cross-reactivity by assaying a larger sample of PS monoclonals, but the results so far are not encouraging, particularly when considering the possibility that useful probes will emerge for non-Drosophilids, where the lack of genetics makes alternative tools for developmental analyses especially desirable.

The relative cross-species stability of the PS2 epitope suggests that there may be significant selective pressure on the particular molecular structure in this part of the PS complex (5). Especially considering the very specific pattern of PS2 antibody binding, it seems likely that the PS2 antigenic site is an important functional site on the PS complex.

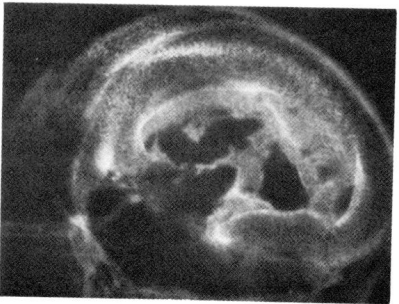

FIGURE 10. PS2 antibody binding to a late third instar wing disc from the mutant ap^{ts78j}/ap^c. The normal dorsoventral pattern of the disc is severely disrupted, with islands of dark dorsal cells surrounded by fluorescing ventral cells in the wing pouch (compare with fig. 2; see text and ref. 16 for details). X130.

In all cases that have been examined, genetic differences that do not significantly alter the morphology or specification of adult pattern in imaginal discs also fail to disrupt the normal pattern of binding of the PS antibodies to discs. As shown here, a strong Minute mutation, which reduces proliferation rate (12) and probably affects cellular metabolic processes, has no detectable effect on the PS antigen distributions. Similarly, wing discs carrying the mutation l(2)gd, which disrupts the normal process of growth control (14), show essentially normal patterns of PS antibody binding. (It should also be mentioned that, in the course of various studies, a number of Drosophila melanogaster stocks bearing sundry mutations for characters such as bristle morphology or cuticle color have been examined without any PS antigen pattern abnormalities being detected.) Moreover, when clear PS antibody binding is detected on wing discs of related species of Drosophila, the patterns are similar to those seen for discs of Drosophila melanogaster. These findings all are consistent with a role for the PS antigens in patterning or morphogenesis (6).

Perhaps more importantly, the data presented here help to validate the usefulness of the PS antibodies as probes for the regional identity of cells in the undifferentiated disc epithelia. A marker for regional character that also varies with proliferation rate is of little use in characterizing patterning in mutant discs where secondary effects

on growth are a possibility. All of our results suggest
that genetic differences that do not affect disc patterning
(as assayed by adult and disc morphology) also fail to
alter the expression of the PS antigens.

From this base, we have used the PS antibodies to
examine pattern abnormalities in discs carrying mutations
at a number of loci. In engrailed wing discs, we found PS2
antigen patterns that indicated extensive posterior to
anterior transformations at relatively early times in disc
development, effectively ruling out some of the alternative
explanations that had been offered for the adult engrailed
phenotype (Figure 9 and reference 15). In discs from
animals carrying mutations at the apterous locus, we found
a disruption of the normal dorsoventral pattern specification, including islands of dorsal cells surrounded by
ventral cells and vice versa (Figure 10 and reference 16).
This finding provides an explanation for the observation
that adults of this genotype display circles and tufts of
"wing margin" bristles throughout the wing blade. Although
the wild type antigen distributions place limits on the
precise types of pattern irregularities that can be investigated, it is clear that the antibodies will prove to be
useful in similar studies of other mutations.

ACKNOWLEDGEMENTS

I would like to thank Bill Heed and Danny Sponaugle
for providing fly stocks, Mary Stevens for help with the
apterous work, and Michel Piovant and Daryl Chan, who
performed preliminary experiments on PS antigens from
different species.

REFERENCES

1. Gehring W (1972). The stability of the determined state
   in cultures of imaginal disks in Drosophila. In
   Ursprung H, Nothiger R (eds): "The Biology of Imaginal
   Discs," Berlin: Springer-Verlag, p 35.
2. Bryant PJ (1975). Pattern formation in the imaginal
   wing disc of Drosophila melanogaster: Fate map,
   regeneration and duplication. J Exp Zool 193:49.

3. Bryant PJ (1978). Pattern formation in imaginal discs. In Ashburner M, Wright TRF (eds): "The Genetics and Biology of Drosophila," London: Academic, p 230.
4. Brower DL, Wilcox M, Piovant M, Smith RJ, Reger LA (1984). Related cell-surface antigens expressed with positional specificity in Drosophila imaginal discs. Proc Natl Acad Sci USA 81:7485.
5. Wilcox M, Brown M, Piovant M, Smith RJ, White RAH (1984). The Drosophila position-specific antigens are a family of cell surface glycoproteins. EMBO J 3:2307.
6. Brower DL, Piovant M, Reger LA (1985). Developmental analysis of Drosophila position-specific antigens. Dev Biol 108:120.
7. Eskens A, Sprey Th, Westra A (1981). Morphological indications for compartmental boundaries in the imaginal wing disc of Drosophila melanogaster. Neth J Zool 31:773.
8. Brower DL, Smith RJ, Wilcox M (1982). Cell shapes on the surface of the Drosophila wing imaginal disc. J Embryol Exp Morphol 67:137.
9. Kuhn DT, Cunningham GN (1977). Aldehyde oxidase compartmentalisation in Drosophila melanogaster wing imaginal discs. Science 196:875.
10. Sprey ThE, Eskens AAC, Kuhn DT (1982). Enzyme distribution patterns in the imaginal wing disc of Drosophila melanogaster and other diptera. Wilhelm Rouxs Arch Dev Biol 191:301.
11. Lindsley DL, Grell EH (1968). "Genetic Variations of Drosophila melanogaster." Carnegie Institution of Washington Publication No 627.
12. Ferrus A (1975). Parameters of mitotic recombination in Minute mutants of Drosophila melanogaster. Genetics 79:589.
13. Bryant PJ, Schubiger G (1971). Giant and duplicated imaginal discs in a new lethal mutant of Drosophila melanogaster. Dev Biol 24:233.
14. Bryant PJ, Levinson P (1985). Intrinsic growth control in the imaginal primordia of Drosophila, and the autonomous action of a lethal mutation causing overgrowth. Dev Biol 107:355-363.
15. Brower DL (1984). Posterior-to-anterior transformation in engrailed wing imaginal disks of Drosophila. Nature 310:496.
16. Stevens ME, Brower DL (1985). Disruption of positional fields in apterous imaginal discs of Drosophila. submitted.

Molecular Determinants of Animal Form, pages 489-519
© 1985 Alan R. Liss, Inc.

MOLECULAR ANALYSIS OF THE INVOLVEMENT OF THE DROSOPHILA engrailed GENE IN EMBRYONIC PATTERN FORMATION[1]

Patrick H. O'Farrell, Claude Desplan
Stephen DiNardo, Judy Kassis, Jerry Kuner[2], Emily Lim
Elizabeth Sher, James Theis and Deann Wright

Dept of Biochemistry & Biophysics
University of California, San Francisco
San Francisco, CA 94143

ABSTRACT

The action of at least 22 genes early in Drosophila development is required to establish the basic segmental body plan of the organism. The spatial regulation of expression of these developmental genes is fundamental to the development of embryonic pattern. We have examined one of these genes, engrailed, at a molecular level. The cloning of the locus, physical mapping of a number of engrailed mutations, and identification of the coding region revealed a surprising discrepancy between the genetic and molecular definitions of the engrailed coding unit. Apparently, the spatial pattern of expression of engrailed, and perhaps other developmental genes, relies on a regulatory region that is dispersed over several tens of kilobases so that mutations reveal that the functional unit is much larger (70 kb) than the primary transcription unit (about 4.5 kb). Preparation of antibodies against the engrailed protein has allowed localization of the engrailed gene product by immunofluorescent staining of whole mount

[1] This work was supported by NSF grant DCB-8418016
[2] Present address: Synergen, 1885 33rd street, Boulder CO 80301

embryos. The immunofluorescence is nuclear and the stained nuclei are restricted to a remarkable pattern of stripes that transect the embryo along its anterior posterior axis. The staining provides an extremely useful aid to observations of the morphogenetic events that first form and then transform the segmental pattern. Segments of the engrailed coding sequence expressed as fusion proteins in E. coli are found to have sequence specific DNA binding activity. The region of the protein having binding activity is within or near the homeobox domain. The engrailed fusion proteins bind near the 5' ends of the engrailed gene and the fushi tarazu gene. These results imply that engrailed and perhaps all homeobox-containing proteins are sequence specific DNA binding proteins and regulators of gene expression and further suggest that the different developmental genes may interact in a complex regulatory network.

INTRODUCTION

At ten hours of development the Drosophila embryo displays its basic segmental body plan. The development of this morphological form requires that at least 22 genes express their products in the developing embryo (1-4). Mutations in these genes lead to ordered abnormalities in the normal segmental periodicities. For example, embryos mutant at the ftz locus develop only half the normal number of segmental units. The seven enlarged segments that do develop in ftz mutants display morphologies that correspond to every alternate segment of the normal pattern (6). This mutant typifies a class of segmentation mutants called "pair-rule mutants" because the mutant defects have a two segment periodicity (1). In contrast, mutants at the engrailed locus produce pattern defects within each segment. In engrailed mutants the bounds of the segments are indistinct and fusions of adjacent segment repeats are observed (7,8). Molecular analyses of the engrailed and ftz genes led to a striking observation--at stages prior to obvious morphological segmentation the ftz gene and the engrailed gene are expressed in a spatially restricted pattern of stripes whose number and position anticipates the position of structures that would be defective in the corresponding mutants. That is, hybridization of DNA probes to ftz RNA in tissue sections detects seven bands of hybridization that transect the blastoderm stage embryo and thus define eight subdivisions of the embryo along the anterior-posterior axis (9). Hybridization with engrailed probes detects a pattern of stripes of expression that develops about 30 minutes later than the ftz pattern and includes roughly twice as many stripes as the ftz pattern (10). These observations support a model that proposes that these genes encode regulators of development and that localized gene expression plays an essential role in defining the positions in which these regulators act. Thus, the basic segmental body plan of the embryo is in part effected by the periodic pattern of expression.

Elegant genetic experiments have traced the clonal fate of cells during Drosophila development, and have examined the requirement for engrailed function later in development (7,8,11-14). These studies have shown that early in development (15) the fate of cells is partially defined--the position that their descendants will occupy is roughly defined but the cells are not specified in terms of their eventual cell type. From this early stage onward, the cells that are specified to develop as the posterior parts of segments are clonally isolated from those cells that develop as anterior parts of segments. These two distinct groups of cells are referred to as developmental compartments (11,13). engrailed gene activity is required only in the posterior compartments and without its function these posterior cells appear to lose their distinct identity. They take on some of the attributes of anterior cells and are no longer clonally isolated (16). These results suggest that early in development engrailed expression is switched on in particular cells and that this regulatory switch is long lasting so that the clonal descendants of these cells continue to express engrailed function. Furthermore, these results support the idea that the engrailed product is a regulator that identifies cells as posterior and defines in part their developmental fate.

Molecular studies of engrailed will allow us to address at a molecular level several issues of pattern formation. We can investigate the mechanism of regulation that defines the spatial program of expression. We can ask how the engrailed gene product acts as a regulator of developmental fate. Finally, we can investigate at a molecular level a determinative event in development--that is, how is the early decision, whether or not a particular cell will express engrailed, propagated clonally to all the daughter cells.

The significance of these molecular studies in Drosophila would be augmented if the processes under scrutiny were universally used mechanisms. A surprising level of sequence homology between several homeotic genes and segmentation genes of Drosophila has been found and implies that the encoded proteins include similar functional domains (17,18). The shared domain, the homeobox, is 60 amino acids long and constitutes only a small portion of each of these regulatory gene products. This shared domain is conserved with extraordinary fidelity in distantly related organisms (19) including human (20), mouse (21), frog (22), and in more degenerate forms, in yeast (23,24). Since this homology implies a common function for this domain, elucidation of the mechanism of action of these genes in Drosophila may have far reaching consequences. Because the yeast genes having distant homology to the homeobox are DNA binding proteins and transcriptional regulators (25, and Johnson and Herskowitz, in prep.), the prevalent hypothesis is that the developmental genes of Drosophila will similarly be regulators of transcription (19,23). The sequence of the homeobox is compatible with sequence features of the DNA binding domains of bacterial and phage transcriptional regulators (23). Since the engrailed gene also contains a homeobox sequence we should be able to test these predictions by examining the function of the engrailed protein.

Although the data are only suggestive, there are a number of diverse indications that the various developmental genes interact so that the expression of each gene is influenced by the others. Expression of one homeotic gene can alter the spatial distribution of a second unlinked homeotic gene (26,27). Additionally, some engrailed mutations produce a transformation of posterior wing structures to anterior haltere structures and the dependence of this phenotype on alleles of the Bithorax complex suggests that normal engrailed function might inhibit bx function (28). This suggestion is nicely compatible with previous suggestions that bx function is normally restricted to the anterior compartment of the third thoracic segment (29). Finally, because the ftz^{Rp1} mutation gives a dominant phenotype resembling that of pbx, it appears that the ftz gene product might also interact with the Bithorax complex (23). Some of the data we present here also suggest regulatory interactions among developmental genes.

Drosophila engrailed Gene in Pattern Formation

Here we summarize studies that document the spatial and temporal regulation of engrailed gene expression, that demonstrate that the engrailed product is a sequence specific DNA binding protein, that define functionally important domains of the engrailed gene and that provide preliminary indications of a regulatory network among the Drosophila developmental genes.

SUMMARIES OF RESULTS

Physical Mapping of the engrailed Gene (Kuner, et al., in prep).

Using a met_2 tRNA as a probe, we isolated a tRNA gene that mapped (48B) near the engrailed locus (48A) and by chromosome walking techniques (30) we isolated 255 kb of DNA that included the engrailed locus and flanking regions. To define the location of the engrailed gene within these clones we mapped the positions of 15 engrailed rearrangement mutations. The map of breakpoint positions shows that interruptions distributed over 70 kb of genomic sequences disturbed engrailed function (Figure 1). This analysis defines the minimum size of the gene as 70 kb.

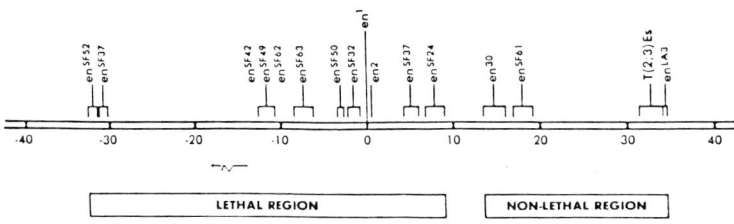

Figure 1. Localization of engrailed Rearrangement Mutations. The physical positions of 15 engrailed mutations were defined by Southern analysis. Their approximate positions are indicated with brackets above a scale in kb. The engrailed transcript is indicated at position -13 to -17½ kb. The zero point of this physical map is the position of an insertion element in the en^1 mutation.

Different parts of the 70 kb engrailed gene do not seem to be equally important for engrailed function. The mutants that were located within the distal 20 kb of this engrailed gene did not give lethal phenotypes and furthermore, breakpoints in the more proximal region, though lethal, do not disturb the morphology of the early embryo as severely as deletions for the engrailed gene. Thus, some aspects of engrailed function do not depend on the integrity of the entire 70 kb. Because all the mutations belong to a single complementation unit, various regions spanning 70 kb must interact in cis for complete engrailed activity.

Although one might expect a fairly close correspondence between the genomic DNA genetically defined as that involved in function, and the position and extent of the primary transcription unit, Northern analysis failed to detect a transcript that spanned the positions of all the engrailed breakpoint mutants (Drees, O'Farrell and Kornberg, in prep). In fact, detailed molecular characterization defines a major embryonic transcript of 2.7 kb that is encoded by about 4.5 kb of genomic sequences (31). Thus, one surprise from these early analyses was that the mutations defined a large complementation unit of 70 kb while the molecular analysis identified a small centrally located transcription unit. Because the small transcription unit appears to encode engrailed function, we presume that the large size of the genetic unit reflects a requirement for extensive flanking sequences for proper regulation of expression (Kuner, et al., in prep).

Isolation of Antibodies Against the engrailed Protein (DiNardo et al., in prep).

To study the engrailed protein and its expression we made antibodies against bacterially synthesized fusion proteins. In these fusions a small segment of engrailed coding sequence was inserted in the N-terminal region of a plasmid β-galactosidase gene (Figure 2). Rabbit antisera were purified by immunoadsorption to remove antibodies directed against β-galactosidase. Subsequently, antibodies reacting with the Drosophila portion of the fusion protein were isolated by immunoaffinity purification. DNA sequencing of the open reading frames used to construct the fusions and studies of the immunological cross reaction of the antibodies to the fusion proteins demonstrated that we had antibodies to two regions of the engrailed protein, en-ORF1 and en-ORF2, that do not overlap or show evident immunological crossreaction.

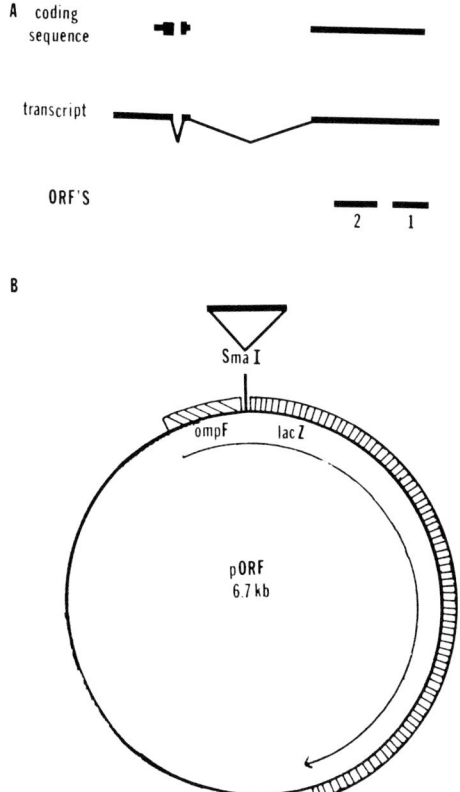

Figure 2. Construction of Fusion Proteins For Use As Antigens.
a) The engrailed transcript and exon structure is schematically represented with the protein coding portions of the sequence indicated above. The homeobox domain is indicated as a thicker region of the line representing the protein coding region. The two regions of this coding sequence that were expressed to produce engrailed antigens, en-ORF1 and en-ORF2, are indicated below the transcript.
b) For expression, en-ORF1 and en-ORF2 were inserted into the Sma I of the E. coli vector pORF (45). Expression initiates at the ompF promoter and proceeds through regions encoding the ompF signal sequence, the engrailed ORF and the lacZ gene. The resulting tribrid protein was used to immunize rabbits.

Immunofluorescent Staining of engrailed Protein (DiNardo et al., in prep).

To study the localization of expression of the engrailed protein we used immunofluorescent staining of whole mount Drosophila embryos. Although we will not be certain until we have examined engrailed mutations, our present results give us considerable confidence that the major immunofluorescent signal is due to the engrailed gene product rather than a crossreacting determinant. engrailed antibodies against en-ORF1 and en-ORF2 give similar results. The fluorescent staining is always nuclear. Embryos at the syncytial stage exhibit faint nuclear staining that is not localized into stripes. Immediately after the beginning of gastrulation we detected the first indications of spatially restricted expression. At this stage, immediately posterior to the developing cephalic furrow we detect six stripes of increased immunofluorescent staining. These stripes are perpendicular to the developing ventral furrow and each stripe consists of a row of highly staining nuclei. When they are first detected, each row is only one nucleus wide and is spaced from the adjacent row by three nuclei. By the germ band elongation stage the embryos exhibit much more intense fluorescence, and 15 bands of brightly staining nuclei are evident. At this stage the stripes are about three cells wide. The change in width is presumably a combined result of cell division and cell movements associated with germ band formation. The space between bands is now 5-6 cells wide. By the time of germ band shortening when segmentation is morphologically evident, we can clearly see that the brightly staining nuclei are in cells at the posterior portion of each evident segment division (Figure 3). This is exactly the distribution of staining that would have been predicted from the genetic analyses (see introduction). Complex patterns of staining are also evident in the head and tail regions of germ band shortened embryos. These patterns parallel a similarly complex pattern of lobulation in these regions. The segmental nature of head and tail regions is known to be complex, and it is apparent that a number of transformations shift the orderly array of early segmental divisions into the more complex patterns seen in these areas. We hope that further analysis of these staining patterns will help illuminate the nature of the complex morphological transformations that occur during gastrulation. At present we would like to emphasize that both spatially and temporally, the pattern of engrailed expression is extremely complex, but orderly, and that the antibody staining provides us with an efficient way of examining this expression. We would like to understand the molecular basis for this orderly pattern of expression. As suggested above, we think that the requirement for extended regions flanking the engrailed transcription unit reflects the regulatory role of these sequences. By examining the pattern of engrailed expression in mutants defective in different parts of the flanking region it might be possible to assign the regulatory responsibility for different aspects of the complex pattern of expression to different regions within the flanking sequences.

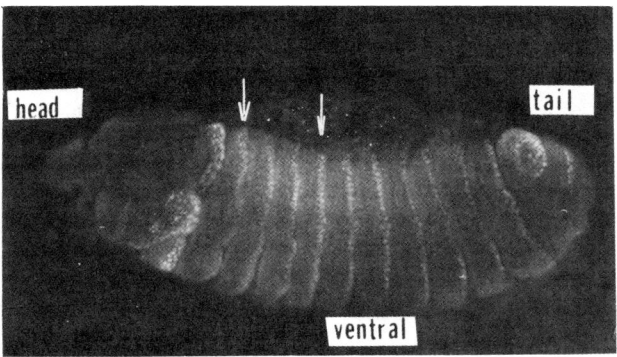

Figure 3. Embryonic Localization of engrailed Protein by Indirect Immunofluorescence. A Drosophila melanogaster embryo after germ band shortening (about 9hr after egg laying) has been fixed, treated with affinity purified anti-engrailed rabbit antisera. After excess antibody was washed out, the preparation was incubated with rhodamine-conjugated goat anti-rabbit antisera, washed and then visualized and photographed under epifluorescent illumination. The arrows mark stripes of engrailed-positive nuclei at the posterior edge of the first thoracic and first abdominal segments (left and right respectively). Also note the complex but orderly patterns of engrailed-positive nuclei in cells of the head and tail region.

Evolutionary Conservation of engrailed Sequences Provides an Assay for Functionally Important Domains (Kassis, Wong and O'Farrell, in prep).

Sequences that have no apparent function diverge during evolution at a rate of roughly 1% per million years (32,33). Divergence of sequences having function is constrained to different degrees by selection (34). The rates of divergence of coding sequences range from rates of about 1% per 100 million years for very conserved sequences, such as the histone genes, down to the unrestrained rate of divergence (1% per 0.7 MY). By comparing the engrailed gene sequence of Drosophila melanogaster with the engrailed gene of a distantly related species it should be possible to identify the more highly conserved sequences. These should correspond to more functionally relevant sequences. Because the pattern of conservation of a sequence with a particular type of function (e.g., a protein coding sequence) has distinctive characteristics, analysis of evolutionarily diverse sequences of a gene can aid functional interpretation. To pursue this approach we cloned 80 kb of genomic sequences representing the engrailed region of D. virilis. Since D. virilis diverged from D. melanogaster over 60 million years ago, only sequences having substantial functional constraints will be recognizable as homologous.

We had three interests in pursuing the evolutionary comparison. First, we wanted to determine whether the requirement for extended flanking sequences for engrailed expression reflects a requirement for a few distant functional sequences or whether functional elements are dispersed throughout the entire flanking region. Second, we wanted to determine whether parts of the protein coding sequence outside the homeobox domain are conserved. Third, we wanted to see whether there were conserved sequences near the 5' end of the engrailed gene that might provide an indication of regulatory sequences involved in controlling the expression of the engrailed transcript.

To detect and localize conserved sequence across the 70 kb engrailed region we made heteroduplexes of the corresponding D. melanogaster and D. virilis clones and examined these by electron microscopy. The positions of duplexed DNA represent sequences that have substantial homology, roughly 70% or greater (the imprecision of the homology estimate is intrinsic to the technique and it reflects the importance of other variables, such as AT

content of the sequence or proximity to a nearby stable
duplex). This analysis revealed numerous regions of
homology dispersed throughout the 70 kb engrailed region
(Figure 4). The blocks of homology indicated that the D.
melanogaster and D. virilis sequences are colinear and that
there are no major differences in the size of the region.
Because the β-globin gene cluster in mammals has not shown
evidence for conservation of size or sequence within the
non-coding sequences, we feel that these observations made
with the engrailed locus indicate the importance of
sequences throughout the large 70 kb engrailed genetic
unit.

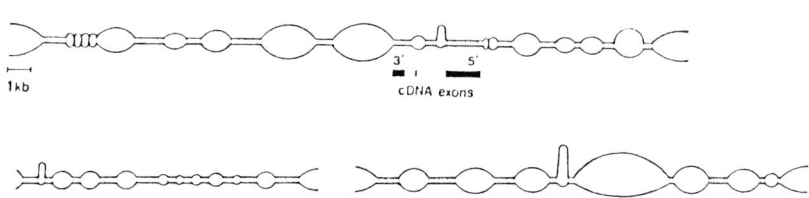

Figure 4. Schematic representation of EM heteroduplex
analysis showing the location of the conserved regions of
the engrailed locus. Single-stranded nonconserved regions
are shown as two curved lines; strands are shown to be the
same length except where known to differ in size. Blank
spaces indicate two short gaps in the data.

To examine the conservation of the engrailed coding
region we have sequenced a portion of D. virilis genomic
DNA that has homology to the transcribed region of D.
melanogaster DNA. We find a comparable protein coding
sequence and predict an exon structure similar to the D.
melanogaster gene (31). The homeobox sequence is
completely conserved at the amino acid level but contains
several nucleotide substitutions. Other regions of the
protein are also well conserved, and overall the pattern of
conservation strongly supports the protein sequence that
was proposed on the basis of conceptual translation of the
engrailed cDNA sequence (note that though the engrailed
protein produced in vivo has not been directly
characterized, the D. virilis sequence data and the data
obtained using antibodies to detect the protein by
immunofluorescence argue strongly that the predicted

sequence is correct). While a few regions of the protein sequence have diverged considerably between D. virilis and D. melanogaster, these regions are small. The major conclusion is that the function of the engrailed coding sequence is not restricted to the highly conserved homeobox domain. This conservation outside the homeobox domain may be useful in identification of the true homolog of the engrailed sequence in very distantly related organisms. When distantly related organisms are screened with a homeobox probe, clones representing a family of genes are detected. Identification of individual members of this family will probably rely on the detection of conserved sequences flanking the homeobox region.

The analysis of 5' flanking sequences is preliminary but encourages us to believe that we will be able to identify conserved regulatory sequences. The EM analysis detected a 0.6 kb region of duplexed sequence just upstream of the engrailed coding unit and recent sequence data detected some sequence homology in regions near the proposed start sites for transcription. Finally, we have detected small regions of sequence homology between the DNA upstream of the D. melanogaster ftz and engrailed genes; the homology between these genes suggests that homologous regulatory interactions control these genes and that we may be able to detect the target sequences for regulatory interactions as conserved sequences upstream of the coding region.

Sequence Specific DNA Binding of engrailed Fusion Proteins (Desplan, Theis and O'Farrell, in prep).

To examine the function of the engrailed protein we first needed a source of the protein. We chose to express the cloned cDNA sequence as a carboxy terminal fusion with β-galactosidase and to express this fusion in E. coli. To test whether the homeobox domain might have an autonomous function we created three fusions, a full length fusion (FL) with all of the engrailed coding sequences included, a homeobox fusion (HB) with just the carboxy terminal quarter of the engrailed coding sequences, and a nonhomeobox fusion (NHB) that included only the first three quarters of the engrailed coding sequences (Figure 5). The fusion proteins were partially purified from E. coli extracts and were tested for DNA binding.

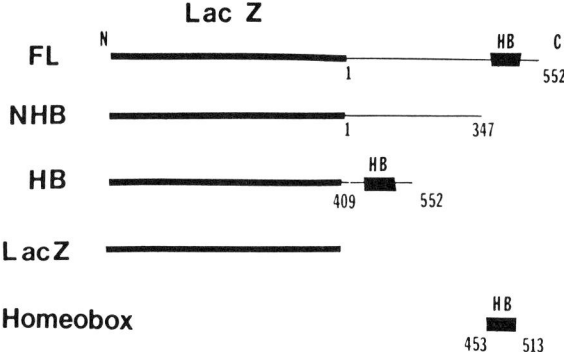

Figure 5. engrailed - lacZ Fusion Proteins
Three different fusion proteins were produced as C-terminal extensions of β-galactosidase in the plasmid pUR (Ruther and Muller-Hill, EMBO J. 2, 1791-94, 1983): The full-length open reading frame at the cDNA (FL), the N-terminal fragment deleted for the homeodomain (NHB), and the C-terminal fragment containing the homeodomain (HB).

Although we expected that the engrailed protein might be a sequence specific DNA binding protein we did not know what the specific target sequence might be. Since all sequence specific DNA binding proteins also exhibit low affinity nonspecific DNA binding we decided to first test the three fusions and a β-galactosidase control for such nonspecific DNA binding. To assay DNA binding, ^{32}P labeled restriction fragments were mixed with the test protein, and bound DNA was isolated by immunoadsorption to an anti-β-galactosidase antibody coupled to fixed Staphylococcus. The β-galactosidase and all of the fusion proteins were precipitated by the antibody with comparable efficiency. At low salt concentrations and in the absence of carrier DNA, weak associations of protein and DNA can be detected, while the addition of either salt or carrier DNA increases the stringency of the assay so that only higher affinity associations are detected. We first looked at nonspecific interaction of labeled Hae III fragments of bacteriophage φX174 with the various proteins. While β-galactosidase and the NHB fusion showed no detectable DNA binding, the FL fusion and the HB fusion bound all of the labeled DNA fragments.

To test whether the fusion proteins could bind DNA by sequence specific interactions without knowing what a target sequence might be, we relied on an approach similar to that used to define the sequence specificity of bacteriophage lambda int protein interaction with DNA (35). We assumed that in a complex mixture of DNA some fragments would contain by chance a sequence that approximated the actual sequence recognized by the protein. Although they might not bind as well as a functional sequence (depending on how well they might approximate the functional sequence), such chance concensus sequences should bind with higher affinity than unrelated sites. To test this, we digested phage lambda DNA to give more than a hundred fragments and then examined how well these were bound to the FL fusion and HB fusion by carrying out the assay at different levels of stringency. Moderately increased salt concentrations or addition of carrier DNA resulted in displacement of the majority of the restriction fragments from the protein, but a number of fragments were retained in the protein-bound fraction with differing efficiencies. Further increases in the stringency of the assay conditions displaced all but four of these restriction fragments from the protein-bound fraction. The FL fusion and the HB fusion gave comparable results. This demonstrates that these fusion proteins are sequence specific DNA binding proteins and that the terminal quarter of the engrailed coding sequence can specify this activity. This is consistent with the idea that the homeobox domain specifies DNA binding and that the engrailed protein might act as a transcriptional regulatory factor. Clearly we would prefer to identify functionally significant sites of engrailed protein binding.

The engrailed Fusion Proteins Bind to Regions Near the 5' End of the engrailed and of the ftz Gene (Desplan et al., in prep).

As discussed in the introduction, the regulation of engrailed gene expression appears to be inherited clonally, as might be expected if a molecular switch had been thrown. One simple way to create such a switch is to have the engrailed gene induce its own expression so that continued expression is guaranteed. Thus, we might propose that the engrailed protein interacts with DNA sequences near its own promoter to induce transcription. Also, as discussed in the introduction, there are suggestions that the various homeotic and segmentation genes might interact in a regulatory network. Perhaps other segmentation or homeotic gene sequences are targets for engrailed protein interaction. We tested these ideas by looking for binding of the fusion proteins with sequences upstream of the engrailed transcription unit and sequences upstream of the ftz transcription unit. As shown in Figure 6 we were able to identify binding upstream of both of these genes.

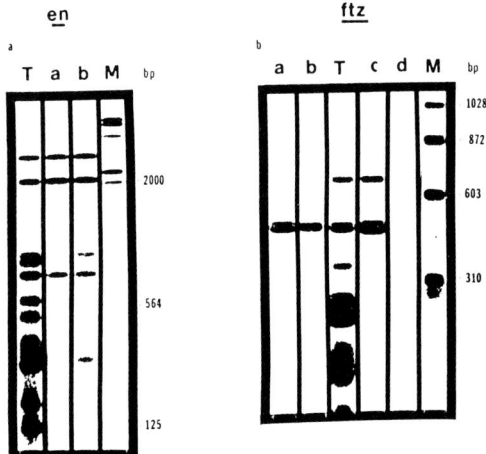

Figure 6. Specific DNA Binding of the Homeodomain Containing Fusion Proteins to Regulatory Region of engrailed and ftz.
en: A plasmid containing the first exon and 2 kb upstream of the engrailed gene was digested with BstN1, end labeled and run on a 2% agarose gel (T). The same digest when complexed to the FL fusion protein was immunoprecipitated with anti-lacZ antibodies; in (a) the incubation was performed in the presence of 200 mM NaCl and in (b) with 4 µg/ml carrier DNA. M is λ Hind III marker. A band of MW 700 bp corresponding to the region upstream of the cDNA was specifically retained.
ftz: A plasmid containing the ftz gene, including its regulatory region, was digested with DdeI, end-labelled and run on a 2% gel (T). Immunoprecipitations were performed with the FL protein in the presence of 100 mM NaCl (a) or 150 mM NaCl (b) or with the fusion protein containing only the homeobox domain with 150 mM NaCl (c) or 250 mM NaCl (d).
The high molecular weight bands present in some experiments are due to nonspecific binding of long fragments.

DISCUSSION

The Sequences Responsible for the Spatial Programming of engrailed Expression Appear to be Dispersed Over a Large Genomic Region.

The comparison of the genetic characterization of the engrailed gene and the molecular characterization of the engrailed transcription unit reveals that complete engrailed function requires the integrity of several tens of kb of genomic sequence that lie outside the characterized transcription unit. Because mutations interrupting these flanking regions produce relatively weak engrailed phenotypes, we suggest that most of the flanking regions are only required for a subset of the functions of the engrailed locus. Furthermore, the mutations that are the most distant from the transcription unit produce the weakest phenotypes; these mutations are nonlethal, and their phenotypes are restricted to different parts of the adult fly in an allele specific pattern. Thus, we would argue that these more distant mutations alter a smaller or less important group of engrailed functions.

Why is it that so many engrailed mutants only partially inactivate the gene? This is not what one would expect of mutations that are due to chromosome breakage and rearrangement. We suggest that only a part of the complex temporal and spatial program of engrailed expression is affected in these mutants and that an extended region flanking the coding unit regulates expression. Different parts of these extensive flanking sequences would be responsible for different aspects of control. Thus, the rearrangement mutations that have breakpoints physically close to the engrailed coding region would be among the most severe because they would dissociate the coding unit from a large number of regulatory elements; more distant breakpoints would affect fewer regulatory functions and would have correspondingly weaker phenotypes. Furthermore, the loss of particular regulatory functions would give the spatially restricted phenotypes observed in the weaker alleles. Figure 7 roughly schematizes these ideas.

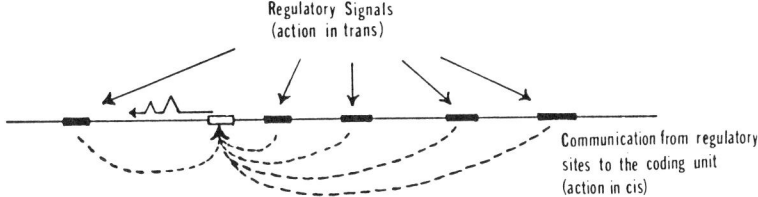

Figure 7. A Dispersed Regulatory Region. The key features of the proposed dispersed regulatory region are illustrated. The line represents genomic DNA. Target sequences for regulatory signals are dispersed over several tens of kb around the coding unit and are indicated here as a thicker line segment. Regulatory factors such as DNA binding proteins would interact with these target sequences. Interaction of regulators with target sequences would alter the expression of the distant coding unit. The mechanism of communication between regulatory regions and the coding unit is not known but because the entire region behaves as a single complementation unit, this regulatory influence must be transmitted predominantly in cis (dashed arrows).

If this is a general phenomenon, many genes having complex spatial patterns of expression during development will also have regulatory sequences dispersed over several tens of kilobases. Loci regulated by these large dispersed sequence domains would exhibit classes of mutant alleles similar to the partially defective alleles of engrailed. Indeed some of the various behaviors of mutants of the Bithorax complex that have been an enigma for many years can be explained in this fashion. For example, the pbx mutations were originally considered defects in a gene that functioned exclusively in the posterior compartment of the third thoracic segment (29). A recent alternative view suggests that these mutants are defective in a control region that regulates expression of the Ubx coding unit (30 kb away) (36,37). The phenotype can be accounted for if this control region is required only for Ubx function in a particular position, the posterior compartment of the third thoracic segment. Several other mutations in the left part of the Bithorax complex might be explained as being defective in other components of a dispersed regulatory region. Some other gene complexes, such as scute, might have similar control regions. scute mutations map to a large region of DNA, but all fall into the same complementation unit. These mutations show extraordinarily varied phenotypes. Each mutant lacks some of the large number of bristles that decorate the adult fly. The curious feature is that the differenti alleles of the "gene" lack different bristles. Again, the defects of mutants at this locus could be explained if scute gene function relied on spatial programming that was somehow encoded in a large regulatory region that was differentially affected by the various scute alleles. Physical mapping of scute mutations and correlation of mutant phenotype and mutant position offers strong support for this model (38).

What type of molecular mechanism could mediate the long range interactions between a coding unit and a distant regulatory site? One simple model might propose that there is a promoter associated with each regulatory unit, that the distant protein coding region is expressed from these promoters as gigantic transcripts, and that multiple control is achieved by the use of multiple alternative primary transcripts. At least for these developmental loci this model seems unlikely. No such large transcripts of the engrailed or scute loci are detected. Some of the "control" mutations in the Bithorax complex suggest that the interaction between the regulatory site and the coding unit can occur in trans if the homologous chromosomes are paired (note that if the regulatory unit and coding unit defined one transcription unit the interaction would be exclusively in cis).

Because of the failure of known molecular interactions to account for these results we have considered a possible regulatory mechanism based on RNA:RNA interactions (O'Farrell, in prep). Regulatory RNA's having regions of homology to a structural transcript might guide RNA processing by hybridizing to regions of the structural transcript, occluding such processing signals as splice junctions and poly A addition signals. This proposed regulation by antisense RNA's might either block or promote production of particular processed messages. Since it is known that RNA splicing systems and poly A addition systems will select alternative sites if the normal site is inactivated, it seems reasonable to propose that the RNA processing route might be guided by regulatory interactions that occlude particular processing signals.

To explain why the regulatory regions act in cis on the physically linked coding element, we need to invoke an additional complexity. The interaction between antisense regulatory RNA's and a structural gene would only occur between nascent transcripts if kinetic and structural constraints did not permit interaction between completed transcripts(39). Clearly such co-transcriptional interaction of RNA's would require close physical positioning of the regulatory coding sequence and the structural gene. Proximity could be assured by having the regulatory sequence closely linked the structural gene. In this configuration the regulatory sequence would primarily interact with the structural gene that is on the same chromosome (cis interaction) because of its proximity. However, when homologs are closely paired the regulatory region should be able to interact with the structural gene on the other chromosome (trans-interaction or transvection) (40).

The above ideas about the regulation of engrailed expression would predict that various rearrangement mutations would affect the pattern or timing of engrailed expression during development. Furthermore, the severity of the defect should be correlated with the proximity of the position of rearrangement to the coding sequence. The positions of all the rearrangement mutations have been localized (Kuner et al., in prep), and now the antibody probe provides us with a powerful way to assess the spatial and temporal pattern of engrailed expression (DiNardo et al., in prep). We are currently using the engrailed antibodies to examine the engrailed rearrangement mutations.

The engrailed Gene Product Might Function as a Pleiotropic Regulator of Transcription.

The observation that engrailed fusion proteins are sequence specific DNA binding proteins has a number of obvious implications of considerable import. Here we summarize these implications and try to extrapolate beyond them because the experimental approaches that are suggested by these speculations may prove important.

The following edifice has been assembled on the basis of several attractive but poorly supported ideas. Because the various homeotic and segmentation genes appeared to act as pivotal regulatory factors in development, it has been proposed that they might function as direct regulators of gene expression (19). The appeal of this proposal is primarily that this mode of regulation is familiar. Years of study of regulation in microbial systems have focused on the actions of such transcriptional regulatory factors. The discovery that the developmental genes of Drosophila all share a region of homology, the homeobox domain, implied that the function of all these proteins involved a very similar physical interaction (17, 18). Again, one of the most attractive speculations is that this region of homology specifies a DNA binding domain and that all these proteins act as sequence specific DNA binding proteins and specific regulators of transcription. This idea received considerable support from observations of homology between this homeobox domain and the yeast proteins a1 and α2 (24). These yeast proteins are sequence specific DNA binding proteins and repressors of transcription (25, and Johnson and Herskowitz, in prep). Further sequence comparisons added support for this idea. The sequence of various homeobox domains is compatible with the sequence constraints of a protein structural motif that characterizes bacterial sequence specific repressors (18, 32).

Our data offer strong support for the previously proposed idea that Drosophila developmental loci encode regulators of transcription. The sequence specific binding of the engrailed fusion protein is not likely to be an artificial consequence of the fusion construction, and therefore we conclude that the natural engrailed protein is a sequence specific binding protein. The sequence specific binding of the HB fusion indicates that the terminal 143 amino acids of the engrailed protein encodes the sequence specific binding activity and is consistent with the proposal that the binding activity is encoded within the homeobox domain. Finally, although we do not yet have a functional assay for the engrailed protein our data strongly suggest a role in modifying transcription. Because the engrailed binding sites are located in the vicinity of the start of transcription of two different genes, engrailed and ftz, the location of the site is presumed to have been preserved because it is functional. The simplest explanation for this result is that this location is conserved because binding of the engrailed protein interacts with other components at the promoter to regulate transcription. Although one might imagine other reasons why these binding sites might have this particular location, they seem contrived. We therefore propose that the engrailed protein exerts at least part of its regulatory functions by altering the rates of transcription of a number of target genes with which it has specific interactions.

One extrapolation of our results is that all the genes with homeobox domains are DNA binding proteins and regulators of gene activity. Although this is obviously not a rigorous conclusion, the speculation seems justified on the basis of the extraordinary conservation of this domain. In fact, the conservation is so extreme that examination of the region of the protein sequence that is proposed to be involved in recognition of specific DNA sequences (23,41) would predict that several different homeobox-containing genes would encode the same DNA binding specificity. If we examine the sequence of the region that ought to determine the sequence binding specificity two groups of homeobox sequences are distinguished; the homeobox sequences originally identified as homologous to Antennapedia and ftz homeobox domains are absolutely identical in this region, while the engrailed homeobox domain and its close relatives have a second type of specificity determining region. The degree of sequence conservation throughout the homeobox regions of these genes shows that the members of each group are much more closely related to each other than members of different groups (Figure 8). We suggest that these groups are functionally different and that within each group all the homeobox domains will specify binding to the same sequence.

If the homeobox domains of one group all specify binding to the same sequence, how can different members of that group express distinct regulatory activities. We are entertaining two classes of explanations. First, the proteins might all bind to the same sequences but interact differently with other components so that the consequence of binding each protein would be different. Secondly, the region of each protein outside the homeobox domain might contribute additional specificity to the DNA binding so that each type of protein binds to distinct but related sequences. We hope to test whether the specificity lies at the level of differing sequence specificities by comparing the binding specificities of different homeobox-containing proteins. At present, because of likelihood of overlapping binding specificities we can not be certain whether the DNA binding sites that we have defined by binding of an engrailed fusion protein are normally targets for engrailed protein interaction. The functional interaction in vivo might involve any member of a group of binding proteins with overlapping specificity.

Several observations lead us to a simple perspective regarding the evolution of sequence specific DNA binding proteins. Our thoughts have been influenced by a very basic argument; the de novo evolution of a new sequence specific DNA binding protein must be a complex and rare event, while gradual and constant divergence of a binding protein and its target site would create new specificities. Thus, we would expect different DNA binding proteins to be related to each other, but as a result of continued divergence their relatedness would be very slight; this is what is observed among prokaryotic DNA binding proteins (41). In contrast, divergence of the homeobox sequence seems to have stalled so that vertebrate and Drosophila sequences are virtually identical (note Hu2 and Antp in Figure 8).

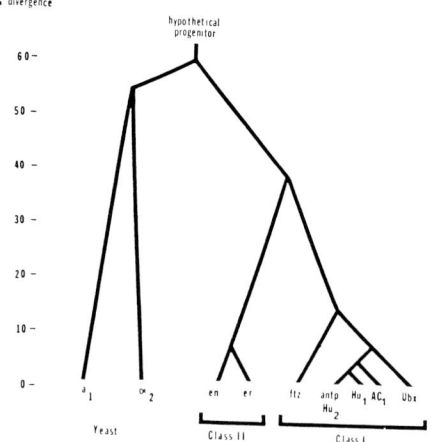

Figure 8. Family Tree of Relatedness of Homeobox Sequences. Because of its high level of conservation a 29 amino acid long portion of the homeobox sequences from position 29 to 57 was used to make pair-wise comparisons of a number of homeodomains. The format of the comparisons was to score each position as one, a half or as zero if occupied by identical, similar or disimilar amino acids, respectively. Relatedness in percent was calculated as the score divided by the number of compared positions (29) times 100 (percent divergence = 100-percent relatedness). The pairwise comparisons were used to assemble a family tree of relatedness. On the basis of degree of homology we have distinguished two classes of homeobox sequences (see text for details).

Two likely consequences of evolution by gene duplication and divergence seem relevant to the extraordinary conservation of the homeodomain. First, homo-oligomeric proteins would tend to evolve toward hetero-oligomeric proteins (42,43). Since most characterized DNA binding proteins are dimeric, gene duplication and divergence might create a family of regulatory proteins that can combine in a variety of homo- and heterotypic associations. Second, if divergence is to occur without loss of DNA binding function, it would most likely occur in sequences outside of those that specify the DNA binding domain. Regulatory genes with the same regions for subunit contact and for DNA binding (obviously this is the initial situation following gene duplication and preceding divergence) will necessarily interact in two ways. They can interact directly by forming heterotypic oligomers that show different specificity than homotypic oligomers (in yeast, the regulation of haploid specific genes by a1 plus α2 versus α2 would appear to be an example of this phenomenon; 17 and Johnson and Herskowitz, in prep). Related regulatory proteins can also interact at a functional level; the pattern of gene regulation can be dramatically affected by a competition between different regulatory proteins that bind to the same sites but have different effects once bound (regulation of bacteriophage lambda lysogeny by cro and lambda repressor is an example of this phenomenon). Thus, we would suggest that evolution of regulatory proteins will have an important tendency to produce regulatory systems whose function relies on interactions among a group of related regulatory proteins. Obviously such networks of regulatory interactions could produce flexible and complex patterns of control. We imagine that evolution of individual sequence specific DNA binding proteins and their sites of interaction is reasonably rapid unless they become members of a regulatory system whose function relies on interaction among related regulatory molecules. In these interacting systems the evolutionary divergence of a number of regulatory proteins is linked; no change that altered subunit association or sequence specificity could occur without coupled changes in all the other members of the group. We think that it is for this reason that evolutionary divergence of the homeobox sequence has virtually stalled. We also suggest that the <u>engrailed</u> class of homeobox sequence (class two homeobox sequence, HBII) though related to the other homeobox sequences (class one, HBI), belong to a distinct

group of interacting regulators that has also become
"fixed" in evolution (Figure 8). Though extremely
speculative these arguments make strong predictions that
can be tested by analysis of the interactions and
relationships among the Drosophila developmental genes.

A Regulatory Network of Interacting Segmentation Genes and
Homeotic Genes Might Play a Fundamental Role in the
Development of Pattern.

As summarized in the introduction, there are diverse
indications that the various developmental genes interact
with each other. Our data are also compatible with the
idea that these genes act to control the expression of each
other. The fact that the engrailed and ftz promoter
regions physically interact with an engrailed fusion
protein suggests that these genes may be ordinarily
regulated by a homeodomain-containing protein.
Furthermore, as discussed above, the sequence homology of
the proposed DNA binding domains of various homeobox
sequences suggests that some level of interaction of the
encoded gene products may be inevitable because of
overlapping sequence specificities. Finally, we are
attracted by these suggestions because, relying on such
interaction, it is possible to propose regulatory networks
that would give rise to ordered pattern formation during
development. We discuss here two proposed types of
regulatory interactions that might be of important
developmental consequences.

The spatially restricted pattern of engrailed
expression is established at early germ band extension
stages (10 and DiNardo et al, in prep). Genetic analysis
tracing the cell lineages suggest that this early decision
to express engrailed is a stable commitment that is
propagated clonally to all daughter cells. At a molecular
level there are two commonly considered mechanisms that are
known to result in a stable regulatory switch. One
mechanism involved gene rearrangement (such as occurs in
phase variation in Salmonella) (44), while another involves
a positive feedback loop that maintains expression of a
gene once it is turned on (e.g., lambda repressor synthesis
during lysogeny). Since no developmental rearrangement of
DNA sequences has been detected in the engrailed region
(or, in fact, in any part of the Drosophila genome) it
seems unlikely that engrailed expression might be
stabilized by such a change. On the other hand, our

observation of engrailed fusion protein binding to the region upstream of the engrailed promoter might be taken as support for the model that engrailed will positively reinforce its own expression. This idea can be tested by examining the expression of engrailed in mutations altered in the engrailed coding region. If the mutant gene encodes a cross-reacting protein we will be able to detect expression immunologically (alternatively we could examine the pattern of expression by in situ hybridization to engrailed RNA). If the engrailed gene product is inactive we can then examine how engrailed activity contributes to its own pattern of expression. The positive feedback model would predict that in these engrailed mutants the stripes would not appear or that they would be faint and transient.

The fundamental question we would like to understand is how does the periodic segmental pattern get established? To guide an experimental approach to this issue we need a conceptual framework. Here I would like to give our view of how the segmental periodicity might be encoded in the cells of the cellular blastoderm, and how this pattern develops from a preceding double segment periodicity. When engrailed expression is first seen as spatially localized, the stripes are a single nucleus wide. Between each row of intensely staining nuclei we see approximately three faintly staining nuclei. Segmentation mutants that produce defects with a single segment periodicity produce defects with fairly evident boundaries. When various segmentation mutants are examined it appears that there might be four boundaries to the various transformations. On this basis we propose that rings of nuclei at the blastoderm are uniquely specified by the localized expression of one or a combination of segmentation genes. According to this hypothesis the positional information that specifies the segmental periodicity along the anterior-posterior axis at the blastoderm stage might be schematically represented as a series of nuclei expressing either gene products a, b, c or d in a repeating array (Figure 9). These gene products would correspond to different segmentation gene products, for example d could be the engrailed gene product.

How might the shell of nuclei be so precisely subdivided into rings of differently specified nuclei? The development of the segmental periodicity seems to be founded on a preceding periodic pattern that is coarser. Thus, a number of genes, the pair-rule genes, are involved in the establishment of a paired-segment pattern, and based on the time at which these genes are expressed it would appear that this periodic pattern precedes the segmental periodicity. We suggest that the paired-segment periodicity is again encoded by the periodic pattern of expression of at least four segmentation genes. The double segment periodicity might then be represented in a cartoon analogous to that used to schematize the segmental pattern (Figure 9). Because cell division is so rapid at the time of establishment of the paired-segment pattern and because the cytoplasm is still syncitial, it does not seem meaningful to try and define the number of nuclei per paired-segment subdivision. Thus, our cartoon for the pattern of expression of pair-rule genes depicts zones of expression that are arbitrary in relation to the position and number of nuclei.

How does the coarse pattern of the pair-rule stage transform into the more detailed pattern of the segmentation stage? Obviously we don't know, but we can propose a fairly straightforward model based on interactions of the segmentation genes. If as schematized in Figure 9b the paired-segment repeat unit is subdivided into successive zones expressing gene products w, x, y and z then the coarse repeat pattern of the paired-segment periodicity could be subdivided into the finer pattern of segmental periodicity if both x and y induced expression of gene a. According to this proposal gene a would then be induced with a periodicity of half the repeat size of the paired-segment period.

There still remains the question of how the fine subdivisions of the segmental unit occur. We can propose a model based on interactions of the segmentation genes. As discussed by Meinhardt and Gierer (46) if genes have local, self-inducing activity and also have a longer range affect to inhibit each others expression, then a two dimensional field will be divided into zones of expression. Expression of a particular gene will dominate in each zone because of local self-induction while others will be excluded because of the inhibitory activity. Explicit patterns of induction and repression can be proposed (Figure 9c) so that the outcome of these interactions is defined pattern. Thus, if one of these genes were expressed in a spatially periodic pattern, interactions among the different regulators could refine this pattern of expression and subdivide the repeat into four zones each with its unique pattern of gene expression.

a) A segmental repeat may be divided into a series of differently specified zones

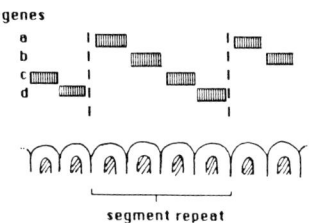

segment repeat

b) A paired-segment repeat pattern also appears to be subdivided

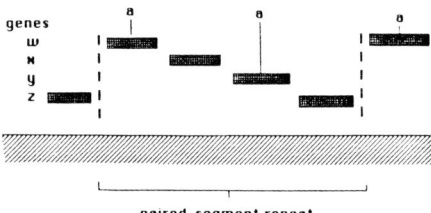

paired-segment repeat

c) Interactions among regulators can produce spatially restricted patterns of expression

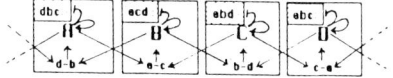

Figure 9. Molecular Pre-patterns Might Direct Formation of the Segmental Pattern.

a) Different segmentation genes are expressed in spatially restricted patterns. Here it is suggested that at cellular blastoderm the expression of four genes, a, b, c and d, might each be restricted to a stripe that is a single cell wide, as we have observed for the engrailed gene. The positions of expression of each of these genes (indicated by the shaded bars) could uniquely specify rings of cells in each segmental repeat.

b) The pair-rule genes are also expressed in a spatially restricted pattern and a set of four of these genes, w, x, y and z, could subdivide the coarser paired segment pattern in to subregions similar to those described for the segmental repeat. The coarser paired-segment pattern could be subdivided into the segment pattern if gene products w and y induced gene a (see 9a) and gene products x and z did not (see text for discussion).

Continued on following page.

c) Stable and accurate subdivision of a segmental unit into zones having a unique pattern of gene expression could be achieved by regulatory interactions among the segmentation genes. The four large squares represent zones within a segmental repeat in the embryo. Each of these zones is characterized by the dominant expression of a particular segmentation gene indicated by the uppercase letters. The arrows describe a set of interactions among these genes that could produce this spatially ordered pattern of expression. The arrows within each box (zone) indicate inductive interactions. Thus, each gene, A, B, C and D, turns itself on. The flatheaded "arrows" indicate repression. Each gene product at high concentration turns off the other three genes that are represented here as lower case letters within the stipled boxes. The arrows between boxes indicate diffusion of the gene products to the adjacent zones. As a result of this transport, each zone has a low level of the gene products that are produced in the zones that flank it (this is indicated by the lower case letters at the bottom of each box). As indicated, the regulators from the flanking regions could combine to produce a positive regulator of the dominant gene in that zone. This specific induction by the flanking regulators is one way to produce an invarient order in the different zones.

We have no confidence that the above proposals accurately predict the particular interactions among segmentation genes. The purpose in making these proposals is to show that an interacting network of regulatory genes could, at a molecular level, account for the progressive development of pattern that is observed. The subdivision of a coarse pattern to a more detailed pattern could account not only for the subdivision of the paired-segmental pattern to the segmental pattern but if successive steps are invoked the process could amplify an original anterior-posterior asymmetry into the complex segmental organization. Most importantly, these considerations make a number of molecular predictions regarding the interaction of the segmentation genes. Using the experimental tools we have developed it should be possible to directly define the regulatory interactions among these genes and hopefully to decipher the molecular algorithm for formation of segmental pattern.

ACKNOWLEDGMENTS

We thank our colleagues for their assistance during the course of this work, particularly Sandy Johnson and Tim Karr. We also thank Harald Biessmann for critically reading the manuscript and Judy Piccini for its preparation. This work was supported by an NSF grant, DCB 8418016.

REFERENCES

1. Nusslein-Volhard, C. and Wieschaus, E. (1980). Nature 287, 795-801.
2. Nusslein-Volhard, C., Wieschaus, E. and Kluding, H. (1984). Roux's Arch. Dev. Biol. 193, 267-282.
3. Simcox, A. A., Sang, J. H. (1983). Dev. Biol. 97 212-221.
4. Jurgens, G., Wieschaus, E., Nusslein-Volhard, C. and Kluding, M. (1984). Wilhelm Roux's Arch. Dev. Biol. 193, 283-295.
5. Wakimoto, B.T., Kaufman, T.C. (1981). Dev. Biol., 81, 51-64.
6. Wakimoto, B.T., Turner, R.F., Kaufman, T.C. (1984). Dev. Biol. 102, 147-172.
7. Kornberg, T. (1981a). Proc. Natl. Acad. Sci. USA 78, 1095-1099.
8. Kornberg, T. (1981b). Devl. Biol. 86, 363-372.
9. Hafen, E., Kuroiwa, A. and Gehring, W.J. (1984). Cell 37, 833-841.
10. Kornberg, T., Siden, I., O'Farrell, P., Simon, M. (1985). Cell 40, 45-53.
11. Garcia-Bellido, A., Ripoll, P. and Morata, G. (1973). Nature New Biol. 245, 251-253.
12. Garcia-Bellido, A. and Santamaria, P. (1972). Genetics 72, 87-104.
13. Lawrence, P.A. and Morata, G. (1976). Dev. Biol. 50, 321-337.
14. Struhl, G. (1981). Dev. Biol. 84, 372-385.
15. Simcox, A. A., Sang, J. H. (1983). Dev. Biol. 97 212-221.
16. Garcia-Bellido, A. and Santamaria, P. (1972). Genetics 72, 87-104.
17. McGinnis, W., Levine, M.S., Hafen, E., Kuroiwa, A. and Gehring, W.J. (1984). Nature 308, 428-433.

18. Scott, M.P. and Weiner, A.J. (1984). Proc. Natl. Acad. Sci. USA 81, 4115-4119.
19. McGinnis, W., Garber, R.L., Wirz, J., Kuroiwa, A and Gehring, W.J. (1984). Cell 37, 403-408.
20. Levine, M., Rubin, G., Tjian, R. (1984) Cell 38, 667-673.
21. McGinnis, W., Hart C.P., Gehring, W.J., Ruddle, F. (1984). Cell 38, 675-680.
22. Carrasco, A.E., McGinnis, W., Gehring, W.J., DeRobertis, E.M. (1984). Cell 37, 409-414.
23. Laughon, A., Scott, M.P. (1984). Nature 310, 25-31.
24. Shepherd, J.C.W., McGinnis, W., Carrasco, A.E., DeRobertis, E.M., Gehring, W.J. (1984). Nature 310, 70-71.
25. Miller, A.M., MacKay, V.L., Nasmyth, K.A. (1985). Nature 314, 598-603.
26. Hafen, E., Levine, M., Gehring, W. (1984). Nature 307, 287-289.
27. Duncan, I.M. Personal Communication.
28. Eberlein, S., and Russell, M. (1983). Dev. Biol. 100, 227-237.
29. Lewis, E.B. (1978). Nature 276, 565-570.
30. Lawrence, P.A. and Morata, G. (1976). Dev. Biol. 50, 321-337.
31. Poole, S.J., Kauvar, L.M., Drees, B., and Kornberg, T. (1985). Cell 40, 37-43.
32. Mayashida, M., Miyata, T. (1983). Proc. Natl. Acad. Sci. USA 80, 2671-2675.
33. Perler, F., Efstratiadis, A., Lomedico, P., Gilbert, W., Kolodner, R., Dodgson, J. (1980). Cell 20, 555-566.
34. Wilson, A.C., Carlson, S.S., White, T.J. (1977). Ann. Rev.. Biochem. 46, 573-639.
35. Ross, W., Landy, A. (1982). Proc. Natl. Acad. Sci. USA, 79, 7724-7728.
36. Ingham, P.W. (1984). Cell 37, 815-823.
37. Beachy, P.A., Helfand, S.L., Hogness, D.S. (1985). Nature 313, 545-551.
38. Campuzano, S., Carramolino, L., Cabrera, C.V., Ruiz-Gomez, M., Villares, R., Boronat, A., Modolell, J. (1985) Cell 40, 327-338.
39. Masukata, M., Tomizawa, J.I. (1984). Cell 36, 513-522.
40. Lewis, E.B. (1963). Am. Zoologist 3, 33-56.
41. Pabo, C.O., Saver, R.T. (1984) Ann. Rev. Biochem. 53, 293-321.

42. Dickerson, R.E., Geis, I. (1980). Proteins: Structure, Function and Evolution (Menlo Park, CA: Benjamin Cummings Publishing) 65-116.
43. Noda, N., Takahashi, H., Tanabe, T., Toyosata, M., Kikyotani, S., Furutani, Y., Hirose, T., Takashima, H., Inayama, S., Miyata, T., Numa, S. (1983) Nature 302, 528-532.
44. Silverman, M., Simon, M. (1980) Cell 19, 845-854.
45. Weinstock, G.M., ap Rhys, C., Berman, M.L., Hampara, B., Jackson, D., Silhavy, T.J., Weisemann, J., Zweig, M. (1983) Proc Natl Acad Sci 80, 4432-4436.
46. Meinhardt, H. and Gierer, A (1980) J. Theor. Biol. 85 429-450.

CELL PATTERNING IN NEURAL MAPS: SPECIFICITY AND DYNAMICS IN RETINOTECTAL MAP FORMATION[1]

Scott E. Fraser and Nancy A. O'Rourke

Department of Physiology and Biophysics,
and Developmental Biology Center
University of California,
Irvine, CA 92717

ABSTRACT The visual system of Xenopus laevis is characterized by the regular, topographic projection formed by the axons from the retinal ganglion cells of the eye onto the optic tectum. The regularity and the regenerative ability of the lower vertebrate retinotectal projection have lead to a view of the system as having specific and relatively static neural connections. In contrast, some experiments indicate that the patterning of neural connections is best viewed as a dynamic, ongoing process. In these experiments, conventional anatomical and electrophysiological techniques have provided static views of developmental processes, which then must be compiled to provide a glimpse of the dynamic behavior of the projection. This paper will present some inter-related attempts to better understand the processes involved in retinotectal map formation and to follow the dynamic interactions that shape this map. Both computer modeling and perturbation experiments using antibodies directed against the adhesive molecule N-CAM demonstrate that adhesive interactions may play a major role in the patterning of the retinotectal projection. Experiments using a new vital-dye fiber-tracing technique offer a means to directly follow the process of tectal innervation in developing animals and provide support for the idea that the projection is shaped by dynamic interactions.

[1]This work was supported by grants from the NSF (BNS80-23638 and BNS84-06307) and by a McKnight Foundation Scholar's Award.

INTRODUCTION

The development of the nervous system parallels the development of non-neural tissues in that it is the product of several morphogenetic mechanisms. The cells demonstrate both the position-dependent differentiation and the patterned cell migrations characteristic of many developing tissues. The nervous system carries these basic phenomena a step further by forming patterned neuronal interconnections, refered to as "projections", from one neural center to another. The groups of cells linked by these projections are often separated by relatively large distances. Long nerve tracts interconnect the centers in a stereotyped pattern via a stereotyped pathway. Two aspects of neural projections make them especially attractive for studies of cell patterning: their precision and the physical separation of the two sets of interconnecting cells. The precision of the pattern and the techniques available to assay the pattern make several different experimental approaches possible. The physical separation of the two sets of cells permits the independent manipulation of each set, permitting experiments that would otherwise be impossible.

This paper will consider the patterning of one such projection, the retinotectal projection of lower vertebrates, and the dynamic cell interactions that may serve to shape it. Though our discussion will center on the patterning of the developing visual system, many of the same interactions appear to play central roles in the development of other neural structures and are likely to be important to the elaboration of many patterns in the embryonic development.

The Retinotectal Projection and Assays for its Order.

In lower vertebrates, the major visual pathway is the projection of the retinal ganglion cells to the contralateral optic tectum (the retinotectal projection). The optic nerve fibers that make up the retinotectal projection are organized such that a stereotypic "map" of the retina is formed over the surface of the optic tectum. Optic nerve fibers from the dorsal part of the eye project to the lateral tectum, from the ventral eye to medial tectum, from the nasal eye to caudal tectum, and from the temporal eye to rostral tectum. The consistent order of the retinotectal projection has made it a popular system for the study of the mechanisms involved in neural patterning. The bulk of the studies have employed fish or amphibians (salamanders, Rana or Xenopus). In these animals, the retina and the tectum are readily accessible to experimental manipulations, and the projection regenerates to nearly-normal order following trauma to the optic nerve.

The typical assays used to study the patterning of the retinotectal projection rely on either anatomical or physiological techniques. The physiological assay employs metal microelectrodes, inserted into the neuropil of the optic tectum to record the responses of the optic nerve terminals (c.f. 1). Stimuli, in the form of spots of light, are presented in the visual field of the frog's eyes; the spots of light are used to determine the regions of visual field that elicit activity from the optic nerve terminals near the electrode tip. The responsive region of the visual field for each electrode position is called the "receptive field" for that position. This procedure is repeated for about twenty-five electrode positions, yielding a set of electrode positions and receptive fields that serve as a "map" of the pattern of optic nerve terminals on the tectum. The technique can also be used to determine the precision of the projection pattern. Larger receptive fields reflect a decreased short-range order in the patterning of the projection (2). The advantages of physiological techniques are that they are a functional assay and that the animal can be revived after the procedure to allow further development. The major disadvantage is that the anatomical substrate for the physiological map can only be inferred from the data.

Anatomical techniques for assaying the pattern of the retinotectal projection use selective labeling of subsets of the neurons with a transportable label. A marker that is transported along individual axons either anterogradely (away from the cell body) or retrogradely (toward the cell body) is introduced into small groups of cells in the visual pathway. Histological techniques are then used to determine the distribution of the marker substance. By reconstructing the site where the marker was applied and the location to which it was transported, a measure of the topography of the neuronal map is obtained. For example, Horseradish Peroxidase (HRP) can be applied to a subset of the optic nerve fibers to determine their termination site in the tectum (c.f. 3). Similarly, retrograde transport of HRP after small injections in the tectum can be used to label the retinal ganglion cells that projected to the injection site (4). By studying the dispersion of the labeled cells, both approaches can give a measure of the precision of the projection pattern. These experiments can be carried to single-cell resolution by intracellularly injecting either HRP or cobalt as tracers, thereby permitting detailed analyses of the anatomy of the nerve fibers and their terminals (5,6). The major advantage of the anatomical techniques is that the anatomical basis of the neural projection is directly determined; the major disadvantages are that only one point (or very few points) in the projection can be assayed in each animal and that the animal must be sacrificed in order to perform the assay.

Cell Patterning in the Retinotectal Projection.

Several recent studies have been designed to investigate the mechanisms that guide the formation of the retinotectal projection. The experiments have investigated the patterning of the optic nerve fibers during the normal development of the projection or during its regeneration. Experimental perturbations of the system can be used to challenge the mechanisms that generate this spatial order. These studies provide evidence for the presence of some form of positional cues in the retina and the tectum that help to guide optic nerve fibers to their tectal "targets".

In developing Xenopus larvae, Holt and Harris (7) have used anatomical methods to show that the initial projection formed by the optic nerve fibers has a proper dorsoventral order on the tectum. That is, the optic nerve fibers are found to project to their "correct" dorsoventral positions when they first arrive at the tectum, presumably before any cues based on the function of the visual pathway could act. Additional experiments have ruled out an absolute requirement for nerve activity by allowing the nerves to grow into the tectum in the presence of a neurotoxin; other studies have ruled out a requirement for a specific order of ingrowth of the optic nerve fibers in this crude dorsoventral ordering by imposing an incorrect order of arrival of the nerve fibers (8,9,10). In a related set of experiments on Xenopus embryos, small wedges of eyebud tissue were grafted to ectopic sites in the eyebud (11). When assayed in the adult, the descendants of the grafted tissue were often found to project to the tectum in a manner appropriate for their position of origin. That is, the cells of the grafted fragments of the eyebud behaved according to their original position in the donor eyebud, reflecting that the cells have some inherent information of their position in the eye rudiment.

Experiments on the regenerating retinotectal projection have also yielded results consistent with a role for positional cues in the patterning of the projection. In experiments on the goldfish retinotectal projection, Meyer has found that surgically-deflected optic nerve fibers can relocate their correct termination sites in the tectum (12). Fugisawa and his colleagues have traced the pathways of optic nerve fibers in the regenerated retinotectal projection of the newt; some of the fibers were observed to take a circuitous route to reach their final correct termination sites, as though they were actively searching out their targets (13,14). In related sets of experiments on Xenopus and goldfish, small pieces of the tectum were surgically excised and then reimplanted in a different position or orientation; the regenerating optic nerve fibers were able to locate and terminate on their proper, though malpositioned, tectal targets (c.f 15, reviewed in 1). The popular interpretation of each of these experiments is that each optic nerve fiber has an inherent knowledge of its position within the eye and uses this knowledge to navigate over positional markers in the tectum. While this interpretation may well be correct, and models with similar assumptions have the ability to fit much of the experimental data (16,17), it remains a strong possibility that other interactions, independent of fixed positional markers, play a significant role in the patterning of the retinotectal projection (18).

Dynamic Events in Retinotectal Development.

Recent experiments on the development and regeneration of the retinotectal projection indicate that the ordered projection between eye and brain is the product of a dynamic process. The present interpretation is that the neurons extend and retract neurites, responding to environmental cues in the tissue, in their search for their termination site. The best evidence for this comes from studies of the regeneration of the retinotectal projection of goldfish. A crudely ordered projection is formed initially during regeneration (2,19,20) by optic nerve terminals much larger than those in mature animals (21). The precision of the projection is gradually refined over time as the terminal arbors reach near-normal sizes. This refinement is believed to be the result of a continued "pruning" of inappropriate terminals and sprouting of more appropriate endings.

Evidence for dynamic interactions in the formation of the map can be obtained by studying the normal development of the retinotectal projection, as well. Several lines of evidence indicate that the retinotectal projection is continually changing during development. The retina and the tectum continue to grow during the life of these lower vertebrates, and the optic nerve terminals appear to shift caudo-medially over the surface of the tectum to compensate for the different growth patterns of the retina and the tectum (3,22,23,24). This constant rearrangement of the projection demonstrates a dynamic aspect of the normal, seemingly-static projection pattern. Due to the growth of the eye and the tectum, what was once an appropriate termination site for a given optic nerve fiber becomes an inappropriate site. As in the refinement of the regenerating projection, this shift is thought to be the result of a continued sprouting of new neurites and "pruning" of less appropriate terminals. Unfortunately, if the normal projection pattern is the product of dynamic processes, neither conventional anatomical nor physiological techniques can completely address the normal developmental sequence. Physiological experiments may be able to follow changes in the projection pattern with time, but fail to yield information on the underlying anatomical changes. Furthermore, they are difficult to apply to younger animals, and, except in a few cases, do not indicate the termination site of defined optic nerve fibers (23). Anatomical experiments can show the morphologies of individual cells, but since the techniques require that the tissue be fixed, cannot give direct information on the changes of the projection. It is only through comparisons between the data from different animals or a retrospective interpretation of cell morphologies that anatomical techniques can yield data on dynamic changes.

Adhesive Interactions in Nerve Patterning.

Little is known about the cues that might be guiding the optic nerve terminals in the formation of the retinotectal projection, though chemical or adhesive cues have long been proposed to play a role. The work of the past decade has refined our knowledge of adhesive interactions between cells. This knowledge now permits a better analysis of the role of cell adhesion in the patterning of the retinotectal projection.

An Adhesive Model for Nerve Patterning.

Modeling of developing systems can give insights into the interactions between cells that may be important in development. The experimental evidence available indicates that the patterning of the retinotectal projection is the end-product of dynamic interactions between thousands of players. It impossible to consider dynamic interactions between so many players through thought experiments alone; misleading predictions of even simple rules of interaction are possible. The simplifications that may be required can dramatically alter the results of thought experiments and would render such an analysis invalid. Computer simulations of proposed mechanisms offer a means to avoid these pitfalls. They can help clarify the implications of different hypotheses and therefore allow the experimenter to more thoroughly explore a set of ideas before designing test experiments.

Our laboratory has been examining the possible role of cell surface adhesions in the patterning of the retinotectal projection. Towards this end, we have contructed and tested a simple model that is based upon adhesive cell interactions (16,17). Computer tests of the model demonstrate its ability to explain most, if not all, of the experimental results obtained from experiments on the amphibian and fish retinotectal projection (17). Because of the model's abilities, it is presently being further refined to ease the design of experimental tests of the mechanisms that underlie it.

Briefly, the model proposes that there are three adhesive

mechansims that operate together to pattern the retinotectal projection. The largest of these, **C**, is a position independent adhesion between all of the optic nerve fibers and the tectum. The model works best if the **C** interaction is mediated by a homophilic adhesive molecule; this enables the same molecule to mediate both the fasiculation of the optic nerve on the way to the tectum and the adhesion of the nerve fibers to the tectum. To this large adhesive interaction, two much-smaller adhesive interactions, **AP** and **DV**, are added. **AP** and **DV** represent position-dependent adhesive interactions that give each nerve fiber a crude "best-fit" site on the tectum. If these adhesive interactions are mediated by homophilic adhesive molecules, then the same position-dependent cues are capable of mediating fiber-tectum affinities and fiber-fiber affinities. In addition to these adhesive interactions, a competition between nerve terminal, termed **R**, is added. It is intermediate in strength between **C** and **AP/DV**. The **R** interaction can take several forms; a simple means to provide the competitive **R** interaction between the terminals is to have the terminals compete for a fixed number of **C** adhesive sites on the tectal cells.

To perform computer simulations of the model, two large arrays are set up, one representing the cells of the retina and the other representing the tectum. The cells in the retinal array are allowed to send axons to terminate in the tectal array. The interactions outlined above are expressed in equation form such that each can be considered as an effect on the adhesive free energy of the neurons. The computer simulation keeps track of each nerve fiber and continually calculates the adhesive and competitive interactions experienced by that fiber. Each nerve fiber is allowed to sprout new neurites and retract old ones in its attempt to maximize its adhesive interactions (**C**, **AP** and **DV**) and minimize the competitive influence of **R**. That is, each nerve fiber strives for its own local minimum energy state by exploring its local environment within the tectum, and sprouts or retracts neurites within that region. The process is allowed to proceed until all of the fibers have found a stable position at which to terminate. The computer simulation displays the order of the projection by performing the equivalent of an electrophysiological map on itself. This allows a rather straightforward comparison of the model and the data.

Several lessons can be learned from computer simulations of the model. First, the model is able to replicate the great regularity of the developing and the regenerating retinotectal projection, even though the position-dependent adhesions **AP** and **DV** are made the weakest interaction. The prediction that weak interactions can give rise to the striking order of the projection is an example of the counterintuitive behavior that can take place in large assemblages of dynamically interacting elements. Second, the model generates the plasticity observed both following surgical interference with the retina or tectum and during the normal growth of the visual system. Thus, the same model is consistent with both major behaviors of the system: neuronal specificity and neuronal plasticity. Third, the model demonstrates mixed results from some experimental manipulations. That is, a single experimental paradigm can lead to two or more stable results, as has been observed in some experiments (see discussion in 17). Slight differences in the details of the experimental design are sufficient to cause quite different results in the model. This should serve as a caution to experimenters in both the design and the interpretation of their experiments. Small differences in experimental conditions or technique may "color" the experimental results generated by the system, such that one or another of the mixed results are exclusively obtained. Finally, the model demonstrates the unexpected finding that a dominant position-independent (unpatterned) adhesion plays a central role in the production of patterned neural connections. The simulations clearly show that the ability of the model is severely compromised if the **C** interaction is weakened or eliminated. This large, unpatterned adhesion is central because it permits the smaller **AP** and **DV** interactions to correctly influence the patterning of the nerve fibers. The size of the **C** interaction may make it more experimentally approachable than the smaller **AP** or **DV** interactions.

A Test of the Role of Adhesive Interactions.

The model proposes that the **C** interaction is a major, homophilic adhesion distributed over the retina and the tectum. The work of the past few years has clearly demonstrated that the adhesive molecule N-CAM shares these characteristics (25). In collaboration with Edelman and his colleagues, we have examined the role of N-CAM in the patterning of the retinotectal projection (26). Antibodies raised against Xenopus N-CAM were introduced, in the form of an agarose implant, into the tectum of either a normal froglet or a froglet undergoing regeneration of its retinotectal projection. Extracellular electrophysiology was then used to follow the pattern and the precision of the retinotectal projection after the introduction of the antibody. The anti-N-CAM antibodies distorted and decreased the precision of the retinotectal projections of both normal and regenerating animals (26). Control antibodies (preimmune, anti-L-CAM, or anti-brain membrane depleted of anti-N-CAM activity) had little or no effect on the projection. The effects required about three days to become apparent and peaked at about one week after the introduction of the antibody. This timecourse is consistent with the idea that disrupting some of the N-CAM based adhesions caused the optic nerve fibers to change their termination sites through the sprouting and retracting of neurites. If the retinotectal projection is the product of a dynamic process of neurite extension and retraction, then a local perturbation of a part of the process could bias the sprouting and retraction of the optic nerve fibers. A distortion in the projection is expected with a time-course consistent with the normal on-going processes. The results obtained agree with the predictions of the model and support the proposal that N-CAM plays a role in the patterning of the optic nerve fibers. This lends support to the ideas developed in the adhesive model for nerve patterning, since the predictions of the model and the experimental results are similar. However, it should be kept in mind that this consistency does not rule out the possibility that N-CAM plays a role in neural patterning different from that proposed by the model.

A New Assay for the Dynamics of Neural Patterns.

We have recently refined an anatomical technique that can be used to non-invasively study neuronal projections in living animals (27). This offers the possibility of directly following the development of the retinotectal projection, and should thereby permit a better grasp of the dynamic cell interactions that play a role in the patterning of neural connections. The technique is based upon some recently devised fluorescent cell lineage tracers, fluorescent dextrans (28). The fluorescent dextrans were microinjected into Xenopus embryos at the one-cell stage, yielding an animal in which all of the cells were fluorescently labeled. Classical grafting techniques were then used to implant labeled eyebud cells into an unlabeled host, creating a chimeric animal with only some eyebud cells containing the marker. The fluorescent dextrans fill neurons completely, including their axons, and since Xenopus larvae are transparent, the growth of the labeled neurons can be followed with an epifluorescence microscope in situ, without damaging the host in any way. The fluorescent dextrans have been rendered fixable by the addition of lysine to the dextrans, thereby allowing the distribution of labeled cells to be studied in tissue sections. This approach thus combines the advantages of a vital-dye fiber-tracing technique with more conventional anatomical techniques since the marker can be localized both in situ and in histological sections. To follow the development of the retinotectal projection, chimeric animals with all or part of their eyebuds labeled were created, and the outgrowth of the labeled optic nerve fibers was then followed with an epifluorescence microscope.

METHODS

Embryos were injected with about 10 nL of either fluorescein-labeled lysinated dextran (LFD) or rhodamine-labeled lysinated dextran (LRD) using glass micropipettes and a pressure-injection apparatus. Following injection, the embryos developed normally, with no larger fraction of abnormalities than their uninjected siblings. The injected animals were protected from light to avoid bleaching of the dye and possible phototoxic effects. The LFD and LRD used in these studies were the generous gifts of J. Braun and R. Gimlich (U.C.Berkeley). These dyes appear to remain diffusible within the cell but cannot escape from the cell because of their size and solubility properties. The cells marked with these dyes remain visible in living animals for at least two weeks.

All embryological grafting was performed in Steinberg's solution, under methylsulfonate anaesthesia (Finquel, Ayerst), using the sharpened tips of watchmakers forceps. Whole eyebuds or subregions of the eyebud could be grafted from labeled embryos into unlabeled hosts; the grafted tissues were held in place with small pieces of glass microscope coverslip for two hours, at which time the embryos were removed from the anaesthetic. Most grafts healed-in within 2 hours and remained healthy thereafter; those that did not were discarded. The grafts remained visible under epifluorescence for about 12 hours after the operation. During that time, the grafts could be seen to maintain their integrity and their position within the eyebud. After that time the grafts become partially obscured by the pigmented retinal epithelium that surrounds the eye; therefore, the distribution of the labeled cells in the eyebud can only be clearly followed by using histological techniques.

The positions of the fluorescent cells and their axonal projections were determined using a Zeiss Universal epifluorescence microscope equipped with an image intensifying video camera (RCA, SIT camera). This camera allowed the animals to be observed with minimum bleaching of the fluorescent dye, thereby avoiding possible phototoxic effects of the dye on the cells. This low light level observation appeared to avoid photoxic effects since the same cells could be observed over several days and observed animals mature normally. The projections of the optic nerve fibers remained clearly visible for more than two weeks. Images were recorded on videotape (Sony, U-Matic) or color slide film for later analysis.

RESULTS

To follow the topography of the developing retinotectal projection, fragments of labeled tissue from various positions in the eyebud were exchanged for identical fragments in unlabeled eyebuds. The development of donor derived tissues can be directly followed using an epifluorescence microscope in the living animal. The initial outgrowth of the optic nerve fibers is obscured in part by the refractile yolk present in the surrounding tissues. Therefore, unhindered observation of the growing axons at early stages requires that parts of the tissues surrounding the eye or the optic stalk be deflected. As development proceeds, the yolk inclusions are cleared from the cells in the head and the tissue becomes very transparent. This permits the labeled optic nerve fibers to be followed by direct observation with an epifluorescence microscope in the intact larvae. Since the fluorescent dextrans mark both the position of the labeled retinal ganglion cells and the position of their labeled terminals in the tectum, this technique permits a direct assay of the projection from the labeled retinal cells.

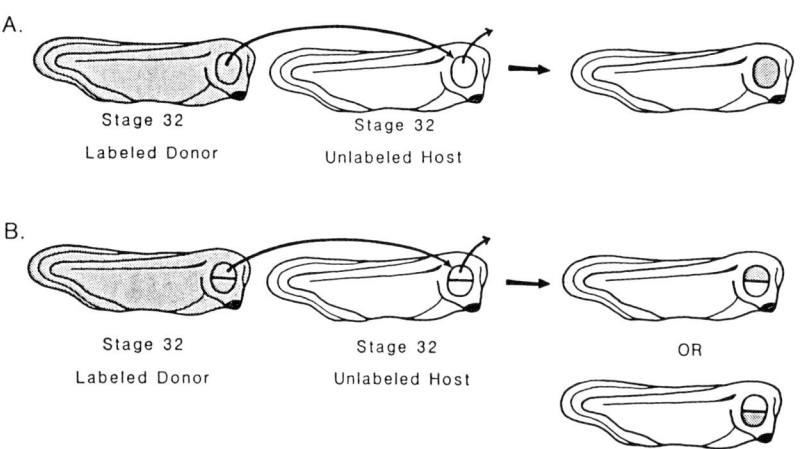

FIGURE 1. Schematic drawing of the operations used to create animals with only some cells labeled with fluorescent dextran.

Dorsoventral Topography.

The dorsoventral order of the retinotectal projection was assayed by implanting either a dorsal half-eyebud or a ventral half-eyebud into an unlabeled host and then determining the position of the optic nerve terminals in the tectum. At stage 47, the tectum has adopted a shape that allows the tectal neuropil to be conveniently viewed from above in an intact larvae. Such observation revealed a clear difference in the projection pattern of optic nerve fibers from the dorsal and ventral halves of the retina. Labeled dorsal eyebud cells sent optic nerve fibers to the lateral region of the optic tectum; labeled ventral eyebud cells projected to the medial tectum (Figure 2). In animals younger than stage 47, the tectal shape made it more difficult to view the tectal neuropil in an intact animal; nevertheless, the fibers showed the same dorsoventral ordering. This pattern of termination is as predicted from the order of the adult and mature larval retinotectal projection, and agrees with anatomical assays of the developing retinotectal projection which show the establishment of a rough dorsoventral order by stage 40 (6,7). This agreement between our vital-dye technique and the more conventional approaches which require that the animal be fixed and processed confirms the accuracy and utility of the technique.

Nasotemporal Topography

The development of the nasotemporal (anteroposterior) order of the projection was explored using a similar paradigm. A labeled nasal or temporal half eyebud was grafted into an unlabeled host animal, and the distribution of these labeled optic nerve fibers was assayed during the next ten days. At stage 47, when a clear dorsoventral order could already be observed, no consistent nasotemporal difference was found (Figure 3). The fibers from the nasal half and the temporal half of the eye overlapped, covering the same area of the tectal surface. That is, the projections formed by eyebuds labeled in entirety or with only their nasal or temporal halves labeled were indistinguishable from one-another. Three to five days later at stage 49, a clear difference became apparent. At this time the projection from the nasal half-eyebud consistently projected more caudally on the tectum than the projection from the temporal half-eyebud (Figure 3). The appearance of this nasotemporal order in the projection can not be the product of new fibers arriving in the tectum, since cell division at the ciliary margin of the eye has diluted the dye in the cells newly arriving at the tectum.

Patterning of Retinotectal Map Formation 533

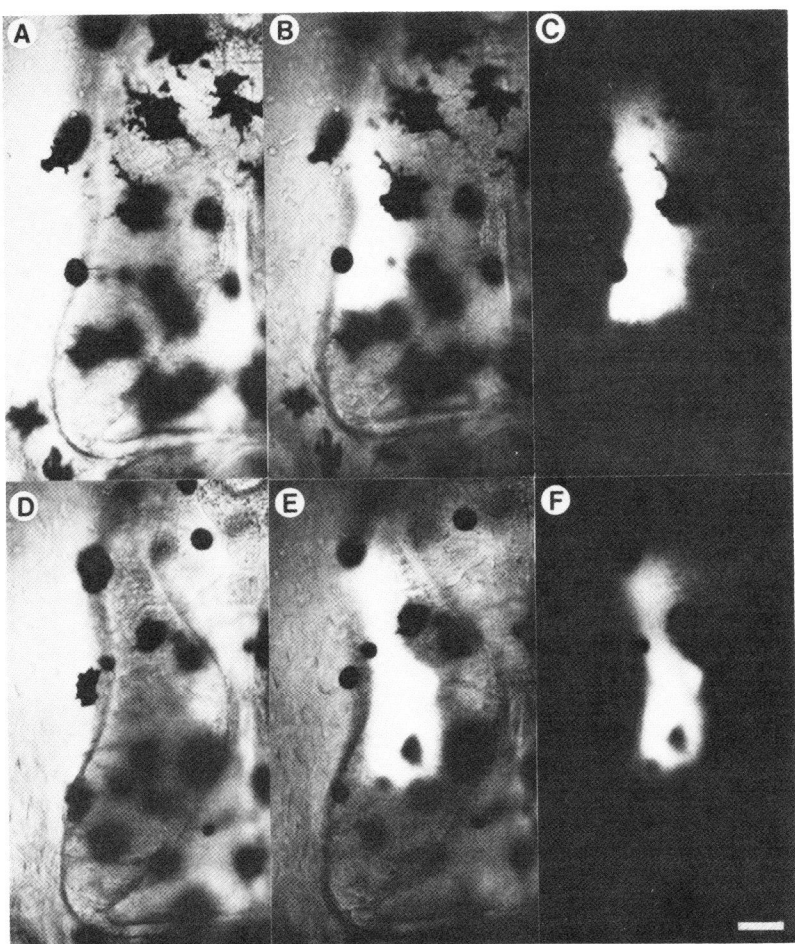

FIGURE 2. Photograph of the projections of eyes with their dorsal or ventral halves labeled. (A,B,C) Photographs of the projection from an animal with its dorsal eye labeled (A:phase, C:fluorescence, B:combination). Note that the labeled region of the neuropil reaches to the lateral (left) edge of the developing tectum. (D,E,F) Photographs of the projection from an animal with its ventral eye labeled, showing a projection that is more medial on the tectum. This polarity is in agreement with that of the adult retinotectal map.

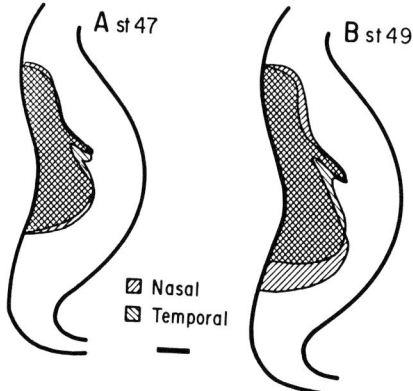

FIGURE 3. Line drawings of the projections from nasal and temporal labeled eyes at stage 47 and 49. The projections demonstrate adult-like topography at stage 49 but not at stage 47.

DISCUSSION

The retinotectal projection offers a rich experimental system for investigations of the interactions that pattern neuronal connections. Several lines of experimentation indicate that the topography of the projection is the product of a dynamic process, in which the optic nerve fibers are continually stiving to refine its order. The cues that guide the optic nerve fibers to their target sites in the tectum, and the manner in which these cues are assigned during development remain largely unknown. Recent work may offer some insight into some of these issues; both theoretical and experimental analyses indicate that adhesive interactions may play a large role in the patterning of the projection. A model based on a simple set of adhesive interactions is capable of fitting not only the basic observations on the development and regeneration of the projection but also the more intriguing findings obtained in experimentally altered animals such as neuronal specificity, neuronal plasticity and ocular dominance columns. The model shows that dynamic interactions among many players can generate a high degree of organization from weak positional cues. Experiments with antibodies directed against N-CAM validate one of the proposed interactions in the model and demonstrate the importance of cell adhesion in the patterning of the projection. Such a combined experimental and theoretical approach may prove fruitful in the search for the cell interactions that serve as cues in the formation of patterned neural connections.

A more detailed analysis of the interactions important in the patterning of the projection may now be possible through the use of the vital-dye fiber-tracing technique described above. The technique is based upon fluorescently-labeled dextran molecules, which are impermeant to the cell membrane. Previously, this fluorescent dye has been injected into individual cells as a marker both for following the descendants of the injected cell (29,30) and for following the course of axons eminating from the descendants of the injected cells (31,32). In our application of these markers, entire embryos were injected with the lineage marker at the one-cell stage; cells from the labeled embryo were then grafted into an unlabeled host. The growth and the pathway of the axons from a defined set of labeled cells could then be followed non-invasively with epifluorescence microscopy. The technique has adequate resolution to observe individual filopodia on growth cones in cultures and in permissive locations in the intact embryo.

To observe the patterning of the developing retinotectal projection, defined halves of labeled eyebud tissue were grafted into unlabeled animals. As development proceeded, the yolk stores in the cells of the head were eliminated, and the tissues surrounding the eye and the tectum became transparent. This permitted the use of an epifluorescence microscope to follow the growth and distribution of the labeled optic nerve fibers. Dorsoventral ordering of the projection could be clearly observed from stage 43 onward: a labeled ventral half-eyebud projected to the medial tectum and a labeled dorsal half-eyebud projected to the lateral tectum. In contrast to this dorsoventral ordering, the fibers did not demonstrate any noticable anteroposterior ordering until about stage 49. Labeled nasal or temporal half-eyebuds projected in an indistiguishable fashion to the developing tectum; the projections then segregated over a three to five day period between stages 47 and 49. Previous investigations had indicated that little or no anteroposterior order was present in the retinotectal projection (6,7). The advantage of the present approach over these previous experiments is that our assay can be performed on living animals; hence, the emergence of the antero-posterior ordering can be directly followed in a group of previously-labeled cells in a single animal. Such observations indicate that the appearance of anteroposterior order is not the result of "pruning" a once-exuberant projection to the tectum. Instead, the appearance of anteroposterior order appears to be the product of the optic nerve fibers from the nasal portion of the eyebud suddenly invading a greater portion of the developing tectum. The reasons for this sudden appearance of polarity remain unknown, but may include the maturation of the retina or tectum, the appearance of cues on the tectum, or the onset of activity. The vital-dye fiber technique, when combined with more conventional embryological techniques should allow the roles of each of these to be tested. The direct observation of the retinotectal projection made possible by this new technique will aid in the design of new experiments and permit the further refinement of the model.

ACKNOWLEDGEMENTS

We thank J. Braun and R. Gimlich for their generous gifts of the fluorescently labeled dextrans used in these experiments, and M. Bronner-Fraser, for her comments on the manuscript.

REFERENCES

1. Fraser SE, Hunt RK (1980). Retinotectal specificity: Models and experiments in search of a mapping function. Ann Rev Neurosci 3:319.
2. Schmidt JT, Edwards DL (1983). Activity sharpens the map during the regeneration of the retinotectal projection in goldfish. Brain Res. 269, 29-40.
3. Easter SS,Jr., Stuermer CAO (1984). An Evaluation of the hypothesis of shifting terminals in goldfish optic tectum. J Neurosci 4:1052.
4. Cook JE, Rankin ECC (1984). Use of a lectin-peroxidase conjugate (WGA-HRP) to assess the retinotopic precision of goldfish optic terminals. Neurosci Lett 48:61.
5. Sterling RV, Merrill EG (1984). How retinal axons grow in the tectum: Branching analysis of axonal arbors of single physiologically identified frog retinal ganglion cells. J Embryol Exp Morph 82:236 (supplement).
6. Sakaguchi DS, Murphey RK (1984). Initial development of the retinotectal projection in Xenopus: an examination of retinal ganglion cell terminal arborizations. Soc Neurosci Abst 10:669.
7. Holt CE, Harris WA (1983). Order in the initial retinotectal map in Xenopus: A new technique for labeling growing nerve fibres. Nature 301:150.
8. Harris WA (1984). Axonal pathfinding in the absence of normal pathways and impulse activity. J Neurosci 4:1153.
9. Ferguson BA (1983). Simultaneous elimination of normal fibre-fibre interactions and impulse activity does not prevent appropriate tectal innervation. Soc Neurosci Abst 9:759.

10. Holt CE (1984). Does timing of axon outgrowth influence initial retinotectal topography in Xenopus? J Neurosci 4:1130.
11. Conway K, Feiock K, Hunt RK (1980). Polyclones and patterns in developing Xenopus larvae. Curr Topics Devel Biol 15:216.
12. Meyer RL (1984). Target selection by surgically misdirected optic fibers in the tectum of goldfish. J Neurosci 4:234.
13. Fugisawa H, Tani N, Watanabe K, Ibata Y (1982). Branching of regenerating retinal axons and preferential selection of appropriate branches for specific neuronal connections in the newt. Devel Biol 90:43.
14. Fugisawa H, Watanabe K, Tani N, Ibata Y (1981). Retinotopic analysis of fiber pathways in the regenerating retinotectal system of the adult newt Cynops pyrrhogaster. Brain Res 206:27.
15. Yoon MG (1975). Topographic polarity of the optic tectum studied by reimplantation of the tectal tissue in adult goldfish. In Cold Spring Harbor Symp Quant Biol 40:503.
16. Fraser SE (1980). A differential adhesion approach to the patterning of nerve connections. Dev Biol 79:453.
17. Fraser SE (1985). "Cell interactions involved in neural patterning: An experimental and theoretecal approach." In Edelman GM, Gall WE, Cowan MW (eds): "Molecular basis of neural development," New York: Wiley, p.481.
18. Willshaw DJ, Von der Malsburg C (1977). How to label nerve cells so that they can interconnect in an orderly fashion. PNAS 74:5176.
19. Meyer R (1983). TTX inhibits the formation of the refined retinotopography in goldfish. Dev Brain Res 6:293.
20. Rankin ECC, Cook JE (1984). Use of WGA-HRP to assess the progress of map refinement after regeneration of the goldfish optic tectum. J Embryol Exp Morph 82:233 (supplement).
21. Schmidt JT, Buzzard MJ, Turcotte J (1984). Morphology of regenerated optic arbors in goldfish tectum. Soc Neurosci Abst 10:667.
22. Gaze RM, Keating MJ, Ostberg A, Chung SH (1979). The relationship between retinal and tectal growth in larval Xenopus: Implications for the development of the retinotectal projection. J Embryol Exp Morphol 53:103.
23. Fraser SE (1983). Fiber optic mapping of the Xenopus visual system: Shift in the retinotectal projection during development. Dev Biol 95:505.
24. Reh TA, Constantine-Paton M (1984). Retinal ganglion cell terminals change their projection sites during larval development of Rana pipiens. J Neurosci 4:442.
25. Edelman GM (1973). Cell Adhesion Molecules. Science 219:450.

26. Fraser SE, Murray BA, Chuong C-M, Edelman GM (1984) Alteration of the retinotectal map in Xenopus by antibodies to neural cell adhesion molecules. PNAS 81:4222.
27. O'Rourke NA, Fraser SE (1985). Dynamic aspects of retinotectal map formation as revealed by a vital-dye fiber-tracing technique. Dev Biol (submitted)
28. Gimlich RL, Braun J (1985). Improved fluorescent compounds for tracing cell lineage. Dev Biol in press.
29. Gimlich RL, Cooke J (1983). Cell lineage and the induction of second nervous systems in amphibian development. Nature 306: 471-473.
30. Kimmel CB, Warga R, Law RD (1984). Cellular lineages forming the zebrafish nervous system. Soc Neurosci Abst 10:1044.
31. Braun J (1984). Dependence of central axon outgrowth on peripheral ectodermal cells in the leech. Soc Neurosci Abst 10: 139.
32. Eisen JS, Myers PZ, Westerfield M (1984). Segmentally specific growth of motor axons in live zebrafish embryos. Soc Neurosci Abst 10:371.

SELECTIVE STABILIZATION OF RETINOTECTAL SYNAPSES BY AN ACTIVITY DEPENDENT MECHANISM[1]

John T. Schmidt

Department of Biological Sciences
and Neurobiology Research Center
State University of New York at Albany
1400 Washington Ave. Albany, NY 12222

ABSTRACT During regeneration of the optic nerve in goldfish, the ingrowing retinal fibers successfully seek out their correct places in the overall retinotopic projection on the tectum. Chemospecific cell surface interactions are sufficient to organize only a crude retinotopic map on the tectum, as assayed with electrophysiological recording or anatomical staining of optic arbors. Precise retinotopic ordering appears to be achieved via an activity dependent stabilization of appropriate synapses, and is based upon the correlated activity of neighboring retinal ganglion cells of the same receptive field type. Four treatments block the sharpening process: blocking activity with intraocular tetrodotoxin, dark rearing, strobe rearing, and blocking retinotectal synaptic transmission with alpha-Bungarotoxin. These experiments support a role for normal, locally correlated visual activity in sharpening the diffuse projection, probably through the summation of EPSP's within the postsynaptic cells. The mature retinotectal map also remains sensitive to activity as a local decrement of synaptic transmission causes a local disruption. Initial projections in development are also often diffuse, and activity dependent synaptic stabilization may be a general mechanism whereby the diffuse projections of early development are brought to the precise, mature level of organization.

[1] This work was supported by NIH grant EY 03736 and a Sloan Foundation Fellowship to the author.

INTRODUCTION

Because much of the brain is taken up with maps of visual and auditory space and of body surfaces, topographic maps and especially the mechanism of their formation have been topics of general interest. The most intensively studied topographic projection is the direct retinotopic map on the optic tectum of fish and frogs. The formation of this projection can be studied both during development in the embryo and during regeneration of the optic nerve in the adult. The central question has always been: 'How does each ingrowing retinal fiber select the correct place in the tectum?'. Initially, the emphasis was on the match between the ingrowing fibers and their postsynaptic target neurons. More recently, the emphasis has been broadened to include cooperative interactions between retinal fibers, as well as the relationship with postsynaptic cells.

The Mechanisms of Map Formation

The major ideas about the mechanisms involved in the formation of maps have centered in four areas: 1) Differential chemospecific adhesion between retinal fibers and tectal cells (1,2), 2) Competition between optic fibers for synaptic space (3,4,5), 3) Fiber self ordering and pathway interactions en route to the tectum (6,7,8,9), 4) Activity dependent stabilization of a retinotopic pattern after a diffuse early innervation (10,11,12,13,14). In this paper, I will briefly review the evidence that the first three mechanisms above are sufficient only to orient a crude map, then present in more detail the evidence for the sharpening, via an activity dependent synaptic stabilization, of both the retinotectal map and other topographic maps, and finally relate these recent studies to findings elsewhere in the nervous system.

Chemoaffinity mechanisms have limited resolution. Historically, the first of the above mechanisms to be considered was that of selective chemoaffinity between retinal fibers and tectal cells, based upon the position of the ganglion cell in the retina and the tectal cell in the tectum. Sperry (1) initially postulated general biochemical gradients across cells that would influence fiber to tectum interactions, but later (2) made his interpretation much more rigid, postulating unique surface markers for each cell. This idea of unique markers appeared to be supported by the anatomical experiments of Attardi and Sperry (15). They removed half of the retina of the goldfish, crushed the optic nerve and studied the pathways and innervation sites of the regenerating optic fibers using a modified silver stain. They reported that the optic fibers grew back selectively along the original pathways and to the original sites previously occupied, even if the regenerating fibers had to bypass open sites en route. More modern methods have not upheld this finding (see below), although other experiments still support the concept of regional differences (or gradients) across retina (16,17) and tectum (18 but see also 19), and differential affinity between retina and tectum (20).

In addition, the notion of unique markers was not consistent with the dramatic plastic rearrangements made by the projection following surgical intervention or during normal development (see reviews by Schmidt (8) and Easter (21)). Briefly, a surgically created half retina eventually expands its projection over the entire tectal surface, including the inappropriate half. Likewise, if half of the tectum is removed, the whole retina forms a compressed map over the remaining half tectum. These expanded and compressed maps generally maintain retinotopic order, even though each retinal fiber then terminates over a different tectal site. The formation of such projections is not consistent with the matching of unique position dependent retinal and tectal markers, but is consistent with the use of general gradients to orient the map. Moreover, even if the optic fibers have only a relative preference for one spot on the gradient, one must postulate competitive interactions to force the spreading of the half retinal projection and the compression of the full projection onto a half tectum. In the half retinal case the fibers move away from the preferred site merely to gain more territory. Schmidt et al. (4) postulated that each optic fiber has an intrinsic tendency to expand its arbor, that this tendency is opposed by that of nearby arbors, and that the tendency becomes weaker as the arbor grows larger. Recent electron microscopic evidence indicates that this competition may be for a fixed number of postsynaptic sites. Murray et al. (5) concluded that each fiber in the compressed projection on a half tectum must occupy approximately half the usual number of synaptic sites.

During development, a similar movement of optic arbors, with a continual changing of postsynaptic partners, occurs because of the disparate geometric growth patterns of the retina and tectum of fish and frog (for a review see 21). The retina grows by adding annuli, but the tectum grows by adding cells in a crescent at the caudal end. The earliest retinal ganglion cells (around the optic nerve head at the center of the retina) initially connect with the most rostral (earliest generated) tectal cells, although this tectal region will eventually receive fibers from far temporal ganglion cells. As more retinal and tectal cells are added, the older central retinal ganglion cells move their arbors progressively more caudally to stay at the center of the tectum.

Such movements, as well as the surgically induced rearrangements, are consistent with weak chemoaffinity gradients, which may in fact be necessary to orient the projections. Indeed, since oriented retinotopic projections can also be formed within nonvisual nuclei following removal of their original inputs (22), these chemoaffinity gradients may be present in other brain areas. However, the experiments outlined above argue strongly against unique or individual chemical positional markers on the tectum. Clearly factors other than chemospecific fiber-target interactions must contribute to the precision of these retinotopic maps.

<u>Fiber self ordering also not precise.</u> A second postulated mechanism is a selective fiber-fiber affinity which could maintain the relative ordering of retinal fibers in the presynaptic array (6,7,8,9). Such an interaction was suggested when Meyer (23) showed that deflection of a few fiber bundles across the midline sometimes resulted in a partial map of opposite polarity within the frame of a normally oriented map. Since the polarity was opposite that to be expected from any tectal polarity cues, the orderly but reversed partial map could only have been formed via fiber-fiber interactions. This result and similar ones (24,25) therefore implied more than mere compression or expansion along one tectal axis. They suggested that selective interfiber affinity based upon the closeness of the two cells in the retina may be another mechanism in organizing a retinotopic map. This type of interaction has recently been included in models of the retinotectal projection (26). This emphasis on selective fiber-fiber affinity led to a careful study (9) of the degree of order in both the normal and the regenerated optic pathway using horseradish peroxidase (HRP) staining. The normal pathway in goldfish was found to be highly ordered, and to undergo several reorganizations along the way to the tectum. These

reorganizations appear to suggest interactions with the pathway, particularly at the level of the diencephalon where the tract splits into dorsomedial and ventrolateral branches. The fibers in the regenerated pathway were also somewhat ordered, but the precision was not as great as in the normal. For example, at the rostral pole of the tectum, where the optic tract bifurcates, the fibers from dorsal retina normally all go ventrally and those from ventral retina go dorsally. After regeneration, however, approximately one fiber in five enters through the wrong branch. This 4 to 1 ratio of correct to incorrect pathways selected shows that fiber self ordering and pathway interactions may play some role in organizing the projection. However, many mistakes are left to be corrected after entry into the tectum.

Early Diffuse Projections In Regeneration

Anatomical studies. While the above experiment demonstrates mistakes along the pathway, many other experiments demonstrate mistakes in the zone of tectal innervation. Meyer (27) studied the order in the regenerated projection, after crushing the nerve and removing part of the retina. Instead of the silver stain employed by Attardi and Sperry (15), he used radioautography to trace the projection from the remaining retina, and determined how much retina would have to be removed to create a denervated zone in the tectum. The normal projection is sufficiently orderly that small nasal retinal lesions produce corresponding denervated areas in caudal tectum. (Caudal tectum was chosen because it does not have as many fibers of passage which could complicate matters.) During the early stages of regeneration, however, at least the nasal half of the retina had to be removed to cause any caudal area to be denervated. Stuermer and Easter (9) showed this same diffuseness by making small punctate injections of HRP both in normal tecta and in tecta after regeneration. Normally the ganglion cells that are retrogradely labelled through their terminals by this technique are spatially confined to a small area within the retina. A few other cells are labelled because axons of passage are broken at the injection site. In the early stages of regeneration, however, labelled cells were scattered over the entire retina with only a slight preponderance in the correct half. Cook and Rankin (28) also made tectal injections, but used wheat germ agglutinin conjugated to HRP (WGA-HRP) which is taken up more by terminals than by broken axons of passage, allowing a clearer picture. Early in regeneration, the punctate injections yielded a similar scattering of retinal ganglion cells labelled through their terminals, but after two months or more, the injection labelled a cluster of cells that was once again fairly compact and resembled the normal pattern.

In the newt, Fujisawa et al. (29) were the first to use anterograde transport of HRP to stain small numbers of regenerated optic fibers from discrete areas of the retina. Using tectal whole mounts, they could see that individual regenerating fibers not only took aberrant paths into and through the tectum, but also made many branches in inappropriate areas of the tectum. Many months after regeneration, the arbors had shrunk back to normal size, although the paths of the regenerated fibers remained abnormal.

Electrophysiological studies. In the case of the goldfish, electrophysiological recordings during this early diffuse phase are not readily feasible because the responses fatigue extremely rapidly. The earliest recordings at 34 days postcrush show maps that are already normal in organization. In the frog, however, the sharpening of the map could be followed with the electrophysiological mapping technique (30,31). Early in regeneration, recording at each tectal point yielded many units (thought to be the arbors of retinal ganglion cells) responding to stimulation of a wide area of the retina or visual field. The inclusive area that was responsive is called a multiunit receptive field, and its large size indicates that only a very crude level of retinotopic organization is present on the tectum. Over the next 20 to 30 days, these large responsive areas shrunk to the normal size, and a normal map emerged. For the Australian tree frog (31), there was a pronounced rostral to caudal progression of the sharpening.

In the frog, visual information, after reaching the contralateral tectum, is relayed through the Nucleus Isthmi to the ipsilateral tectum, where it is easily recorded. This relay projection allowed Adamson et al. (30) to assess the postsynaptic effects of the crude topography of the direct retinotectal projection. Early in regeneration, the receptive fields of single relay fibers in the ipsilateral tectum were grossly enlarged, reflecting the crude map on the opposite tectum. Thus, the misdirected arbors or branches of arbors within the crude map appeared to have established effective synaptic connections, in spite of the fact that they were in the 'wrong' places. We will see below in the studies of goldfish regeneration that synaptic transmission appears to play a role in the sharpening process.

The experiments in the studies presented below were designed to demonstrate that the final level of retinotopic organization depends not only upon the presence of activity during regeneration, but also upon the spatial correlations in that activity, and upon the effective synaptic transmission of that activity to the tectal cells. Both electrophysiological unit recordings and anatomical fiber staining techniques have been employed to demonstrate these effects.

METHODS

Goldfish 10-13cm in overall length were anaesthetized before surgery or recording by immersion in a 0.1% solution of tricaine methanesulfonate. The optic nerve was crushed in the orbit behind the eye using fine jewelers forceps (32).

Five groups of fish with regenerating optic nerves were studied. The fish of Group 1 either received no injections or received control injections of vehicle solutions, and all regenerated under normal visual conditions. The fish of Group 2 received intraocular injections of tetrodotoxin every second day to achieve a continuous block of activity during various periods of the regeneration (13). The fish of Group 3 were placed in light tight boxes to regenerate in complete darkness (33). The fish of Group 4 were placed in similar light tight boxes with strobe light illumination (xenon strobe unit) operating at one flash per second continuously (33). The fish of Group 5 were fitted with intraventricular cannulas that delivered approximately 3 ul of 1 uM alpha-Bungarotoxin (BTX) per day from an osmotic minipump attached above their heads (35). The BTX was purified from crude venom as described previously (35).

The electrophysiological unit recordings were performed with the fish's eye in water to avoid any problems with the refraction of the light rays at the air cornea interface(4). The apparatus consisted of a water filled plastic hemisphere with a holder for the fish attached to the flat side (13). The fish's eye viewed the entire surface of the hemisphere, and the position of the eye was carefully monitored using the projection of the optic disc onto the surface of the hemisphere. Units recorded from the synaptic layers of the tectum were driven and mapped with movements across the hemisphere of small black or white disks from 3 to 8 degrees in diameter. Intraretinal recordings were made from some fish using the method of Macy (36), and were mapped on a tangent screen with the same black and white disks.

Field potentials were recorded from the tecta receiving regenerated optic projections using methods previously described(32,37). Briefly, the eye was removed, a concentric bipolar suction electrode was placed over the stump of the optic nerve, and shocks were administered at low frequency from an optically isolated stimulator. The field potentials were recorded (DC) in the tectum with a Ringer filled pipette, relative to a silver-chloride ground electrode behind the cerebellum. Analogue-to-digital conversions, the averaging of traces, and the second differencing calculations were performed with an LSI 11/03 based microcomputer.

The anatomical staining of both normal and regenerating optic axons and arbors was accomplished via the anterograde transport of Horseradish peroxidase (HRP) in broken axons, giving a solid filling of the axon interior (38). Micropipettes tipped in dried HRP were inserted into the optic tracts of the fish, which after 2-3 days survival were sacrificed and perfused. The tecta were fixed lightly and removed, reacted with o-dianisidine, cleared in methyl salicylate and mounted in Canada Balsam for light microscopic examination. The yield of stained axons per tectum ranged from single axons to several hundred. Individual optic axons and arbors from the best filled cases were drawn from the tectal whole mounts using a camera lucida drawing tube on a Zeiss microscope.

RESULTS

Staining of Optic Arbors

Normal optic arbors. One hundred and fifteen optic axons and arbors were traced in their entirety (38). The optic arbors have on the average about 15 to 25 branches which either give off terminal swellings as tiny branchlets or simply terminate as a series of swellings. The number of such swellings per arbor is variable (probably depending to some degree on the intensity of the staining), but in well filled cases the swellings are very numerous, often more than 100. The primary purpose of the study was to quantify the spatial extent of the arbors for comparison with the arbors regenerated after optic nerve crush. Optic arbors in normal goldfish tectum fall roughly into three classes: small arbors range approximately from 80um to 150um across, medium arbors are approximately 150 to 250 um across, and large arbors are 250 to 400um across. These figures are similar to those reported recently by Stuermer(39). The small arbors tended to ramify in a more superficial plane than the medium and large ones (50-120 versus 120-200um corrected depth). The numbers of each type in our study were approximately 55% small, 40% medium and 5% large. However, these numbers cannot be taken to indicate the exact distribution because of the possibility of sample bias, either in the staining process or in the selection procedure for drawing the axons. It is safe to say that the small arbors are by far the most numerous and the large ones the least numerous. If anything, the percentage of large arbors may be overestimated because they stain very darkly and are more noticeable in the whole mounted tecta. They arise from larger caliber axons and these axons are a very small percentage of the total number of stained axons, again supporting our conclusion that they represent only a few percent of the total.

Regenerated arbors. More than 150 regenerated optic arbors were traced under camera lucida (38). Regenerated arbors could be filled as early as two weeks after nerve crush. At these early stages of regeneration, they were much larger than normal in their spatial extent although they were sparsely branched. At 2 weeks, they averaged 1200 to 1500um across. At 3 weeks, still larger arbors 2000 to 2500um across were noted (Figure 1), although by this time some smaller arbors only 200-300um across were also present. At 4 to 5 weeks postcrush, the normal sized arbors began to predominate with only a few of the grossly enlarged arbors still present. At 6 to 7 weeks postcrush, only occasionally were arbors found to have wider than normal extents. Late regenerates, stained after 6 months or more (37 cases), showed a distribution of sizes that closely paralled those of the normal. Their distribution with depth also paralleled that of the normals, with the small arbors more superficial than the medium and large ones. Other than the axonal trajectories, which remained tortuous and indirect, there was little that differed from normal projections.

In summary, the regenerating optic axons initially make

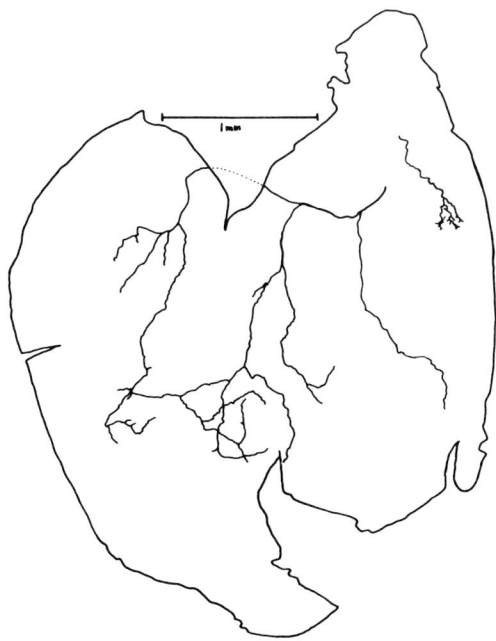

FIGURE 1. Camera lucida drawing of a widely branched optic arbor stained with HRP early in regeneration (21 days postcrush). The heavy lines around the outside mark the boundaries of the tectum, which had to be slit rostrally (top) and caudally (bottom) to allow it to lie flat. A normal optic arbor (large class) is shown at the upper right for comparison. The scale bar is 1 mm.

widely branched arbors that later become restricted. Most of this restriction occurs over the first 5 weeks postcrush but some of it takes months to complete. It is not known whether all of the regenerating optic arbors go through this widely branched phase or only a subset of the total. Likewise we do not know from this data whether the regenerating optic arbors arrive back at the tectum simultaneously or whether certain groups regenerate more slowly than others. At the extremes, one might postulate either that all arbors (whether fast or slow to regenerate) are initially large, or alternatively that only a small number of fast regenerating arbors go through this phase. Because some large arbors are present at 5 to 7 weeks postcrush, the former possibility seems more likely.

Electrophysiological recordings of the retinotopic map.

Lack of sharpening in TTX blocked fish. The electrophysiological recording of retinotectal projections in goldfish in which visual activity had been blocked with intraocular TTX during regeneration showed that the diffuse map had failed to sharpen (13). In control fish, maps recorded at 35 days were already normal both in organization and in the size of the multiunit receptive fields, which averaged 11 degrees (Figure 2). In projections blocked for

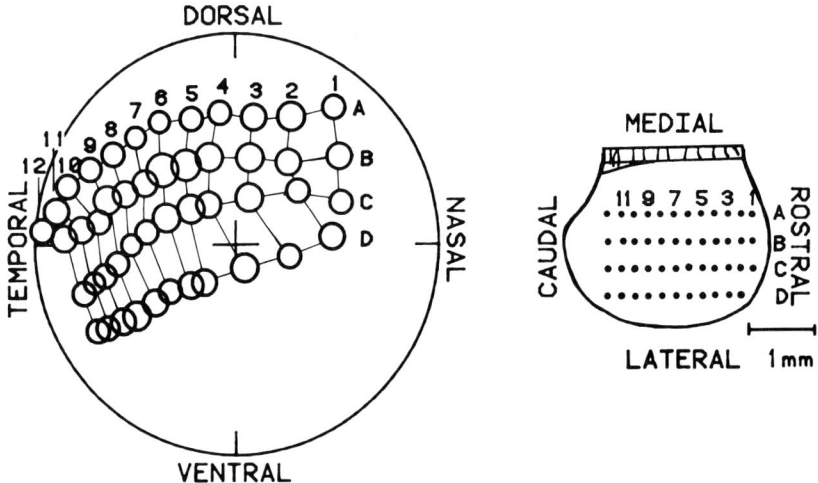

FIGURE 2. Map of a control regenerated retinotectal projection recorded 35 days postcrush. On the left, the large circle represents the hemispherical visual field, and to the right is a drawing of the tectal surface viewed from above. Each point in the tectal array is an electrode penetration. The receptive fields recorded at those points fall into an orderly array in the visual field numbered to match the array of tectal penetrations. Circles give the approximate size of each receptive field. For convenience, the drawing of the tectum has been inverted about one axis so that the arrays are oriented in the same direction: rostral on the tectum corresponds to nasal in the visual field, medial to dorsal, lateral to ventral, and caudal to temporal. Reproduced with permission from reference 13.

the first 28 days, however, the multiunit receptive fields recorded at each tectal point were greatly enlarged, averaging around 30 degrees (Figure 3). As in the study of Adamson et al. (30), the multiunit receptive fields were defined as the inclusive areas of the visual field that yielded any clear responses. The centers of these enlarged fields were in the retinotopically appropriate region of the visual field, indicating that the gross organization of the map was correct.

The enlarged multiunit receptive fields probably reflect the convergence onto each tectal point of arbors from retinal ganglion cells distributed over a wide area of retina. This interpretation is supported by other evidence.

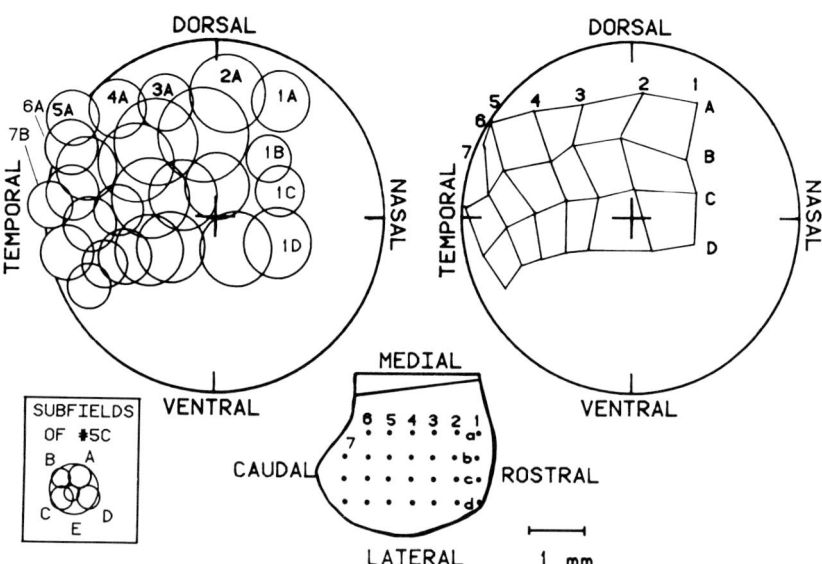

FIGURE 3. Retinotectal map recorded 63 days postcrush in a fish blocked intraocularly with TTX from 0 to 28 days. Two representations of the visual field are shown containing the outlines of the receptive fields (left) and the positions of their centers (right). Other conventions, including the numbered array of tectal penetrations, are the same as in Figure 1. The inset at the lower left diagrams the relationship between the multiunit receptive field at position 5C in the array, and several of its single unit components, isolated with a spike height discriminator. Reproduced with permission from reference 13.

First, the fact that the units persisted during a postsynaptic block with BTX demonstrated that the tectal recordings were of presynaptic origin (13,40). Second, the receptive fields of single ganglion cells recorded in the retina of these same fish were not enlarged (13). Finally, the tectal recordings yielded many units at each site, which when isolated by amplitude, were found to have receptive fields of normal size (13; Figure 3). The simplest interpretation is that the enlarged multiunit receptive fields reflect inherent errors in the targeting of the many regenerated arbors converging on each point. These errors might be due either to entire arbors of normal size misplaced within the retinotectal map, or to dispersed branches of greatly enlarged arbors that are more or less centered on the correct area. These errors are normally corrected when activity is present in the optic fibers. Anterograde HRP staining of the arbors from TTX blocked fish are underway to distinguish between these two possibilities.

Sensitive period. The period of sensitivity corresponds to the period of synaptogenesis (Figure 4). Synapses, as assessed by recording field potentials, were first detectable on day 20 (32). Blocking before this time (0-14 days) did not prevent the sharpening (13), and the maps recorded at 35 days were normal as in Figure 2. Blocking from 14 to 34 days, however, was extremely effective, producing enlarged fields averaging approximately 40 degrees. Synaptogenesis continues at a declining rate until 80 to 100 days postcrush as determined by counts of total synaptic density (5), by the restoration of the levels of acetylcholine receptor (BTX binding, 41) and choline

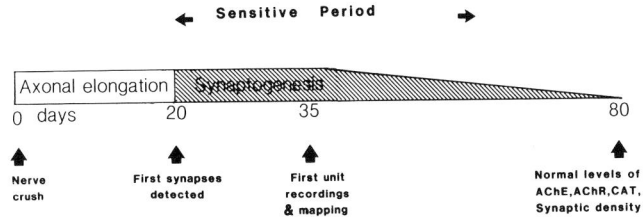

FIGURE 4. Schematic diagram showing the sensitive period for activity dependent sharpening following regeneration. This period coincides roughly with the period of synaptogenesis.

acetyltransferase (42). Schmidt and Eisele (unpublished) have now assessed the effects of blocking activity for two weeks starting at later times (35, 50, 65 or 80 days postcrush). Blocks imposed at 35 days cause the multiunit receptive fields, which had already sharpened, to become enlarged again. The size of this effect, however, becomes progressively smaller at later times, and no effect is seen in the mature projection, where synaptogenesis should be minimal. The results suggest a parallel between the rate at which synapses are added and the degree of disorder produced by blocking activity.

Activity and synapse formation. TTX does not appear to cause a decreased rate of synaptogenesis as judged by the field potentials elicited by optic nerve shock (32). A similar lack of effect of TTX was noted concerning neuromuscular synapse formation in vitro (43). In the fish tectum, the amplitude of the field potentials from eyes blocked during regeneration was not decreased when they were recorded just after the block wore off (32). For the small field potentials in early regeneration, amplitude is likely to be a reasonably sensitive measure of the number of synapses formed. This result suggests that the block of activity may not affect the quantity of synapses made, but instead may interfere with the deployment of the synapses in the retinotopically correct order.

Lack of sharpening in dark reared fish. To control for any possible side effects of TTX beyond the block of spike activity, Schmidt and Eisele (33) tested whether the removal of visually evoked activity also prevented sharpening of the retinotopic map. Placing the fish in complete darkness during the regeneration of the optic nerve allowed only spontaneous activity to occur. Projections regenerated in total darkness also had enlarged multiunit receptive fields, averaging 29 degrees in diameter. This was slightly smaller than the average of 40 degrees for TTX blocks during the equivalent period of 14-34 days, and this difference may reflect the presence of spontaneous activity.

Lack of sharpening in strobe reared fish. An important question at this point was whether the lack of sharpening depended merely upon the amount of activity, or upon some feature of the visually driven activity such as the pattern of activity, which should be correlated between neighboring but not distant ganglion cells. In order to test this possibility, fish were reared under stroboscopic illumination which caused correlated firing in all ganglion cells (33). Single unit recordings from the retinas during regeneration showed that both 'OFF' cells and 'ON' cells were effectively entrained by the strobe light. When all cells, not just near neighbors, fire in synchrony, correlated firing can no longer be a cue for finding neighbors from the retina. Projections regenerated under stroboscopic illumination, like those regenerated without

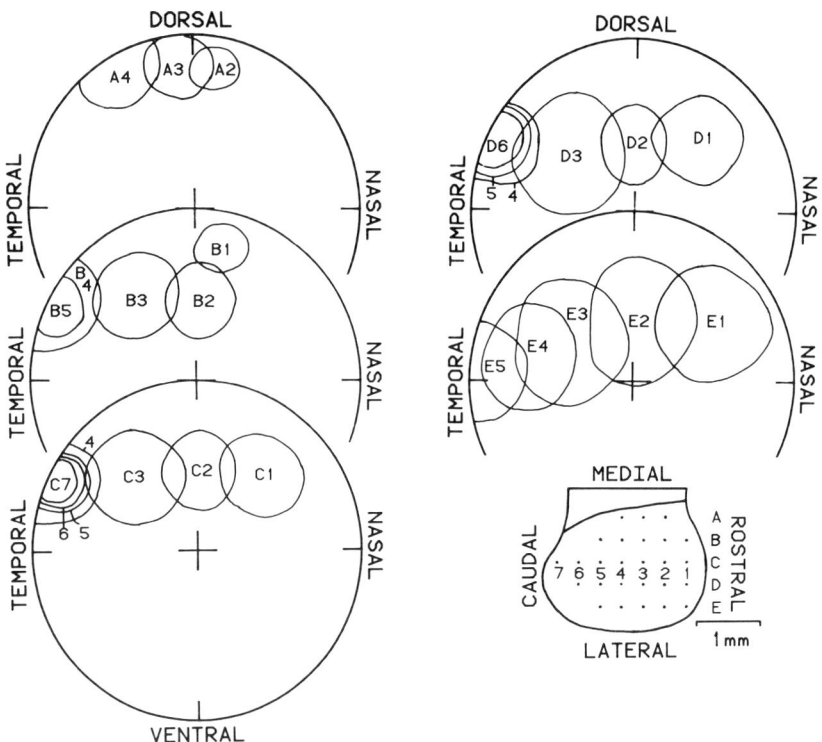

FIGURE 5. Map of a retinotectal projection regenerated under stroboscopic illumination and recorded 55 days postcrush. The multiunit receptive fields recorded at each point on the tectum, although enlarged fall into a rough array similar to the array on the tectal surface and numbered accordingly. For clarity, five separate representations of the visual field are presented, one for each row of points A through E on the tectum. Conventions as in Figure 2. Reproduced with permission from reference 33.

activity, had enlarged multiunit receptive fields that averaged 33 degrees in diameter (Figure 5). This seems to imply that the spatial patterns of correlated activity are important in sharpening the retinotopic map. As with TTX, nonregenerating projections were not sensitive to strobe illumination and remained sharply retinotopic. Finally, the sensitive period for strobe illumination was very similar to that for TTX block of activity.

<u>Syanptic transmission involved in sharpening.</u> If locally correlated activity is used to stabilize the appropriate synapses, it is likely that this correlated activity must be transmitted to the postsynaptic tectal cell to allow summation of the EPSP's. To test this prediction, synaptic transmission was blocked during the period of early synaptogenesis (34). An osmotic minipump was attached to the fish's head to deliver a continuous infusion of BTX, previously shown to block retinotectal synaptic transmission (35,44,45). The maps from fish that had been infused from 20 to 35 days postcrush had enlarged multiunit receptive fields similar to those seen in the TTX blocked, strobe blocked and dark reared fish, averaging 29 degrees in diameter. This experiment suggests that fibers with correlated activity do not interact directly, but must interact through the transmission of their signals to the postsynaptic tectal cells, probably through the summation of EPSP's.

DISCUSSION

Activity-dependent Sharpening of the Retinotopic Map

<u>Major findings.</u> The retinotectal projection that regenerates after optic nerve crush in the adult goldfish is initially only crudely retinotopic but later sharpens in an activity dependent manner. Four lines of evidence have been presented. 1) Anatomically, this sharpening can be seen as the transition from the initially enlarged arbors that explore vast areas of the tectal surface to the spatially restricted arbors characteristic of normal projections. 2) Electrophysiological evidence from the TTX blocked and dark reared fish shows that this sharpening is dependent upon visually driven activity. 3) The lack of sharpening in strobe reared fish shows that the sharpening is driven by the spatiotemporal pattern of activity. 4) The ability of BTX to prevent sharpening suggests that the correlated

activity interacts via the summation of EPSP's within the postsynaptic tectal cells.

Correlation between electorphysiology and anatomy. The period during which the arbors become restricted compares reasonably well with the earliest time that one can record a normally ordered map on the tectum. This time is around 5 weeks postcrush when the restricted arbors first predominate. At times earlier than 5 weeks, reliable recordings cannot usually be made in goldfish simply because the responses to visual stimuli fatigue very rapidly so that the exact determination of receptive field size is very difficult. Previous attempts at recording earlier than 5 weeks (13) suggest however that the multiunit receptive fields are greatly enlarged, on the order of 60 to 90 degrees. A similar transient appearance of enlarged fields has been noted in the frog (30,31) as a normal stage in regeneration. Thus, the enlarged receptive fields seen here after blocking activity, rearing in the dark or under strobe illumination, or with BTX infusion can readily be interpreted as interfering with the usual sharpening process. Based upon the results of the initial TTX experiments, Schmidt and Edwards (13) presented a model for the activity dependent sharpening.

A Model Based on Correlated Activity. The proposed model (Figure 6) for sharpening the map is based upon the correlated firing of neighboring ganglion cells and the resultant summation of their EPSP's in the postsynaptic tectal cells (13,14). Neighboring ganglion cells, which view the same part of the visual world, are likely to fire with a high degree of correlation if they are of the same type (eg., ON or OFF; 46). In fact, both Arnett (47) and Ginsberg et al. (46) have demonstrated such correlations, even in absolute darkness when only spontaneous activity is present. The model also assumes that the tectal arbors in the initial diffuse projection are large (as demonstrated anatomically), and have a high degree of overlap. For neighboring ganglion cells having some overlap of their arbors and firing with a high degree of correlation, there would be a summation of the postsynaptic EPSP's. Finally, if the most effective synapses are differentially stabilized and retained (10,48), the correlated activity, resulting in larger EPSP's, would stabilize their synapses in the region of overlap. Of course, arbors from distant ganglion cells would also overlap, but without correlated firing there would be no summation and no stabilization of those synaptic connections. Such a cue for convergence would, therefore, be specific to the arbors of neighboring ganglion cells, because they would have correlated activity.

<u>Testing the model.</u> The experiments reported here have upheld two of the major features of this model. First, the model proposes that the sharpening would still be disrupted even if the ganglion cells were allowed activity as long as the strictly local correlation in firing of neighboring ganglion cells is disrupted. Two methods were used : stroboscopic illumination and dark rearing. The strobe illumination caused a universal rather than local correlation in activity, thereby removing the cue that two fibers in the tectum having correlated activiy would come from nearby ganglion cells. Since this was also very effective in preventing the sharpening, the cue must reside in the spatiotemporal pattern of correlated activity generated under normal conditions. For the dark reared fish, the effect was somewhat smaller than that produced under strobelight illumination or by the equivalent TTX block. This may reflect a slight degree of sharpening due

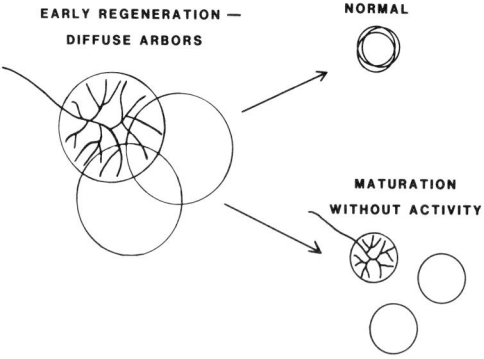

FIGURE 6. Schematic diagram of the model for activity dependent sharpening of the retinotectal map. On the left are the arbors from three neighboring retinal ganglion cells that are large and inaccurately targeted. Within the area of overlap, the three cells synchronous firing would lead to summation of their EPSP's, and to the stabilization of their synapses moreso than outside the area. This might lead to a concentration of branches within this area and the retraction of the others, progressing to a situation with smaller, nearly completely overlapping arbors. Reproduced with permission from reference 13.

to the correlation in the spontaneous activity of neighboring ganglion cells (47). The rate of spontaneous activity in goldfish is much lower than in the cat(47), where spontaneous activity could conceivably play a larger role.

The strobe and dark rearing experiments also make unlikely the possibility that the TTX effect upon sharpening was caused by the deficit in axonal transport in regenerating optic fibers that was demonstrated by Edwards and Grafstein (49). The high and low levels of activity in the strobe and dark reared fish are unlikely to be associated with similar decrements in axonal transport.

A second prediction from the model is that, in order for the synapses to be stabilized, the correlated activity must be transmited to the postsynaptic tectal cell to allow summation of EPSP's. When synaptic transmission was disrupted by the infusion of BTX, the regenerated maps had similarly enlarged multiunit receptive fields as those seen in the other three groups. The fibers with correlated activity apparently interact through a cholinergic mechanism, probably via the transmission of their signals to common postsynaptic cells allowing the summation of EPSP's. At the present time, we cannot determine exactly what form of activity in the postsynaptic cells (whether EPSP or spike activity) may be important in the stabilization of synaptic contacts, nor do we know anything about the nature of the feedback to stabilize the presynaptic contact.

In the above model, we have assumed that arbors maturing without cues from activity patterns must still retract many of the branches of the initially enlarged arbors. The reason for this assumption is that upon release from the TTX block (or from the dark, the strobe, or the BTX block), the projection is much less efficient at using normal visual activity in sharpening the map. Normal exposure for more than a month results in very little sharpening compared to that usually accomplished in two weeks early in regeneration. After a year or more of normal visual exposure, however, significant sharpening occurs, although some areas of the maps show vestiges of the crude projection. The simplest explanation for this phenomenon is that the ability to use correlated activity to sharpen the projection is related to the spatial extent and degree of overlap of the axonal arbors in the tectum. If the arbors are forced to retract randomly in the absence of cues based upon activity, then the resulting small arbors in inappropriate areas would then have greatly lessened chances of finding the appropriate areas because of their lessened overlap with other arbors.

Meyer (12), using intraocular TTX injections in small goldfish, has also demonstrated that the sharpening of the retinotopic map is activity dependent, but he found that the diffuse map is more readily sharpened upon release from TTX block. He used the technique of retinal lesioning and radioautographic tracing to show that a block from 32 to 80 days postcrush prevented the return to a normal topography. However, if the fish were allowed a further 24 days of activity after the TTX was discontinued, the map was able to sharpen. Thus, small goldfish apparently retain the ability to use activity to sharpen the map several months after the fibers reached the tectum. The difference between his results and ours may hinge on the size of the arbors relative to the tectal surface and the overlap factor mentioned above. It is well known that smaller goldfish demonstrate a higher degree of plasticity in other tests. For example, when half of the tectum is surgically removed, small 5cm goldfish take only 30 days to compress the projection onto a half tectum (50), while large 15cm fish take approximately 6 months (51).

Diffuse Projections in Development. The diffuse projections found during regeneration raise the question of whether there are any direct parallels in development. Instances of diffuse retinotectal maps have been documented in several developing animals, but the role of activity has not been studied.

In early larval Xenopus, several groups are studying single cobalt stained optic axons. Sakaguchi (52) reported that the ingrowing fibers usually took very direct paths toward the appropriate quadrant of tectum, thus initially forming a crude retinotopic map. However, the early arbors at stages 40-45 were often large relative to the total tectal neuropil, covering approximately 75% of the rostrocaudal extent, and approximately 33 to 50% of the mediolateral extent. At later stages (St. 45-50), the arbors were more restricted, in that they covered a somewhat smaller percentage of the neuropil. The arbors had not shrunk, but the tectum had grown in size, as it continues to add new cells(21). Piper, Steedman and Stirling (53) found the arbors to be approximately 250 um across at larval stages 50 to 55, still quite large with respect to the tectal dimensions.

In chick embryos, McLoon (54) used quadrantal retinal lesions combined with HRP staining of the remaining projection to demonstrate a diffuse projection at 10 days of incubation. The tectal quadrant that corresponded topographically to the retinal lesion was not denervated but still received retinal innervation at 10 days. By 14 days, the map had sharpened and the corresponding tectal quadrant was denervated. Thus, a gradual sharpening of the projection begins several days after the fibers arrive at the tectum. This sharpening occurs before hatching, so that any role for activity would depend upon correlated spontaneous activity.

In rodents, which are born in an altricial state, the innervation of the tectum occurs around the time of birth or just after. Several studies have shown that the initial retinotectal projection is diffuse. In hamsters, Schneider et al. (55), using anterograde staining with HRP, found that optic arbors transiently made widespread branches with sparse arborizations but later concentrated into the dense, restricted arbors that are characteristic of the mature projection. In the rat, O'Leary et al. (56,57) used Fast Blue to retrogradely label the retinal ganglion cells on the day of birth. Initially, injections into caudal tectum labelled many ganglion cells from all over the retina. By day 12, however, the cells that were labelled on the day of birth were largely restricted to the appropriate nasal retina. Because this dye (Fast Blue) is known to persist in cells, the vast majority of the cells from inappropriate retinal quadrants appeared to be eliminated. This may indicate that cell death rather than, or in addition to, synapse elimination may mediate much of the retinotopic sharpening. It is tempting to suggest that the cells that die may be those that lose all of their terminal branches in the tectum during an activity dependent sharpening (8,57).

Comparison with Other Studies of Synapse Elimination.

The segregation of ocular dominance patches. Recent evidence suggests that the segregation of visual afferents into eye specific patches or stripes may also be driven by an activity dependent stabilization of synapses. This segregation normally occurs during the development of the visual cortex in cat and monkey, as a pattern overlaid on the retinotopic map (58), but a similar segregation of direct retinal afferents occurs when two eyes of fish or frog (59,60,61) innervate a single tectum (Figure 7A). Each of the two projections initially covers the entire tectum, but later retracts into patches occupied almost exclusively by fibers from the same eye. If binocular TTX injections

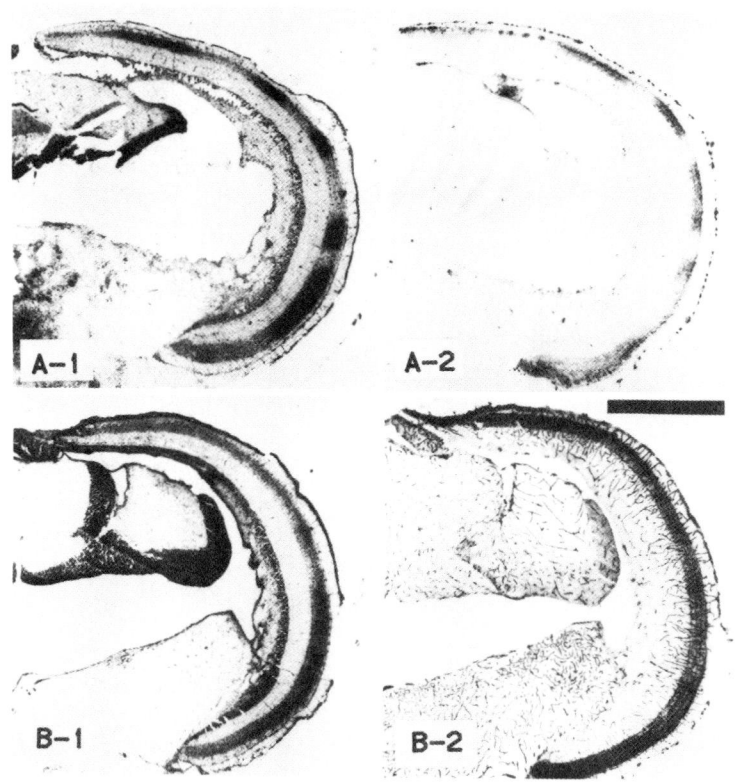

FIGURE 7. Tectal cross sections from a control (vehicle injected) fish (above) and a TTX injected fish (below) demonstrating the segregation of the projections into complementary ocular dominance patches and its prevention with TTX. Top: Adjacent sections from the control fish. A-1: Ipsilateral projection radioautographically labelled and counterstained with neutral red. A-2: Contralateral projection labelled with HRP-TMB reaction and no counterstain. Bottom: Adjacent section from the TTX injected fish. B-1: Ipsilateral projection radioautographically labelled, neutral red counterstain. B-2: Contralateral projection stained with HRP-DAB reaction, no counterstain. Calibration bar: 1mm. Reproduced with permission from reference 60.

are used to block activity, the segregation is prevented (59,60,61,62; Figure 7B). The differential stabilization of synapses may be driven by the correlated activity of neighboring ganglion cells within each eye and a probable lack of correlation in the activity of ganglion cells from different eyes. The evidence was recently reviewed by Schmidt and Tieman (63).

Retraction of Ineffective Synapses. The selective stabilization of the most effective synapses implies that ineffective ones are destabilized. Such an effect has been reported to occur in both the marine toad (64) and the goldfish (40) after application of alpha-Bungarotoxin to block postsynaptic responses in a small localized region of the tectum. One week after such a block, there was a local disturbance in the retinotopic maps, suggesting that the optic terminals normally terminating within the blocked area withdrew their arbors to make synapses outside. The disturbance was shown to be activity dependent, as no retraction from the toxin blocked area was found to occur if activity in the optic fibers was simultaneously blocked with intraocular TTX. Thus fibers with activity appear to be able to respond to differences in responsiveness of postsynaptic cells.

Activity and other maps. In addition to the direct retinotectal projection, several other maps are known to be sharpened, organized or otherwise altered by activity dependent mechanisms. One of these is the indirect retinotectal pathway in Xenopus. Keating (65) has shown electrophysiologically that this pathway, which relays information from the ipsilateral eye, can adjust itself, following rotation of one eye, to bring the maps from the two eyes into register. Udin (66,67) has reported anatomical changes in the isthmotectal fibers and arbors corresponding to the changes noted in the electrophysiologically recorded maps.

A second example of a map sharpened by activity is the tectal map of auditory space. In the barn owl, this map apparently aligns itself to the visual map already in place on the tectum and can be shifted relative to the visual map by the placement of a plug in one ear during the early life of the owl (68).

A third example is the set of visual callosal connections that connects the two representations of the visual midline in the respective half field maps of the two sides of the brain (69). The callosal fibers in the cat initially arise from a large set of cortical neurons, but after synapse elimination only a small subset of these neurons continue to send their fibers across the callosum. The selection of these neurons may depend upon patterned visual activity, as a different subset remain connected when the animal is reared in the dark (69,70).

Finally, the lateral geniculate nucleus of the cat also has a retinotopic map that is apparently sharpened by activity (71). Blocking activity with intraocular TTX during the first three postnatal weeks resulted in the appearance of enlarged single unit (postsynaptic) receptive fields, forming a close parallel with the findings in goldfish. In addition, the individual cells of the geniculate normally respond selectively to 'ON' or 'OFF', but not to both. Without activity during development, the majority end up with mixed 'ON' and 'OFF' input. Lack of activity also causes abnormal mixing of the X and Y inputs. If the initial innervation is not selective by receptive field type, the TTX may be interfering with synapse elimination, preventing the capture of geniculate cells by 'ON' or 'OFF' fibers exclusively. Such a segregation of parallel streams of incoming information may represent another important role of activity dependent stabilization in the development of highly ordered neural connections found in the mature nervous system.

Activity and synapse elimination outside the CNS. The production of excess synaptic connections and the subsequent elimination of many of these connections is a familiar pattern in neural development, occurring in the peripheral and autonomic nervous systems as well as in the central nervous system. In the autonomic ganglia, there is a loss of excess preganglionic inputs during development, although the role of activity is not easily studied and only indirectly implicated (72). At the neuromuscular junction, the elimination of polyneuronal innervation requires that presynaptic activity be transmitted to the postsynaptic muscle fiber (73,74,75,76,77). However, synchronous activation of all nerve fibers does not promote stable convergence of those inputs, but only hastens the elimination process (78,79). This apparently reflects a major difference between the neuromuscular junction and many synapses within the central nervous system.

Concluding remarks. Activity dependent mechanisms for the selective stabilization or elimination of the initially diffuse synaptic connections may represent a widespread phenomenon, occurring at all levels from the neuromuscular junction to the callosal connections between the cortices.

They may be relatively efficient mechanisms in development, particularly if the excess initial synaptic branches do not degenerate but become resorbed, as has been suggested at the neuromuscular junction.

Activity dependent stabilization does not act alone, but in concert with the differential cell adhesion mechanisms commonly known as chemoaffinity. Activity dependent stabilization of totally random synapses would not be able to generate the reproducibly oriented maps that one finds in the nervous system. Instead a random mix of retinal fibers would result either in randomly oriented maps or in 'mosaic' maps with occasional discontinuities (14). Rather, the reproducible polarity must stem from the ability of differential affinity mechanisms to bring a greater number of temporal versus nasal retinal fibers to rostral tectum, and nasal fibers to caudal tectum, etc.

On the other hand, a completely rigid developmental process (rigid chemoaffinity alone) might not be sufficiently flexible to succeed. Small mistakes occurring at one site would not be able to be corrected at later points along the pathways. Factors such as the variable geometry of the head, the separation between the eyes, between the ears, etc., might demand flexibility in the process of aligning the visual maps from the two eyes or aligning the auditory map with the visual maps.

We know very little about how an activity dependent stabilization of synapses might occur; we have as yet only a very rough outline. The exact type of postsynaptic activity that is important is unknown (EPSP's or action potentials). In addition, we do not yet know the spatial area over which facilitation occurs. Purves and Hume (80) have pointed out the implications that this might have for postsynaptic dendritic geometry. Finally, we do not know what form the feedback signal for stabilization might take. It will suffice to point out that the signal need not be an actual factor emitted postsynaptically, but might instead be a recruitment or anchoring of additional intercellular adhesion molecules in the postsynaptic density which would serve to stabilize the mechanical linkage between the pre- and postsynaptic cells.

ACKNOWLEDGEMENTS

Leslie Eisele, Valerie Boss, Joanne Turcotte, Louise Edwards and Marlene Buzzard participated in some of the work described herein. I thank Dr. David Tieman for helpful comments on the manuscript.

REFERENCES

1. Sperry RW (1943). Effect of 180 degree rotation of the retinal field on visuomotor coordination. J Exptl Zool 92: 263.
2. Sperry RW (1963). Chemoaffinity in the orderly growth of nerve fiber patterns and connections. Proc Nat Acad Sci USA 50: 703.
3. Prestige M, Willshaw D (1975). On a role for competition in the formation of patterned neural connections. Proc Roy Soc B 190:77.
4. Schmidt JT, Cicerone CM, Easter SS (1978). Expansion of the half retinal projection to the tectum in goldfish: An Electrophysiological and anatomical study. J Comp Neurol 177: 257.
5. Murray M, Sharma SC, Edwards MP (1982). Target regulation of synaptic density in the compressed retinotectal projection of goldfish. J Comp Neurol 209: 374.
6. Horder TJ, Martin KAC (1977). Morphogenetics as an alternative to chemospecificity in the formation of nerve connections. In Curtis ASG (ed): "Cell-cell recognition (32nd Symposium Soc. Exp. Biol.)," Cambridge: Cambridge Univ. Press, p 275.
7. Bodick N, Levinthal C (1980). Growing optic nerve fibers follow neighbors during embryogenesis. Proc Nat Acad Sci USA 77: 4374.
8. Schmidt JT (1982). The formation of retinotectal projections. Trends in NeuroSci 5: 111.
9. Stuermer C, Easter SS (1984). A comparison of the normal and regenerated retinotectal pathways of goldfish. J Comp Neurol 223: 57.
10. Changeux JP, Danchin A (1976). Selective stabilisation of developing synapses as a mechanism for the specification of neuronal networks. Nature 264:705.
11. Keating MJ (1975). The time course of experience dependent synaptic switching of visual connections in Xenopus laevis. Proc R Soc Edinburgh B189: 603.
12. Meyer RL (1983). Tetrodotoxin inhibits the formation of refined retinotopography in goldfish. Dev Brain Res 6: 293.

13. Schmidt JT, Edwards DL (1983). Activity sharpens the map during the regeneration of the retinotectal projection in goldfish. Brain Res 269: 29.
14. Willshaw DJ, von der Malsburg C (1976). How patterned neural connections cbe set up by self-organization. Proc Roy Soc Lond B194: 431.
15. Attardi DG, Sperry RW (1963). Preferential selection of central pathways by regenerating optic fibers. Exp Neurol 7: 46.
16. Straznicky C, Gaze RM (1980). Stable programming for map orientation in fused eye fragments in Xenopus. J Embryol Exp Morph 55: 123.
17. Conway K, Feiock K, Hunt RK (1981). Polyclones and patterns in growing Xenopus eye. Current Topics in Developmental Biology 15: 217.
18. Straznicky C (1978). The acquisition of tectal positional specification in Xenopus. Neurosci Lett 9: 177.
19. Chung S-H, Cooke J (1978). Observations on the formation of the brain and of nerve connections following embryonic manipulation of the amphibian neural tube. Proc R Soc Lond B201: 335.
20. Bonhoeffer F, Huf J (1982). In vitro experiments on axon guidance demonstrating and anterior-posterior gradient on the tectum. EMBO J 1: 427.
21. Easter SS (1983). Postnatal neurogenesis and changing neuronal connections. Trends in NeuroSci 6: 53.
22. Frost D (1981). Orderly anomalous retinal projections to the medial geniculate, ventrobasal and lateral posterior nuclei of the hamster. J Comp Neurol 203: 227.
23. Meyer RL (1979). Retinotectal projection in goldfish to an inappropriate region with a reversal in polarity. Science 205: 819.
24. Horder TJ, Martin KAC (1983). Some determinants of optic terminal localization and retinotopic polarity within fibre populations in the tectum of goldfish. J Physiol 333: 481.
25. Sharma SC (1975). Visual projection in surgically created 'compound' tectum in adult goldfish. Brain Res 93: 497.
26. Fraser SE (1985). Cell interactions involved in neuronal patterning: An experimental and theoretical approach. In Cowan MW (ed): "Molecular Bases of Neural Development," New York: John Wiley and Sons, (In press).
27. Meyer RL (1980). Mapping the normal and regenerating retinotectal projection of goldfish with

autoradiographic methods. J Comp Neurol 189:273.
28. Cook JE, Rankin ECC (1984). Use of a lectin-peroxidase conjugate (WGA-HRP) to assess the retinotopic precision of goldfish optic terminals. Neurosci Lett 48: 61.
29. Fujisawa H, Tani N, Watanabe K, Ibata Y (1982). Branching of regenerating retinal axons and preferential selection of appropriate branches for specific neuronal connection in the newt. Dev Biol 90:43.
30. Adamson JR, Burke J, Grobstein P (1984). Reestablishment of the ipsilateral oculotectal projection after optic nerve crush in the frog: evidence for synaptic remodeling during regeneration. J Neurosci 4: 2635.
31. Humphrey MF, Beazley LD (1982). An electrophysiological study of early retinotectal projection patterns during optic nerve regeneration in *Hyla moorei*. Brain Res 239: 595.
32. Schmidt JT, Edwards DL, Stuermer CAO (1983). The Reestablishment of synaptic transmission by regenerating optic axons in goldfish: Time course and effects of blocking activity by intraocular injection of tetrodotoxin. Brain Res 269: 15.
33. Schmidt JT, Eisele LE (1985). Stroboscopic illumination and dark rearing block the sharpening of the retinotectal map in goldfish. Neuroscience 14: 535.
34. Schmidt JT, Eisele LE (1983). Goldfish stroboscopic illumination: regeneration in a flash fails to sharpen the retinotectal map. Soc Neurosci Abstr 9: 858.
35. Freeman JA, Schmidt JT, Oswald RE (1980). Effect of a-bungarotoxin on retinotectal synaptic transmission in the goldfish and the toad. Neurosci 5: 929.
36. Macy, A (1981). Growth related changes in the receptive field properties of retinal ganglion cells in goldfish. Vision Res 21: 1491.
37. Schmidt JT (1979). The laminar organization of optic nerve fibers in the tectum of goldfish. Proc Roy Soc London B205: 287.
38. Schmidt JT, Buzzard M, Turcotte J (1984). Morphology of regenerated optic arbors in goldfish tectum. Soc Neurosci Abstr 10: 667.
39. Stuermer CAO (1985). Rules for retinotectal arborizations in goldfish optic tectum: a whole mount study. J Comp Neurol 229: 214.
40. Schmidt JT (1985). Apparent movement of optic terminals out of a local postsynaptically blocked region in goldfish tectum. J Neurophysiol 53: 237.
41. Schechter N, Francis A, Deutsch DG, Gazzaniga MS (1979).

Recovery of tectal nicotinic-cholinergic receptor sites during optic nerve regeneration in goldfish. Brain Research 166: 57.
42. Francis A, Schechter N (1979). Activity of choline acetyltransferase and acetylcholinesterase in the goldfish tectum after disconnection. Neurochem Res 4: 547.
43. Obata K (1977). Development of neuromuscular transmission in culture with a variety of neurons and in the presence of cholinergic substances and tetrodotoxin. Brain Res 119: 141.
44. Oswald RE, Schmidt JT, Norden JJ, Freeman JA (1980). Localization of bungarotoxin binding sites to the goldfish retinotectal projection. Brain Res 187: 113.
45. Schmidt JT, Oswald RE, Freeman JA (1980). Electrophysiologic evidence that the retinotectal projection in the goldfish is nicotinic cholinergic. Brain Res 187:129.
46. Ginsberg KS, Johnsen JA, Levine MW (1984). Common noise in the firing of neighboring ganglion cells in goldfish retina. J Physiol (London) 351: 433.
47. Arnett DW (1978). Statistical dependence between neihboring retinal ganglion cells in goldfish. Exp Brain Res 32:49.
48. Hebb DO (1949). "Organization of Behavior." New York: John Wiley and Sons.
49. Edwards DL, Grafstein B (1983). Intraocular tetrodotoxin in goldfish hinders optic nerve regeneration. Brain Res 269: 1.
50. Yoon MG (1976). Progress of topographic regulation of the visual projection in the halved optic tectum of adult goldfish. J Physiol (London) 257: 621.
51. Schmidt JT (1983). Regeneration of the retinotectal projection following compression onto a half tectum in goldfish. J Embryol Exp Morph 77: 39.
52. Sakaguchi DS (1984). The development of the retinotectal projection in Xenopus laevis. Ph D Thesis, State Univ of New York at Albany.
53. Piper EA, Steedman JG, Stirling RV (1979). Three-dimensional computer reconstruction of cobalt stained optic fibers in whole brains of Xenopus tadpoles. J Physiol (London) 286: 13P.
54. McLoon S (1982). Alterations in precision of the crossed retinotectal projection during chick development. Science 218: 1418.
55. Schneider GE, Rava L, Sachs GM, Jhaveri S (1981).

Widespread branching of retinotectal axons: Transient in normal development and anomalous in adults with neonatal lesions. Soc Neurosci Abstr 7: 732.
56. O'Leary DDM, Fawcett JW, Cowan MW (1984). Elimination of topographical targeting errors in the retinocollicular projection by ganglion cell death. Soc Neurosci Abstr 10: 464.
57. Cowan MW, Fawcett JW, O'Leary DDM, Stanfield BB (1984). Regressive events in neurogenesis. Science 225: 1258.
58. LeVay S, Stryker MP, Shatz CJ (1978). Ocular dominance columns and their development in layer IV of the cat's visual cortex: A quantitative Study. J Comp Neurol 179: 223.
59. Meyer RL (1982). Tetrodotoxin blocks the formation of ocular dominance columns in goldfish. Science 218:589.
60. Boss V, Schmidt JT (1984). Activity and the formation of ocular dominance patches in dually innervated tectum of goldfish. J Neuroscience 4: 2891.
61. Reh T, Constantine-Paton, M (1985). Eye-specific segregation requires neural activity in three-eyed Rana pipiens. J Neurosci 5: (In press).
62. Stryker MP (1981). Late segregation of geniculate afferents to the cats' visual cortex after recovery from binocular impulse blockade. Soc Neurosci Abstr 7: 842.
63. Schmidt JT, Tieman SB (1985). Eye-specific segregation of optic afferents in mammals, fishes and frogs: The role of activity. Cell and Molec Neurobiol (In press).
64. Freeman JA (1977). Possible regulatory function of acetylcholine receptor in maintenance of retinotectal synapses. Nature 269: 218.
65. Keating MJ (1975). Time course of experience dependent synaptic switching of visual connections in Xenopus laevis. Proc Roy Soc Edinburgh B189: 603.
66. Udin S (1983). Abnormal visual input leads to development of abnormal axon trajectories in frogs. Nature 301: 336.
67. Udin S (1985). The role of visual experience in the formation of binocular visual projections in frogs. Cell and Molec Neurobiol (In press).
68. Knudsen EI (1983) Early auditory experience aligns the auditory map of space in the optic tectum of the barn owl. Science 222: 939.
69. Innocenti GM (1981). Growth and reshaping of axons in the establishment of visual callosal connections. Science 212: 8246.
70. Innocenti GM, Frost DO, Illes J (1985). Maturation of

visual callosal connections in visually deprived kittens: A challenging critical period. J Neurosci 5: 255.
71. Archer SM, Dubin MW, Stark LA (1982). Abnormal development of kitten retinogeniculate connectivity in the absence of action potentials. Science 217: 743.
72. Purves D, Lichtman JW (1980). Elimination of synapses in the developing nervous system. Science 210: 153.
73. Thompson WJ (1985). Activity and synapse elimination at the neuromuscular junction. Cell and Molec Neurobiol (In press).
74. Thompson W, Kuffler DP, Jansen JKS (1979). The effect of prolonged, reversible block of nerve impulses on the elimination of polyneuronal innervation of new-born rat skeletal muscle fibers. Neurosci 4: 271.
75. Sohal GS, Creazzo TL, Oblak TG (1979). Effects of chronic paralysis with alpha-Bungarotoxin on development of innervation. Exp Neurol 66: 619.
76. Magchielse T, Meeter E (1982). Reduction of polyneuronal innervation of muscle cells in tissue culture after long term indirect stimulation. Dev. Brain Res 3: 130.
77. Duxson MJ (1982). The effect of postsynaptic block on development of the neuromuscular junction in postnatal rats. J Neurocytol 11: 395.
78. O'Brien RA, Ostberg AJC, Vrbova G (1978). Observations on the elimination of polyneuronal innervation in developing mammalian skeletal muscle. J Physiol (London) 282: 571.
79. Srihari T, Vrbova G (1978). The role of muscle activity in the differentiation of neuromuscular junctions in slow and fast chick muscles. J Neurocytol 7: 529.
80. Purves D, Hume RI (1981). The relation of postsynaptic geometry to the number of presynaptic axons that innervate autonomic ganglion cells. J Neurosci 1: 441.

SELECTION MECHANISMS IN NEURAL MAPPING

Leif H. Finkel

The Rockefeller University
New York, New York 10021

ABSTRACT The theory of neuronal group selection
is applied to the problem of the organization of
neural maps. Selection in the adult nervous system
bears certain similarities to various aspects of
embryonic induction, and these are discussed.
Evidence for anatomical and functional variability in
the nervous system is reviewed and is used to support
the existence of a selective process in the nervous
system. Formation of neural maps is tied to three
processes (group confinement, group selection, and group
competition) involving neuronal groups -- local sets of
tightly coupled neurons which act collectively.
Finally, a model of synaptic modification is presented
to account for group selection. This model proposes
independent presynaptic and postsynaptic modifications,
and investigates how the two are coupled through the
anatomy of the network by means of a continuing
developmental process.

INTRODUCTION

The premise of this paper is that a process of selection occurs in the developing and mature nervous system, and the object is to show how this process is responsible for the formation and maintainance of ordered neural maps. Selection is the primary force responsible for creating taxonomies in evolution, and we will argue that a process of selection, analogous to natural selection acting on organisms in evolution, acts on groups of neurons during the lifetime of the animal. As originally formulated (Edelman, 1978;

1981), the theory of neuronal group selection proposed that there are two phases of selection: a developmental phase which gives rise to a degenerate repertoire of neuronal groups, and an experiential phase in which certain groups are selected over others as a result of biochemical and biophysical changes occuring at synapses.

Selective systems cannot operate unless there is variability in the population, and a requirement of our theory of neuronal group selection is that a large degree of variability will exist in neural structure and function. We shall review the evidence for such variability including recent experimental findings that the functional maps found in cortex of vertebrates are not static and fixed, but are dynamically modifiable by changes in afferent input.

The underlying molecular basis for anatomical variability lies in the interaction of cell adhesion molecules (CAM's) with the primary processes of development. Developmental control of the prevalence, distribution, and binding strengths of various CAM's allows for dramatic changes in the neural structures that arise. As has been discussed at this meeting, the dynamic, symmetry-breaking properties characteristic of the modulation of cell adhesion lead to variability in connections in an obligate fashion (Edelman, 1984).

The immune system, which constitutes the other great biological recognition system, uses another primary process of development as its basis for memory. Clonal selection (Edelman, 1975) uses cell division as its mechanism of differential enhancement of selected cells in the lymphocyte population. But the immune system can only respond over a time course of minutes to hours, orders of magnitude too slow for the nervous system which must rapidly respond to environmental challenges. Together with other considerations to be discussed below, this implies that selection in the nervous system must occur through non-developmental mechanisms. These mechanisms nevertheless reflect a certain developmental character which will become apparent in our discussion of synaptic modification.

The parallel between development and the functioning of the adult nervous system also holds in the other direction. The phenomenology of embryonic induction, in particular, the classic work of Paul Weiss, bears a distinct resemblance to the descriptions of neuronal group selection to be advanced here. These similarities reflect the population properties of biological systems and the power of selective systems to dynamically adapt to the local environment.

The concept of degeneracy -- that neural structure and function are not related in a one-to-one fashion and thus, that isofunctional groups need not be isomorphic -- has its counterparts in development. It is well documented that the prospective potency of an embryonic region is not the same as its prospective fate (Weiss, 1939). Weiss discusses the "selection of potencies" during the course of development, "the cellular material of the early germ could produce a practically infinite variety of morphogenetic and histogenetic results. However, they do produce only one definite pattern from among the many. This pattern is the standard pattern of the species. It implies definite localization, correct timing, and proper intensity of the morphogenetic and histogenetic processes." (Weiss, 1939) We will see that these same factors -- localization, timing, and intensity -- will be crucial to the mechanism of neuronal group selection.

Weiss' notion of a developmental "field" shares features with our concept of a neuronal group. He states (Weiss, 1939) that "the fate of an individual cell depends upon the particular local field within which it happens to lie ... A group of cells acts collectively through its field ...." The competitive effects between groups which we shall argue are responsible for the organization of neural maps also have their parallels in development. Again from Weiss (1939) "A cell group lying in the area of overlap of two fields, unable to produce some sort of comprimise structure, can obey only either one or the other ... the decision ... depends solely on the relative strength of the various fields in the locality: the strongest will prevail."

We will discuss some recent results on the organization and reorganization of neural maps which well exemplify these types of competitive effects. Before that, however, we begin by reviewing evidence for anatomical variability in the networks comprising the maps.

ANATOMICAL VARIABILITY

The presence of anatomical variability in the nervous system is a necessary requirement of the theory of neuronal group selection. Variability is also a strong argument in favor of selection since any information processing theory is seriously hampered by variability in the system, and furthermore, can not cogently explain its presence. It is difficult to cite examples of anatomical variability as few studies have been done to examine the populational or numerical aspects of neuroanatomy in well-controlled situations. In vertebrates, it is known that complex variants of chemical synapses can arise during development and can persist in even such highly regular regions as the cerebellum (Chan-Palay, et al 1981; 1982). Variations related to sexual dimorphism may also be significant, for example, in the form- ation of callosal connections (de Lacoste-Utamsing and Holloway, 1982). The callosal connections are, in fact, a paradigmatic case, as the exact pattern of the callosal inputs differs to such a degree among individuals, that the termination pattern has been compared to a finger-print (van Essen, et al, 1982).

Studies of variability in vertebrates are problematic due to the complexity of the system. An exception to this is the Mauthner cell, which when examined by intracellular fills with horseradish peroxidase does show variability in the number of terminals present at neuromuscular junctions (Korn, et al, 1984). Nevertheless, the most dramatic studies of variability have utilized identified neurons in invertebrates.

Pearson and Goodman (1979) studied the morphology of the descending contralateral movement detector neuron in the locust (Locusta migratoria). Although identified neurons in invertebrates are commonly thought to be identical in all members of the species, the neurons were found to vary extensively. In fact, the position, size, and number of branches varied to such an extent that the authors were unable to define a "normal" form.

Macagno et al (1973) examined neurons from the visual system of Daphnia magna. All the individuals studied were genetically identical having been produced by parthenogenesis, and all were reared under similar environmental conditions. Nevertheless, there was a large degree of variability between individuals, as well as between corresponding neurons from the left and right optic lobes of the same individual.

Stent and his colleagues (Kramer, et al, 1985) have recently reported that the receptive fields and the neuronal arborizations of mechanosensory neurons in the leech (Haementeria ghilianii) show a suprising amount of variability. This is a particularly poignant example given the developmental lineage relationships of the neurons of the leech.

These anatomical differences reflect the epigenetic control of normal development. Given the dynamic nature of the primary processes of development, in particular, cell adhesion mediated by CAM's, the generation of anatomical variability appears to be unavoidable. The key question, however, is whether this variability is in any way related to the function of the nervous system. It could be, and in fact is, universally dismissed as functionally inconsequential. For example, as long as the same cell is contacted, the exact course followed by the axon to its target does not matter. We will now discuss evidence, however, that anatomical variability is not only functionally relevant, but in fact is superseded by an overlying functional variability.

FUNCTIONAL VARIABILITY

The experimental literature of the last 150 years has documented the finding that the brain, and in particular, the cortex is partitioned into distinct anatomical areas, each with distinct physiological and functional properties. However, several experiments which have been performed over the last decade demonstrate that anatomy and function are not related in a one-to-one fashion in which a given anatomical area is always constrained to perform the same function. These recent experiments have been carried out by a number of investigators, primarily on the somatosensory system of cats and monkeys. Among the first were Wall and his colleagues (Dostrovsky, et al, 1976; Devor and Wall, 1981) who showed that transection of the dorsal spinal roots leads to the emergence of new representations in the area of the dorsal column nuclei (cuneate) formerly devoted to the severed roots. These changes occured with short latencies (they were present within 2 hours) and if the spinal roots were blocked by cold rather than being cut, the changes were reversible. On a broader level, Hyvarinen (1984) blinded monkeys at birth, and showed that Area 7, which in normal monkeys is a visual area (with a slight degree of multimodality to somatosensory responses), in the blind monkeys was totally devoted to somatosensory responses with rather normal receptive fields.

The most extensive and elegant set of physiological studies have been done by Merzenich and his collaborators (Kaas, et al, 1981; 1983; Merzenich, et al, 1983). Their studies represent a detailed analysis of changes in areas 3b and 1 of the somatosensory cortex of adult owl and squirrel monkeys. The paradigm involved making extremely fine maps of these areas in normal monkeys (electrode penetrations every 100-150 microns), and then remapping the same cortical areas at subsequent times after various procedures had been performed. These procedures included: (1) Transection of peripheral nerves such as the median nerve. In some experiments the transected nerve was ligated to prevent regeneration, and in other cases the nerve sheath was re-sutured to promote regeneration. (2) Amputation of single digits or of two adjacent digits (usually digits 2 and 3). (3) Alterations of stimulation to the hand without transection achieved by bandaging the hand, applying casts, or various protocols in which the monkey was trained to tap his finger against a moving object. (4) Local cortical ablations.

The results of these studies compel a reassesment of cortical mapping. The key experimental observation is that cortical maps rapidly reorganize following nerve transection or digit amputation in adult monkeys. The reorganization involves changes in both the micro-properties of the map (the size of receptive fields and the amount of overlap between the receptive fields of nearby neurons), and in the gross properties of the representational map (the peripheral loci represented at each cortical site, the locations of representational discontinuites, etc.). Equally compelling is the observation that each individual monkey has a unique cortical map which varies by a greater extent than, for example, the difference in facial characteristics; and that following nerve transection or digit amputation, map reorganization takes place in an idiosyncratic manner in each monkey. This reorganization involves lesion-induced territorial expansions of the representations of the regions innervated by the remaining peripheral nerves, but the changes are not restricted to the areas represented by the transected nerves.

Immediately after transection of the median nerve, an incomplete new representation arises within the cortical area that formerly represented the median nerve field (Merzenich, et al, 1983a). This supports the notion that the anatomical basis for this new representation was always present but was effectively suppressed from organizing. During the weeks and months following peripheral nerve transection, the new cortical representation gradually changes. The location on the skin which is represented at each cortical site is altered, as is the organization of the representation. These findings indicate that the receptive field recorded at any cortical site is but one of the many possible receptive fields that, under different conditions, could be expressed at that cortical site. The receptive field found at a given cortical site in the new map may have been found at a site up to a millimeter away in the original map. Thus, the same receptive field can be expressed anywhere over a relatively wide area of cortex. As the maps are restructured, basic topography is always maintained, so that at all times during the reorganization, both representational continuity and some degree of global somatotopy are preserved. Despite the fact that the location on the cortex of the representational border can move substantial distances, the borders themselves remain sharp. For example, cells on one side of a border might respond exclusively to glabrous inputs, while those on the other side will respond exclusively to stimulation of the the dorsum of the hand.

During reorganization, different temporal sequences of expansions and contractions occur in the representations of those nearby regions of the hand which are innervated by the remaining peripheral nerves (Merzenich, et al, 1984). Within the first few weeks after median nerve transection, the radial nerve captures the greater share of the former median nerve representational territory in Area 3b, whereas in Area 1, the ulnar nerve captures the greater share of the territory.

Merzenich also made several interesting quantitative observations concerning receptive field size and overlap, and of the distance limits on reorganization. The larger the area of cortex that is devoted to a representation (the greater the magnification factor), the smaller are the associated receptive fields. Conversely, smaller cortical representations are characterized by larger receptive fields. Of particular significance is the fact that the percentage overlap of the areas of the receptive fields of two cortical cells is a monotonically decreasing function of the cortical separation of the cells, reaching a value of zero percent overlap at separations of approximately 600 microns in normal animals (Sur, et al, 1981).

Finally, there appears to be a distance limit of approximately 600 microns in any direction, over which expansion of a representation can take place. After amputation, a silent area of cortex that extends over more than 600 microns in radius can never be completely reoccupied. Together with the observation that the site of representations can effectively move millimeter distances, this is another indication that the anatomical basis for these translocated representations must have preexisted in a suppressed form in the near vicinity.

It is important to note that there is no evidence for the occurence of sprouting of nerve terminals following transection and ligation of the median nerve, either in the hand, or centrally in cortex. In fact, even if sprouting were to occur, it could not account for these observations. Changes in the map occur immediately after transection and continue for months; whereas sprouting would follow an intermediate time course.

We have previously shown that the theory of neuronal group selection can account for the findings on the overlap of receptive fields, the distance limit on map reorganization, and the area of cortex occupied by a representation. In addition, the central conclusion compelled by these results, is directly supportive of the theory. Namely, that a dynamic process selects particular neuronal groups from the degenerate anatomical substrate to form the normal functional adult map. This dynamism is indicated by the movements of representations and by the reestablishment of receptive field structure after transection. In other words, following the critical periods of development, there remains a degenerate anatomical substrate upon which selection operates competitively to create a functional map from a manifold of possible maps. The major determinant of the competitive selection process appears to be significant neuronal activity resulting from coactivation of overlapping afferents. In the next section, we turn our attention to the limits and mechanisms of this competition, and we propose a group selection model to account for the formation of neural maps.

MECHANISMS OF GROUP CONFINEMENT, GROUP SELECTION, AND GROUP COMPETITION

We propose that there are three processes that give rise to the organization of cortical maps. The first concerns the limitation of group size, which we call group-confinement. The second process, group selection, is largely responsible for the determination of the receptive field of the group. Selection arises from the distribution of afferent arbors, each of which spreads over a limited cortical region and extensively overlaps with neighboring arbors. Across these degenerate arbors, the play of coactivated versus uncorrelated stimuli leads to selection of groups according to a synaptic modification rule. The final and highest-level process, group competition, concerns the competitive interactions among those groups that arise from confinement and are then selected.

In our model, the expressed map emerges as a result of the three processes acting upon a degenerate anatomical substrate capable of giving rise to numerous possible maps.

The fundamental challenge to the model is to explain how a map is selected out of the degenerate primary repertoire. Towards this end, we now consider, in order, the processes of confinement, selection, and competition.

Group Confinement

A neuronal group in the cerebral cortex is functionally defined, in this model, as an ensemble of cohesively interconnected cells which together determine their individual receptive field properties. The concept of a neuronal group differs from that of a "cell assembly" (Hebb, 1949) or the various ephemeral multi-neuronal assemblies which have been proposed (e.g. von der Malsberg, 1981), in that a neuronal group is a stable, distinguishable, localizable, anatomical entity. From this aspect, neuronal groups also differ (in sensu strictu) from the minicolumns proposed by Mountcastle (1978) which are ontogenetic rather than anatomic entities. Thus, the neurons belonging to a group continue to belong to that group indefinitely unless a supervening process (see group competition below) overrides that membership. Moreover, we propose that each neuron belongs to only one group (or in some cases a small number of groups). Neuronal groups are not, in general, composed of neurons distributed widely throughout a large region of the brain. However, the inter-group connections which link widely separated groups can be modified; so the associative power of a cell-assembly model is not lost.

The fundamental operation of neuronal groups is to compete for domination of cell activity. Domination is acheived by strengthening the synaptic connections between the particular cell to be captured and the cells in the group. However, as new cells are incorporated, the receptive field properties of the group may be altered. As long as significant activation of the group is maintained, the group can maintain its control over the activity of its constituent cells. But other groups are constantly competing for control of the same cells, and any weakening of connections puts the group at risk either of losing a few cells, or in the extreme case, of being "divided and conquered."

As discussed below, neuronal groups extend vertically through all cortical laminae. We conjecture that the average diameter of a group in the hand representation of somatosensory area I (SI) is in the range of 50 - 100 microns. Given these dimensions, a group would contain between 250 and 1500 cell bodies based on Rockel et al's (1980) data on

cell density in the macaque somatosensory cortex. We choose
the value 50 - 100 microns because it represents the upper
bound for separated locations in the cortex with identical
receptive fields (Sur, et al, 1980). Group size may be
somewhat different in other brain regions; but in any case,
a group extends over only a fraction of the extent of the
arborization of a thalamic afferent. The size of a group
depends on all three of the processes mentioned, but before
selection of groups and competition between them can occur,
the size of the group must be confined to a limited region.
Confinement reflects the existence of a stable size range
for groups due to the interplay of excitatory and inhibitory
corticocortical connections in different laminae. To appreciate these phenomena, we must first briefly review some of
the cytoarchitectonics of the somatosensory cortex.

The majority of synapses in SI, as in all of cortex,
are from intrinsic fibers (Jones and Powell, 1970; White and
Hersch, 1982). The vast majority of these intrinsic fibers
are oriented vertically, and serve to link cells in different laminae. The basic conception concerning the operation
of a cortical unit is shown in Figure 1. Thalamic afferents

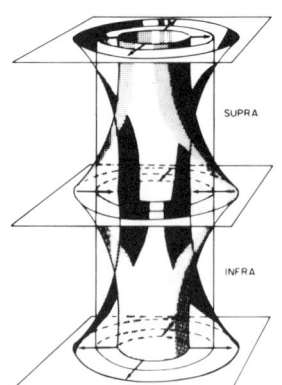

Figure 1 -- Schematic conceptualization of the hypothesized
process of group confinement. Three different configurations of a neuronal group in the cerebral cortex are demarcated by the three surfaces. A group will tend to expand in
the supra-granular layers due to excitatory horizontal connections. This expansion leads to increased inhibition in
layer IV leading to constriction of the group. Conversely,
constriction of the group in supra-granular layers leads to
expansion in layer IV. The intermediate cylinder represents
an equilibrium configuration for the group.

excite nonspiny stellate cells and also spiny stellate cells
in layer IV. These afferents also excite pyramidal cells in
other layers.

Pyramidal cells, particularly those in the supra-granular layers, are thought to receive excitatory connections
from other pyramidal cells and from the layer IV spiny stellate cells (Szentagothai, 1975; Winfield, et al, 1981). In
conjunction with the thalamic afferents, these excitatory
interconnections could provide the "cell-catching" mechanism
of the group. Pyramidal cells that receive significant
excitation (from thalamic, commissural, and associational
afferents as well as from other nearby pyramidal cells and
from spiny stellates directly below) are activated and
strengthen their connections to the pyramidal cells that
they excite in turn. Local inhibitory cells sharpen the
dynamic response by lateral inhibition. As depicted in
Figure 1, the supragranular and infragranular layers are
sites of excitatory group expansion. However, layer IV, the
predominant thalamic recipient zone, is dominated by
inhibitory nonspiny stellate cells. As larger regions of
layer IV are excited by thalamic inputs and by the
increasing amount of excitation in other cortical layers,
increased inhibition is generated, which leads to a tamponade of the source of excitation. The dynamic equilibrium
that occurs between granular level "contraction" and supraand infra-granular "expansion", inextricably linked by the
predominant vertical connectivity, results in the formation
and confinement of the group.

This interpretation of the operation of the cortical
anatomy is consistent with the models which have been advanced by Szentagothai (personal communication) and Jones
(1981).

Group Selection

Group confinement is an intrinsic property of the cortex, but group selection depends upon both the anatomy and
the spatiotemporal properties of the input. The temporal
overlap of inputs depends mostly on the pattern of stimulation, but the spatial overlap of the inputs requires convergence of stimulated afferents to a local region.

Thalamic afferents to the somatosensory cortex in the
cat terminate in large arborizations which densely cover
about 0.5-1.0 mm^2 of cortex (Landry and Deschenes, 1981).
Gilbert and Wiesel (1979) have found similar arborizations
in afferents to cat visual cortex. No experimental evidence

exists on the degree of anatomic overlap between separate afferent arbors. However, based on degeneration studies (Jones, 1981; Kosar and Hand, 1981), and reported data on the innervation of the barrel fields of the mouse (Pasternak and Woolsey, 1975; Lee and Woolsey, 1975), we have predicted (Finkel and Edelman, 1984) that afferent arbors will be extensively overlapped with nearby arbors, perhaps to the extent of 90% of their area.

One result of this large degree of afferent overlap is that most cortical cells receive synapses from a large number of afferents, especially considering the possible interactions of the widespread axonal arbors with large dendritic trees. Moreover, despite the fact that cortical cells receive hundreds to thousands of synaptic inputs, the vast majority of these inputs are not represented in the receptive field of the cortical neurons. This phenomena is usually dismissed as due to the existence of subthreshold inputs and "silent synapses". However, this leaves the control mechanism a mystery and the functional significance an enigma. From the point of view of group selection, it is this degenerate, variable, anatomical substrate which serves as the battleground for an ongoing "Darwinian" competition among neuronal groups that occurs in somatic time. To explain how a map is selected against the background of this extensively degenerate anatomical substrate, we must consider a model of synaptic modification.

The fundamental property of any synaptic model is that some classes of inputs lead to synaptic strengthening and other classes lead to weakening. Most recent models are constructed so that intense, coactivated barrages of inputs strengthen and weak, uncorrelated firing patterns weaken synaptic strength. This property is demonstrated by the synaptic model we will propose later. In order to explain the mechanism of group selection, we must consider how the synaptic rule operates within the constraints of the anatomy of the neural network.

Figure 2 is an attempt at this explanation. The Y-shaped structures represent a highly idealized set of overlapping cortical afferents. The branches marked "X" receive coactivated stimulation from peripheral receptors, leading to coactivation of neurons all across the hatched area. Adjacent afferents marked "O" are not coactivated; thus all the neurons outside the darkened region receive uncorrelated activity, Despite the wide arborization of the afferent terminals, there is only a small cortical region

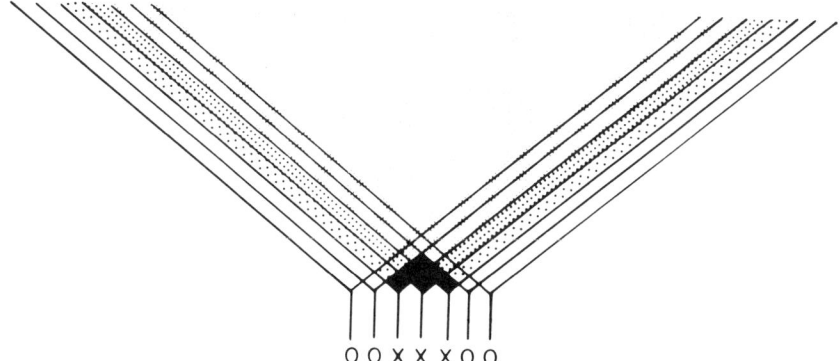

Figure 2 -- Highly idealized representation of overlapping thalamic afferent arbors. Afferents marked "X" receive coactive stimulation, while those marked "O" receive uncorrelated stimulation. The 3 densities of stippling indicate regions in which 1, 2, or 3 coactive afferents overlap. The darkened region receives coactivated but no uncorrelated input.

that experiences maximal coactivation but minimal uncorrelated activity. If the synaptic modification rule is such that this condition, maximal coactivation with minimal uncorrelated activity, leads to differential synaptic strengthening, then only groups situated within this relatively small region will be selected by the coactive stimuli. We will examine this property of the selection mechanism in our model of synaptic modification. But first we briefly consider the last of the three processes responsible for map organization, the process of group competition.

Group Competition

Group competition is perhaps the most important single process in determining what the actual map will look like. As different stimuli are successively encountered, we suggest that, in additon to group confinement and selection, a Darwinian competition occurs between various groups for cortical representation space. In general, within the constraints of peripheral innervation density, groups that receive coactivated stimulation most frequently have the

competitive edge. We have proposed that there must be a set
of hierarchy rules according to which groups can compete
with other groups (Finkel and Edelman, 1984). A self-
consistent set would include, for example, the following:
(1) the most competitive groups are those that are
associated with the most frequently stimulated peripheral
locations; (2) Groups whose dimensions are within a certain
size range are the most competitive; (3) groups whose
receptive fields overlap within certain limits with those of
neighboring groups are favored; (4) extant groups have an
advantage over incipient groups.

The basic argument developed in this section is that
group selection within a degenerate anatomical substrate
leads to the dynamic expression of maps corresponding to
environmentally relevant structures. A cortical group is a
cooperative, self-organizing unit whose mechanims of formation constrains all cells within it to share a common receptive field. Group size can vary, but is limited by the
vertically mediated dynamic balance between a propensity to
contract in layer IV and a tendency to expand in other layers (group confinement). The choice of exactly which afferents will be expressed in the receptive field of a group
depends on the temporal coactivation of peripheral inputs
stimulating these afferents through the various relays, and
it operates under the constraints of previously established
corticocortical synaptic efficacies (group selection).
Every stimulus acts, in some small way, to alter the competitive balance between groups (group competition). The
functional map represents the combined effects of group
confinement, selection, and competition.

SYNAPTIC MODIFICATION MODEL

In the previous section we argued that some form of
synaptic modification based on coactivation of afferent
input was necessary for group selection. In this section a
detailed model for such a process will be put forward. The
fundamental assumption of the model is that both presynaptic
and postsynaptic modifications occcur, but that each is controlled by a separate and independent mechanism. The presynaptic mechanism provides stability and homeostasis for
the neuronal group while the postsynaptic mechanism provides
the specificity according to which individual synapses are
modified under appropriate context-dependent conditions. As
a consequence, presynaptic and postsynaptic modifications
can be coupled at the neuronal and network levels, but

remain independent at the level of each synapse.
The two proposed rules governing the separate modifications operate concurrently in parallel at each synapse, and jointly contribute to the net modification of the individual synapse. An important consequence of this idea is that at the level of the individual synapse, presynaptic and postsynaptic modifications will be functionally indistinguishable. However, the two rules differ markedly with regard to how the modifications are anatomically distributed within a network, and the two types of modifications may be distinguished by considering the distribution of these changes over the population of synapses. For this reason, the anatomical and pharmacological details of a given network assume major importance in determining the interactions among synapses.

We shall first discuss the postsynaptic rule which applies, in general, to short-term changes at specific individual synapses. In contrast, the presynaptic rule, which we discuss next, regulates long-term changes at all presynaptic terminals of the neuron. The presynaptic rule thus affects large numbers of synapses defined by the connectivity of that neuron and distributed non-specifically over the population.

The central aim of this discussion, however, is an analysis of the properties of the two modification rules acting within a network with group structure. The analysis specifically addresses the fundamental problem of how short-term and long-term modifications evoked by the same history of activation can both occur in the same network without interference and loss of information. Indeed, we show that not only can the two processes operate concurrently over the same population of synapses, but that in doing so significant new emergent properties can be gained. We suggest that independent presynaptic and postsynaptic mechanisms acting together over the same period can account for: (1) the existence of heterosynaptic as well as homosynaptic modifications, (2) the occurrence of stable network changes over several different time scales while producing a continual source of variance in the synaptic couplings of neurons, (3) the coexistence of long and short-term changes distributed within the same network, and based upon modifications of the same population of synapses, and (4) the property that these long and short-term modifications are functionally related to each other despite independent biochemical mechanisms of pre- and postsynaptic change.

THE POSTSYNAPTIC RULE

The postsynaptic rule regulates a family of biochemical modifications of receptors, channels, and various other postsynaptic structures and describes how these modifications are modulated by other synaptic inputs to the cell. These modifications alter the postsynaptic response to both the presynaptic release of neurotransmitter and to heterosynaptically conducted voltages.

There are several possible mechanisms which could carry out the postsynaptic rule (see Table 1, below). In the specific example of the postsynaptic rule to be discussed here, we assume that voltage-sensitive channels are present at or near the postsynaptic processes. Homosynaptic inputs lead to the production of a substance which modifies some of these channels at the synapse. But the susceptibility of these channels to modification can be altered by heterosynaptic inputs (Figure 3).

The main assumption of the proposed mechanism is that the local biochemical modifications are state-dependent: the probability of modifying a channel depends upon its functional state (e.g. open, closed, inactivated). Due to steric properties of different channel conformations, one particular conformational state, for example the "inactivated" conformation, could be more susceptible to biochemical modification (such as phosphorylation by a particular protein kinase). We propose that the functional state of voltage-dependent channels can be transiently altered by conducted voltages from other synaptic inputs to the postsynaptic cell. This is a population phenomena since only a fraction of the channels at any synapse will be so affected. Nevertheless, the susceptibility of these channels to modification will be altered according to the state-dependent assumption. Appropriately timed heterosynaptic inputs can thereby modulate the degree of local synaptic modification. Without a state-dependent assumption, heterosynaptic effects would be obligately non-specific. Conducted voltages would give rise to modifications at all or most postsynaptic sites (within several space constants) regardless of the state of activity of those synapses.

One possible action of the biochemical modification is a change in the voltage-dependent probability of the channel switching between functional states, e.g. open, closed, inactivated. However, other possible mechanisms could involve a change in the effective lifetime of the membrane channel, the maximum open-channel conductance, the spectrum of kinetic states the channel exhibits, or the actual kinetics of these states. In any case, the modification alters

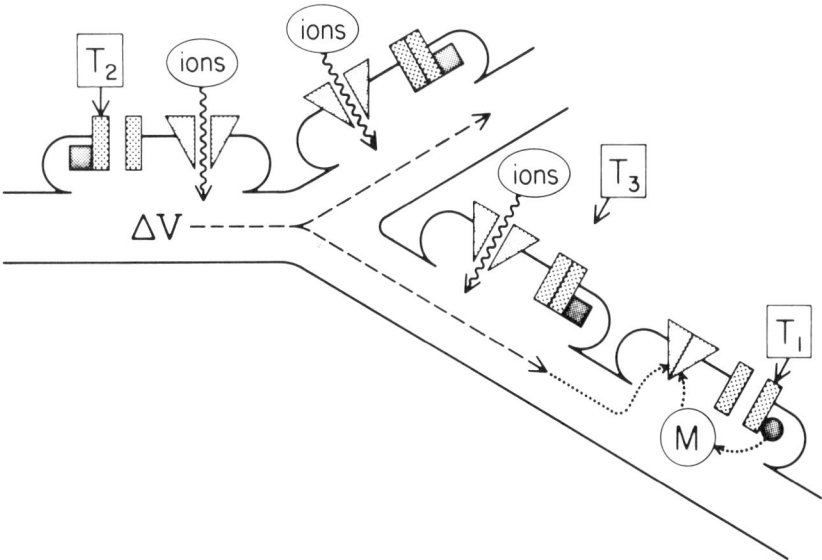

Figure 3 -- Operation of Postsynaptic Mechanism. Schematic of four synapses on a branching dendritic tree. Shaded triangles represent voltage-sensitive channels (VSC's) and shaded rectangles represent receptor-operated channels (ROC's). Transmitter (T_1) has bound to synapse at lower right leading to opening of the ROC's and also to the production of modifying substance, M, through activation of a membrane associated protein (small shaded circle). M modifies VSC's which are in the modifiable state. At a different time, a potentially different transmitter, T_2, has bound to leftmost synapse opening local ROC's and VSC's. The resulting change in membrane potential (ΔV) is propagated through the dendritic tree (dashed line), changing the states of VSC's as they are reached. If the potential reaches the bottom right synapse (dotted line) at a time when the concentration of M is still high, the fraction of VSC's in the modifiable state at the synapse will be altered, and consequently, so will the number of channels modified by M. Synapses which have not yet bound transmitter (e.g. T_3), and synapses not yet reached by the potential, will not be affected.

both the postsynaptic potential evoked by subsequent homosynaptic inputs, and the sensitivity to subsequent heterosynaptic inputs.

To summarize, in the present model the degree of postsynaptic modification will depend upon (1) the timing of the different synaptic inputs to a neuron, (2) the number and intensity of heterosynaptic inputs occuring during the modification period, (3) the spatial distribution of synapses on the post-synaptic cell (conduction delays), and (4) the types of transmitters, receptors, ion channels, and modifying substances present.

Formal Model of the Post-Synaptic Mechanism

In all likelihood, the several conformational states of a channel will be differentially modifiable; for simplicity, let us assume that only a single state is modifiable. Suppose the modifiable state is I, the inactivated state. We will not treat all the complexities of the microkinetics of

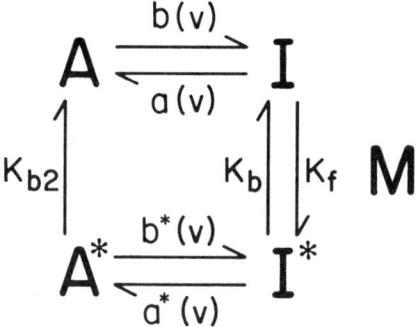

Figure 4 -- Kinetics of State-Dependent Modification. Two-state model of channel. "A" represents lumped activated states, "I" represents lumped inactivated states, which we take to be the modifiable states. Channels in state I can be modified, in the presence of modifying substance, M, to modified state, I^*. Forward and backward rate constants for modification are k_f and k_b. Modification is short-term, and decays from I^* according to k_b and by potentially different constant k_{b2} from activated state.

state transitions; instead we consider only a 2-state model, consisting of a lumped active state, A, and a lumped inactivated state, I.
Such a two-state model is shown in Figure 4 where M is the concentration of modifying substance. The primed quantities in the lower tier represent the modified channels. a(v) and b(v) are voltage-dependent parameters -- one or both of which is modified by the biochemical mechanism. The decay of the modification may also be state-specific, but, for simplicity, we assume that it is not; thus, $k_b = k_{b2}$.

It also seems reasonable to assume that the time constants for state transitions, $(a+b)^{-1}$ and $(a^*+b^*)^{-1}$, are small with respect to the time constant for modification, $(K_f+K_b)^{-1}$. Thus,

$$dN^*(t)/dt = K_f \cdot I(t) \cdot M(t) - K_b \cdot (I^*(t) + A^*(t)) \quad (1)$$

where N^* is the fraction of channels in the modified state; I and I^* are the fraction of normal and modified inactivated channels, respectively; A^* is the fraction of modified activated channels, and M is the amount of modifying substance present locally. The concentration of M depends upon the pattern of local inputs (see formulation below); and I, I^*, and A^* are voltage-dependent functions.

The local voltage at synapse i is given by

$$V_i(t) = \delta \cdot V_i(t-1) + \sum_j A_{ij} \cdot \xi_j \cdot \eta_{ij} \cdot S_j(t) \cdot f_{open} \quad (2)$$

where δ is the decay constant of the voltage, A_{ij} is the voltage attenuation between synapses i and j, ξ is the presynaptic efficacy, η is the postsynaptic efficacy, $S_j(t)$ corresponds to the firing of cell j, and f_{open} is the fraction of channels in the open state.

It is difficult to find experimental values for the various parameters and variables in equation 1, however, we have argued (Finkel and Edelman, 1985) that the modifications produced by such a mechanism would be measurable and would have a significant effect on the subsequent behavior of the synapse in question.

We next consider the time window within which the homosynaptic and heterosynaptic inputs must occur in order to produce a modification. Suppose, as shown in Figure 5, that homosynaptic inputs lead to production of some modifying substance which persists for a time t_M after a lag time of t_L, and that conducted heterosynaptic inputs produce a local change in membrane potential which persists for a time t_V

after a conduction delay of t_D. Then the time window for modification requires that the heterosynaptic inputs occur no earlier than a time $(t_V + t_D - t_L)$ before the occurence of the homosynaptic inputs, and no later than a time $(t_M + t_L - t_D)$ after them. It is possible that one or both of these quantities will be negative for a given synapse. In that case, the presentation of the inputs in that particular order will not lead to a modification.

Multi-synaptic circuits with dense reentrant connections can receive a complex set of inputs distributed both temporally and anatomically over the network. Only those

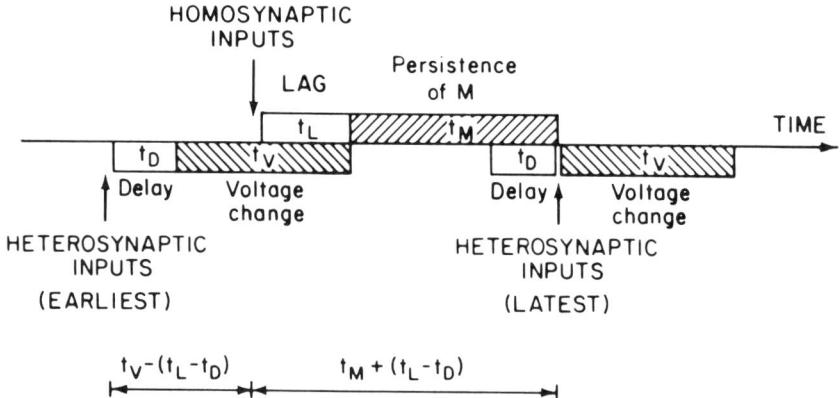

Figure 5 -- Timing Constraints for Postsynaptic Modification. Time axis showing the time window before and after a homosynaptic input during which the occurrence of heterosynaptic inputs can lead to modification. After barrage of homosynaptic inputs, a biochemical cascade of duration t_L, leads to production of modifying substance, M, which persists for a time t_M. Heterosynaptic inputs lead to a depolarization (or hyperpolarization) at the local synapse which begins after a conduction delay, t_D, and then persists for t_V. Arrows indicate the earliest and latest time at which the heterosynaptic inputs can occur with respect to the homosynaptic inputs so that the effect of the conducted voltage temporally overlaps with the prevalance of M. Depending on the values of the time constants, t_V, t_M, etc. (see text) heterosynaptic inputs may be constrained to either exclusively follow or preceed the homosynaptic inputs in order to achieve modification.

heterosynaptic inputs which are appropriately synchronized will be able to strengthen each other. Viewed in another way, only those circuits whose time delays and patterns of activation are such that the reentrant input returns at the appropriate time will be the circuits which are differentially selected.

Postsynaptic Rule Computer Results

The postsynaptic rule was simulated with a Fortran program run on an IBM 4331. The postsynaptic strengths, η, of all the synapses onto a single cell were computed as a function of stimulation over time.

The local voltage at each synapse (see equation 2) is used to calculate the fraction of channels at that synapse which are in the modifiable state (assumed to be the inactivated state). The postsynaptic rule involves a number of kinetic processes governing channel transitions and modifying substance production, the fraction of inactivated channels being one specific example. These fractions should be calculated using a kinetic scheme similar to that shown in Figure 4. However, to increase computational speed, we assumed that the state transitions occur instantaneously (compared to the duration of a iteration step), and are determined by a sigmodial function of voltage.

A second sigmoidal curve which was shifted to the right was used to calculate the fraction of modified channels in the inactivated state·

We also made the assumption that the production of modifying substance increases in a sigmoidal fashion with the amount of transmitter released by the local presynaptic input. The concentration of modifying substance was then determined by the following equation:

$$M(t+1) = m_1 \cdot M_p \cdot S - m_2 \cdot M(t) \qquad (3)$$

This equation represents the accumulation of modifying substance at a rate proportional to the production rate, M_p, whenever the local afferent input fires, together with an exponential decay of the substance with rate constant, m_2 (see discussion of equation 5, below).

The postsynaptic efficacy, η, was given by

$$\eta = \eta^0 \cdot [\, (1-N^*)(1-f_{inact.}) \cdot g + N^*(1-f^*_{inact.}) \cdot g^* \,] \qquad (4)$$

where η^0 is the baseline postsynaptic efficacy, and g and g^*

are the conductances of the normal and the modified channels, respectively, f_{inact} and f^*_{inact} are the fraction of normal and modified channels in the inactivated state, and the fraction of channels modified, N^*, is determined from equation 1.

Figure 6 shows the basic operation of the postsynaptic rule. We consider the case of afferents from group A and group B synapsing onto cell C. The mean postsynaptic efficacy, η_{CA}, of the inputs from group A is plotted versus time (cycle number). The run began with no channels in the modified state and with $\eta^0 = 0.25$. An initial burst of 5 spikes (cycles 3-7) gives rise to a modest increase in η. This follows from the local production of modifying substance concurrently with inactivation of channels. The modification of postsynaptic efficacy is short-term and thus starts to decay back to its baseline value, η^0, with a decay constant determined by k_b and k_{b2}. A large stimulus burst from the group B (heterosynaptic) inputs (cycles 17-26) produced no detectable change in η. This is because despite the inactivation of channels due to the conducted voltage, no modifying substance is being locally produced at the CA synapse, and that which was present has already decayed away. In contrast, simultaneous bursts from both the group A and group B inputs (cycles 42-46) leads to a larger change in η. In this case production of M and inactivation were concurrent. However, if the group B inputs are given a large burst and at the appropriate time, the group A inputs are stimulated, the increase in η is dramatically greater (cycles 67-76). This result demonstrates the basic property of the postsynaptic rule, namely, that appropriately timed heterosynaptic inputs can dramatically modulate local postsynaptic modifications.

In other simulations (not shown) we found that a large homosynaptic burst is equally effective in increasing postsynaptic efficacy (Finkel and Edelman, 1985). The locally generated voltage is able to inactivate channels at least as well as can be done by the attenuated voltage from heterosynaptic inputs. With the particular choice of parameters used in Figure 6, the increase in η is greatest when the heterosynaptic inputs precede the onset of the homosynaptic inputs. This is because, given the time constants of M and inactivation chosen here, if the homosynaptic inputs occur too early, the modifying substance all decays away before enough channels inactivate to produce significant modification. As shown in Figure 5, by adjusting these time constants, we could obtain various timing relationships

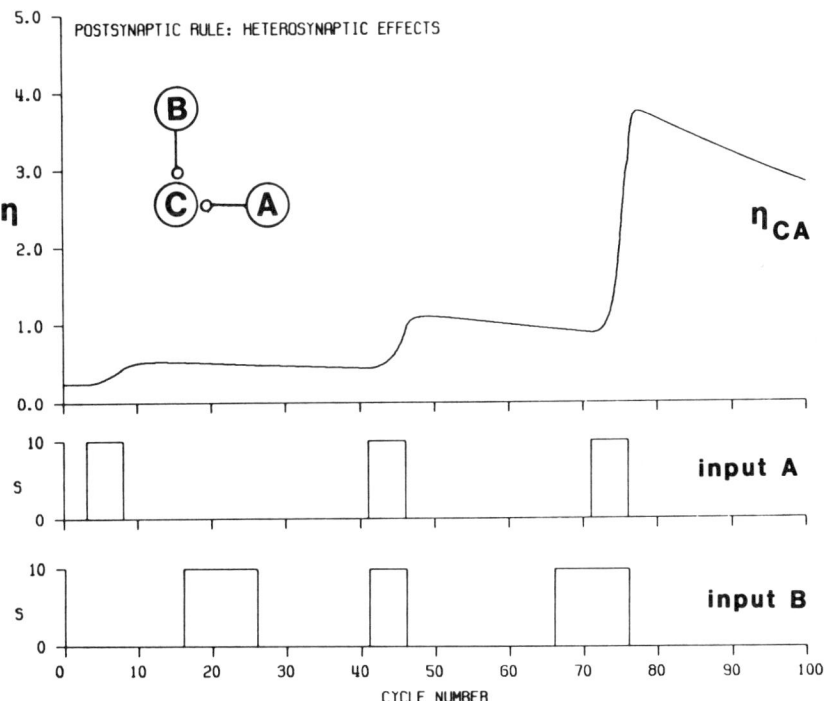

Figure 6 -- Simulation of the Postsynaptic Rule. Postsynaptic efficacy, η_{CA}, of input from cell A to cell C as a function of stimuli to cell A and to heterosynaptic input (cell B). Stimulus burst to cell A increases η_{CA}, lone heterosynaptic burst has no effect, paired homosynaptic and heterosynaptic bursts produce larger increase in η_{CA}. However, large heterosynaptic burst with appropriately timed homosynaptic burst (final pair) produce much larger increase. Shows that heterosynaptic inputs can modulate local postsynaptic modifications.

The following values were used for the input parameters: Number substeps per iteration = 10, number of synaptic inputs = 50, mean intragroup voltage attenuation = 0.5, mean intergroup voltage attenuation = 0.25, coefficient of variation of voltage attenuations = 0.0, $\delta=0.9$, $m_1=5.0$, $m_2=0.5$, $k_f=0.1$, $k_b=k_{b2}=0.986$, $g=1.0$, $g^*=9.0$, no channels are initially modified, and presynaptic parameters are the same as in Figure 8.

between the classes of inputs required to produce large modifications. What is important in this case, is that given the appropriate timing relationship, a dramatic increase in postsynaptic efficacy can be achieved.

The postsynaptic rule can account for heterosynaptic changes and for the differential strengthening or weakening of certain subsets of synapses. While the geometry and pharmacology of the synapses on a given neuron are fixed, depending upon the play of inputs over the cell, different combinations of synapses will be modified.

Certain cohorts of synapses may be in privileged anatomical positions to influence or to be influenced by others. The degree of voltage attenuation between two synaptic sites depends upon the neuronal geometry. For example, the voltage attenuation in the retrograde direction (i.e., going from proximal to distal sites) is far less than that in the orthograde direction (Rall and Rinzel, 1973; Koch, et al, 1982).

This suggests that instead of dividing inputs into only two classes - excitatory and inhibitory - and considering neuronal operations in Boolean terms, it may be more valuable to consider a kind of "transmitter logic" in which each transmitter (in association with its post-synaptic partners) can lead to characteristic modifications of synapses receiving only certain other transmitters and located on only certain other parts of the dendritic tree. Such a transmitter logic would generate a great diversity of classes of synaptic modifications, all specific, but varying with respect to magnitude, time course, origin, and target.

Biochemical Basis of the Modification Mechanism

As discussed above, the particular version of the postsynaptic rule presented here may be viewed as one example of the more general mechanism. In general, the target of the modification can be either voltage-sensitive channels or receptor-operated channels. At present, the main candidates for modifying substances are Ca^{++} and the seven known protein kinases which include cAMP-dependent and Ca/calmodulin-dependent kinases (Nestler and Greengard, 1984). The modifying substances can be produced either by the binding of specific transmitters or by conducted voltages in the case of Ca-dependent processes. A number of different mechanisms (see table 1) given by the various possible combinations of modifying substances and targets of modification may come into play.

While there is, at present, no direct experimental support for state-dependent modifications, Catteral (1979) has shown that the mechanism by which scorpion toxins block the inactivation of sodium channels is voltage dependent. He argues that the toxin binds with different specificities to sodium channels in different conformational states (active, inactive, etc.). There is also some evidence that the drug quinacrine only binds to acetylcholine receptors when the receptor is in its "open" state (Changeux, 1981). While these results concern foreign toxins as opposed to internal second messengers they nonetheless support the possibility of the proposed state-dependent modifications.

Table 1
POSSIBLE MECHANISMS FOR POSTSYNAPTIC MODIFICATIONS

(Each mechanism would involve one or more choices from each of the following categories)

Local Modifier
* Receptor-associated second messenger
* Ca^{++}
* Voltage
* pH
* direct structural interaction between target molecules

Heterosynaptic Mediator
* External paracrine diffusion of transmitter
* Internal second messenger diffusion
* Ca^{++} diffusion
* Voltage (electronic or active conduction)
* Global cell-surface modulation

Target
* Receptors
* Ionophores
* Other ultrastructural or regulatory proteins

Effect
* Change in effective number of channels or receptors
* Change in maximal conductance of channels or binding of receptors
* Change in kinetics of state transitions for channels or receptors

Developmental Aspects of the Postsynaptic Rule

The specificity of the postsynaptic rule which allows for the independent modification of individual synapses, suggests that it may play a role during the formation and removal of connections during development. It has been found (Witzemann, et al, 1983; Llinas and Sugimori, 1980; Thiele, et al, 1982) that there are different complements of channels at different cellular sites, and that these channels may have site-specific properties. For example, certain neurons have been shown to have K^+ channels at the soma that shunt antidromic stimuli from reaching the dendrites (Llinas, 1980). Ca^{++} channels in the distal dendrites have been shown (Llinas, 1979) to give rise to long plateau depolarizations allowing more time for other events to overlap temporally. It would be interesting to determine whether these inhomogeneities in channel distribution are programmed into the cell and aid afferent contact, or whether they actually result from contact by afferents. At present, it is only known that the distribution of channels changes with development.

Given that these inhomogeneities may be controlled by post-translational protein modifications intrinsic to the neuron (Bloebel, 1983), the model may explain how afferents contact and remain at the appropriate region of the postsynaptic cell during development. It would postulate that contacts are initially made with many different regions of the neuron, but only those contacts that are strengthened remain. As we have discussed, the ability of conducted voltages to affect the local membrane potential at other regions of the neuron depends upon the neuronal geometry. This geometry, as well as other neuronal properties, changes as the neuron develops. The only synapses which will persist are those which are strengthened by virtue of their position relative to other synaptic inputs. Classes of connections to certain cell types may either be lost or inappropriately retained if an abnormal complement of afferents is received. Possible mechanisms of synaptic elimination are beyond the purview of this paper, however, they might involve a coupling between either voltage-dependent effects or second messengers and cell-adhesion molecules. The most dramatic effects should be observed if the abnormal afferents received release a different transmitter from the normal complement of afferents. For example, noradrenergic afferents may greatly influence the formation or removal of additional synapses depending upon the types of connections that have already been made. As originally shown by Kasamatsu and Pettigrew (1976), intraventricular or local microperfusion of 6-OHDA which destroys monoaminergic cells protects a cortical region from the effects of monocular deprivation -- i.e., the region remains binocular. However, if the injection is made in very young animals (Paradiso, et al, 1983), the region can still become monocular.

The postsynaptic rule also raises the possibility that a result found in several developmental systems (Schmidt, 1982) namely, that TTX blocks synapse dissolution but not formation, may have less to do with non-specific properties of "activity" per se than with TTX and Na channels. Perhaps Ca-blockers or K-blockers have very different effects.

Finally, the model may be able to account for the kinds of competitive exclusion effects that Purves and others have described (Hume and Purves, 1981). Consider a set of afferents synapsing upon a dendritic branch, and furthermore, suppose that the effect of the firing of these afferents is to produce a voltage change which leads to the weakening of the synaptic strengths of nearby afferents. Then, when the density of afferents exceeds a certain threshold, the voltage-induced synaptic weakening would be sufficient to eliminate a synapse. In this situation, afferents within the same class would compete while those in different classes would cooperate. It is clear, however, that there are many other possible schemes based upon this type of model that could account for such competitive exclusion.

THE PRESYNAPTIC RULE

The underlying premise of our proposed presynaptic rule is that long-term changes in synaptic efficacy occur, that they are manifested at the cellular and network levels rather than at the synaptic level, and that they reflect a change in the homeostatic set-level of transmitter release by the cell in response to large sustained or repeated fluctuations in release.

The presynaptic rule describes the amount of neurotransmitter released by a presynaptic terminal in response to a given depolarization. Transmitter release is a complicated event and its control involves multiple cell-biological processes. At present, insufficient data exists about these individual processes to base the model at the molecular level. Therefore, we have pitched the present model (Finkel and Edelman, 1985) at the macroscopic level. However, we have previously investigated a simple model

framed at the molecular level (Edelman and Finkel, 1984).
Evidence from a variety of preparations including avian ciliary ganglion (Martin and Pilar, 1964), Mauthner cells (Korn, et al, 1984), frog neuro-muscular junction (Mallart and Martin, 1968; Magleby and Zengel, 1982), crayfish neuromuscular junction (Parnas, et al, 1984), Aplysia ganglia (Hawkins, et al, 1983), and squid giant synapse (Smith, et al, 1985), demonstrates that the level of transmitter release from a pre-synaptic terminal can be transiently increased or decreased after various stimulation sequences. In general, there appear to be several components of increased release (Mallart and Martin, 1968; Magleby and Zengel, 1982), as well as of decreased release (Bryan and Atwood, 1981) each with characteristic time constants. In perhaps the most elegantly analysed system, the frog neuromuscular junction, Magleby and Zengel (1982) have described four separate components of increased transmitter release, each decaying with a different time course. The frequency and duration of stimulation necessary to maximally evoke each of these processes differs.

Synaptic depression is less well understood. Various mechanisms have been proposed including depletion of a finite store of releasable transmitter, or transient inactivation of release sites involved in vesicular exocytosis. At the macroscopic level of the model proposed below, these two alternatives are formally similar.

In order to keep the model simple, we will use only a single component of increased release, a generic "facilitation", and only a single component of "depression". Facilitation can be described by an equation, similar to that used by Magleby and Zengel (1982) and others,

$$dF_i/dt = \varepsilon \cdot S_i(t) - \lambda \cdot F_i(t) \tag{5}$$

where $F_i(t)$ is the degree of facilitation in the presynaptic terminal of neuron i; λ is the decay time constant; $S_i(t)$ is the firing rate of the neuron at time t; and ε is the amount of increase in facilitation per spike. We assume, for simplicity, that the increase in facilitation (ε) is a constant corresponding to a constant bolus of calcium entering with every spike. Equation (5) represents the simplest formal description of a facilitatory process -- it increases in constant increments each time a spike occurs, and decays exponentially with a single time constant.

Synaptic depression is described by a similar equation,

$$dD_i(t)/dt = \kappa \cdot \xi_i(t) \cdot S_i(t) - \beta \cdot D_i(t) \quad (6)$$

where $D_i(t)$ is the degree of depression in the pre-synaptic terminal; β is the time decay constant; and κ is the constant of proportionality between release and depression. The first term describes the increase in depression due to release of transmitter. The $S(t)$ in the first term indicates that substantial (i.e. non-spontaneous) levels of release only occur when significant depolarization has occurred due to the rapid diffusion of calcium away from the release site. The first term as a whole indicates that depression increases linearly with the amount of release. It would perhaps be more realistic to assume a non-linear dependence, with significant levels of depression only arising after a substantial amount of release has already occurred. The second term represents the decay of depression due to, as discussed above, replenishment of depleted transmitter, return of previously inactivated release sites, or return to equilibrium of whatever molecular process is actually involved.

A final equation relates the amount of release to the degree of facilitation and depression,

$$\xi_i(t) = \xi_i^0 \cdot (1+F_i(t))^3 \cdot (1-D_i(t)) \quad (7)$$

where ξ^0 is the baseline amount of release that would occur in response to an isolated action potential in the absence of facilitation or depression. F is defined to range between 0 and some maximum value while D ranges from 0 to 1. Equation (7) resembles those studied by Magleby and Zengel (1982) and could be generalized to include products of all the components of facilitation and depression. Magleby and Zengel found that raising the facilitatory term to the third power resulted in a better fit to their data. We adopt this non-linearity which also agrees with the results of Smith and colleagues (1985) on the dependence of PSP magnitude on calcium current.

The pattern of activity over time will determine the relative changes in facilitation and depression. The specific changes depend upon the values of the kinetic constants. For the case when the time constant of depression (β) is greater than that for facilitation (λ), short, high frequency bursts favor facilitation; longer, low frequency bursts may favor depression, and lack or low levels of activity will leave transmitter release unchanged. This property has been observed in many experimental preparations. For instance, stimulation of the right pleural connective in Aplysia at 5 Hz for 30 seconds leads to several minutes of post-tetanic potentiation, but stimulation at 0.1 Hz leads only to homosynaptic depression (Kandel, 1976).

Thus far we have considered short-term changes in transmitter release from individual terminals in response to cell activity. We now assume that the long-term change in transmitter release occurs through a shift in the baseline amount of transmitter release, ξi^0. Our assumption is based on the idea that the long-term modification in ξ^0 results from a total cell response to time-averaged fluctuations, both facilitatory and depressive, in the pre-synaptic strength, ξ.

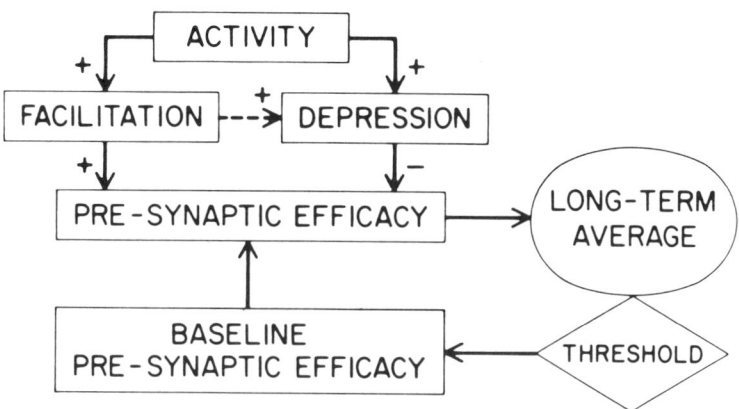

Figure 7 -- Flowchart of the presynaptic rule. Activity increases the degree of both facilitation and depression, which are also coupled as greater facilitation increases depression (see eq. 5). Facilitation increases and depression decreases presynaptic efficacy, ξ, which also depends on the baseline presynaptic efficacy, ξ^0. The presynaptic rule states that a long-term average is kept of ξ as it fluctuates due to the temporal pattern of activity. If this average reaches a threshold, the baseline value of presynaptic efficacy is reset to a new value. This changes the response of the cell to future inputs.

The basic idea is shown in Figure 7. The presynaptic efficacy of a neuron, ξ, depends upon three factors as given by equation 7: the degree of facilitation, the degree of depression, and the baseline presynaptic efficacy. Firing of the neuron leads to both facilitation and depression in a manner dependent on the temporal pattern of activity, as described by equations 5-7. The presynaptic rule says that if the long-term average (perhaps on the order of a second) of the instantaneous value of the presynaptic efficacy exceeds a threshold, then the baseline presynaptic efficacy, ξ^0, is reset to a new value.

The shift in ξ^0 will alter the dynamical properties of the neuron with respect to the amount of transmitter released for a given temporal pattern of stimulation. It is important to observe that even though the baseline presynaptic efficacy has been increased, the actual efficacy can still be driven to zero by transient depressions, or can be still further increased by transient facilitations.

Presynaptic Rule Computer Results

The presynaptic rule acting at one terminal was simulated on an IBM 4331 computer. The firing level of the cell as a function of time, $S_i(t)$, was given as input to the program. The program simulated equations 5, 6, and 7 with the four parameters ϵ, λ, κ, and β specified as input parameters.

Long-term changes in ξ were effected whenever the running average of a parameter (L) surpassed a threshold, θ_s. L was determined by the equation:

$$L(t+1) = \alpha \cdot L(t) + \gamma \cdot \frac{(\xi(t) - \xi^0)}{\xi^0} \qquad (8)$$

Thus L represents a running average of the deviation of from its steady-state value, with α the averaging time constant and γ a constant of proportionality.

Figure 8 shows the results of two runs which demonstrate the operation of the presynaptic rule. First a run was done with two bursts of stimulation separated by 5 cycles of rest (the first 2 bursts in Figure 8). The threshold, θ_s, was chosen to be 4.0, and L never reached this threshold. During the next 65 cycles, no stimuli were applied, so that ξ and L both returned towards resting values. A test burst at cycle 85 yielded a change in similar to that observed with the initial burst. The next

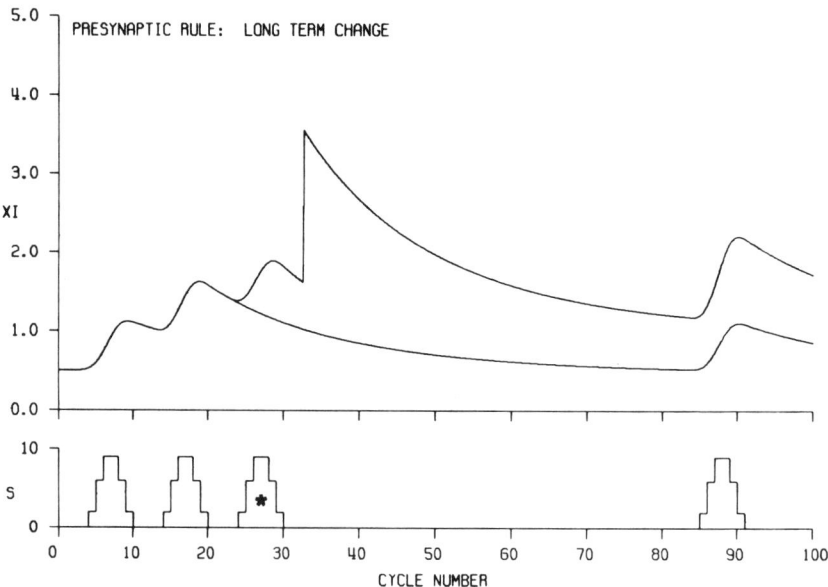

Figure 8 -- Simulation of the Presynaptic Rule. Two stimulus bursts produce lower curve plotted which does not achieve a long-term change, and which shows a response to test burst at cycle 85 which is virtually identical to initial response. Three stimulus bursts (including the burst marked with *) give rise to long-term change (upper curve) manifested as increased baseline presynaptic strength. Test burst at cycle 85 produces an enhanced response. Values of input parameters are: number of substeps per iteration = 10, $\lambda=0.04$, $\beta=0.01$, $\varepsilon=0.10$, $\kappa=0.02$, $\xi^0=0.5$, $\alpha=0.10$, $\gamma=0.20$.

run (plotted as upper curve in Figure 8) involved 3 initial bursts of stimuli (the extra "third" burst is distinguished by an asterisk). This was sufficient to allow L to surpass θ_S and a long-term change resulted. The long-term change is seen as the sharp rise in ξ at cycle 32. As discussed below, such long-term changes are proposed to result from changes in gene expression and would require minutes to hours to be fully manifested. For simplicity, we opted to make the change take place instantaneously. Such discontinuities could potentially introduce spurious oscillations into the network, but in our simulations, we never observed such effects. The change in ξ^0 was 0.6, and ξ underwent a proportionate increase. During the next 55 cycles of rest, ξ, decreased with the same decay constant as before the long-term change, since λ and β are unchanged. The test pulse at cycle 85 reveals a much larger response than was initially produced -- this is the result of the long-term change. The increase in ξ would give rise to greater stimulation of any follower cells. It might be objected that the increase in ξ seen at cycle 85 is due to the shorter recovery time form the preceeding bursts. This, however, is not the case; the same result is obtained with identical resting times between the last stimulus and the test burst.

Basis for the Presynaptic Rule

There are two reasons for postulating that the long-term modification occurs on a total cell basis rather than independently at individual synapses. The first reason is that most of the terminals given off by a given axon will experience a similar history of activation. The series of facilitatory and depressive fluctuations will thus be similar at the various terminals of an axon. Thus, even if the modifications were carried out locally, they would largely be identical across all presynaptic terminals of the cell.

We do not wish to diminish the possibility that variation occurs among axonal terminals -- indeed, the raison d'être of a selective theory is to emphasize such variability and to utilize it for selective amplifications. The sources for such variation are abundant: axonal branch conduction block (safety factor) has been observed in several species (Parnas, 1972; Henneman, 1984), autoreceptors or receptors to other transmitters (Chesselet, 1984) influence presynaptic transmitter release, colocalized transmitters may interact differently at each terminal, and the initial distribution of pre-synaptic strengths must itself have a finite variance. In addition, influences of local circuits will give each axonal terminal an essentially unique environment.

The second reason for assuming that presynaptic modification must occur across the entire cell rather than at individual synapses concerns the physical limitations of any synaptic mechanism. We are implicitly supposing, as was originally proposed by Greengard (Greengard and Kuo, 1970) that these long-term modifications involve a change in gene expression, potentially in response to second messengers produced at synapses and retrogradely transported (Nestler and Greengard, 1984). In this case, the net effect would reflect an average of the individual synaptic short-term fluctuations. Given the inherent time lags in the production and transport of newly synthesized gene products, there is no currently known way to selectively route the new material to those individual synapses which originally experienced the short-term changes. In any case, a long-term modification will reflect a fairly long temporal average (otherwise the synaptic state would thrash about unstably) and such an average will necessarily involve a loss of specificity of the contributions of individual synapses.

It has been shown (Edelman and Finkel, 1984) that this presynaptic mechanism is sufficient to account for group selection within a neural network. Neurons within a group receive similar extrinsic inputs and are densely connected (see Figure 10). Many of the neurons in a group will fire in a coactive fashion, particularly because intrinsic inhibition quickly dampens all activity that does not manage to excite a large fraction of the neurons in the group.

If the extrinsic inputs to a group are infrequent or not highly correlated temporally, most of the neurons in the group will be only weakly activated. Individual neurons all modify their presynaptic efficacy to a new, lower set-point, and the result is a weakening of the group. This is manifested by a decreased probability that neurons in the groups will fire in a correlated fashion with each other. Conversely, if the extrinsic inputs are strong and coactive, most of the neurons in the group may be activated to a sufficient degree to re-set their presynaptic efficacies upward; hence, the group will be strengthened. In general then, coactivation strengthens a neuronal group, while uncorrelated activation weakens the group. This was the condition required of the synaptic rule to account for group selection (see discussion of Figure 2). It is worth noting that the Hebb rule (Hebb, 1949) also shares this property of coactivation leading to synaptic strengthening. However, the Hebb rule achieves this property by definition; in contrast, we have provided a biophysically based mechanism for the rule.

INTERACTION OF PRESYNAPTIC AND POSTSYNAPTIC MODIFICATIONS

The assumption that the mechanisms regulating presynaptic and postsynaptic modifications can operate independently raises the important question of how the two types of changes interact within the same population of synapses to produce functional alterations in network behavior. In considering this question, we will consider synaptic modifications occuring on only two time scales, short-term postsynaptic modifications and long-term presynaptic modifications. However, both presynaptic and postsynaptic sites probably undergo changes over several different time scales.

Long-term synaptic modifications originating in a particular group have very different effects on that group versus other groups. In general, the group originating the changes becomes more cohesive and more tightly connected, while neighboring groups are weakened. Since synaptic changes are occurring in all groups in parallel, there will be an ongoing competition between groups to maintain a set of synaptic states.

We have previously shown that the relationship between short and long-term changes has four main characteristics (Finkel and Edelman, 1985): (1) Short-term changes in a group lead to long-term changes primarily within that same group. (2) The presence of group structure is sufficient to ensure that long-term changes arising from short-term changes in a particular group differentially affect future short-term changes in that particular group. (3) Long-term presynaptic changes increase the variance of the distribution of subsequent short-term postsynaptic strengths. (4) Although short-term changes are differentially affected in the group originating the long-term change, the group still retains a significant capacity to respond to novel input with novel short-term modifications. In other words, the repertoire of possible short-term modifications does not become entrained, stereotyped or overgeneralized.

We have examined these characteristics of the interaction in a formal analytical model (Finkel and Edelman, 1985a; 1985b). Given the inter-group connection scheme shown in Figure 9, we supposed that a long-term presynaptic modification has occured in group I. We asked what conditions are sufficient to guarantee that the internal connections, n_{II}, of group I experience the largest change in subsequent short-term modifications as a result of the long-term change.

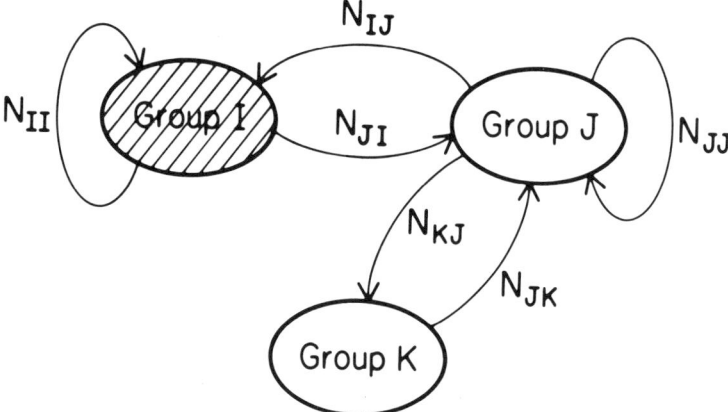

Figure 9 -- Classes of Group Connections. Ellipses represent groups, arrows represent all connections between the groups. There are N_{IJ} connections from group J to group I, N_{JI} from group I to group J, and N_{II} from group I to itself, etc. The analysis assumes a long-term change has occured in group I (cross-hatching), and examines the effect on subsequent short-term changes in the various classes connections between the groups.

We found that there are three jointly sufficient conditions; namely, that (1) connectivity within groups is stronger than between groups, (2) neurons in the same group fire together more often than neurons in different groups, and (3) on the average, input from different groups is uncorrelated.
 The key point of this formal analysis was that within the context of this simple model, these three sufficient conditions define a neuronal group (Finkel and Edelman, 1985). The analysis thus indicates that the organization of a network into neuronal groups provides a sufficient condition for long-term modifications to differentially affect Given the group condition, we can rank the classes of connections according to the magnitude of the effected change in short-term modifications. Group I itself always

undergoes the greatest change, connections between unaffected groups (n_{JK}) are always the least affected; and the three remaining classes of connections (n_{IJ}, n_{JI}, n_{JJ}) may be affected in any relative order depending upon the number and strength of the connections and on the degree of coactivation of the various groups.

Computer Simulations of Combined Model

We now turn to computer simulations of the dual pre- and postsynaptic rules operating within a network. We have simulated the interaction in two separate programs (Finkel and Edelman, 1985b); one based on the simplified analytical model just presented, and a second which uses the more detailed individual formulations of the rules presented earlier. Both give concordant results, and here we will only present results from the first program. However, the fact that both programs give similar results argues for the robustness of the conclusions concerning the interactions.

A typical example of the network used is shown in Figure 10. Neurons within each group are interconnected randomly with a fixed average density of connections, and each connection is randomly assigned an initial presynaptic and postsynaptic strength. Each cell is also randomly connected to neurons in a fixed number of other groups. Each cell in the network also receives extrinsic fibers with fixed synaptic strengths to provide inputs. These connections were not generated randomly, but consisted of an orderly topographic mapping onto the network. These inputs were stimulated in patterns designed to elicit activity in various regions of the network.

Figure 11 shows the results of one computer simulation. Two groups, called I and J for convenience, were arbitrarily chosen; and the first group, I, was stimulated via its extrinsic inputs to induce short-term postsynaptic changes in its intrinsic connections. The actual stimulation used was a burst of 5 cycles of stimulation to half the cells in the group, followed by a 5 cycle rest, followed by 5 cycles of the same stimulation. After allowing time (50 cycles) for the activity in the system to return to its resting state, the same stimulation protocol was applied to the other group, J. The magnitude of the short-term changes, averaged over all group connections is shown as the solid lines in Figure 11. Note that the second barrage of stimulation induced a larger synaptic change than the first

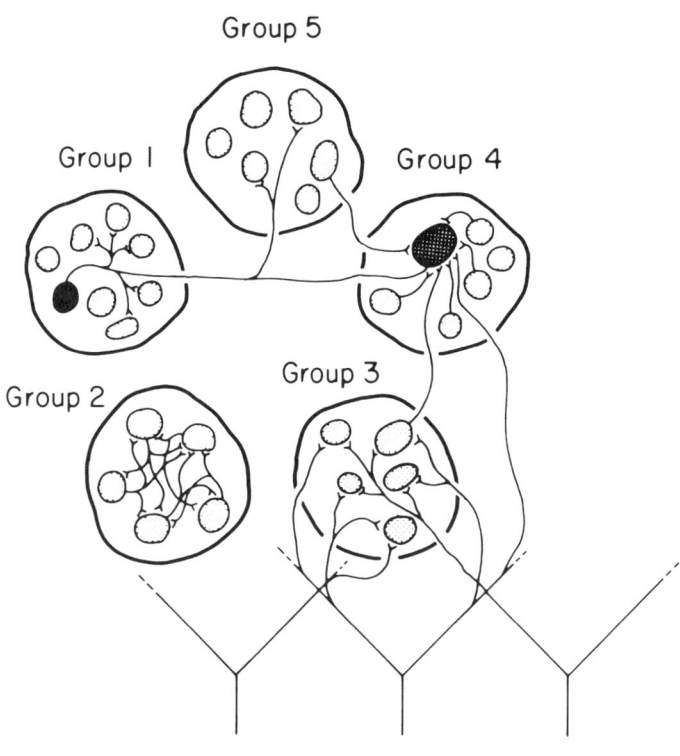

extrinsic inputs

Figure 10 -- Schematic network connectivity used in computer simulations. Five groups (dark outlines) are shown with some of their cells indicated. Each group is used to illustrate one aspect of the connectivity. Group 1 shows that each cell contacts cells in its own group and in other groups. Group 2 shows the dense internal connectivity of groups. Group 3 shows that each group also receives inputs from a set of overlapped extrinsic inputs which can be selectively stimulated. Group 4 shows that each cell thus receives inputs from cells in its own group, from cells in other groups, and from extrinsic sources. The computer simulation allows observation of changes in presynaptic and postsynaptic efficacies of the various connections after different stimulation paradigms.

Selection Mechanisms in Neural Mapping

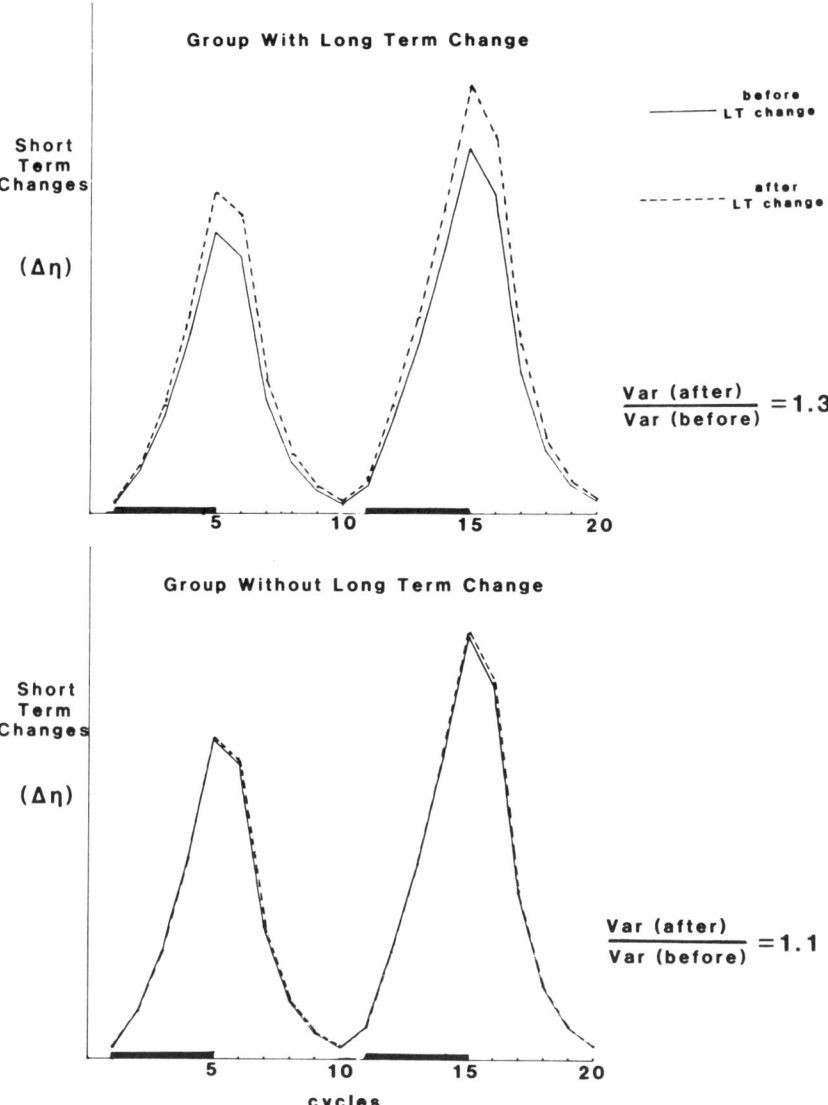

Figure 11 -- Legend appears on following page.

Figure 11 -- Results of Computer Simulation. Plots of magnitude of change in postsynaptic efficacy, $\Delta\eta$, versus time (iterations). Two groups were chosen at random and were stimulated, in turn, (darkened bar on abcissa) for 5 cycles, allowed to rest for 5 cycles, then stimulated again for 5 cycles. The solid line shows the resulting short-term changes. One group (top graph) was then repeatedly stimulated for 50 cycles to produce a long-term presynaptic change in some of the cells in the group. After allowing 50 cycles of rest, the initial stimulation sequence (5 cycles stimulation/ 5 cycles rest/ 5 cycles stimulation) was given again. The short-term changes are shown by the dashed line. Only the group which received the long-term change, shows an alteration in subsequent short-term changes. Also shown is the ratio of the variance in the η of the group (measured at cycle 15) for before and after the long-term change. The variance increases in both groups.

barrage. If the barrages are repeated, eventually the threshold for a presynaptic modification is surpassed. We thus repeated the stimulus barrrages to group I 10 times, without rest, to produce a long-term change in several of the cells in group I. After 50 cycles of rest, the initial stimulation protocol (2 barrages of 5 cycles of stimulation separated by 5 cycles of rest) was repeated, first to group I; then, after a rest, to group J. The dashed curve in Figure 11 shows the resultant short-term postsynaptic changes in both group I and group J subsequent to the long-term change induced in group I.

It is clear from Figure 11 that there is no significant change in subsequent short-term modifications of the group which did not receive the long-term presynaptic modification (solid and dashed curves are practically superposed). However, in the group with the long-term change, repetition of the stimuli which induced the long-term change gives rise to an amplified pattern of short-term changes.

We also examined the change in the variability of short-term modifications as a result of a long-term modification. The definition of variability is elusive and in the present study we examined the effect of modifications upon the variance of the postsynaptic strengths. As shown in Figure 11, the variance of postsynaptic strengths in a group increases after a long-term change in that group (here by 30%); but significantly, it also increases in groups which

did not experience the long-term change (here by 10%).
Thus, groups which do not directly experience a long-term
change themselves, but which receive inputs from a group
that did, are still affected as a result of the increased
variance in their short-term modifications. Because each
group contacts many other groups, the variability-generating
property of long-term presynaptic changes will therefore
have wide effects upon the network. These effects provide a
continuing range of repertoire variation that will be the
basis for future selective events.

CONCLUSIONS

There are several conclusions which we can draw from
the present model. The large amount of anatomical variability present in the nervous systems of vertebrates and invertebrates argues for and provides a substrate for a process
of selection. The functional plasticity of maps in the nervous system, as documented by the work of Merzenich and his
colleagues, goes further to suggest that selective mechanisms are responsible for the reorganization, and in all
likelihood the organization of neural maps.

The synaptic rules discussed here have been previously
used (Finkel and Edelman, 1985b) to generate an extensive
set of experimental predictions, most of which can be tested
using current technology. The main point of the present
model is the consideration of how the synaptic rules interact with each other and with the network anatomy. The
general theory is supported by the finding that networks
organized into neuronal groups allow short and long-term
modifications to be linked in a meaningful fashion.

Synaptic modifications regulated by the proposed rules
provide the mechanism of selection responsible for the
organization of neural maps. Short-term modifications can
temporarily alter the properties of the extant map through
changes in the coordination of different groups as well as
through direct changes in the receptive fields. These
short-term changes may put the group at a competitive advantage or disadvantage, but these changes are not stable.
Consistent with population views of the nervous system,
there will be a competition between groups to retain the
integrity of their short-term modification repertories --
each group acts to increase the likelihood of its own set of
short-term responses while acting to decrease the likelihood
of those of surrounding groups. Clearly this requires a
delicate balance to maintain a state intermediate between

completely rigid stereotyped response and amnesic de novo trial and error. This balance may shift with input history and experience.

Persistent or repeated short-term changes lead to stable long-term changes in the groups. Long-term changes in several neurons of a group are probably necessary to lead to selection of that group. Together with the underlying anatomy, these long-term modifications give rise to the modal properties of the map. Short-term changes allow adaptability to small scale environmental fluctuations without changing the group structure. Larger environmental fluctuations can cause selection of new groups, and consequently, give rise to larger changes in the map.

It is convenient to think of the functions of the nervous system as a kind of ongoing development. It is therefore of interest that the synaptic rules presented here have a certain developmental "flavor". The rules reflect the operation of complex biochemical and biophysical mechanisms, while they depend upon interactions among large populations of cells. The same rules presumably operate in different contexts throughout the nervous sytem. In this way, the synaptic rules are not unlike epigenetic sequences in development.

It would be naive to jump from synaptic modifications to cognitive processes: no such link can be made at present; and in any case, such higher level processes most likely reflect the combined influences of many brain regions operating in parallel. Nevertheless, this model of short and long-term modifications interacting over synaptic populations does have useful memorial properties that might be reflected by higher level cognitive processes. For example, the heterosynaptic properties of the postsynaptic rule demonstrate how context-dependent associations may be mediated; and as a result of the anatomical properties of the synaptic rules the model helps to refine the notion of a distributed memory.

The foregoing analysis provides detailed mechanisms and substance to the assumptions of the theory of neuronal group selection. It provides a basis upon which to analyze further the various levels of cortical organization and the suprising functional responses of neural maps.

REFERENCES

1. Blobel G (1983). Regulation of intracellular protein traffic. Prog Br Res 58:77.
2. Bryan JS, and HL Atwood (1981). Two types of synaptic depression at synapses of a single crustacean motor axon. Mar Behav Physiol 8:99.
3. Catteral WA (1979). Binding of scorpion toxin to receptor sites associated with sodium channels in frog muscle -- correlation of voltage-dependent binding with activation. J Gen Physiol 74:375.
4. Chan-Palay V, Nilaver G, Palay SL, Beinfeld MC, Zimmerman EA, Wu JY, O'Donohue TL (1981). Chemical heterogeneity in cerebellar Purkinje cells: Existence and coexistence of glutamic acid decarboxylase-like and motilin-like immunoreactivities. Proc Natl Acad Sci USA 78:7787.
5. Chan-Palay, V, Palay SL, Wu JY (1982). Sagittal cerebellar microbands of taurine neurons: immunocytochemical demonstration by using antibodies against the taurine-synthesizing enzyme cysteine sulfinic acid decarboxylase. Proc Natl Acad Sci USA 79:4221.
6. Changeux J-P (1981). The acetylcholine receptor: an "allosteric" membrane protein. The Harvey Lectures, Ser 75:85.
7. Chesselet MF (1984). Presynaptic regulation of neurotransmitter release in the brain -- facts and hypothesis. Neurosci 12:347.
8. de Lacoste-Utamsing C, Holloway RL (1982). Sexual dimorphism in the human corpus callosum. Science 216:431.
9. Devor M, PD Wall PD (1981). Effects of peripheral nerve injury on receptive fields of cells in the cat spinal cord. J Comp Neurol 199:227.
10. Dostrovsky JO, Millar J, Wall PD (1976). The immediate shift of afferent drive of dorsal column nucleus cells following deafferentiation: a comparison of acute and chronic deafferentiation in gracile nucleus and spinal cord. Exp Neurol 52:400.
11. Edelman GM (1975). The shock of molecular recognition. In Smith EE (ed): "Molecular Approaches to Immunology," New York: Academic Press.
12. Edelman GM (1978). Group selection and phasic reentrant signalling: a theory of higher brain function. In Edelman GM, Mouncastle VB (eds): "The Mindful Brain: Cortical Organization and the Group-Selective Theory of

Higher Brain Function," Cambridge: MIT Press, p 55.
13. Edelman GM (1981). Group selection as the basis for higher brain function. In Scmitt FO, et al (eds): "Organization of the Cerebral Cortex," Cambridge: MIT Press, p 535.
14. Edelman GM (1984). Modulation of cell adhesion during induction, histogenesis, and perinatal development of the nervous system. Ann Rev Neurosci 7:339.
15. Edelman GM, Finkel LH (1984). Neuronal group selection in the cerebral cortex, In Edelman GM, Gall WE, Cowan WM (eds): "Dynamic Aspects of Neocortical Function," New York: John Wiley, p 653.
16. Edelman GM, and Reeke GN Jr. (1982). Selective networks capable of representative transformations, limited generalizations, and associative memory. Proc Nat Acad Sci USA 79:2091.
17. Finkel LH, Edelman GM (1985a). Interaction of synaptic modification rules within populations of neurons. Proc Nat Acad Sci USA 82:1291.
18. Finkel LH, Edelman GM (1985b). Population rules for synapses in networks. In Edelman GM, Gall WE, Cowan, WM (eds): "New Aspects of Synaptic Function," New York: John Wiley, (in press).
19. Gilbert CD, Wiesel TN (1979). Morphology and intracortical projections of functionally characterized neurons in the cat visual cortex. Nature 280:120.
20. Greengard P, Kuo JF (1970). On the mechanism of action of cyclic AMP. Adv Biochem Psychopharmacol 3:287.
21. Hawkins RD, Abrams TW, Carew TJ, Kandel ER (1983). A cellular mechanism of classical conditioning in Aplysia-activity dependent amplification of pre-synaptic facilitation. Science 219:400.
22. Hebb DO (1949). "The Organization of Behavior." New York: John Wiley.
23. Henneman E, Luscher HR, and Mathis J (1984). Simultaneously active and inactive synapses of single Ia fibers on cat spinal motoneurones. J Physiol Lond 352:147.
24. Hume RI, Purves D (1981). Geometry of neonatal neurons and the regularity of synapse elimination. Nature 293:469.
25. Hyvarinen J (1984). Modification of parietal association cortex by early blindness. Int J Neuro 22:177.
26. Jones EG (1981). Anatomy of cerebral cortex: columnar input-output organization. In Schmitt FO, et al (eds): "The Organization of the Cerebral Cortex," Cambridge:

MIT Press, p 199.
27. Jones EG, and Powell TPS (1970). An electron microscopic study of the laminar pattern and mode of termination of afferent fibre pathways in the somatic sensory cortex of the cat. Philos Trans Roy Soc Lond [Biol] 257:45.
28. Kaas JH, Merzenich MM, Killackey HP (1983). The reorganization of somatosensory cortex following peripheral-nerve damage in adult and developing mammals. Ann Rev Neurosci 6:325.
29. Kaas JH, Nelson RJ, Sur M, Merzenich MM (1981). Organization of somatosensory cortex in primates. In Schmitt FO et al (eds): "The Organization of the Cerebral Cortex," Cambridge: MIT Press, p237.
30. Kandel ER (1976). "Cellular Basis of Behavior." San Francisco: W.H. Freeman & Co.
31. Kasamatsu T, Pettigrew JD (1976). Depletion of brain catecholamines: failure of ocular dominance shift after monocular occlusion in kittens. Science 194:206.
32. Koch C, Poggio T, Torre V (1982). Retinal ganglion cells: a functional interpretation of dendritic morphology. Phil Trans Roy Soc Lond B298:227.
33. Korn H, Faber DS, Burnod Y, Triller A (1984). Regulation of efficacy at central synapses. J Neurosci 4:125.
34. Kosar E, Hand PJ (1981). First somatosensory cortical columns and associated neuronal clusters of nucleus ventralis posterolateralis of the cat: An anatomical demonstration. J Comp Neurol 198:515.
35. Kramer AP, Goldman JR, Stent GS (1985). Developmental arborization of sensory neurons in the leech Haementeria ghilanii 1. Origin of natural variations in the branching pattern. J Neurosci 5:759.
36. Landry P, Deschenes M (1981). Intracortical arborizations and receptive fields of identified ventrobasal thalamocortical afferents to the primary somatic sensory cortex in the cat. J Comp Neurol 199:345.
37. Lee KJ, Woolsey TA (1975). A proportional relationship between peripheral innervation density and cortical neuron number in the somatosensory system of the mouse. Brain Res 99:349.
38. Llinas R (1979). The role of calcium in neuronal function. In Schmitt FO, et al (eds): "The Neurosciences 4th Study Program," Cambridge: MIT Press, p. 555.
39. Llinas R, Sugimori M (1980). Electrophysiological properties of *in vitro* Purkinje cell somata in mammalian

cerebellar slices. J Physiol 305:171.
40. Macagno ER, Lopresti V, Levinthal C (1973). Structure and development of neuronal connections in isogenic organisms: variations and similarities in the optic system of Dapnia magna. Proc Nat Acad Sci USA 70:57.
41. Magleby KL, Zengel JE (1982). Quantitative description of stimulation-induced changes in transmitter release as the frog neuromuscular junction. J Gen Physiol 80:613.
42. Mallart A, Martin AR (1968). An analysis of facilitation of transmitter release at the neuromuscular junction of the frog. J Physiol 193:679.
43. Martin AR, Pilar G (1964). Presynaptic and postsynaptic events during post-tetanic potentiation and facilitation in the avian ciliary ganglion. J Physiol Lond 175:17.
44. Merzenich MM, Kaas JH (1982). Reorganization of mammalian somatosensory cortex following peripheral-nerve injury. Trends Neurosci 5:434.
45. Merzenich MM, Jenkins WM, Middlebrooks JC (1984). Observations and hypotheses on special organizational features of the central auditory nervous system. In Edelman GM, Gall WE, Cowan WM (eds): "Dynamic Aspects of Neocortical Function," New York: John Wiley, p 397.
46. Merzenich MM, Kaas JH, Sur M, and Lin CS (1978). Double representation of the body surface within cytoarchitectonic areas 3b and 1 in "SI" in the owl monkey (Aotus trivirgatus). J Comp Neurol 181:41.
47. Merzenich MM, Kaas JH, Wall JT, Nelson RJ, Sur M; Felleman DJ (1983a). Topographic reorganization of somatosensory cortical areas 3b and 1 in adult monkeys following restricted deafferentation. Neurosci 8:33.
48. Merzenich MM, Kaas JH, Wall JT, Sur M, Nelson RJ, Felleman DJ (1983b). Progression of change following median nerve section in the cortical representation of the hand in areas 3b and 1 in adult owl and squirrel monkeys. Neurosci 10:639.
49. Merzenich MM, Nelson RJ, Stryker MP, Cynader M, Schoppmann A, Zook AM (1984b). Somatosensory cortical map changes following digit amputation in adult monkeys. J Comp Neurol 224:591.
50. Meyer RL (1980). Mapping the normal and regenerating retino-tectal projection of goldfish with autoradiographic methods. J Comp Neurol 189:273.
51. Miledi R (1973). Transmitter release induced by injection of calcium ions into nerve terminals. Proc Roy Soc Lond B183:421.
52. Mountcastle VB (1978). An organizing principle for

cerebral function: The unit module and the distributed system. In Edelman GM, Mouncastle VB (eds): "The Mindful Brain: Cortical Organization and the Group-Selective Theory of Higher Brain Function," Cambridge: MIT Press, p 7.
53. Nestler EJ, Greengard P (1984). "Protein Phosphorylation in the Nervous System." New York: John Wiley.
54. Paradiso MA, Bear MF, Daniels JD (1983). Effects of intracortical infusion of 6-hydroxydopamine on the response of kitten visual-cortex to monocular deprivation. Exp Br Res 51:413.
55. Parnas I (1972). Differential block at high frequency of branches of a single axon innervating two muscles. J Neurophysiol 35:903.
56. Parnas I, Dudel J, Parnas H (1984). Depolarization dependence of the kinetics of phasic transmitter release at the crayfish neuromuscular junction. Neurosci Lett 50:157.
57. Pasternak JF, Woolsey TA (1975). The number, size, and spatial distribution of neurons in lamina IV of the mouse SmI neocortex. J Comp Neurol 160:291.
58. Pearson KG, Goodman CS (1979). Correlation of variability in structure with variability in synaptic connections of an identified interneuron in locusts. J Comp Neurol 184:141.
59. Rall W, Rinzel J (1973). Branch input resistance and steady state attenuation for input to one branch of a dendritic model. Biophys J 13:648.
60. Reeke GN, Edelman GM (1984). Selective networks and recognition automata. Ann NY Acad Sci 426:181-201.
61. Rockel AJ, Hiorns RW, Powell TPS (1980). The basic uniformity in structure of the neocortex. Brain 103:221.
62. Schmidt JT (1982). The formation of retinotectal projections. Trends Neurosci 46:111.
63. Shinoda Y, Yokota J, Futami T (1981). Divergent projection of individual corticospinal axons to motoneurons of multiple muscles in the monkey. Neurosci Lett 23:7.
64. Shinoda Y, Yokata J, Futami T (1982). Morphology of physiologically identified rubrospinal axons in the spinal cord. Brain Res 242:321.
65. Smith SJ, Augustine GJ, Charlton MP (1985). Transmission at voltage-clamped giant synapse of the squid - evidence for cooperativity of presynaptic calcium action. Proc Nat Acad Sci USA 82:622.

66. Sur M, Merzenich MM, Kaas JH (1980). Magnification, receptive field area, and "hypercolumn" size in areas 3b and 1 of somatosensory cortex in owl monkeys. J Neurophysiol 44:295.
67. Sur M, Sherman SM (1982). Retinogeniculate terminations in cats--morphological differences between X-cell and Y-cell axons. Science 218:338.
68. Sur M, Wall JT, Kaas JH (1981). Modular segregation of functional cell classes within the postcentral somatosensory cortex of monkeys. Science 212:1059.
69. Szentagothai J (1975). The module-concept in cerebral cortex architecture. Brain Res 95:475.
70. Thiele J, Klumpp S, Schultz JE, Bardele CF (1982). Differential distribution of voltage-dependent calcium channels and guanylate-cyclase in the excitable ciliary membrane from Paramecium-tetraurelia. Eur J Cell Biol 28:3.
71. Van Essen DC, Newsome WT, Bixby JB (1982). The pattern of interhemispheric connections and its relationship to extrastriate and visual areas in the macaque monkey. J Neurosci 2:265.
72. von der Malsburg C (1981). The correlation theory of brain function. Internal Report 81-2, Dept. Neurobiology, Max Planck Institute for Biophysical Chemistry, Goettingen.
73. Weiss P (1939). "Principles of Development." New York: Henry Holt.
74. White EL, Hersch SM (1982). A quantitative study of thalamocortical and other synapses involving the apical dendrites of corticothalamic projection cells in mouse Sml cortex. J Neurocytol 11:137.
75. Winfield DA, Brooke RNL, Sloper JJ, Powell TPS (1981). A combined Golgi-electron microscopic study of the synapses made by the proximal axon and recurrent collaterals of a pyramidal cell in the somatic sensory cortex of the monkey. Neurosci 6:1217.
76. Witzemann V, Richardson G, Bousted C (1983). Characterization and distribution of acetylcholine-receptors and acetyl-cholinesterase during electric organ development in Torpedo-marmorata. Neurosci 8:333.

Index

AChE, motorend plate morphogenesis, 395, 396, 403, 409, 411–416
AChR and AChE aggregating factor, 386–389
AChR, motorend plate morphogenesis, 395, 396, 402, 403, 406, 409, 412, 414–416
Actin
  filaments, epithelial-mesenchyme transformation, 297–299
  α-, induction, *Xenopus laevis*, cytoplasmic localization, 30–31
  -rich filopodia, epithelial-mesenchyme transformation, 301–303
  see also Microvilli formation, actin mechanical properties in
Actin genes, muscle-specific, *Xenopus laevis* embryo, 37–43
  activation
    without cell-cell contact, 39–41
    by experimental induction mechanism, 40–41, 43
    by materials localized in uncleaved egg, 39–40
  induction, 37, 39, 41–43
    vs. cytoplasmic localization, 37–38, 41–43
    directive vs. permissive, 43
    mechanism, 43
  transcription, normal, 38
α-Actinin, 241
Action cf. reaction systems, 63–65
Activity-dependent stabilization. *See* Retinotectal synapses, activity-dependent stabilization, goldfish
Actomyosin gel, fibrous network, and microvilli formation, 367–369; *see also* Cortical tractor model, epithelial folding *entries*
Afferent arbor overlap, neural mapping, 581–583

Aldehyde oxidase, 479
Amanitin, 32, 77
Ambystoma, 135
  embryo, 5
  maculatum, gastrula cell adherence to FN-Sepharose beads, 272, 273, 276, 279
  mexicanum (Axolotl), 180, 272
Amphioxus, 68–69
Angiogenesis, kidney, 357–359
Animal-vegetal axis, 61–62
antennapaedia homeobox sequence, *Drosophila*, 509
Antibodies. *See also* Monoclonal antibodies
Anti-GP40, and cell-substratum attachment glycoprotein complex, 244, 245
Anti-GP80, cell-CAM 128/80, 236–238, 240
Anti-gp 105 antiserum, and cell-CAM 105, hepatocellular carcinoma, 256–258, 261
  anti-gp 105-2, 258–260
Anti-N-CAM in motor endplate morphogenesis, 408
Apical seal, cortical tractor model, epithelial folding, 148, 149
  newt neural plate, 152–153, 157, 159
  preservation while cells interdigitate, 149–150
Arborization, axon terminal, motor endplate morphogenesis, 397
Archenteron
  convergent extension by cell intercalation, *Xenopus laevis*, 115, 122
  gastrula cell adherence to FN-Sepharose beads, 280, 285
aristapaedia gene, 432
Auditory space map, barn owl, 561
Autocatalytic process
  actin contractility, microvilli formation, 372

hydra, head effect on body column, 449
Axis, *Xenopus laevis* induction
   deficiency, induction, 21, 23
   formation, induction, 18–22, 27, 30–31
   schema, 19
   structure determinants, 16–17
   truncation, induction, 16
Axon, motor endplate morphogenesis
   early contacts with myotube, 396
   terminal arborization, 397

Basal lamina (basement membrane)
   epithelial-mesenchyme transformation, 294, 296, 298
   hydra body column pattern formation (mesoglea), 437
   and skin morphogenesis, ECM in, 321
   synaptic
     motor endplate morphogenesis, 395, 396, 402, 403, 411, 412, 415
     synaptogenesis, regenerating neuromuscular junction, 385–389
Blastocoelic roof
   and cell migration, gastrulation, 173–175
   gastrula cell adherence to FN-Sepharose beads, 272, 280, 282, 285, 288–290
   gastrulation, *Xenopus laevis*, 129–130, 134–135
   gastrulation without, 118–126
Blastocyst, cell lineages and body plan, mammalian embryo, 73–74, 76, 80, 92
   inner cell mass, 74, 76–78
   trophectoderm, 74, 76, 77
Blastoderm, avian, morphogenetic movements and fate maps, 99–107
   cinemicrography, 100
   endoblast, 101, 103–105
   incubation temperature, 102
   mesoblast, 101, 103–106
   pharyngula, 106
   prechordal plate, 105
   primitive streak, 103–107
   quail-chick xenograft technique, 100, 107
   symmetrization, 100, 106–107
   TEM, SEM, 107

   vital dye and iron oxide markers, 100
Blastopore closure, *Xenopus laevis*, 112, 116, 121, 129–130
Body column. *See* Hydra, body column pattern formation
Body plan. *See* Cell lineages and body plan, mammalian embryos
Bottle cells
   cortical tractor model, epithelial folding, newt neural plate, 153
   gastrulation, convergent extension by cell intercalation, *Xenopus laevis*, 112, 116, 119–120
Brain development, CAMs, 211, 212
Breast cancer cell line MCF-7, 237
Bullous pemphigoid antigen, and skin morphogenesis, 321, 322, 325, 326, 343
α-Bungarotoxin
   fluorescein-labeled, motor endplate morphogenesis, 407, 409, 411
   ^{125}I-, motor endplate morphogenesis, 396, 402, 403
   retinotectal synapses, activity-dependent stabilization, 545, 551, 555, 557, 561
   rhodaminated, synaptogenesis, regenerating neuromuscular junction, 386, 388
bx function, *Drosophila*, 492

Cadherins, 202, 237, 239, 242
Cadherins, Ca^{2+}-dependent selective cell adhesion, 223–231
   E-cadherin, 224, 228–231
     cf. L-CAM, 230–231
     lens and neural tube, 230, 231
     MAb ECCD-1, 224–226, 229, 254
     PCC3 cell line, 228
     see also L-CAM
   N-cadherin, 224, 228, 229
     G26-20 glioma cell line, 229
     MAb NCD-1, 224–227, 229, 230
     cf. N-CAM, 231
     see also N-CAM
   cell-type specificity, 226–228
   developmental change in subclass expression, 230
   fluorescent bead cell labeling, 226, 228

Index 621

segregation of E- and N-cadherin-positive cells, 228
unidentified cadherin, 229–230
Calcium
L-CAM, 199, 202, 203, 207
-free medium, 273
and microvilli formation, actin role, 367, 368, 374
neuronal selection mechanisms, neural mapping, 594–596
triggering in cortical tractor model, epithelial cells, 146–147
see also Cadherins, Ca^{2+}-dependent selective cell adhesion
Calpodes, 442
CAM(s) (cell adhesion molecules), 167–170, 184
kidney morphogenesis, 354–355
and neural crest cell migration, 179, 184, 185
neuronal selection mechanisms, neural mapping, 572, 574
phosphorylation, 204, 264
see also specific CAMs
L-CAM, 168–170
Ca^{++}, 199, 202, 203, 207
cf. cadherin, 202
cf. cell-CAM 105, 254, 262–265, 267
cf. cell-CAM 128/80, 202, 237, 239, 242
cDNA probes for mRNA, 201
epithelial-mesenchyme transformation, 295
feather histogenesis, 213, 215, 216
molecular structure and binding, 198–199, 201–203, 207
and neural crest cell migration, 179, 184, 185
in neural development, 208, 209
tunicamycin, 202
cf. uvomorulin, cadherin, and cell-CAM 120/80, 202
see also Cadherins entries; Cell-CAM 105, rat hepatocellular carcinoma; Cell-CAM 120/80; Cell-CAM 128/80; Uvomorulin
N-CAM, 168–171
A and E forms, 211–213, 217, 415
cf. N-cadherin, 231
cDNA probes for mRNA, 200–202
feather histogenesis, 213, 215, 216
homophilic, 205
molecular structure and binding, 198–203, 205–207
and neural crest cell migration, 179, 184, 185
in neural development, 208–213, 217
retinotectal map formation, Xenopus laevis, 528–529, 534
sialic acid, 198, 200, 202, 206
tunicamycin, 198, 200
vesicles, 206–207
see also Cadherins entries; Motor endplate morphogenesis, mouse
Ng-CAM, 168–170
heterophilic, 207
molecular structure and binding, 198, 203–205, 207
and neural crest cell migration, 185
in neural development, 208, 210, 211
polarity modulation, 211
phosphorylation and glycosylation, 204
CAMs in feather histogenesis, 213–218
CAM couples, 215
listed, cf. other tissues, 217
and keratin genes, 215
stages, 213, 214
CAMs in neural development, immunofluorescence, 208–213
brain development, 211, 212
neural tube formation, 208, 209
oligodendroglia, 211
polarity modulation, 211
primary cf. secondary CAMs, 208
primitive streak stage, 209
spinal cord, 208, 210–212
see also Motor endplate morphogenesis, mouse
Carcinoma, breast, MCF-7 cells, 237; see also Cell-CAM 105, rat hepatocellular carcinoma
Cartilaginous elements, positional information and pattern formation, chick limb bud, 424–427
Cell adhesion
cortical tractor model, epithelial cells, 146, 148, 157

folding, newt neural plate, 157
hydra body column pattern formation, 442–444
retinotectal map formation, 526–529, 534–535, 540
see also Cadherins, Ca^{2+}-dependent selective cell adhesion; CAMs; Gastrula cell adherence to fibronectin-Sepharose beads, amphibians; specific CAMs
Cell attachment. See Cell-substratum attachment glycoprotein complex (CSATag; GP140 glycoprotein)
Cell-CAM 120/80
cf. cell-CAM 105, 254, 262–265, 267
cf. L-CAM, 202
see also L-CAM
Cell-CAM 128/80, early mouse development, 236–244
anti-GP80, monospecific antiserum, 236–238, 240
identification, glycoprotein, 236, 237
localization, role in intracellular junctions, 238–243
regulation of expression, 243–244
cf. uvomodulin, cadherin, and L-CAM, 237, 239, 242
see also L-CAM
Cell-CAM 105, rat hepatocellular carcinoma, 253–267
acidity, 257–259
cf. adult rat hepatocyte reaggregation, 255–258
anti-gp105 antiserum, 256–258, 261
anti-gp105-2, 258–260
cell interactions, role in, 267
characterization, cf. non-malignant hepatocytes, 261–262
one-dimensional peptide maps, 261
sialic acid, 262
immunologic assay, 254
cf. L-CAM, ECCD-1 antigen, cell-CAM 120/80, uvomodulin, 254, 262–265, 267
endoglycosidase sensitivity, 262–263
immunofluorescence analysis, various rat tissues, 263
phosphorylation, 264

MAb 362.50, 255, 259–264
localization in intermediate junctions, nonmalignant cells, 265–267
PAGE, two-dimensional, 257, 258, 260
see also L-CAM
Cell density, and skin morphogenesis, ECM in, 329
Cell lineages and body plan, mammalian embryos, 73–93
α-amanitin treatment, 77
amniotic cavity, three types, 74–75
blastocyst, 73–74, 76, 80, 92
inner cell mass, 74, 76–78
trophectoderm, 74, 76, 77
chimeras, 80, 87
compaction, 76, 77
cytochalasin D treatment, 77
cytoplasmic reorganization, 76–77
direct cf. indirect analysis, 80–81, 84, 89
DNA probes, 76, 80
early nutritive connections, 73–74
epiblast development, 78–81, 92
kinetics and mesoderm, 78, 79
mitosis and cell proliferation, 78–80
primitive streak, 78
mitomycin C cytotoxic studies, 91–92
Mus musculus–Mus caroli in situ hybridization, 76
primitive streak, microsurgery on, mouse, 80–89
distal and posterior cells, 88–89
ectoderm grafting, 85–89, 92
fate maps, 81, 82, 84, 87
heart development, 84, 89
orthotopic posterior graft, 88
teratoma studies, 80, 91
tracing in situ, 81
X-inactivation mosaics, 77, 80, 89–92
PGK activity, 90
Cell migration, gastrulation, amphibians, 173–177; see also Neural crest cell migration
Cell patterning and skin morphogenesis, ECM in, 329
Cell proliferation
and epiblast development, 78–80
kidney morphogenesis, 356–357
Cell shape, gastrula cell adherence to FN-Sepharose beads, 285

Cell-substrate adhesion molecules, ECM, 170–173, 184
Cell-substratum attachment glycoprotein complex (CSATag; GP140 glycoprotein), 236
 anti-GP140, 244, 245
 cell surface proteolysis effect, 247
 CSAT complex, avian cells, 245–249
 CSAT MAb, 245, 249
 FN, 246, 247
 laminin, 246, 247
 monoblasts, 245–246
 talin, 249
 see also Extracellular matrix
Chemoaffinity, limited resolution, retinotectal map formation, 540–542, 563
Chick
 gastrula cell adherence to FN-Sepharose beads, 289
 retinotectal synapses, activity-dependent stabilization, 558
 and skin morphogenesis, ECM in, 320–322, 325, 326, 330, 334–342
 see also Synaptogenesis, regenerating neuromuscular junction, chick
Chimeras, cell lineages and body plan, mammalian embryo, 80, 87
Chondroitin sulfate, and neural crest cell migration, 179–181
Chordamesoderm
 cortical tractor model, epithelial folding, newt neural plate, 151
 gastrulation, convergent extension by cell intercalation, *Xenopus laevis*, 132–133
 induction, *Xenopus laevis*, cytoplasmic localization, 17, 25–27, 29–32
Chorio-allantoic membrane, quail, 358, 359
Cinemicrography. *See under* Microscopy/ micrography
Circulatory system, retrovirus insertional mutagenesis, 53
Circus or limnicola movement, gastrulation, *Xenopus laevis*, 134
Cobalt labeling, retinotectal map formation, *Xenopus laevis*, 523
Cockroach, larval, positional information, polar coordinate model, 458, 460, 465

Collagen
 -binding peptide of FN, gastrula cell adherence to FN-Sepharose beads, 290
 epithelial-mesenchyme transformation, 296, 299, 300, 306–307, 312
 gel suspension, and epithelial-mesenchyme transformation, 294, 296, 298, 301, 302–310, 312
 and neural crest cell migration, 179, 181, 183
 and skin morphogenesis, ECM in, 321–327, 343
 dermal cell cultures, 328–330, 332, 333
 see also under Retroviruses, insertional mutagenesis, mouse
Compaction, cell lineages and body plan, mammalian embryo, 76, 77
Competence, 63, 64, 66
Computer models/simulations
 neuronal selection mechanisms, neural mapping
 postsynaptic, 591–594
 presynaptic, 601–603
 presynaptic-postsynaptic modification interaction, 607–611
 retinotectal map formation, *Xenopus laevis*, 527–528
Contact inhibition, cortical tractor model, epithelial folding, 147, 157
Contractility and microvilli formation, actin role, 367, 370, 372
 autocatalytic, 372
Convergent extension. *See* Gastrulation, *Xenopus laevis*, convergent extension by cell intercalation
Corneal epithelium, avian, epithelial-mesenchyme transformation, 297–300
Cortical tractor model, epithelial folding, 143–150
 adhesion, 146, 148, 157
 apical seal, 148, 149
 preservation while cells interdigitate, 149–150
 contact inhibition, 147
 cortical cycling by motile cells, 145–146
 cortical cytogen flow, 145–146

dynamic nature of epithelial structures, 147–149
epithelial cf. mesenchymal cells, 143–145, 147, 161
microfilament bundles, 148
other applicable phenomena, 144
plasma membrane, 146
trigger by ionic stimuli, 146–147
see also Hydra, body column pattern formation; Microvilli formation, actin mechanical properties in
Cortical tractor model, epithelial folding, neural plate, newt, 151–163
apical seal, 152–153, 157, 159
bottle cells, 153
chordamesoderm, 151
contact inhibition, 157
cytogel, mechanical properties, 159–160
differential adhesion, 157
cf. early and late gastrula, 151
epidermis as organizing boundary, 158–160
keyhole stage, 152, 153, 158
neural crest, 155, 158
neural folds, 151, 154, 155, 158–162
neural tube, 151–152, 155, 158, 161–162
rolling, 161–163
notoplate, 150, 151, 154, 156, 161–162
elongation, 153, 156–159
interdigitation (lamellipodia), 153, 156–158
simulations, 159, 161–163
Cortico-cytoplasmic reorganization, induction, *Xenopus laevis*, cytoplasmic localization, 15–16, 29
Cricket, supernumerary regeneration, polar coordinate model, 465–467, 469–471
Crystallins, 304, 307, 312
Cutaneous appendages, morphogenesis, ECM in, 321, 326, 343
Cytochalasin D, 77
Cytodifferentiation, 296, 297, 299, 312
Cytogel
cortical tractor model, epithelial folding, newt neural plate, 159–160
spatially unstable, and microvilli formation, actin role, 370–373
Cytoplasm/cytoplasmic

localization, actin genes, muscle-specific, *Xenopus laevis* embryo, 37–38, 41–43
reorganization, cell lineages and body plan, mammalian embryo, 76–77
segregation, yellow ooplasm, 68
see also Induction, embryonic, *Xenopus laevis*, cytoplasmic localization

Danilchik's medium, 113, 121, 123, 128, 129, 140–141
Daphnia magna, 574
Dark-reared fish, retinotectal synapses, 552
Dermis-epidermis junction, 321–323, 325–326; *see also* Skin morphogenesis, ECM in
Desmogleins, 241
Determination. *See* Inductive interaction and determination, overview
Developmental field concept, cf. neuronal selection mechanisms, neural mapping, 573
Developmental genes. *See Drosophila* embryo, *engrailed* gene, segmental pattern formation
Dictyostelium discoideum, 150
Dispersed regulatory region, *Drosophila engrailed* gene, 504–507
DNA
binding, *Drosophila engrailed* gene product, 500–502, 508–511
microinjection, and retrovirus insertional mutagenesis, 48, 49, 54
cf. recombinant DNA, 48, 53, 54
probes
cell lineages and body plan, mammalian embryo, 76, 80
L-CAM, 201
N-CAM, 200–202
DNase I hypersensitive regions, retrovirus insertional mutagenesis, 51, 55
Drosophila
Antennapedia homeobox sequence, 509
leg segments, 444
mutations
pbx and *Ubx*, 505
scute, 505–506
positional information, polar coordinate model

Index 625

imaginal disc, 457, 467, 468, 471–473
  leg disc, 461
  see also Position-specific MAbs,
    Drosophila melanogaster imaginal
    wing disc
Drosophila embryo, engrailed gene,
  segmental pattern formation, 489–516
  antibodies vs. protein, 494–495
    immunofluorescent staining, 496–497
  and bx function, 492
  developmental compartments, 491
  dispersed regulatory region, 504–507
    RNA:RNA interactions, 506–507
  engrailed fusion proteins, 500, 507
    binding to 5' end engrailed and ftz
      genes, 502–503, 508, 509
    sequence-specific DNA binding,
      500–502, 508, 510
  evolutionary conservation, 492,
    498–500, 510–512
    assay for functionally important
      domains, 498–500
    D. virilis and D. melanogaster,
      498–500
  and ftz promoter gene, 490–492,
    500, 502–503, 512
  gene mapping, 493–494
  homeobox sequence shared domain,
    492, 495, 499–503, 508, 511, 512
  "pair-rule genes," 490, 513–515
  pleiotropic transcription regulator,
    507–512
    DNA binding, homeobox domain,
      508–511
  regulatory network of homeotic and
    segmentation genes, 512–516
  molecular prepatterns, 515
  periodicity encoding, 513–514
  positional information, 513
  regulatory role, 491, 492, 497, 504

Ectoderm
  gastrula cell adherence to FN-Sepharose
    beads, 280, 288–289
  grafting, microsurgery on mouse
    primitive streak, 85–89, 92
  hydra body column pattern formation,
    436–437

cell rearrangements, sphere-cylinder
  conversion, 439–441
Electrophoresis, two-dimensional gel, cell-
  CAM 105, 257, 258, 260
Electrophysiology. See under Retinotectal
  synapses, activity-dependent
  stabilization, goldfish
Endodoerm
  hydra body column pattern formation,
    436–437
  subblastoporal, gastrulation, Xenopus
    laevis, 114, 115, 120–122
Endoglycosidase sensitivity, various CAMs,
  262–263
Endothelium, kidney morphogenesis,
  350, 351
engrailed gene. See Drosophila embryo,
  engrailed gene, segmental pattern
  formation
Environmental effects, gene expression,
  epithelial-mesenchyme
  transformation, 294
Epiblast development, mammalian embryo,
  78–81, 92
Epidermis
  and epithelial folding, newt neural plate,
    158–160
  positional information, polar coordinate
    model, 457–460, 469
Epifluorescence microscopy, retinotectal map
  formation, Xenopus laevis, 529, 533,
  535
Epigenetic cf. preformistic view, 59, 61, 69
Epithelial cell(s)
  cell-CAM 128/80, 236, 238, 239, 242
  -specific antigen, hydra, MAb TS19 to,
    447, 448
  see also Cortical tractor model, epithelial
    folding entries
Epithelial-mesenchyme interaction
  kidney morphogenesis, as model system,
    350–352; see also Kidney
    morphogenesis
  and retrovirus insertional mutagenesis, 52
  and skin morphogenesis, 320
Epithelial-mesenchyme transformation, ECM
  effect, 293–313
  cell polarity, 295, 312

collagen gel suspension, 294, 296, 298, 301, 302
cytodifferentiation, 296, 297, 299, 312
electron micrography, 299, 303, 307, 308
environmental effects on gene expression, 294
epithelial-mesenchyme transformation in collagen gels, 302-310, 312
epithelial-mesenchyme transformation in embryo, 310-313
  hyaluronic acid, 312
  vimentin, primitive streak, 312
epithelium, characterization, 294-300
mesenchyme, characterization, 294-296, 300-302
SEM, 296
sialic acid, 295
EPSPs, retinotectal synapses, activity-dependent stabilization, 554-556, 563
Equatorial region developmental autonomy, induction, *Xenopus laevis*, 22-26, 30-31
  rescue capacity increases during cleavage, 27
  transplant, schema, 24
Equilibrium, mechanical, actin role in microvilli formation, 369
Evolutionary conservation, *Drosophila engrailed* gene, 492, 498-500, 510-512
  assay for functionally important domains, 498-500
Extracellular matrix, 52
  chemical inductors, 60, 65, 67
  kidney morphogenesis, 354
  and neural crest cell migration, 178, 181, 183-185
  see also Epithelial-mesenchyme transformation, ECM effect; Skin morphogenesis, ECM in; specific components

Fate maps, 81, 82, 84, 87, 115; *see also* Blastoderm, avian, morphogenetic movements and fate maps
Feather morphogenesis, ECM in, 321, 322, 327, 343; *see also* CAMs in feather histogenesis

Fiber self-ordering, retinotectal map formation, 542-543
Fibronectin, 171-173
  anti-FN antibodies prevent gastrulation, 272
  cell migration, gastrulation, 173-175
    receptors, 175, 176
  cell-substratum attachment glycoprotein complex, 246, 247
  neural crest cell migration, 179-183, 185
    receptors, 183
  receptors, 171
  RGDS peptide, 172-173
  and skin morphogenesis, ECM in, 321, 322, 325, 343
    dermal cell cultures, 328-330, 332, 333, 343
  see also Gastrula cell adhesion to fibronectin-Sepharose beads, amphibians
Fibrous network, actomyosin gel, 367-369
Field potentials, retinotectal synapses, activity-dependent stabilization, 546
Filopodia
  actin-rich, epithelial-mesenchyme transformation, 301-303
  hydra body column pattern formation, 442
  and neural crest cell migration, 182
Fluorescein
  -dextran-amine
    gastrulation, *Xenopus laevis*, 113, 123
    induction, *Xenopus laevis*, cytoplasmic localization, 16, 21, 25, 26
    retinotectal map formation, *Xenopus*, 529-532, 535
  -labeled α-bungarotoxin, motor endplate morphogenesis, 407, 409, 411
  motor endplate morphogenesis, 407, 409, 411
Fluorescent cell lineage tracers, retinotectal map formation, *Xenopus laevis*, 529-532, 535
Force balance equation, microvilli formation, actin role, 379-380
ftz and *engrailed* genes, *Drosophila*, 490-402, 500, 502-503, 508, 509, 512

GAGs
  epithelial-mesenchyme transformation, 300
  and skin morphogenesis, 321, 322, 324, 327, 343
Gap junctions, 32, 43, 60, 239, 241
Gastrula cell adhesion to fibronectin-Sepharose beads, amphibians, 271–290
  Ambystoma maculatum, 272–273, 276, 279
  anti-FN antibodies prevent gastrulation, 272
  archenteron, 280, 285
  blastocoelic roof, 272, 280, 282, 285, 288–290
  cf. blastula cells, 278–281
  cell-binding cf. collagen-binding peptide of FN, 290
  cell shapes, 285
  cf. chick, 289
  kinetics, 276–277
  lamellipodia, 285–288
  methods, 273–275
    Ca^{2+}-, Mg^{2+}-free medium, 273
    light microscopy, 275, 282–283, 285
    SEM, 275, 284–285, 286–287
    time-lapse cinemicrography, 275, 287–288
  cf. neural crest cell migration, 289
  normal cf. hybrid embryos, 273, 276–277, 279–282, 288, 289
  cf. other types of beads, 281
  prospective ectoderm and mesoderm, 280, 288–289
  Rana
    catesbiana, 273, 280, 288
    pipiens, 272–273, 276, 279, 280, 288
    sylvatica, 272, 273, 276, 279
  various developmental stages, 277–278
Gastrulation
  cell migration, amphibians, 173–177
  morphogenetic movements during, 62
  neural induction during, 1, 2–6, 8, 10
  signal substances, 9
  schema, 151
Gastrulation, *Xenopus laevis*, convergent extension by cell intercalation, 111–136, 140–141
  archenteron, 115, 122
  blastocoelic roof, 129–130, 134–135
  gastrulation without, 118–126
  blastopore closure, 112, 116, 121, 129–130
  bottle cells, 112, 116, 119, 120
  chordamesoderm, 132–133
  classical evidence for, 128–131
  function of cell intercalation, 133–134
    limnicola or circus movement, 134
    radial cf. circumferential intercalation, 133
  function of convergent extension, 131–133
  function of specific regional processes, 111–112
  involuting marginal zone, 114–136
  mechanism of convergent extension, 122–128
    deep cells, 122–123, 133
    intercalation by protrusion, 126, 128, 130, 133, 135–136
  mesodermal cell migration, 112, 116, 134–135
  microsurgery, cell labeling, explant culture, 113
    Danilchik's medium, 113, 121, 124, 128, 129, 140–141
    FDA, 113, 123
    "sandwich" explants, 113, 117, 119, 121, 124, 126, 129, 132, 133
  morphogenetic movements, 112, 114–116
  fate map, 115
  neural plate, 132
  neurula stage, 118–120, 122, 124–127, 129
  SEM, 114, 119
  subblastoporal endoderm, 114, 115, 120–122
  time-lapse videomicrography and cinemicrography, 114, 118, 124, 130
Gene mapping, *Drosophila engrailed* gene, 493–494
Germ cells, primordial, and germ plasm, 67–69
Glioma cell line G26-20, 229
Glycoprotein

cell-surface. *See under* Position-specific MAbs, *Drosophila melanogaster* imaginal wing disc
identification, cell-CAM 128/80, 236, 237
myelin-associated, 204–205
see also CAMs *entries*; Cell-substratum attachment glycoprotein complex
Glycosaminoglycans
epithelial-mesenchyme transformation, 300
and skin morphogenesis, 321, 322, 324, 327, 343
Glycosylation, Ng-CAM, 204
Gray code, insect embryo, positional information, 428, 429
Growth cones, motor endplates morphogenesis, 402

Haementeria ghilianii, 574
Hair cells, inner ear, cf. microvilli formation, actin role, 365, 374
Hair morphogenesis, ECM in, 321–325, 327, 343
Head, hydra. *See* Hydra, body column pattern formation
Heart development, microsurgery on mouse primitive streak, 84, 89
Hebb rule, 604
Hepatocellular carcinoma. *See* Cell-CAM 105, rat hepatocellular carcinoma
Homeoboxes. *See under Drosophila* embryo, *engrailed* gene, segmental pattern formation
HRP labeling
retinotectal map formation, 523, 542
retinotectal synapses, activity-dependent stabilization, 546, 548, 558, 560
HSPG
epithelial-mesenchyme transformation, 299, 301
motorend plate morphogenesis, 394
Hyaluronate
neural crest cell migration, 179, 181, 183
skin morphogenesis, ECM in, 321–322
Hyaluronic acid, epithelial-mesenchyme transformation in embryo, 312

Hybridization in situ, *Mus musculus–Mus caroli*, 76
Hydra, body column pattern formation, 435–452
head effect on body column dimensions, 448–449
autocatalytic process, 449
inhibition, 449
prepattern model, 449–451
head morphogenesis, 444–448
head as organizer, 445, 450
MAb CP8 to head-specific antigen, 445–448
MAb TS19 tentacle-epithelial-cell-specific antigen, 447, 448
prepattern model, 435–436, 441, 445, 449–451
regeneration, 437–438
sphere-cylinder conversion, 438–441, 450
ectodermal cell rearrangements, 439–441
head-body proportions, 440, 451
size of original piece, 440, 451
sphere-cylinder conversion, nonprepattern model, 441–444
differential forces, adhesion and strain, 442–444
filopodia and contraction, 442
two-layered structure (endoderm and ectoderm), 436–437
basement membrane, 437
hypostome and tentacles, 437
Hyla regilla, 118
Hypostome, hydra body column pattern formation, 437

Imaginal disc. *See* Position-specific MAbs, *Drosophila melanogaster* imaginal wing disc
Immune system, cf. neural system, recognition function, 572
Immunoelectrophoresis, N-CAM in motor endplate morphogenesis, 405, 407, 409
Immunofluorescence
Drosophila engrailed gene, 496–497
indirect
N-CAM in motor endplate morphogenesis, 406–407, 410, 411

Index   629

and skin morphogenesis, ECM in, 320–323, 325, 330–331, 333
position-specific MAbs, *Drosophila melanogaster* imaginal wing disc, 480, 482
  see also CAMs in neural development, immunofluorescence
Immunohistology, kidney morphogenesis, 351, 359
Index of axis deficiency, induction, *Xenopus laevis*, cytoplasmic localization, 21, 23
Induction, embryonic
  kidney morphogenesis, 354
  cf. neuronal selection mechanisms, neural mapping, 572
  search for chemical inductors, 59–60
  see also Actin genes, muscle-specific, *Xenopus laevis* embryo
Induction, embryonic, *Xenopus laevis*, cytoplasmic localization, 15–32
  accumulation of α-actin mRNA, 30–31
  axial structure determinants, 16–17
  axis formation, rescue by vegetal cell transplantation, 18–22, 27, 30–31
    schema, 19
  axis truncation, 16
  cell lineage tracers, 16
  fluorescein-dextran-amine, 21, 25, 26
  cortico-cytoplasmic reorganization, 15–16, 29
  embryonic organizer (chordamesoderm), 17, 25–27, 29–32
  equatorial region, developmental autonomy, 22–26, 30–31
  rescue capacity increases during cleavage, 27
  transplant, schema, 24
  index of axis deficiency, 21, 23
  injection of RNA synthesis inhibitor, amanitin-based, 32
  neural plate formation, 30
  rescue activity, complete distribution in donor embryos, 27–29
  role of intercellular communication, 32
  sperm entry point, 19, 24
Inductive interaction and determination, overview, 1–10, 59–69

  action cf. reaction system, 63–65
  animal-vegetal axis, 61–62
  competence, 63, 64, 66
  directive cf. permissive, 3, 4
  Fucus egg, 61
  gap junctions, 32, 43, 60
  gastrulation, morphogenetic movements during, 62
  germ plasm and primordial germ cells, 67–69
  heterotypic inductors, listed, 8, 10
  meso-endoderm, 62–65
  messages or signals, release and recognition, 5–9, 60–61, 64
  metanephron into kidney tubules, 1, 4–6, 8–10
  molecular determinants, terminological problems, 66–68
  implication of specificity, 67–68
  mosaic eggs, 61–62, 69
  neural crest, 1–6, 8–10, 62, 64, 65
  preformistic cf. epigenetic views, 59, 61, 69
  pluripotentiality, 2, 63, 65, 68
  reciprocal interactions, 61
  search for chemical inductors, 59–60
    ECM, 60, 65, 67
    cf. specificity in responding system, 60
  supporting interactions, 63
  yellow ooplams, cytoplasmic segregation, 68
Inner ear hair cells, cf. microvilli formation, actin role, 365, 374
Intercalary regeneration, positional information, polar coordinate model, 459–461, 463; *see also* Gastrulation, *Xenopus laevis*, convergent extension by cell intercalation
Intercellular junctions, cell-CAM 128/80, 238–243; *see also* Junctions
Interdigitation, newt notoplate, 153, 156–158
Intermediate filaments
  cyto(alpha)keratin, 295, 310, 312
  vimentin, epithelial-mesenchyme transformation, 296, 310
Interphase between mesenchyme and inducer, kidney morphogenesis, 352–353

Involuting marginal zone, gastrulation,
    Xenopus laevis, 114–136
Ion channels, 414, 586, 587, 591, 592,
    595–597; *see also* Calcium
Ionic stimuli, cortical tractor model,
    epithelial cells, 146–147
Iron oxide markers, avian blastoderm, 100

Junctions
  gap, 32, 43, 60, 239, 241
  intercellular, cell-CAM 128/80, 238–243
  intermediate, and cell-CAM 105,
    hepatocellular carcinoma, 265–267
  see also Synaptogenesis, regenerating
    neuromuscular junction, chick

Keratin, 215, 295, 310, 312
Keyhole stage, newt neural plate,
    152, 153, 158
Kidney morphogenesis, 349–360
  angiogenesis, 357–360
  transplantation into quail chorio-
    allantoic membrane, 358, 359
  CAMs, 238–240, 354–355
  cell-cell interaction, 350
  cell-ECM adhesion, 354
  cell proliferation, 356–357
    transferrin-dependent, 356, 357
  endothelium, 350, 351
  epithelial ureter bud, 350, 351
  immunohistology, 351, 359
  interphase between mesenchyme and
    inducer, 352–353
  as model system, epithelial-mesenchyme
    interaction, 350–352
  nephrogenic mesenchyme, 350–352,
    357, 358
  response to induction, 354
  tubule formation, 1, 4–6, 8–10, 351, 352

Lamellipodia, gastrula cell adherence to FN-
    Sepharose beads, 285–288
Laminin, 170–171
  and cell-substratum attachment
    glycoprotein complex, 246, 247
  epithelial-mesenchyme transformation,
    296, 299, 301, 307
  and neural crest cell migration, 179, 181,
    183, 184

and skin morphogenesis, ECM in, 321
synaptogenesis, regenerating
    neuromuscular junction, 388–389
Lens
  avian anterior, epithelial-mesenchyme
    transformation in collagen gels,
    305, 306, 309
  and cadherins, 230, 231
Limb
  bud, chick. *See under* Positional
    information and pattern formation
  fore- and hind-, divergence, 431–432
Limnicola or circus movement, gastrulation,
    Xenopus laevis, 134
Linear analysis and pattern selection,
    microvilli formation, actin role,
    381–384
Liver. *See* L-CAM; Cell-CAM 105, rat
    hepatocellular carcinoma
Locusta migratoria, 574

Mapping, neural. *see* Retinotectal map
    formation *entries*
Mass balance equations, microvilli
    formation, actin role, 377–379
Mathematical models
  and microvilli formation, actin role, 367,
    377–384
  neuronal selection mechanisms, neural
    mapping, 588–591
  see also Computer models/simulations
MCF-7 breast cancer cell line, 237
Mesenchyme
  cells, cortical tractor model, cf. epithelial
    cells, 143–145, 147, 161
  and inducer, interphase between, kidney
    morphogenesis, 352–353
  see also Epithelial-mesenchyme *entries*
Mesoderm
  cells, and cell migration, gastrulation,
    173, 175
  and epiblast development, 78, 79
  gastrula cell adherence to FN-Sepharose
    beads, 280, 288–289
  gastrulation, *Xenopus laevis*, cell
    migration, 112, 116, 134–135
Meso-endoderm, 62–65
Mesoglea, hydra body column pattern
    formation, 437

Index 631

Metanephric induction into kidney tubules, 1, 4–6, 8–10
Methylation, *de novo*, retrovirus insertional mutagenesis, 50, 55
Microelectrodes, retinotectal map formation, *Xenopus laevis*, 523
Microfilament(s)
 bundles, cortical tractor model, epithelial cells, 148
 and skin morphogenesis, 338
Microscopy/micrography
 cinemicrography
  avian blastoderm, 100
  gastrula cell adherence to FN-Sepharose beads, 275, 287–288
  gastrulation, *Xenopus laevis*, 114, 118, 124, 130
 electron
  blastoderm, avian, morphogenetic movements and fate maps, TEM and SEM, 107
  epithelial-mesenchyme transformation, ECM effect, SEM, 296
  gastrula cell adhesion to FN-Sepharose beads, amphibians, SEM, 275, 284–287
  gastrulation, *Xenopus laevis*, SEM, 114, 119
  motor endplate morphogenesis, mouse, genetic abnormalities, 398
 epifluorescence, retinotectal map formation, *Xenopus laevis*, 529, 533, 535
 light, gastrula cell adhesion to FN-Sepharose beads, amphibians, 275, 283–283, 285
 videomicrography
  gastrula cell adhesion to FN-Sepharose beads, amphibians, 275, 287–288
  gastrulation, *Xenopus laevis*, 114, 118, 124, 130
Microsurgery. *See* Cell lineages and body plan, mammalian embryos; Gastrulation, *Xenopus laevis*, convergent extension by cell intercalation

Microvilli formation, actin mechanical properties in, 365–374, 377–384
 calcium, 367, 368, 374
 constant polymerization/depolymerization, 370
 contractility, 367, 370, 372
  autocatalytic, 372
 cytogel spatially unstable, 370–373
 fibrous network of actomyosin gel, 367–369
 hexagonal packing, 365–367, 373, 374
 mathematical model, 367, 377–384
  force balance equation, 379–380
  linear analysis and pattern selection, 381–384
  mass balance equations, 377–379
  model equation, 377
  simplified model system, 380–381
 mechanical equilibrium, 369
 mechanochemical properties, viscoelastic, 369–370
 osmotic pressures initiate, 370–372, 374
 sliding filament, 369
 cf. stereocilia, inner ear hair cells, 365, 374
 see also Cortical tractor model, epithelial folding *entries*
Mitomycin C, 91–92
Mitosis and epiblast development, 78–80
Molecular determinants, terminological problems, 66–68
 implication of specificity, 67–68
Molecular specificity, neural crest cell migration, 184–185
Moloney leukemia provirus, 48, 55
Monoclonal antibodies
 362.50, and cell-CAM 105, hepatocellular carcinoma, 255, 259–264
 and cell-substratum attachment glycoprotein complex, 245, 249
 CP8, to hydra head-specific-antigen, 445–448
 6D4 and 2F6, synaptogenesis, regenerating neuromuscular junction, 387–389
 ECCD-1, and cadherins, 224–226, 229, 254

NCD-1, and cadherins, 224–227, 229, 230
TS19, to hydra tentacle-epithelial-cell-specific antigen, 447, 448
see also Position-specific MAbs, *Drosophila melanogaster* imaginal wing disc
Mosaic(s)
  eggs, 61–62, 69
  X-inactivation, 77, 80, 89–92
  PGK activity, 90
Motoneurone, 401, 414
Motor endplate morphogenesis, mouse, 393–416
  AChE, 395, 396, 403, 409, 411–416
  AChR, 395, 396, 402, 403, 406, 409, 412, 414–416
  basal lamina, synaptic, 395, 396, 402, 403, 411, 412, 415
  ^{125}I-α-bungarotoxin, 396, 402, 413
  N-CAM role, 403–411, 415–416
  cf. adult motor endplate, 404–407
  expression, nerve control, 409–411
  properties in neuron and muscle listed, 414, 415
  central issues in nerve-muscle interactions, listed, 412
  migration and differentiation, 412
  synapse adhesion, 412, 413
  early neuromuscular system development, 395–398
  axon-myotube, early contacts, 396
  axon terminal arborization, 397
  FITC, 407, 409, 411
  genetic abnormalities, 398–403
  motor endplate disease *(med)*, 398–414
  muscular dysgenesis *(mdg)*, 399, 401–403, 413–414
  HSPG, 394
  motoneurone, 401, 414
  muscle inactivity, 414
  Na$^+$ channels, voltage-dependent, 414
  node of Ranvier, 397, 398, 401
  polarity modulation, 416
  Schwann cell, 394, 401, 413
  silver nitrate impregnation technique, 397, 399, 401

Mov-13 mutant mouse strain, 48–50
Multipotent target cells, 2, 63, 65, 68
Muscle. *See* Actin genes, muscle-specific, *Xenopus laevis* embryo; Motor endplate morphogenesis, mouse
Muscular dysgenesis *(mdg)*, motor endplate morphogenesis, 399, 401–403, 413–414
Mus musculus–Mus caroli in situ hybridization, 76
Mutagenesis. *See* Retroviruses, insertional mutagenesis, mouse
Mutations, developmental. *See under* Motor endplate morphogenesis, mouse; Position-specific MAbs, *Drosophila melanogaster* imaginal wing disc
Myelin-associated glycoprotein, 204–205
Myoblasts and cell-substratum attachment glycoprotein complex, 245–246

NCD-1 MAb, 224–227, 229, 230
Nephrogenic mesenchyme, 350–352, 357, 358
Neural crest, 62, 64, 65
  cortical tractor model, newt, 155, 158
Neural crest cell migration, 178–184
  CAMs, 179, 184–185
  chondroitin sulfate, 179–181
  collagens I and III, 179, 181, 183
  directional migration mechanisms, 182–183
  ECM, 178, 181, 183–185
  final localization and aggregation, 183–184
  FN, 179–183, 185
  receptors, 183
  gastrula cell adherence to FN-Sepharose beads, 289
  hyaluronate, 179, 181, 183
  laminin, 179, 181, 183, 184
  molecular specificity in regulating, 184–185
  numerous filopodia, 182
  patterns, migratory, 179–180
  plasminogen activator synthesis, 179
  separation from neural tube, 178–179
Neural folds, cortical tractor model, newt, 151, 154, 155, 158–162

Index   633

Neural induction during gastrulation,
    1, 2–6, 8, 10
  signal substances, 9
Neural mapping. *See* Neuronal selection
    mechanisms in neural mapping;
    Retinotectal map formation *entries*
Neural plate
  formation, induction, *Xenopus laevis*,
    cytoplasmic localization, 30
  gastrulation, convergent extension by cell
    intercalation, *Xenopus laevis*, 132
  see also Cortical tractor model, epithelial
    folding, neural plate, newt
Neural tube
  and cadherins, 230, 231
  cortical tractor model, newt, 151, 152,
    155, 158, 161–162
  rolling, 161–163
  formation and CAMs in neural
    development, 208, 209
Neuromuscular junction. *See* Motor endplate
    morphogenesis, mouse;
    Synaptogenesis, regenerating
    neuromuscular junction, chick
Neuronal selection mechanisms in neural
    mapping, 571–612
  afferent arbor overlap, 581–583
  CAMs, 572, 574
  degeneracy, 573
  cf. developmental field concept, 573
  cf. embryonic induction, 572
  group
    competition, 578, 583–584
    confinement, 578–581
    selection, 578, 581–583
  cf. immune system, recognition function,
    572
  layer IV spiny stellate cells, 580–581,
    584
  postsynaptic rule, 586–597
    biochemistry of modification,
      594–595
    Ca^{2+}, 594–596
    computer results, 591–594
    developmental aspects, 596–597
    efficacy, 591–592
    cf. excitatory/inhibitory division, 594
    formal mathematical model, 588–591
    ion channels, 586–587, 591, 592,
      595–597
    kinetics, 588, 591
    protein kinases, 594
    state-dependent modifications,
      586, 588, 594
    timing constraints, 590, 592
  presynaptic-postsynaptic modification
    interaction, 605–611
    classes of group connections,
      606, 608
    computer simulations, 607–611
    formal analysis, 605–606
    short- and long-term changes,
      605–612
  presynaptic rule, transmitter release,
    597–604
    basis for, 603–604
    computer results, 601–603
    facilitation cf. depression, 598–600
    Hebb rule, 604
    long-term modification, 600–601,
      603, 604
  pyramidal cells, 581
  silent synapses, 582
  somatosensory area I, 579–580
  somatosensory system, cats and monkeys,
    575–577
    receptive field size and overlap,
      577–578
  synaptic modification model, 584–585
  variability required, 572
    anatomic, 572–575
    functional, 575–578
Neurula stage, convergent extension by cell
    intercalation, *Xenopus laevis*,
    118–120, 122, 124–127, 129
Neurulation, 62, 64, 65; *see also* Cortical
    tractor model, epithelial folding,
    neural plate, newt
Newt. *See* Cortical tractor model, epithelial
    folding, neural plate, newt
Nitroblue tetrazolium stain, 480
Node of Ranvier, 397, 398, 401
Notoplate, cortical tractor model, epithelial
    folding, newt, 150, 151, 154, 156,
    161–162
  elongation, 153, 156–159

interdigitation, 153, 156–158
Nutritive connections, early, mammalian embryo, 73–74

Ocular dominance patches, 559–561
Oligodendroglia and CAMs in neural development, 211
Ooplasm, yellow, 68
Optic arbor stainng, normal cf. regenerated, 546–549, 554
Osmotic pressure and microvilli formation, actin role, 370–372, 374
Owl, auditory space map, 561

"Pair-rule genes," *Drosophila* embryo, 490, 513–515
Pattern formation. See *Drosophila* embryo, *engrailed* gene, segmental pattern formation; Hydra, body column pattern formation; Positional information *entries*; Retinotectal map formation, cell patterning, *Xenopus laevis*
Pattern selection, microvilli formation, actin role, 381–384
pbx mutation, *Drosophila*, 505
PCC3 cell line and cadherins, 228
Peptide maps, cell-CAM 105, 261
Preformistic cf. epigenetic view, 59, 61, 69
Pericanalicular area and cell-CAM 105, 265
Periodicity encoding, *Drosophila engrailed* gene, 513–514
PGK activity and X-inactivation mosaics, 90
Pharyngula, 106
Phosphorylation
  CAMs, 204, 264
  and encoding of positional information, 427
  protein kinases, neural mapping, 594
Pigment, avian, 429, 431
Plasma membrane, cortical tractor model, 146
Plate morphogenesis. See Motor endplate morphogenesis, mouse
Pleiotropic transcription regulation, *Drosophila engrailed* gene, 507–512
  DNA binding, homeobox domain, 508–511
Pleurodeles waltlii, 272

Pluripotent target cells, 2, 63, 65, 68
Polarity
  and epithelial-mesenchyme transformation, ECM effect, 295, 312
  modulation
  and CAMs in neural development, 211
  motor endplate morphogenesis, 416
  positional information and pattern formation, chick limb bud, 424–426
  and skin morphogenesis, ECM in, 338
  see also Positional information, polar coordinate model, insect leg
Polymerization/depolymerization, actin, microvilli formation, 370
Positional information
  Drosophila engrailed gene, 513
  retinotectal map formation, *Xenopus laevis*, 524
Positional information and pattern formation, 423–432
  chick limb bud, 424–427
    cartilaginous elements, 424–427
    morphogen, diffusible, 424, 425, 427
    polarizing region, 424–426
    retinoic acid effect, 425, 427
  divergence of similar structures (vertebrae, fore- and hindlimbs), 431–432
  encoding, 427–428
    insect embryo, Gray code, 428, 429
    phosphorylation, 427
  feather patterns, 431
  interpretation, positional values, 428–431
    avian pigment formation, 429, 431
  modes, 426–427
  wave-like pattern, 424
Positional information, polar coordinate model, insect leg, 456–473
  different coordinate systems, 463–465
    cf. Cartesian, 464–465
  distal regeneration, 461–463
    amputation, 461, 462
  Drosophila imaginal disc, 457, 461, 467, 468–473
  epidermis, central role, 457–460, 469

intercalary regeneration, 459-461, 463
  larval cockroach, schema, 458, 460, 465
  supernumerary regeneration, 465-467,
      469-471
    cricket, left and right grafts, 466, 467
    temperature, 467
    Tenebrio, 455, 468
  and thoracic segment, 467-472
    compartments, 471
Position-specific MAbs, *Drosophila
    melanogaster* imaginal wing disc,
    477-487
  aldehyde oxidase, 479
  cell-surface glycoproteins, 478-479, 483
  PS1 MAb CF.5E5, 479, 480,
      482-484
  PS2 MAb CF.2C7, 479, 480,
      482-486
  PS3 MAb CF.6G11, 479, 483
  developmental mutations, 478, 481
    apterous, 486, 487
    engrailed, 485, 487
    lethal (2) giant discs (l(2)gd), 479,
        481, 482, 486
    Minute, 479, 481, 482, 486
  immunofluorescence, 480, 482
  nitroblue tetrazolium stain, 480
  cf. other *Drosophila* species, 479, 480,
      482-484
Posterior cells, microsurgery on mouse
    primitive streak, 88-89
Postsynaptic rule. *See under* Neuronal
    selection mechanisms in neural
    mapping
Prechordal plate, avian blastoderm, 105
Prepattern
  Drosophila engrailed gene, 515
  hydra body column pattern formation,
      435-436, 441, 445, 449-451
    head effect, 449-451
Presynaptic rule. *See under* Neuronal
    selection mechanisms in neural
    mapping
Primitive streak
  avian blastoderm, 103-107
  and CAMs in neural development, 209
  epithelial-mesenchyme transformation in
      embryo, 312

  see also under Cell lineages and body
      plan, mammalian embryos
Primordial germ cells, 67-69
Protein kinases, neuronal selection
    mechanisms, neural mapping, 594
Proteoglycans, 296, 299, 321; *see also*
    specific proteoglycans
Proteolysis, cell-surface, 247
Pyramidal cells, neuronal selection
    mechanisms, neural mapping, 581

Quail
  -chick xenograft technique, 100, 107
  chorio-allantoic membrane, 358, 359

Rana, gastrula cell adherence to FN-
    Sepharose beads
  R. catesbiana, 273, 280, 288
  R. pipiens, 272, 273, 276, 279, 280, 288
  R. sylvatica, 272, 273, 276, 279
Reaction cf. action systems, 63-65
Reciprocal interactions, induction, 61
Recombinant DNA and retrovirus insertional
    mutagenesis, 48, 53, 54
Regeneration
  hydra body column pattern formation,
      437-438
  optic arbor staining, retinotectal
      synapses, activity-dependent
      stabilization, 546-549, 554
  retinotectal map formation, *Xenopus
      laevis*, 524
  see also Synaptogenesis, regenerating
      neuromuscular junction, chick;
      under Positional information,
      polar coordinate model, insect leg
Regulatory region, dispersed, *Drosophila
    engrailed* gene, 504-507
RNA:RNA interactions, 506-507
Rescue, induction, *Xenopus laevis*,
    cytoplasmic localization, 27-29
Retinoic acid, 425, 427
Retinotectal map formation, 540-545
  adhesion, 540
  early diffuse projections in regeneration,
      543-545
  anatomy, 543-544
  electrophysiology, 544-545
  fiber self-ordering, 542-543

HRP, 542
limited resolution of chemoaffinity, 540–542, 563
WGA-HRP, 544
Retinotectal map formation, cell patterning, *Xenopus laevis*, 521–536, 558
assays for retinotectal projection, 522–523
cell adhesive interactions in nerve patterning, 526–529, 534–535
dorsoventral topography, 532, 533
dynamic process, 525–526
nasotemporal topography, 532, 534
noninvasive anatomic assay, 529, 530
epifluorescence microscopy, 529, 533, 535
fluorescent cell lineage tracers, 529–532, 535
grafts, 530, 531
positional information, 524
regenerating, 524
Retinotectal synapses, activity-dependent stabilization, goldfish, 545–563
α-bungarotoxin, 545, 551, 555, 557, 561
cf. cat, 557, 561–562
correlated activity model, 555–558
diffuse projections in development, 558–559
cf. rodents, 559
cf. *Xenopus* and chick, 558
electrophysiology, 545–546, 549–554
correlation with anatomy, 555
dark-reared fish, 552
sensitive period, 551–552
strobe-reared fish, 552–554
synapse formation, 552
synaptic transmission, 554
TTX-blocked fish, 549–552, 554
EPSPs, 554–556, 563
field potentials, 546
HRP, 546, 548, 558, 560
optic arbor staining, normal cf. regenerated, 546–549, 554
cf. other activity maps, 561–562
cf. peripheral mechanisms, 562
retraction of ineffective synapses, 561
segregation of ocular dominance patches, 559–561

TTX, 555, 557, 558, 560–562
Retroviruses, insertional mutagenesis, mouse, 47–55
collagen I alpha 1 gene, 48, 50–51, 54
de novo methylation, 50, 55
DNase I hypersensitive regions, 51, 55
molecular analysis of mutation, 50–51
collagen I, role in early embryonic development, 52–53
microinjection of DNA, 48, 49, 54
cf. recombinant DNA, 48, 53, 54
Mov-13 mutant mouse strain, 48–50, 55
mutations, listed, 54
RGDS peptide, fibronectin, 172–173
RNA
messenger (mRNA)
α-actin, induction, *Xenopus laevis*, cytoplasmic localization, 30–31
cDNA probes for L-CAM and N-CAM, 200–202
:RNA interactions, *Drosophila engrailed* gene, 506–507
synthesis inhibitor, induction, *Xenopus laevis*, cytoplasmic localization, 32
Rous sarcoma virus, src gene, 54, 55

Sandwich culture
gastrulation, *Xenopus laevis*, 113, 117, 119, 121, 124, 126, 129, 132, 133
and skin morphogenesis, ECM in, 339
Scale morphogenesis, ECM in, 321, 327, 343
Schwann cell, motor endplate morphogenesis, 394, 401, 413
scute mutation, *Drosophila*, 505–506
Segmental pattern. See *Drosophila* embryo, *engrailed* gene, segmental pattern formation
Sensitive period, retinotectal synapses, 551–552
Sepharose. See Gastrula cell adhesion to fibronectin-Sepharose beads, amphibians
Sialic acid
and cell-CAM 105, hepatocellular carcinoma, 262
and epithelial-mesenchyme transformation, ECM effect, 295

Index 637

N-CAM, 198, 200, 202, 206
Signals, induction, 5-9, 60-61, 64
Silver nitrate impregnation technique, motor endplate morphogenesis, 397, 399, 401
Skin morphogenesis, ECM in, 319-344
  chick, 320-322, 325, 326, 330, 334-342
  cutaneous appendages, 321, 326, 343
  dermal cell cultures, three-dimensional hydrated collagen gel, 320, 327,, 333-344
    cell locomotion, 339, 342-343
    dissociated cells vs. explants, 334, 336, 339, 341
    fibronectin, 343
    microfilaments, 338
    polarization, 338
    sandwich culture, 339
    ultrastructure, 336-338
  dermal cell cultures, two-dimensional, 327-333, 340, 344
    cell locomotion, 328, 332
    cell patterning and density, 329
    dissociated cells vs. explants, 327, 333
    fibronectin and collagen, 328-330, 332, 333
  dermis-epidermis junction, 321-323, 325-326
  distribution of ECM components, 320-327
  epithelial-mesenchyme interaction, 320
  feather, 321, 322, 327, 343
  hair, 321-325, 327, 343
  indirect immunofluorescence, 320-323, 325, 330-331, 333
  mouse, 320-324
  scale, 321, 327, 343
Sliding filament and microvilli formation, actin role, 369
Sodium channels, voltage-dependent, motor endplate morphogenesis, 414
Somatosensory area I, neuronal selection mechanisms, neural mapping, 579-580
Somatosensory system, cats and monkeys, neuronal selection mechanisms, neural mapping, 575-577

receptive field size and overlap, 577-578
Sperm entry point, induction, *Xenopus laevis*, 19, 24
Spinal cord development, 208, 210-212
Spiny stellate cells, 580-581, 584
Stabilization. *See* Retinotectal synapses, activity-dependent stabilization, goldfish
State-dependence of modification, neuronal selection mechanisms, 586, 588, 594
Stereocilia cf. microvilli formation, actin role, 365, 374
Strobe-reared fish, retinotectal synapses, 552-554
Subblastoporal endoderm, gastrulation, *Xenopus laevis*, 114, 115, 120-122
Symmetrization, avian blastoderm, 100, 106-107
Synaptogenesis, regenerating neuromuscular junction, chick, 385-389
  AChR and AChE aggregating factor, 386-389
  MAbs 604 and 2F6, 387-389
  α-bungarotoxin, rhodaminated, 386, 388
  laminin, 388-389
  synaptic basal lamina, *Torpedo californica*, 385-389
  see also Motor endplate morphogenesis, mouse; Neuronal selection mechanisms in neural mapping; Retinotectal *entries*

Talin, 249
Temperature
  incubation, avian blastoderm, 102
  positional information, polar coordinate model, 467
Tenebrio, positional information, 455, 468
Tentacle, hydra. *See* Hydra, body column pattern formation
Teratoma studies, cell lineages and body plan, mammalian embryo, 80, 91
Thyroid follicles, epithelial-mesenchyme transformation in collagen gels, 309-310
Torpedo californica, 385-389
Transcription regulation, *Drosophila engrailed* gene, 507-512

DNA binding, homeobox domain,
    508–511
Transferrin-dependent cell proliferation,
    kidney morphogenesis, 356, 357
Transfilter experiments, induction,
    mechanism, 7
Transmitter release. *See under* Neuronal
    selection mechanisms in neural
    mapping
Triton
    cristatus, 151
    taeniatus, 152
TTX, 552, 555, 557–558, 560–562, 597
Tunicamycin, 198, 200, 202

Ubx mutation, *Drosophila*, 505
Ureter bud, epithelial, kidney
    morphogenesis, 350, 351
Uvomorulin, 202, 237, 239, 242, 254,
    262–265, 267, 355; *see also* L-CAM

Vegetal cells
    actin genes, *Xenopus laevis* embryo,
        activation, 40, 43
    transplantation, axis formation rescue,
        Xenopus laevis, 18–22, 27, 30–31
    schema, 19
Vertebrae, 431–432
Vesicles, N-CAM, 206–207
Videomicrography. *See under* Microscopy/
    micrography

Vimentin, 296, 310, 312
Vinculin, 241
Viruses. *See* Retroviruses, insertional
    mutagenesis, mouse
Viscoelastic properties, actin, 369–370
Vital dye, avian blastoderm, 100
Voltage-dependent sodium channels, motor
    endplate morphogenesis, 414

WGA
    binding, cell-CAM 105, hepatocellular
        carcinoma, 255–256, 258
    -HRP, retinotectal map formation, 544

Xenograft, quail-chick, 100, 107
Xenopus, 272; *see also* Actin genes, muscle-
    specific, *Xenopus laevis* embryo;
    Gastrulation, *Xenopus laevis*,
    convergent extension by cell
    intercalation; Induction, embryonic,
    Xenopus laevis, cytoplasmic
    localization; Retinotectal map
    formation, cell patterning, *Xenopus
    laevis*
X-inactivation mosaics, 77, 80, 89–92
    PGK activity, 90

Yellow ooplasm, 68

Zonula adherens, 241, 242